Université Joseph Fourier

Les Houches

Session LXXXI

2004

Nanophysics: Coherence and Transport

Contributors to this volume

J.P. Eisenstein
D. Estève
L.I. Glazman
F.W.F. Hekking
A. Kamenev
V.E. Kravtsov
A. Levy Yeyati
I. Lerner
T. Martin
D. Maslov
Y. Meir
M. Pustilnik
B. Reulet
J. van Ruitenbeek
D. Vion
I.V. Yurkevich

ÉCOLE D'ÉTÉ DE PHYSIQUE DES HOUCHES

SESSION LXXXI, 28 JUNE – 30 JULY 2004

EURO SUMMER SCHOOL
NATO ADVANCED STUDY INSTITUTE
ÉCOLE THÉMATIQUE DU CNRS

NANOPHYSICS: COHERENCE AND TRANSPORT

Edited by

H. Bouchiat, Y. Gefen, S. Guéron, G. Montambaux and J. Dalibard

ELSEVIER

2005

Amsterdam – Boston – Heidelberg – London – New York – Oxford
Paris – San Diego – San Francisco – Singapore – Sydney – Tokyo

ELSEVIER B.V.   ELSEVIER Inc.   ELSEVIER Ltd   ELSEVIER Ltd
Sara Burgerhartstraat 25   525 B Street, Suite 1900   The Boulevard, Langford Lane   84 Theobalds Road
P.O. Box 211, 1000 AE   San Diego, CA 92101-4495   Kidlington, Oxford OX5 1GB   London WC1X 8RR
Amsterdam,   USA   UK   UK
The Netherlands

First edition 2005

Library of Congress Cataloging in Publication Data
A catalog record is available from the Library of Congress.

British Library Cataloguing in Publication Data
A catalogue record is available from the British Library.

ISBN: 0-444-52054-6
ISSN: 0924-8099

⊗ The paper used in this publication meets the requirements of ANSI/NISO Z39.48-1992 (Permanence of Paper).
Printed in The Netherlands.

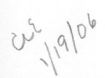

ÉCOLE DE PHYSIQUE DES HOUCHES

Service inter-universitaire commun
à l'Université Joseph Fourier de Grenoble
et à l'Institut National Polytechnique de Grenoble

Subventionné par le Ministère de l'Éducation Nationale,
de l'Enseignement Supérieur et de la Recherche,
le Centre National de la Recherche Scientifique,
le Commissariat à l'Énergie Atomique

Previous sessions

Publishers:
- Session VIII: Dunod, Wiley, Methuen
- Sessions IX and X: Herman, Wiley
- Session XI: Gordon and Breach, Presses Universitaires
- Sessions XII–XXV: Gordon and Breach
- Sessions XXVI–LXVIII: North Holland
- Session LXIX–LXXVIII: EDP Sciences, Springer
- Session LXXIX-LXXX: Elsevier

1	D. Tanaskovic	23	S. Lal	45	T.-C. Wei
2	Y. Fominov	24	A. Halbritter	46	S. Braig
3	J. Salomez	25	A. Romito	47	R. Zitko
4	J.P. Morten	26	C. Texier	48	A. Ghosal
5	G. Fève	27	M. Zvonarev	49	V. Kravtsov
6	N. Chtchelkatchev	28	F. Portier	50	L. Glazman
7	R. Egger	29	D. Ralph	51	Y. Gefen
8	A. Bender	30	A. Grishin	52	A. Braggio
9	G. Catelani	31	C. Schönenberger	53	M. Houzet
10	A. Ratchov	32	B. Michaelis	54	J. Dufouleur
11	M. Ferrier	33	T. Martin	55	D. Meidan
12	M. Turek	34	S. Csonka	56	V. Rytchkov
13	H. Bouchiat	35	C. Sloggett	57	B. Huard
14	J. Segala	36	S. Guéron	58	M. Chauvin
15	M. Cholascinski	37	Y. Utsumi	59	M. Marganska
16	J. Hagelaar	38	S. Florens	60	D. Basko
17	J. Koch	39	A. Bobkov	61	R. Saha
18	H.-S. Sim	40	A. de Martino	62	G. Montambaux
19	A. Ossipov	41	A. Tajic	63	J. Dubi
20	N. Sedlmayr	42	I. Bobkova	64	C. Seoanez Erkell
21	M. Ancliff	43	O. Tsyplyatyev	65	P. Vitushinsky
22	M. Stenberg	44	L. Goren		

Lecturers

EISENSTEIN Jim P., California Institute of Technology, mail code 114-36, Pasadena, CA 91125, USA

ESTÈVE Daniel, SPEC/CEA Quantronics Group, Orme des Merisiers, 91191 Gif sur Yvette, France

GLAZMAN Leonid, Theoretical Physics Institute, University of Minnesota, 116 Church St SE, Minneapolis, MN 55455, USA

HEKKING Frank, Laboratoire de Physique et Modélisation des Milieux Condensés, C.N.R.S, B.P. 166, 38042 Grenoble Cedex, France

KAMENEV Alex, Dept. Physics, University of Minnesota, 116 Church St SE, Minneapolis, MN 55455, USA

KRAVTSOV Vladimir, ICTP, 11 strada Costiera, POB 586, 34100 Trieste, Italy

MARTIN Thierry, Centre de Physique Théorique, Université de la Méditerranée, case 907, 13288 Marseille Cedex 09, France

MASLOV Dmitrii, Department of Physics, University of Florida, PO. Box 118440, Gainesville, FL 32611-8440, USA

RALPH Dan, 536 Clark Hall, Cornell University, Ithaca, NY 14853, USA

VAN RUITENBEEK Jan, Atomic and Molecular Conductors, Leiden Institute of Physics, Leiden University, Niels Borhweg 2, 2333 CA Leiden, The Netherlands

Seminar speakers

APRILI Marco, ESPCI, 10 rue Vauquelin, 75231 Paris cedex 05, France

EGGER Reinhold, Institute für Theoretische Physik, Heinrich-Heine Univ., Universitaetsstrasse 1, D-40225 Duesseldorf, Germany

ENSSLIN Klaus, Solid State Physics Laboratory, ETH Zurich, CH 8093, Switzerland

FAZIO Rosario, Scuola Normale Superiore, Classe di Scienze, Piazza dei Cavalieri 7, 56126 Pisa, Italy

FEIGELMAN Mikhail, Russian Academy of Sciences, Landau Institute for Theoretical Physics, Kosiginstr.2, 119334 Moscow, Russian Federation

IMRY Yoseph, Weizmann Institute of Science, Department of Condensed Matter, 76100 Rehovot, Israel

KASUMOV Alik, Laboratoire de Physique des Solides, CNRS, Université Paris-Sud, 91405 Orsay, France

KOENIG Juergen, Institut für Theoretische Physik III, Ruhr-Universität Bochum, D-44780 Bochum, Germany

LERNER Igor, University of Birmingham, School of Physics and Astronomy, Edgbaston, Birmingham B15 2TT, UK

LEVY YEYATI Alfredo, Department de Fisica de la Materia Condensada, Centro U/O, Faculdad de Ciencias, Campus de Cantoblanco, km 15, 28049 Madrid, Spain

MAKHLIN Yuriy, Institute für Theoretische Festörperphysik, Univ. Karlsruhe, D-76128 Karlsruhe, Germany

MEIR Yigal, Ben Gurion University, Department of Physics, 87109 Beer Sheva, Israel

REULET Bertrand, Applied Physics Dept, BCT 411, Yale Univ., New Haven, CT 06520, USA

SHKLOVSKII Boris, Theoretical Physics Institute, 116 Church St SE, Minneapolis, MN 55455, USA

SCHOENENBERGER Christian, Nanocenter Basel and Institute of Physics, Univ. Basel, Klingelbergstrasse 82, CH-4056 Basel, Switzerland

STERN Ady, Weizmann Institute of Science, Department of Condensed Matter, 76100 Rehovot, Israel

YACOBY Amir, The Joseph and Belle Centre for Submicron Research, Weizmann Institute, 76100 Rehovot, Israel

Participants

ANCLIFF Mark, Dept. Physics, Univ. of Warwick, Coventry CV4 7 AL, UK

BASKO Denis, ICTP, Strada Costiera 11, 34014 Trieste, Italy

BENDER Angelika, LPM2C, Maison des Magistères, BP 166, 38042 Grenoble cedex, France

BOBKOV Alexander, Inst. of Solid State Physics, 142432, Chernogolovka, Moscow region, Russia

BOBKOVA Irina Inst. of Solid State Physics, 142432, Chernogolovka, Moscow region, Russia

BRAGGIO Alessandro Inst. für Theoretische Physik III, Ruhr-Universitaet Bochum, D-44780 Bochum, Germany

BRAIG Stephan, Cornell University, Lab. of Atomic and Solid State Physics, 117 Clark Hall, Ithaca, NY 14853, USA

CATELANI Gianluigi, Physics dept., Mail Code 5262, Columbia Univ., New York, NY 10027, USA

CHAUVIN Martin, Quantronics Group, SPEC, CEA Saclay, 91191 Gif-sur-Yvette cedex, France

CHOLASCINSKI Mateusz, Inst. für Theoretische Festkorperphysik, Univ. Karls-ruhe, W. Gaede strasse 1, D-76128 Karlsruhe, Germany

CHTCHELKATCHEV Nikolai, Landau Inst. for Theoretical Physics, Kosygina str. 2, 119334 Moscow, Russia

CSONKA Szabolcs, Budapest Univ. of Technology, Dept. Physics, Electron Transport Research Group, Budafoki ut 8, 1111 Budapest, Hungary

DE MARTINO Alessandro, Inst. Theoretical Physics IV, Heinrich Heine University, Universitaetsstrasse 1, Gebaeude 25.32, D-40225 Duesseldorf, Germany

DIMITROVA Olga, Landau Institute for Theoretical physics, Chernogolovka, Moscow region, 142432, Russia

DUBI Jonathan, Physics dept., Ben-Gurion Univ. of the Negev, Beer-Sheva, 84105, Israel

DUFOULEUR Joseph, LPN, route de Nozay, 91460 Marcoussis, France

FERRIER Meydi, Laboratoire de Physique des Solides, Université Paris-Sud, 91405 Orsay, France

FEVE Gwendal, Lab. Pierre Aigrain, ENS, 24 rue Lhomond, 75231 Paris cedex 05, France

FLORENS Serge, Inst. für Theorie der Kondensierten Materie, Univ. Karlsruhe, Postfach 6980, 76128 Karlsruhe, Germany

FOMINOV Yakov, Landau Inst. for Theoretical Physics, ul. Kosygina, 2, 119334 Moscow, Russia

GHOSAL Amit, Physics dept., Duke Univ., Box 90305, Science drive, Durham, NC 27708, USA ghosal@phy.duke.edu

GOREN Liliach, Weizmann Inst. of Science, 76100 Rehovot, Israel

GRISHIN Alexander, School of Physics and Astronomy, Univ. Birmingham, Edgbaston, Birmingham B15 2 TT, UK

HAGELAAR Joris, Eindhoven Univ. Technology, Dept. Applied Physics, Spectrum 1.80, Den Dolech 2, PO Box 513, 5612 AZ Eindhoven, The Netherlands

HALBRITTER Andras, Budapest Univ. of Technology, Dept. Physics, Electron Transport Research Group, Budafoki ut 8, 1111 Budapest, Hungary

HOUZET Manuel, CEA DRFMC/SPSMS Bat C1, 17 avenue des Martyrs, 38054 Grenoble cedex 9, France

HUARD Benjamin, Groupe Quantronique, SPEC, CEA Saclay, 91191 Gif sur Yvette cedex, France

KOCH Jens, Freie Univ. Berlin, Inst. Theoretische Physik, Arnimallee 14, D-14195 Berlin, Germany

LAL Siddhartha, Inst. for Theoretical Physics, Univ. Cologne, Zuelpicher strasse 77, 50937 Koeln, Germany

MARGANSKA Magdalena, Dept. Theoretical Physics, Univ. Silesia,ul. Uniwersytecka 4, 40007 Katowice, Poland

MEIDAN Dganit, Weizmann Inst. of Science, 76100 Rehovot, Israel

MICHAELIS Björn, Inst. Lorentz, PO Box 9506, Nl-2300 RA Leiden, The Netherlands

MORTEN Jan Petter, Norwegian Univ. of Science and Technology, Dept. Physics, 7491 Trondheim, Norway

OSSIPOV Alexander, ICTP, Condensed Matter Section, Strada Costiera 11, I-3014 Trieste, Italy

PORTIER Fabien, SPEC, DRECAM, CEA Saclay, 91191 Gif sur Yvette, France

RATCHOV Alexandre, LPM2C, Maison des Magistères, 25 avenue des martyrs, BP 166, 38042 Grenoble cedex, France

ROMITO Alessandro, Scuola Normale Superiore and NEST-INFM, piazza dei Cavalieri, 7, I-56126 Pisa, Italy

RYTCHKOV Valentin, Univ. Geneva, Dept. Theoretical Physics, 24 quai E. Ansermet, 1211 Geneva, Switzerland

SAHA Ronojaoy, Dept. Physics, Univ. Florida, Gainesville, FL 32611, USA

SALOMEZ Julien, LPM2C, Maison des Magistères, 25 avenue des martyrs, BP 166, 38042 Grenoble cedex, France

SEDLMAYR Nicholas, School of Physics and Astronomy, Univ. Birmingham, Edgbaston, Birmingham B15 2TT, UK

SEGALA Julien, SPEC, DRECAM, CEA Saclay, 91191 Gif sur Yvette, France

SEOANEZ ERKELL Cesar, Oscar ICMM-CSIC, Cantoblanco, 28049 Madrid, Spain

SIM Heung-Sun, School of Physics, Korea Inst. for Advanced Study, 207-43 Cheongryangri-dong, Dongdaemun-gu, Seoul 130-722, Korea

SLOGGETT Clare, School of Physics, Univ. New South Wales, Sidney NSW 2052, Australia

STENBERG Markku, Materials Physics Lab., Helsinki Univ. of Technology, PO Box 2200, FIN-02015 HUT, Finland

TAJIC Alireza, Leiden Univ., Inst. Lorentz, PO Box 9506, 2300 RA Leiden, The Netherlands

TANASKOVIC Darko, Nat. High Magnetic Field Lab., Florida State Univ., 1800 E. Paul Dirac dr, Tallahassee, Florida 32310-3706, USA

TEXIER Christophe, LPTMS, Univ. Paris Sud, Bât. 100, 91405 Orsay cedex, France

TSYPLYATYEV Oleksandr, Lancaster Univ., Lancaster, LA1 4YB, UK

TUREK Marko, Univ. Regensburg, Inst. Theoretical Physics, D-93040 Regensburg, Germany

UTSUMI Yasuhiro, MPI of Microstructure Physics, Weinberg 2, 06120 Halle, Germany

VITUSHINSKY Pavel, CEA DRFMC/SPSMS, 17 rue des martyrs, 38054 Grenoble, France

WEI Tzu-Chieh, Dept. Physics, Univ. Illinois at Urbana-Champaign, 1110 West Green Street, Urbana, IL 61801-3080, USA

ZITKO Rok, Jozef Stefan Inst., Jamova 39, 1000 Ljubljana, Slovenia

ZVONAREV Mikhail, Orsted Lab., Niels Bohr Inst., Universitetsparken 5, DK-2100 Copenhagen, Denmark

PREFACE

The Les Houches summer school of July 2004 devoted to "Nanophysics: Coherence and Transport" can be viewed as a continuation of the series devoted to condensed matter physics. It began with "Ill Condensed Matter" (1978), followed by "Chance and Matter" (1986) and by "Mesoscopic Quantum Physics" (1994). The aim of the present session was to review the some of the main developments that took place over the last decade, with an eye towards new open directions in the field of mesoscopic and nanoscopic physics.

The developments of nanofabrication in the past years have enabled the design of electronic systems, that exhibit spectacular signatures of quantum coherence. Experimental results gave rise to numerous initially unexpected (and at times theoretically unresolved) surprises.

Nanofabricated quantum wires and dots containing a small number of electrons are ideal experimental playgrounds for probing electron-electron interactions and their interplay with disorder. Going down to even smaller scales, molecules such as carbon nanotubes, fullerenes or hydrogen molecules can now be inserted in nanocircuits. Measurements of transport through a single chain of atoms have been performed as well. Much progress has also been made in the design and fabrication of superconducting and hybrid nanostructures, be they normal/superconductor or ferromagnetic/superconductor. Quantum coherence is then no longer that of individual electronic states, but rather that of a superconducting wavefunction of a macroscopic number of Cooper pairs condensed in the same quantum mechanical state. Beyond the study of linear response regime, the physics of non-equilibrium transport (including non-linear transport, rectification of a high frequency electric field as well as shot noise) has received much attention, with significant experimental and theoretical insights. All these quantities exhibit very specific signatures of the quantum nature of transport which cannot be obtained from basic conductance measurements.

Participants in the School were exposed to the basic concepts and analytical tools needed to understand this new physics. This was presented in a series of theoretical fundamental courses, in parallel with more phenomenological ones where physics was discussed in a less formal way and illustrated by numerous experimental examples.

Dmitri Maslov gave a long series of lectures emphasizing the importance of electron-electron interactions in one-dimensional quantum transport. Starting from the Fermi liquid description of transport, he showed that electron-electron interactions, whose signature shows up already at higher dimensions, dominate the physics of transport in 1D. Technical aspects concerning the subtleties involved in the re-summation of perturbative diagrammatic expansions, the renormalisation group in 1D and bosonisation were all discussed in great detail. Emphasis was given to the underlying physics. Particular attention was given to the various signatures of interactions in the physical world, including the density of energy states as well as the electric and thermal conductances of 1D wires and their specific sensitivity to the presence of reservoirs or tunneling barriers. This last important point was also treated in detail in the seminar of Igor Lerner. Concrete illustrations of these concepts were given in the seminars of Amir Yacoby on cleaved-edge semiconducting quantum wires and of Reinhold Egger on carbon nanotubes which various physical properties were also reviewed by Christian Schoenenberger whereas Alik Kasumov focused on their superconducting properties (both intrinsic and proximity induced.)

Electron-electron interactions are also responsible for the spectacular transport properties of 2D electron systems confined at the interface between GaAs and GaAlAs in a doped heterostructure, giving rise in particular to the fractional quantum Hall effect at a strong magnetic field. Specifically, the course of Jim Eisenstein was devoted to recent experiments in double layer systems addressing (depending on the strength of the inter-layer coupling) the physics of Coulomb drag or the formation of a completely coherent state involving strong tunneling between the two layers. This spectacular formation of a macroscopic Bose–Einstein condensate of excitons was demonstrated and discussed in the seminar of Ady Stern. Boris Shklovskii presented interesting extensions of the concepts of Coulomb interactions and screening to the physics of biological molecules.

The description of transport and noise out of equilibrium in nanoscale electronic systems has received growing attention in recent years, proving to be an efficient diagnostic tool (experimental and theoretical) in characterizing disorder, interactions and correlations in electronic systems. The course of Alex Kamenev reviewed the theoretical tools (Keldysh theory combined with a non-linear sigma model field theory) required for an efficient treatment of this physics.

Following similar lines Vladimir Kravstov focused on the perturbative aspects of real time Keldysh theory, specifically addressing non-linear ac transport in mesoscopic systems. He made the connection between high frequency rectification of an Aharonov Bohm ring and dephasing in the presence of an ac electric field or a non-equilibrium noise. He also showed that energy absorption gives rise to localization in energy space (dynamic localization).

Non-equilibrium shot noise in conductors contains inter alia specific signatures of the quantum statistics of the charge carriers involved. This was discussed in detail by Thierry Martin in various physical contexts, covering both ballistic and diffusive conductors and showing that in certain strongly interacting systems shot noise constitutes a unique tool to characterize fractional charge excitations. Special emphasis was put on noise in hybrid superconducting/normal devices and multi-terminal structures where signatures of the entanglement of quantum states can be revealed. Bertrand Reulet demonstrated the possibility to measure higher moments of noise which contain different information from that included in the commonly discussed second moment. It turns out that such higher moments depend in a non-trivial way on the electromagnetic environment at hand.

The course of Frank Hekking described the physics of Andreev reflection at Normal/Superconductor interfaces which produces rich sub-gap energy dependent features, both in transport and noise through hybrid nanostructures. These strongly depend on the nature of the barrier at the NS interface and on disorder on the normal side, giving rise to long-range coherent scattering. The more recently investigated Ferromagnetic/Superconducting hybrid systems, also presented in the seminar of Marco Aprili, lead to new spectacular phenomena with oscillations of the phase of the superconducting order parameter in the ferromagnetic region and the possibility to make π junctions. The seminar of Michael Feigelman also showed how Andreev scattering on small superconducting particles can strongly affect quantum transport in a 2D disordered metallic plane.

Leonid Glazmann gave a review of electronic transport through quantum dots in the Coulomb blockade regime. He started from the physics of resonant tunneling and its treatment within the simple-minded picture of the addition spectrum. Attention was given to the role of fundamental symmetries and interactions characterising dots weakly coupled to reservoirs. He went on to investigate more subtle limits, first an off-resonance scenario giving rise to cotunneling (inelastic or elastic at very low temperature), and then, as the effective dot-lead coupling increases, Kondo physics comes into play. Here the magnetic moment involving the highest occupied energy level of the dot combines into a coherent many-body state with the electrons in the reservoirs. There is a broad spectrum of signatures of Kondo physics on the conductance of the dot as a function of temperature and magnetic field. This course was complemented by seminars of Klaus Ensslin on spectroscopy experiments on quantum dots, Yigal Meir on Kondo physics in quantum point contacts and Juergen Koenig on the possibility to observe Fano-like resonances in quantum dots as well as relating dephasing to spin-flip processes.

The course of Dan Ralph, devoted to tunneling measurements of individual quantum states in metallic nanostructures, also constituted a beautiful illustration of these concepts with a particular emphasis on spin effects and superconductivity in nanoparticles, going to the limit of single-molecule transistors.

Atomic point contacts both in the normal and superconducting states constitute the ultimate object for the investigation of quantum transport and requires both specific experimental techniques and theoretical concepts as presented in the course of Jan van Ruitenbeek and Alfredo Levy Yeyati. The notion of conductance channels for a single atom determines the physical properties of these atomic-sized conductors out of equilibrium such as shot noise and superconducting sub-gap structures.

Finally, the lectures of Daniel Estève were devoted to one of the great promises of nanophysics– the possibility to realize solid state quantum bits (in the context of quantum computation) by coherent manipulation of charge, flux or spin variables in a nanocircuit. Selecting the right observables, and controlling the sources of decoherence by designing an optimum electromagnetic environment, are outstanding problems in the field. The crucial question of dephasing by an electromagnetic environment was also addressed in the seminar of Joe Imry. Adding to this broad spectrum of lectures were the seminars by Rosario Fazio and Yuriy Makhlin, who addressed the possibility to investigate and control types of topological Berry phase factors in qubits.

These lectures were complemented by informal sessions or tutorials organized on the spot by the lecturers, as well as by a number of seminars and poster sessions given by the students. They contributed by their unabated enthusiasm and curiosity to the great atmosphere of the school.

The summer school and the present volume have been made possible by the financial support of the following institutions, whose contribution is gratefully acknowledged:

-The "Marie Curie Conferences and Training Course" program of the European Union through the Consortium of European Physics School (CEPS).

-The "Advanced Scientific Institute" program of the Scientific and Environmental Affairs division of NATO.

-The "Lifelong learning" program of the Centre National de la Recherche Scientifique (France).

-The Université Joseph Fourier, the French Ministry of Research and the Commissariat à l'Energie Atomique, through their constant support to the School.

The staff of the School, especially Brigitte Rousset and Isabelle Lelièvre, have been of great help for the preparation and development of the school, and we would like to thank them warmly on behalf of all students and lecturers.

H. Bouchiat, Y. Gefen, S. Guéron, G. Montambaux and J. Dalibard

CONTENTS

Contents

Course 7. *Low-temperature transport through a quantum dot, by Leonid I. Glazman and Michael Pustilnik* 427

Contents

Course 1

FUNDAMENTAL ASPECTS OF ELECTRON CORRELATIONS AND QUANTUM TRANSPORT IN ONE-DIMENSIONAL SYSTEMS

Dmitrii L. Maslov

Department of Physics, University of Florida, P.O. Box 118441, Gainesville, FL 32611-8440, USA

H. Bouchiat, Y. Gefen, S. Guéron, G. Montambaux and J. Dalibard, eds.
Les Houches, Session LXXXI, 2004
Nanophysics: Coherence and Transport

Contents

Abstract

Some aspects of physics on interacting fermions in 1D are discussed in a tutorial-like manner. We begin by showing that the non-analytic corrections to the Fermi-liquid forms of thermodynamic quantities result from essentially 1D collisions embedded into a higher-dimensional phase space. The role of these collisions increases progressively as dimensionality is reduced and, finally, they lead to a breakdown of the Fermi liquid in 1D. An exact solution of the Tomonaga-Luttinger model, based on exact Ward identities, is reviewed in the fermionic language. Tunneling in a 1D interacting systems is discussed first in terms of the scattering theory for interacting fermions and then via bosonization. Universality of conductance quantization in disorder-free quantum wires is discussed along with the breakdown of this universality in the spin-incoherent case. A difference between charge (universal) and thermal (non-universal) conductances is explained in terms of Fabry-Perrot resonances of charge plasmons.

1. Introduction

The theory of interacting fermions in one dimension (1D) has survived several metamorphoses. From what seemed to be a purely mathematical exercise up until the 60s, it had evolved into a practical tool for predicting and describing phenomena in conducting polymers and organic compounds–which were *the* 1D systems of the 70s. Beginning from the early 90s, when the progress in nanofabrication led to creation of artificial 1D structures–quantum wires and carbon nanotubes, the theory of 1D systems started its expansion into the domain of mesoscopics; this trend promises to continue in the future. Given that there is already quite a few excellent reviews and books on the subject [1–10]. I should probably begin with an explanation as to what makes this review different from the others. First of all, it is not a review but–being almost a verbatim transcript of the lectures given at the 2004 Summer School in Les Houches–rather a tutorial on some (and definitely not all) aspects of 1D physics. A typical review on the subject starts with describing the Fermi Liquid (FL) in higher dimensions with an aim of emphasizing the differences between the FL and its 1D counter-part –Luttinger

5

Liquid (LL). My goal–if defined only after the manuscript was written–was rather to highlight the *similarities* between higher-D and 1D systems. The progress in understanding of 1D systems has been facilitated tremendously and advanced to a greater detail, as compared to higher dimensions, by the availability of exact or asymptotically exact methods (Bethe Ansatz, bosonization, conformal field theory), which typically do not work too well above 1D. The downside part of this progress is that 1D effects, being studied by specifically 1D methods, look somewhat special and not obviously related to higher dimensions. Actually, this is not true. Many effects that are viewed as hallmarks of 1D physics, *e.g.,* the suppression of the tunneling conductance by the electron-electron interaction and the infrared catastrophe, do have higher-D counter-parts and stem from essentially the same physics. For example, scattering at Friedel oscillations caused by tunneling barriers and impurities is responsible for zero-bias tunneling anomalies in all dimensions [11, 16]. The difference is in the magnitude of the effect but not in its qualitative nature. Following the tradition, I also start with the FL in Sec. 2, but the main message of this Section is that the difference between $D = 1$ and $D > 1$ is not all that dramatic. In particular, it is shown that the well-known non-analytic corrections to the FL forms of thermodynamic quantities (such as a venerable $T^3 \ln T$-correction to the linear-in-T specific heat in 3D) stem from rare events of essentially 1D collisions embedded into a higher-dimensional phase space. In this approach, the difference between $D = 1$ and $D > 1$ is quantitative rather than qualitative: as the dimensionality goes down, the phase space has difficulties suppressing the small-angle and $2k_F$–scattering events, which are responsible for non-analyticities. The crucial point when these events go out of control and start to dominate the physics happens to be in 1D. This theme is continued in Sec. 5, where scattering from a single impurity embedded into a 1D system is analyzed in the fermionic language, following the work by Yue, Matveev, Glazman [11]. The drawback of this approach–the perturbative treatment of the interaction–is compensated by the clarity of underlying physics. Another feature which makes these notes different from the rest of the literature in the field is that the description goes in terms of the original fermions for quite a while (Secs. 2 through 5), whereas the weapon of choice of all 1D studies–bosonization– is invoked only at a later stage (Sec. 6 and beyond). The rationale–again, a post-factum one–is two-fold. First, 1D systems in a mesoscopic environment–which are the main real-life application discussed here– are invariably coupled to the outside world via leads, gates, etc. As the outside world is inhabited by real fermions, it is sometimes easier to think of, *e.g.,* both the interior and exterior a quantum wire coupled to reservoirs in terms of the same elementary quasi-particles. Second, after 40 years or so of bosonization, what could have been studied within a model of fermions with *linearized* dispersion and not too strong interaction–and this is when bosonization works–was probably studied. (As all statements of this kind, this is one is also an exaggeration.) The last couple of years are characterized by a growing interest in either the effects that do not occur in a model with linearized dispersion, *e.g.*, Coulomb drag due to small-momentum transfers [17], energy relaxation, and phase breaking [18] (the last two phenomena also require three-body processes in 1D) or situations when strong Coulomb repulsion does not permit linearization of the spectrum at any energies [19, 20, 21]. Experiment seems to indicate that the Coulomb repulsion is strong in most systems of interest, thus the beginning of studies of a truly strongly-coupled regime is quite timely. Once the assumption of the linear spectrum is abandoned, the beauty of a bosonized description is by and large lost, and one might as well come back to original fermions. Sec. 6 is devoted to transport in quantum wires, mostly in the absence of impurities. The universality of conductance quantization is explained in some detail, and is followed by a brief discussion of the recent result by Matveev [19], who showed that incoherence in the spin sector leads to a breakdown of the universality at higher temperatures (Sec. 7.4). Also, a difference in charge (universal) and thermal (non-universal) transport–emphasized by Fazio, Hekking, and Khmelnitskii [22]– in addressed in Sec. 7.5. What is missing is a discussion of transport in a disordered (as opposed to a single-impurity) case. However, this canonically difficult subject, which involves an interplay between localization and interaction, is perhaps not ready for a tutorial-like discussion at the moment. (For a recent development on this subject, see Ref. [18].)

Even a brief inspection of these notes will show that the choice between making them comprehensive or self-contained was made for the latter. It is quite easy to see what is missing: there is no discussion of lattice effects, bosonization is introduced without Klein factors, he sine-Gordon model is not treated properly, chiral Luttinger liquids are not discussed at all, and the list goes on. The discussion of the experiment is scarce and perfunctory. However, the few subjects that are discussed are provided with quite a detailed–perhaps somewhat excessively detailed– treatment, so that a reader may not feel a need to consult the reference list too often. For the same reason, the notes also cover such canonical procedures as the perturbative renormalization group in the fermionic language (Sec. 4) and elementary bosonization (Sec. 6), which are discussed in many other sources and a reader already familiar with the subject is encouraged to skip them.

Also, a relatively small number of references (about one per page on average) indicates once again that this is *not* a review. The choice of cited papers is subjective and the reference list in no way pretends to represent a comprehensive bibliography to the field. My apologies in advance to those whose contributions to the field I have failed to acknowledge here.

$\hbar = k_B = 1$ through out the notes, unless specified otherwise.

2. Non-Fermi liquid features of Fermi liquids: 1D physics in higher dimensions

One often hears the statement that, by and large, a Fermi liquid (FL) is just a Fermi gas of weakly interacting quasi-particles; the only difference being the renormalization of the essential parameters (effective mass, $g-$ factor) by the interactions. What is missing here is that the similarity between the FL and Fermi gas holds only for leading terms in the expansion of the thermodynamic quantities (specific heat $C(T)$, spin susceptibility χ_s, etc.) in the energy (temperature) or spatial (momentum) scales. Next-to-leading terms, although subleading, are singular (non-analytic) and, upon a closer inspection, reveal a rich physics of essentially 1D scattering processes, embedded into a high-dimensional phase space.

In this chapter, I will discuss the difference between "normal" processes which lead to the leading, FL forms of thermodynamic quantities and "rare", 1D processes which are responsible for the non-analytic behavior. We will see that the role of these rare processes increases as the dimensionality is reduced and, eventually, the rare processes become the norm in 1D, where the FL breaks down.

In a Fermi gas, thermodynamic quantities form regular, analytic series as function of either temperature, T, or the inverse spatial scale (bosonic momentum q) of an inhomogeneous magnetic field. For $T \ll E_F$, where E_F is the Fermi energy, and $q \ll k_F$, where k_F is the Fermi momentum, we have

$$C(T)/T \quad = \quad \gamma + aT^2 + bT^4 + \ldots; \tag{2.1a}$$

$$\chi_s(T, q = 0) \quad = \quad \chi_s^0(0) + cT^2 + dT^4 + \ldots; \tag{2.1b}$$

$$\chi_s(T = 0, q) \quad = \quad \chi_s^0(0) + eq^2 + fq^4 + \ldots, \tag{2.1c}$$

where γ is the Sommerfeld constant, χ_s^0 is the static, zero-temperature spin susceptibility (which is finite in the Fermi gas), and $a \ldots f$ are some constants. Even powers of T occur because of the approximate particle-hole symmetry of the Fermi function around the Fermi energy and even powers of q arise because of the analyticity requirement [1].

Our knowledge of the interacting systems comes from two sources. For a system with repulsive interactions, one can assume that as long as the strength of the interaction does not exceed some critical value, none of the symmetries

---

[1] The expressions presented above are valid in all dimensions, except for $D = 2$ with quadratic dispersion. There, because the density of states (DoS) does not depend on energy, the leading correction to the $\gamma T-$ term in $C(T)$ is exponential in E_F/T and χ_s does not depend on q for $q \leq 2k_F$. However, this anomaly is removed as soon as we take into account a finite bandwidth of the electron spectrum, upon which the universal (T^{2n} and q^{2n}) behavior of the series is restored.

(translational invariance, time-reversal, spin-rotation, etc.), inherent to the original Fermi gas, are broken. In this range, the FL theory is supposed to work. However, the FL theory is an asymptotically low-energy theory by construction, and it is really suitable only for extracting the leading terms, corresponding to the first terms in the Fermi-gas expressions (2.1a-2.1c). Indeed, the free energy of a FL as an ensemble of quasi-particles interacting in a pair-wise manner can be written as [25]

$$F - F_0 = \sum_k (\epsilon_k - \mu)\, \delta n_k + \frac{1}{2} \sum_{k,k'} f_{k,k'} \delta n_k \delta n_{k'} + O\left(\delta n_k^3\right), \qquad (2.2)$$

where F_0 is the ground state energy, δn_k is the deviation of the fermion occupation number from its ground-state value, and $f_{k,k'}$ is the Landau interaction function. As δn_k is of the order of T/E_F, the free energy is at most quadratic in T, and therefore the corresponding $C(T)$ is at most linear in T. Consequently, the FL theory–at least, in the conventional formulation–claims only that

$$C^*(T)/T \;=\; \gamma^*;$$
$$\chi_s^*(T, q) \;=\; \chi_s^*(0),$$

where γ^* and $\chi_s^*(0)$ differ from the corresponding Fermi-gas values, and does not say anything about higher-order terms [2].

Higher-order terms in T or q can be obtained within microscopic models which specify particular interaction and, if an exact solution is impossible–which is always the case in higher dimensions– employ some kind of a perturbation theory. Such an approach is complementary to the FL: the former nominally works for weak interactions [3] but at arbitrary temperatures, whereas FL works both for weak and strong interactions, up to some critical value corresponding to an instability of some kind, *e.g.*, a ferromagnetic transition, but only in the low-temperature limit. In the {temperature, interaction} plane, the validity regions of these two approaches are two strips running along two axes (Fig. 1). For weak interactions and at low temperatures, the regions should overlap.

Microscopic models (Fermi gas with weak repulsion, Coulomb gas in the high-density limit, electron-phonon interaction, paramagnon model, etc.) show that the higher-order terms in the specific heat and spin susceptibility are non-analytic functions of T and q [26, 27, 28, 29, 30, 31, 32, 33, 34, 35, 36, 37, 38].

---

[2]Strictly speaking, non-analytic terms in $C(T)$ can be obtained from the free energy (2.2) by taking into account the temperature dependence of the quasi-particle spectrum, see Ref. [29]b.

[3]Some results of the perturbation theory can be rigorously extended to an infinite order in the interaction, and most of them can be guaranteed to hold even if the interactions are not weak.

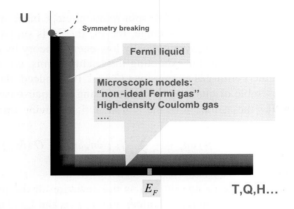

Fig. 1. Combined "diagram of knowledge". x-axis: energy scale (given by temperature T, bosonic momentum Q, magnetic field H) in appropriate units. y-axis: interaction strength. Fermi liquid works for not necessarily weak interactions but smaller than the critical value for an instability of the ground state (grey dot) but at the lowest energy scales. Microscopic models work for weak interactions but an arbitrary energy scale.

For example,

$$
\begin{align}
C(T)/T &= \gamma_3 - \alpha_3 T^2 \ln T \text{ (3D)}; & \text{(2.3a)} \\
C(T)/T &= \gamma_2 - \alpha_2 T \text{ (2D)}; & \text{(2.3b)} \\
\chi_s(q) &= \chi_s(0) + \beta_3 q^2 \ln q^{-1} \text{ (3D)}; & \text{(2.3c)} \\
\chi_s(q) &= \chi_s(0) + \beta_2 |q| \text{ (2D)}, & \text{(2.3d)}
\end{align}
$$

where all coefficients are positive for the case of repulsive electron-electron interaction [4]. Recently, it has been shown that in 2D one can go beyond the perturbation theory and to express the coefficients of the non-analytic terms via an exact scattering amplitude of quasi-particles in the FL.

As seen from Eqs. (2.3a-2.3d), the non-analyticities become stronger as the dimensionality is reduced. The strongest non-analyticity occurs in 1D, where–at

---

[4]Notice that not only the functional forms but also the **sign** of the $q-$ dependent term in the spin susceptibility is different for free and interacting systems. "Wrong" sign of the $q-$ dependent corrections has far-reaching consequences for quantum critical phenomena. For example, it precludes a possibility of a second-order, homogeneous quantum ferromagnetic phase transition in an itinerant system [39]. What is possible is either a first-order transition or ordering at finite q (spin-density wave). In 1D, a homogeneous magnetized state is forbidden anyhow by the Lieb-Mattis theorem [46] which states that the ground state of 1D fermions interacting via spin-independent, but otherwise arbitrary forces, is non-magnetic. One could speculate that the non-analyticities in higher dimensions indicate an existence of a higher-D version of the Lieb-Mattis theorem.

least as long as single-particle properties are concerned–the FL breaks down:

$$C(T)/T = \gamma_1 + \alpha_1 \ln T \quad (1D);$$
$$\chi(q) = \chi_0 + \beta_1 \ln |q| \quad (1D).$$

It turns out that the evolution of the non-analytic behavior with the dimensionality reflects an increasing role of special, almost 1D scattering processes in higher dimensions. Thus non-analyticities in higher dimensions can be viewed as precursors of 1D physics for $D > 1$.

It is easier to start with the non-analytic behavior of a single-particle property–self-energy, which can be related to the thermodynamic quantities via standard means [23] (see also appendix A). Within the Fermi liquid,

$$\text{Re}\Sigma^R(\varepsilon, k) = -A\varepsilon + B\xi_k + \dots \tag{2.4a}$$
$$-\text{Im}\Sigma^R(\varepsilon, k) = C(\varepsilon^2 + \pi^2 T^2) + \dots \tag{2.4b}$$

Expressions (2.4a) and (2.4b) are equivalent to two statements: i) quasi-particles have a finite effective mass near the Fermi level

$$m^* = m_0 \frac{A+1}{B+1},$$

and ii) damping of quasiparticles is weak: the level width is much smaller than the typical quasi-particle energy

$$\Gamma = -2\text{Im}\Sigma^R(\varepsilon, k) \propto \max\left\{|\varepsilon|^2, T^2\right\} \ll |\varepsilon|, T.$$

Landau's argument for the ε^2 (or T^2) behavior of $\text{Im}\Sigma^R$ relies on the Fermi statistics of quasiparticles and on the assumption that the effective interaction is screened at large distances [23]. It requires two conditions. One condition is obvious: the temperature has to be much smaller than the degeneracy temperature $T_F = k_F v_F^*$, where v_F^* is the renormalized Fermi velocity. The other condition is less obvious: it requires inter-particle scattering to be dominated by processes with large (generically, of order k_F) momentum transfers. Once these two conditions are satisfied, the self-energy assumes a universal form, Eqs. (2.4a) and (2.4b), *regardless of a specific type of the interaction (e-e, e-ph) and dimensionality.* To see this, let's have a look at $\text{Im}\Sigma^R(\varepsilon, k)$ due to the interaction with some "boson" (Fig. 2).

D.L. Maslov

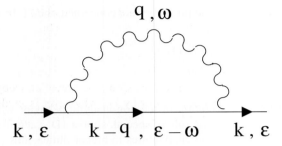

Fig. 2. Self-energy to first order in the interaction with a dynamic bosonic field.

The wavy line in Fig.2 can be, *e.g.*, a dynamic Coulomb interaction, phonon propagator, etc. On the mass shell ($\varepsilon = \xi_k$) at $T = 0$ and for $\varepsilon > 0$, we have [5]

$$\mathrm{Im}\Sigma^R\left(\varepsilon\right) = -\frac{2}{(2\pi)^{D+1}} \int_0^{\varepsilon} d\omega \int d^d q \, \mathrm{Im} G^R\left(\varepsilon - \omega, \mathbf{k} - \mathbf{q}\right) \mathrm{Im} V^R\left(\omega, \mathbf{q}\right).$$

$$(2.5)$$

The constraint on energy transfers ($0 < \omega < \varepsilon$) is a direct manifestation of the Pauli principle which limits the number of accessible energy levels. In real space and time, $V(r, t)$ is a propagator of some field which has a classical limit (when the occupation numbers of all modes are large). Therefore, $V(r, t)$ is a real function, hence $\mathrm{Im} V$ is an odd function of ω. I will make this fact explicit writing $\mathrm{Im} V$ as

$$\mathrm{Im} V^R\left(\omega, q\right) = \omega W\left(|\omega|, q\right).$$

Now, suppose that we integrate over q and the result does not depend on ω. Then we immediately get

$$-\mathrm{Im}\Sigma^R\left(\varepsilon\right) \sim C \int_0^{\varepsilon} d\omega \omega \sim C\varepsilon^2,$$

where C is the result of the $q-$ integration which contains all the information about the interaction. Once we got the ε^2-form for $\mathrm{Im}\Sigma^R\left(\varepsilon\right)$, the ε-term in $\mathrm{Re}\Sigma^R\left(\varepsilon\right)$ follows immediately from the Kramers-Kronig transformation, and we have a Fermi-liquid form of the self-energy regardless of a particular interaction

---

[5]To get Eq. (2.5), you can start with the Matsubara form of diagram Fig. 2, convert the Matsubara sums into the contour integrals, use the dispersion relation

$$D^R(\varepsilon) = \frac{1}{\pi} \int_{-\infty}^{\infty} d\varepsilon' \frac{\mathrm{Im} D^R\left(\varepsilon'\right)}{\varepsilon' - \varepsilon - i0^+},$$

which is valid for any retarded function, and take the limit $T \to 0$.

and dimensionality. Thus a sufficient condition for the Fermi liquid is the *separability* of the frequency and momentum integrations, which can only happen if the energy and momentum transfers are decoupled.

Now, what is the condition for separability? As a function of q, W has at least two characteristic scales. One is provided by the internal structure of the interaction (screening wavevector for the Coulomb potential, Debye wavevector for electron-phonon interaction, etc.) or by k_F, whichever is smaller. This scale (let's call it Q) does not depend on ω. Moreover, as $|\omega|$ is bounded from above by ε, and we are interested in the limit $\varepsilon \to 0$, one can safely assume that $Q \gg |\omega|/v_F$. The role of Q is just to guarantee the convergence of the momentum integral in the ultraviolet, that is, to ensure that for $q \gg Q$ the integrand falls off rapidly enough. Any physical interaction will have this property as larger momentum transfer will have smaller weight. The other scale is $|\omega|/v_F$. Now, let's summarize this by re-writing $\text{Im} V$ in the following scaling form

$$\text{Im} V^R (\omega, q) = \omega \frac{1}{Q^D} U \left(\frac{|\omega|}{v_F Q}, \frac{q}{Q} \right),$$

where U is a dimensionless function, and I have extracted Q^{-D} just to keep the units right.

In the perturbation theory, the Green's function in (2.5) is a free one. Assuming free-electron spectrum $\xi_k = (k^2 - k_F^2)/2m$,

$$\text{Im} G^R \left(\varepsilon - \omega, \vec{k} - \vec{q} \right) = -\pi \delta \left(\varepsilon - \omega - \xi_k + \vec{v}_k \cdot \vec{q} - q^2/2m \right).$$

On the mass shell,

$$\text{Im} G^R \left(\varepsilon - \omega, \vec{k} - \vec{q} \right) |_{\varepsilon = \xi_k} = -\pi \delta \left(\omega - \vec{v}_k \cdot \vec{q} + q^2/2m \right).$$

The argument of the delta-function simply expresses the energy and momentum conservation for a process $\varepsilon \to \varepsilon - \omega, \vec{k} \to \vec{k} - \vec{q}$. The angular integral involves only the delta-function. For any D, this integral gives

$$\langle \delta (\dots) \rangle_\Omega = \frac{1}{v_F q} A_D \left(\frac{\omega + q^2/2m}{v_F q} \right),$$

where I have replaced v_k by v_F because all the action takes place near the Fermi surface. For $D = 3$ and $D = 2$,

$$
\begin{aligned}
A_3 (x) &= 2\theta(1 - |x|); \\
A_2 (x) &= \frac{2\theta (1 - |x|)}{\sqrt{1 - x^2}}.
\end{aligned}
$$

The constraint on the argument of A_D is purely geometric: the magnitude of the cosine of the angle between $\vec{k}$ and $\vec{q}$ has to be less then one. For power-counting purposes, function A_D has a dimensionality of 1. Therefore, its only role is to provide a lower cut-off for the momentum integral. Then, by power counting

$$\mathrm{Im}\Sigma^R(\varepsilon) \sim \frac{1}{v_F Q^D} \int_0^\varepsilon d\omega\omega \int_{q \geq |\omega|/v_F} dq q^{D-2} U\left(\frac{|\omega|}{v_F q}, \frac{q}{Q}\right). \tag{2.6}$$

Now, if the integral over q is dominated by $q \sim Q$ and is convergent in the *infrared*, one can put $\omega = 0$ in this integral. After this step, the integrals over ω and q decouple. The $\omega-$ integral gives ε^2 regardless of the nature of the interaction and dimensionality whereas the $q-$ integral supplies a prefactor which entails all the details of the interaction

$$\mathrm{Im}\Sigma^R(\varepsilon) = C_D \frac{\varepsilon^2}{v_F Q}.$$

For example, for a screened Coulomb interaction in the weak-coupling (high-density) limit $Q = \kappa$, where κ is the screening wavevector, we have in 3D

$$-\mathrm{Im}\Sigma^R(\varepsilon) = \frac{\pi^2}{64} \frac{\kappa}{k_F} \frac{\varepsilon^2}{E_F}.$$

Now we can formulate a sufficient (but not necessary) condition for the Fermi-liquid behavior. It will occur whenever if kinematics of scattering is such that typical momentum transfers are determined by same internal and, what is crucial, $\omega-$ independent scale, whereas energy transfers are of order of the quasi-particle energy (or temperature). Excluding special situations, such as the high-density limit of the Coulomb interaction, Q is generically of order of the ultraviolet range of the problem $\sim k_F$. In other words, isotropic scattering guarantees a ε^2-behavior. Small-angle scattering with typical angles of order $\varepsilon/v_F \ll Q \ll k_F$ gives this behavior as well.

The ε^2- result seems to be quite general under the assumptions made. When and why these assumptions are violated?

A long-range interaction, associated with small-angle scattering, is known to destroy the FL. For example, transverse long-range (current-current [44] or gauge [45]) interactions, which–unlike the Coulomb one–are not screened by electrons, lead to the breakdown of the Fermi liquid. However, the current-current interaction is of the relativistic origin and hence does the trick only at relativistically small energy scales, whereas the gauge interaction occurs only under special circumstances, such as near half-filling or for composite fermions. What about a generic case when nothing of this kind happens? It turns out that even if the

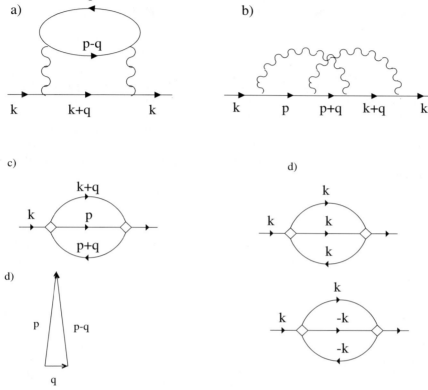

Fig. 3. a) and b) Non-trivial second order diagrams for the self-energy. c) Same diagrams as in a) and b) re-drawn as a single "sunrise" diagram. d) Diagrams relevant for non-analytic terms in the self-energy. e) Kinematics of scattering in a polarization bubble: the dynamic part $\Pi \propto \omega/v_F q$ comes from the processes in which the internal fermionic momentum ($\vec{p}$) is almost perpendicular to the external bosonic one ($\vec{q}$).

bare interaction is of the most benign form, *e.g.*, a delta-function in real space, there are deviations from a (perceived) FL behavior. These deviations get amplified as the system dimensionality is lowered, and, eventually, lead to a complete breakdown of the FL in 1D.

A formal reason for the deviation from the FL-behavior is that the argument which lead us to the ε^2-term is good only in the leading order in ω/qv_F. Recall that the angular integration gives us q^{-1} factors in all dimensions, and, to arrive at the ε^2 result we put $\omega = 0$ in functions A_D and U. If we want to get a next term in ε, then we need to expand U and A in ω. Had such expansions generated regular series, $\text{Im}\Sigma^R$ would have also formed regular series in ε^2: $\text{Im}\Sigma^R =$

$a\varepsilon^2 + b\varepsilon^4 + c\varepsilon^6 + \dots$. However, each factor of ω comes with q^{-1}, so that no matter how high the dimensionality is, at some order of $\omega/v_F q$ we are bound to have an infrared divergence.

2.1. Long-range effective interaction

Let's look at the simplest case of a point-like interaction. A frequency dependence of the self-energy arises already at the second order. At this order, two diagrams in Fig. 3 are of interest to us. For a contact interaction, diagram b) is just -1/2 of a) (which can be seen by integrating over the fermionic momentum $\vec{p}$ first), so we will lump them together. Two given fermions interact via polarizing the medium consisting of other fermions. Hence, the effective interaction at the second order is just proportional to the polarization bubble, which shows how polarizable the medium is

$$\mathrm{Im}V^R(\omega, q) = -U^2\mathrm{Im}\Pi^R(\omega, q).$$

Let's focus on small angle-scattering first: $q \ll 2k_F$. It turns out that in all three dimensions, the bubble has a similar form (see Appendix A for an explicit derivation of this result)

$$-\mathrm{Im}\Pi^R(\omega, q) = \nu_D \frac{\omega}{v_F q} B_D\left(\frac{\omega}{v_F q}\right), \qquad (2.7)$$

where $\nu_D = a_D m k_F^{D-2}$ is the DoS in D dimensions [$a_3 = (2\pi)^{-2}$, $a_2 = (2\pi)^{-1}$, $a_1 = 1/2\pi$] and B_D is a dimensionless function, whose main role is to impose a constraint $\omega \leq v_F q$ in 2D and 3D and $\omega = v_F q$ in 1D. Eq. (2.7) entails the physics of *Landau damping*. The constraint arises because collective excitations– charge- and spin-density waves– decay into particle-hole pairs. Decay occurs only if bosonic momentum and frequency (q and ω) are within the particle-hole continuum (cf. Fig. 4). For $D > 1$, the boundary of the continuum for small ω and q is $\omega = v_F q$, hence the decay takes place if $\omega < v_F q$. The rest of Eq. (2.7) can be understood by dimensional analysis. Indeed, Π^R is the retarded density-density correlation function; hence, by the same argument we applied to $\mathrm{Im}V^R$, its imaginary part must be odd in ω. For $q \ll k_F$, the only combination of units of frequency is $v_F q$, and the frequency enters as $\omega/v_F q$. Finally, a factor ν_D makes the overall units right. In 1D, the difference is that the continuum shrinks to a single line $\omega = v_F q$, hence decay of collective excitations is possible only on this line. In 3D, function B_3 is simply a $\theta-$ function

$$\mathrm{Im}\Pi^R(\omega, q) = -\nu_3 \frac{\omega}{v_F q}\theta\left(q - |\omega/v_F|\right).$$

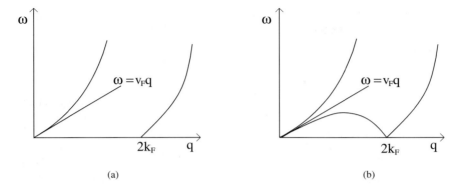

Fig. 4. Particle-hole continua for $D > 1$ (left) and $D = 1$ (right). For the 1D case, only half of the continuum ($q > 0$) is shown.

Next-to-leading term in the expansion of $\mathrm{Im}\Sigma^R$ in ε comes from retaining the lower limit in the momentum integral in Eq. (2.6), upon which we get

$$-\mathrm{Im}\Sigma^R \ \sim \ U^2 m k_F \int_0^\varepsilon d\omega \int_{\omega/v_F}^{Q \sim k_F} dq\, q^2 \frac{1}{v_F q} \frac{\omega}{v_F q}$$

$$\sim \ U^2 \frac{m k_F}{v_F^2} \int_0^\varepsilon d\omega\, \omega \left[\underbrace{k_F}_{\text{FL}} - \underbrace{\frac{\omega}{v_F}}_{\text{beyond FL}} \right]$$

$$\sim \ a\varepsilon^2 - b\,|\varepsilon|^3 \,.$$

The first term in the square brackets is the FL contribution that comes from $q \sim Q$. The second term is a correction to the FL coming from $q \sim \omega/v_F$. Thus, contrary to a naive expectation an expansion in ε is *non-analytic*. The fraction of phase space for small-angle scattering is small–most of the self-energy comes from large-angle scattering events ($q \sim Q$); but we already start to see the importance of the small-angle processes. Applying Kramers-Kronig transformation to a non-analytic part ($|\varepsilon|^3$) in $\mathrm{Im}\Sigma^R$, we get a corresponding non-analytic contribution to the real part as

$$\left(\mathrm{Re}\Sigma^R\right)_{\text{non-an}} \propto \varepsilon^3 \ln|\varepsilon| \,.$$

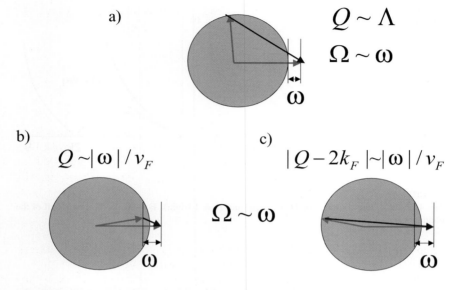

Fig. 5. Kinematics of scattering. a) "Any-angle" scattering. Momentum transfer Q is of order of the intrinsic scale of the interaction, Λ, and is independent of the energy transfer, Ω, which is of order of the initial energy ω. This process contributes regular terms to the self-energy. b) Dynamic forward scattering: $Q \sim |\omega|/v_F$. c) Dynamic backscattering: $|Q - 2k_F| \sim |\omega|/v_F$. Processes b) and c) are responsible for the non-analytic terms in the self-energy.

Correspondingly, specific heat which, by power counting, is obtained from $\mathrm{Re}\,\Sigma^R$ by replacing each ε by T, also acquires a non-analytic term [6]

$$C(T) = \gamma_3 T + \beta_3 T^3 \ln T.$$

This is the familiar $T^3 \ln T$ term, observed both in He^3 [40] and metals [41] (mostly, heavy-fermion materials) [7].

---

[6]One has to be careful with the argument, as a general relation between $C(T)$ and the single-particle Green's function [23] involves the self-energy on the mass shell. In 3D, the contribution to Σ from forward scattering, as defined in Fig. 6, vanishes on the mass shell; hence there is no contribution to $C(T)$ [50]. The non-analytic part of $C(T)$ is related to the backscattering part of the self-energy (scattering of fermions with small total momentum), which remains finite on the mass shell. That forward scattering does not contribute to non-analyticities in thermodynamics is a general property of all dimensions, which can be understood on the basis of gauge-invariance [42].

[7]The $T^3 \ln T$-term in the specific heat coming from the electron-electron interactions is often referred in the literature as the "spin-fluctuation" or "paramagnon" contribution [27, 28] Whereas it is true that this term is enhanced in the vicinity of a ferromagnetic (Stoner) instability, it exists even far way from any critical point and arises already at the second order in the interaction [29].

In 2D, the situation is more dramatic. The $q-$ integral diverges now logarithmically in the infrared:

$$-\text{Im}\Sigma^R(\omega) \quad \sim \quad \frac{U^2}{v_F^2}m \int_0^\varepsilon d\omega\omega \int_{\sim|\omega|/v_F}^{\sim k_F} \frac{dq}{q}$$

$$\sim \quad \frac{U^2}{v_F}m\varepsilon^2 \ln\frac{E_F}{|\varepsilon|},$$

Now, dynamic forward-scattering (with transfers $q \sim \omega/v_F$) is not a perturbation anymore: on the contrary, the ε dependence of $\text{Im}\Sigma^R$ is dominated by forward scattering (the $\varepsilon^2 \ln|\varepsilon|$-term is larger than the "any-angle" ε^2-contribution). Correspondingly, the real part acquires a non-analytic term $\text{Re}\Sigma \propto \varepsilon|\varepsilon|$, and the specific heat behaves as [8]

$$C(T) = \gamma_2 T - \beta_2 T^2.$$

A non-analytic T^2-term in the specific heat has been observed in recent experiments on monolayers of He^3 adsorbed on solid substrate [43] [9].

Finally, in 1D the same power-counting argument leads to $\text{Im}\Sigma^R \propto |\varepsilon|$ and $\text{Re}\Sigma^R \propto \varepsilon \ln|\varepsilon|$ [10]. Correspondingly, the "correction" to the specific heat behaves as $T \ln T$ and is larger than the leading, $T-$ term. This is the ultimate case of dynamic forward scattering, whose precursors we have already seen in higher dimensions [11].

---

[8] Again, only processes with small total momentum contribute.

[9] If a T^2 term in $C(T)$ does not fit your definition of non-analyticity, you have to recall that the right quantity to look at is the ratio $C(T)/T$. Analytic behavior corresponds to series $C(T)/T = \gamma + \delta T^2 + \sigma T^4 + \ldots$, whereas we have a $T^2 \ln T$ and T terms as the leading order corrections to Sommerfeld constant γ for $D = 3$ and $D = 2$, correspondingly.

[10] Special care is required in 1D as in the perturbation theory one gets a strong divergence in the self-energy corresponding to interactions of fermions of the same chirality (Fig. 8a,c). This point will be discussed in more detail in Section 2.3 (along with a weaker but nonetheless singularity in 2D). For now, let focus on a regular part of the self-energy corresponding to the interaction of fermions of opposite chirality (Fig. 8b).

[11] Bosonization predicts that $C(T)$ of a fermionic system is the same as that of 1D bosons, which scales as T for $D = 1$ [10]. This is true only for spinless fermions, in which case bosonisation provides an asymptotically exact solution. For electrons with spins, the bosonized theory is of the sine-Gordon type with the non-Gaussian $(\cos\phi)$ term coming from the backscattering of fermions of opposite spins. Even if this term is marginally irrelevant and flows down to zero at the lowest energies, at intermediate energies it results in a multiplicative $\ln T$ factor in $C(T)$ and a $\ln \max\{q, T, H\}$ correction to the spin susceptibility (where H is the magnetic length, and units are such that q, T, and H have the units of energy). The difference between the non-analyticities in $D > 1$ and $D = 1$ is that the former occurs already at the second order in the interaction, whereas the latter start only at *third* order. Naive power-counting breaks down in 1D because the coefficient in front of $T \ln T$ term in $C(T)$ vanishes at the second order, and one has to go to third order. In the sine-Gordon model, the third order in the interaction is quite natural: indeed, one has to calculate the correlation function

Even if the bare interaction is point-like, the effective one contains a long-range part at finite frequencies. Indeed, the non-analytic parts of Σ and $C(T)$ come from the region of small q, and hence large distances. Already to the second order in U, the effective interaction $\tilde{U} = U^2 \Pi(\omega, q)$ is proportional to the *dynamic* polarization bubble of the electron gas, $\Pi(\omega, q)$. In all dimensions, $\mathrm{Im}\Pi^R$ is universal and singular in q for $|\omega|/v_F \ll q \ll k_F$

$$\mathrm{Im}\Pi^R(\omega, q) \sim v_D \frac{\omega}{v_F |q|}.$$

Although the effective interaction is indeed screened at $q \to 0$ –and this is why the FL survives even if the bare interaction has a long-range tail–it has a slowly decaying tail in the intermediate range of q. In real space, $\tilde{U}(r)$ behaves as ω/r^{D-1} at distances $k_F^{-1} \ll r \ll v_F/|\omega|$.

Thus, we have the same singular behavior of the bubble, and the results for the self-energy differ only because the phase volume q^D is more effective in suppressing the singularity in higher dimensions than in lower ones.

There is one more special interval of q : $q \approx 2k_F$, *i.e.*, Kohn anomaly. Usually, the Kohn anomaly is associated with the $2k_F$- on-analyticity of the *static* bubble and its most familiar manifestation is the Friedel oscillation in electron density produced by a static impurity (discussed later on). Here, the static Kohn anomaly is of no interest for us as we are dealing with dynamic processes. However, the dynamic bubble is also singular near $2k_F$. For example, in 2D,

$$\mathrm{Im}\Pi^R(q \approx 2k_F, \omega) \propto \frac{\omega}{\sqrt{k_F(2k_F - q)}}\theta(2k_F - q).$$

Because of the one-sided singularity in $\mathrm{Im}\Pi^R$ as a function of q (divergent derivative), the $2k_F$–effective interaction oscillates and falls off as a power law in real space. By power counting, if a static Friedel oscillation falls off as $\sin 2k_F r/r^D$, then the dynamic one behaves as

$$\tilde{U} \propto \frac{\omega \sin 2k_F r}{r^{(D-1)/2}}.$$

Dynamic Kohn anomaly results in the same kind of non-analyticity in the self-energy (and thermodynamics) as the forward scattering. The "dangerous" range of q now is $|q - 2k_F| \sim \omega/v_F$–"dynamic backscattering". It is remarkable that the non-analytic term in the self-energy is sensitive only to strictly forward or backscattering events, whereas processes with intermediate momentum transfers contribute only to analytic part of the self-energy. To see this, we perform the analysis of kinematics in the next section.

---

of the $\cos\phi$ term, which already contains two factors of the interaction; the third factor occurs by expanding the exponent to leading (first) order. For more details, see [47],[48],[49].

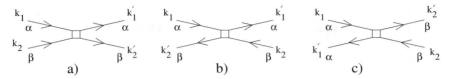

Fig. 6. Scattering processes responsible for divergent and/or non-analytic corrections to the self-energy in 2D. a) "Forward scattering"–an analog of the "g_4"-process in 1D. All four fermionic momenta are close to each other. b) Backscattering–an analog of the "g_2"-process in 1D. The net momentum before and after collision is small. Initial momenta are close to final ones. Although the momentum transfer in such a process is small, we still refer to this process as "backscattering" (see the discussion in the main text). c) $2k_F-$ scattering.

2.2. *1D kinematics in higher dimensions*

The similarity between non-FL behavior in 1D and non-analytic features in higher dimensions occurs already at the level of kinematics. Namely, one can make a rather strong statement: *the non-analytic terms in the self-energy in higher dimensions result from essentially 1D scattering processes.* Let's come back to self-energy diagram 3a. In general, integrations over fermionic momentum $\vec{p}$ and bosonic $\vec{q}$ are independent of each other: one can first integrate over $(\vec{p}, \varepsilon)$, forming a bubble, and then integrate over $(\vec{q}, \omega)$. Generically, $\vec{p}$ spans the entire Fermi surface. However, the non-analytic features in Σ come not from generic but very specific $\vec{p}$ which are close to either to $\vec{k}$ or to $-\vec{k}$.

Let's focus on the 2D case. The $\varepsilon^2 \ln|\varepsilon|$ term results from the product of two q^{-1}-singularities: one is from the angular average of $\text{Im}G$ and the other one from the dynamic, $\omega/v_F q$, part of the bubble. In Appendix A, it is shown that the $\omega/v_F q$ singularity in the bubble comes from the region where $\vec{p}$ is almost perpendicular to $\vec{q}$. Similarly, the angular averaging of $\text{Im}G$ also pins the angle between $\vec{k}$ and $\vec{q}$ to almost 90°.

$$\text{Im}G^R(\varepsilon - \omega, \vec{k} - \vec{q}) = -\pi\delta\left(\varepsilon - \omega - qv_F \cos\theta'\right) \rightarrow$$

$$\cos\theta' = \frac{\varepsilon - \omega}{v_F q} \sim \frac{\omega}{v_F q} \ll 1 \rightarrow \theta' \approx \pi/2.$$

As $\vec{p}$ and $\vec{k}$ are almost perpendicular to the same vector $(\vec{q})$, they are either almost parallel or anti-parallel to each other. In terms of a symmetrized ("sunrise") self-energy (cf. Fig. 3), it means that either all three internal momenta are parallel to the external one or one of the internal one is parallel to the external whereas the other two are anti-parallel [12]. Thus we have three almost 1D processes:

---

[12]In 3D, conditions $\vec{p} \perp \vec{q}$ and $\vec{k} \perp \vec{q}$ mean only that $\vec{p}$ and $\vec{k}$ lie in the same plane. However, it is still possible to show that for a closed diagram, *e.g.*, thermodynamic potential, $\vec{p}$ and $\vec{k}$ are

- all four momenta (two initial and two final) are almost parallel to each other;
- the total momentum of the fermionic pair is near zero, whereas the transferred momentum is small;
- the total momentum of the fermionic pair is near zero, whereas the transferred momentum is near $2k_F$.

These are precisely the same 1D processes we are going to deal with in the next Section–the only difference is that in 2D, trajectories do have some angular spread, which is of order $|\omega|/E_F$. The first one is known as "g_4" (meaning: all four momenta are in the same direction) and the other one as "g_2" (meaning: two out of four momenta are in the same direction). Both of these processes are of the forward-scattering type as the transferred momentum is small. In 1D, these processes correspond to scattering of fermions of same (g_4) or opposite chirality (g_2). The last ($2k_F$) process is known "g_1" in 1D.

It turns out (see next Section) that of these two processes, the g_2-one and $2k_F-$ ones are directly responsible for the $\varepsilon^2 \ln \varepsilon$ behavior. The g_4-process leads to a mass-shell singularity in the self-energy both in 1D and 2D, discussed in the next section, but does not affect the thermodynamics, so we will leave it for now.

What about a $2k_F-$ scattering? Suppose electron $\vec{k}$ scatters into $-\vec{k}$ emitting an electron-hole pair of momentum $2\vec{k}$. In general, $2\vec{k}$ of the e-h pair may consist of any two fermionic momenta which differ by $2\vec{k}$: $\vec{p}$ and $\vec{p} + 2\vec{k}$. But since $|2\vec{k}| \approx 2k_F$, the components of the e-h pair will be on the Fermi surface only if $\vec{p} \approx -\vec{k}$ and $\vec{p} + 2\vec{k} \approx \vec{k}$. Only in this case does the effective interaction –bubble have a non-analytic form at finite frequency. Thus $2k_F$-scattering is also of the 1D nature for $D > 1$.

What we have said above, can be summarized in the following pictorial way. Suppose we follow the trajectories of two fermions, as shown in Fig. 7. There are several types of scattering processes. First, there is a "any-angle" scattering which, in our particular example, occurs at a third fermion whose trajectory is not shown. This scattering contributes regular, FL terms both to the self-energy and thermodynamics. Second, there are dynamic forward-scattering events, when $q \sim |\omega|/v_F$. These are *not* 1D processes, as fermionic trajectories enter the interaction region at an arbitrary angle to each other. In 3D, a third order in such processes results in the non-analytic behavior of $C(T)$–this is the origin of the "paramagnon" anomaly in $C(T)$. In 2D, dynamic forward scattering does not lead to non-analyticity. Finally, there are processes, marked by "g_1", "g_2",

---

either parallel or anti-parallel to each other. Hence, the non-analytic term in $C(T)$ also comes from the 1D processes. In addition, there are dynamic forward scattering events (marked with a star in Fig. 7) which, although not being 1D in nature, do lead to a non-analyticity in 3D. Thus, the $T^3 \ln T$ anomaly in $C(T)$ comes from both 1D and non-1D processes [50]. The difference is that the former start already at the second order in the interaction whereas the latter occur only at the third order. In 2D, the entire $T-$ term in $C(T)$ comes from the 1D processes.

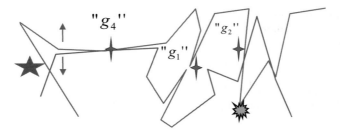

 dynamic forward scattering

"any-angle" scattering event
→ Regular (FL) contribution

 1D forward or backscattering

Fig. 7. Typical trajectories of two interacting fermions. Explosion: "any-angle" scattering at a third fermion (not shown) which leads to a regular (FL) contribution. Five-corner star: dynamic forward scattering $q \sim |\omega|/v_F$. This process contributes to non-analyticity in 3D (to third order in the interaction) but not in 2D. Four-corner star: 1D dynamic forward and backscattering events, contributing to non-analyticities both in 3D and 2D.

and "g_4", when electrons conspire to align their initial momenta so that they are either parallel or antiparallel to each other. These processes determine the non-analytic parts of Σ and thermodynamics in 2D (and also, formally, for $D < 2$). A crossover between $D > 1$ and $D = 1$ occurs when all other processes but g_1, g_2, and g_4 are eliminated by a geometrical constraint.

We see that for non-analytic terms in the self-energy (and thermodynamics), large-angle scattering does not matter. Everything is determined by essentially 1D processes. As a result, if the bare interaction has some q dependence, only two Fourier components matter: $U(0)$ and $U(2k_F)$. For example, in 2D

$$\mathrm{Im}\Sigma^R(\varepsilon) \propto \left[U^2(0) + U^2(2k_F) - U(0)U(2k_F) \right] \varepsilon^2 \ln |\varepsilon| \,;$$

$$\mathrm{Re}\Sigma^R(\omega) \propto \left[U^2(0) + U^2(2k_F) - U(0)U(2k_F) \right] \varepsilon |\varepsilon| \,;$$

$$C(T)/T = \gamma^* - a \left[U^2(0) + U^2(2k_F) - U(0)U(2k_F) \right] T \,;$$

$$\chi_s(Q, T) = \chi_s^*(0) + bU^2(2k_F) \max \{ v_F Q, T \} \,;$$

where a and b are coefficients. These perturbative results can be generalized for the Fermi-liquid case, when the interaction is not necessarily weak. Then the leading, analytic parts of $C(T)$ and χ_s are determined by the angular harmonics of the *Landau interaction function*

$$\hat{F}\left(\vec{p}, \vec{p}'\right) = F_s\left(\theta\right) \hat{I} + F_a\left(\theta\right) \vec{\sigma} \cdot \vec{\sigma}',$$

where θ is the angle between $\vec{p}$ and $\vec{p}'$. In particular,

$$\begin{aligned}
\gamma^* &= \gamma_0 \left(1 + \langle \cos\theta F_s \rangle\right); \\
\chi_s^*\left(0\right) &= \chi_s^0 \frac{1 + \langle \cos\theta F_s \rangle}{1 + \langle F_a \rangle},
\end{aligned}$$

where γ_0 and χ_s^0 are the corresponding quantities for the Fermi gas. Because of the angular averaging, the FL part is rather insensitive to the details of the interaction. As generically F_s and F_a are regular functions of θ, the whole Fermi surface contributes to the FL renormalizations. Vertices $U(0)$ and $U(2k_F)$, occurring in the perturbative expressions, are replaced by *scattering amplitude* at angle $\theta = \pi$

$$\hat{A}\left(\vec{p}, \vec{p}'\right) = A_s\left(\theta\right) \hat{I} + A_a\left(\theta\right) \vec{\sigma} \cdot \vec{\sigma}',$$

Beyond the perturbation theory [37],

$$\begin{aligned}
C(T)/T &= \gamma^* - \bar{a}\left[A_s^2\left(\pi\right) + 3A_a^2\left(\pi\right)\right]T; \\
\chi_s(Q, T) &= \chi_s^*\left(0\right) + \bar{b}A_a^2\left(\pi\right)\max\left\{v_F Q, T\right\}.
\end{aligned}$$

Non-analytic parts are not subject to angular averaging and are sensitive to a detailed behavior of $A_{s,a}$ near $\theta = \pi$.

2.3. Infrared catastrophe

2.3.1. 1D
By now, it is well-known that the FL breaks down in 1D and an attempt to apply the perturbation theory to 1D problem results in singularities. Let's see what precisely goes wrong in 1D. I begin with considering the interaction of fermions of opposite chirality, as in diagram Fig. 8b. Physically, a right-moving fermion emits (and then re-absorbs) left-moving quanta of density excitations (same for left-moving fermion emitting/absorbing right-moving quanta). Now, instead of the order-of-magnitude estimate (2.7), which is good in all dimensions but only for power-counting purposes, I am going to use an exact expression for the bubble, Eq. (B.4), formed by left-moving fermions. We have

$$\begin{aligned}
-\mathrm{Im}\Sigma_{+-}^R\left(\varepsilon\right) &\sim U^2 \nu_1 \int_0^\varepsilon d\omega \int dq \,\mathrm{Im}G_+^R\left(\varepsilon - \omega, p - q\right)\mathrm{Im}\Pi_-^R \\
&\sim U^2 \nu_1 \int_0^\varepsilon d\omega \int dq\,\delta\left(\varepsilon - \omega + v_F q\right)\left(\omega/v_F\right)\delta\left(\omega + v_F q\right) \\
&\sim g^2 \left|\varepsilon\right|,
\end{aligned}$$

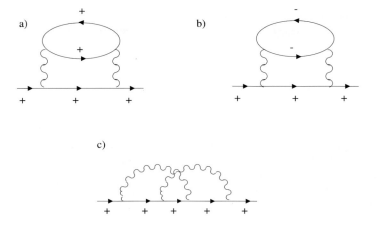

Fig. 8. Self-energy in 1D. $\pm$ refer to right (left)-moving fermions.

where $g \equiv U/v_F$ is the dimensionless coupling constant. The corresponding real part behaves as $\varepsilon \ln |\varepsilon|$. What we got is bad, as $\mathrm{Im} \Sigma^R$ scales with ε in the same way as the energy of a free excitation above the Fermi level and $\mathrm{Re} \Sigma^R$ increases faster than ε (which means that the effective mass depends on ε as $\ln |\varepsilon|$), but not too bad because, as long as $U/v_F \ll 1$, the breakdown of the quasi-particle picture occurs only at exponentially small energy scales: $\varepsilon \lesssim E_F \exp(-(v_F/U)^2)$. Now, let's look at scattering of fermions of the same chirality. This time I choose to be away from the mass shell.

$$-\mathrm{Im} \Sigma^R_{++} \quad \propto \quad \int_0^\varepsilon d\omega \int dq \underbrace{\delta \left(\varepsilon - \omega - v_F (k - q)\right)}_{\mathrm{Im} G^R_+} \underbrace{\omega \delta \left(\omega - v_F q\right)}_{\mathrm{Im} \Pi^R_+} \qquad (2.8)$$

$$= \quad \varepsilon^2 \delta \left(\varepsilon - v_F k\right). \qquad (2.9)$$

It is not difficult to see that the full (complex) self-energy is simply

$$\Sigma^R_{++} \propto -\frac{\varepsilon^2}{\varepsilon - v_F k + i0^+}. \qquad (2.10)$$

On the mass shell ($\varepsilon = v_F k$) we have a strong–delta-function–singularity. This anomaly was discovered by Bychkov, Gor'kov, and Dzyaloshinskii back in the 60s [52], who called it the "infrared catastrophe". Indeed, it is similar to an infrared catastrophe in QED, where an electron can emit an infinite number of soft photons. Likewise, since we have linearized the spectrum, a 1D fermion can emit an infinite number of soft bosons: quanta of charge- and spin-density excitations. The point is that in 1D there is a perfect match between momentum and energy conservations for a process of emission (or absorption) of a boson with energy and momentum related by $\omega = v_F q$:

$$k' = k - q$$
$$\varepsilon' = \varepsilon - \omega = v_F k - v_F q = v_F (k - q).$$

On the mass-shell, the energy and momentum conservations are equivalent. Imagine that you want to find a probability of certain scattering process using a Fermi Golden rule. Then you have a product of two $\delta-$ functions: one reflecting the momentum and other energy conservation. But if the arguments of the delta-functions are the same, you have an essential singularity: a square of the delta-function. As a result, the corresponding probability diverges.

A pole in the self-energy [Eq. (2.10)] indicates an essentially non-perturbative and 1D effect: spin-charge separation. Indeed, substituting Eq. (2.10) we get two poles corresponding to excitations propagating with velocities $v_F (1 \pm g)$ (recall that $g \ll 1$). This peculiar feature is confirmed by an exact solution (see Section 3): already the g_4-interaction leads to a spin-charge separation (but not to anomalous scaling). What we did not get quite right is that the velocities of both–spin- and charge-modes–are modified by the interactions. In fact, the exact solution shows that the velocity of the spin-mode remains equal to v_F, whereas the velocity of the charge mode is modified.

Obviously, there is no spin-charge separation for spinless electrons. Indeed, in this case diagram Fig. 8a does not have an additional factor of two as compared to Fig. 8c (but is still of opposite sign), so that the forward-scattering parts of these two diagrams cancel each other. As a result, there is no infrared catastrophe for spinless fermions.

2.3.2. 2D

What we considered in the previous section sounds like an essentially 1D effect. However, a similar effect exists also in 2D (more generally, for $1 \leq D \leq 2$). This emphasizes once again that the difference between $D = 1$ and $D > 1$ is not as dramatic as it seems.

In 2D, the self-energy also diverges on the mass shell, if one linearizes the electron's spectrum, albeit the divergence is weaker than in 1D–to second order, it is logarithmic [13]. The origin of the divergence can be traced back to the form of the polarization bubble at small momentum transfer, Eq. (A.1). Integrating over

---

[13] In 3D, there is no mass-shell singularity to any order of the perturbation theory.

the angle in 2D, we get

$$\mathrm{Im}\Pi^R(\omega, q) = -\left(\frac{m}{2\pi}\right) \frac{\omega}{\sqrt{(v_F q)^2 - \omega^2}} \theta\left(v_F q - |\omega|\right). \quad (2.11)$$

$\mathrm{Im}\Pi^R(\omega, q)$ has a square-root singularity at the boundary of the particle-hole continuum, *i.e.*, at $\omega = v_F q$. (This is a threshold singularity of the van Hove type–the band of soft electron-hole pairs is terminated at $\omega = v_F q$, but the spectral weight of the pairs is peaked at the band edge.) On the other hand, expanding $\epsilon_{\mathbf{k+q}}$ in $G^R(\varepsilon + \omega, \mathbf{k+q})$ as $\xi_{\mathbf{k+q}} = \xi_k + v_F q \cos\theta$ and integrating over θ, we obtain another square-root singularity

$$\int d\theta \mathrm{Im} G^R = -2\pi \left[(v_F q)^2 - (\varepsilon + \omega - \xi_k)^2\right]^{-1/2}. \quad (2.12)$$

On the mass shell ($\omega = \xi_k$), the arguments of the square roots in Eqs. (2.11) and (2.12) coincide, and the integral over q diverges logarithmically. The resulting contribution to $\mathrm{Im}\Sigma^R$ diverges on the mass shell ($\varepsilon = \xi_k$) [53, 54, 55],[34],[37]

$$\mathrm{Im}\Sigma_{g_4}^R(\varepsilon, k) = -\frac{u^2}{8\pi} \frac{\varepsilon^2}{E_F} \ln \frac{E_F}{|\varepsilon - \xi_k|},$$

where $\Delta \equiv \varepsilon - \xi_k$ and $u \equiv mU/2\pi$. The process responsible for the log-singularity is the "g_4" process in Fig. 6. On the other hand, g_2 and g_1 processes give a contribution which is finite on the mass shell

$$\mathrm{Im}\Sigma_{g_1+g_2}^R(\varepsilon, k) = -\frac{u^2}{4\pi} \frac{\varepsilon^2}{E_F} \ln \frac{E_F}{|\varepsilon + \xi_k|}.$$

(a divergence at $\varepsilon = -\xi_k$ is spurious and is removed by going beyond the log-accuracy [34],[37]). We see therefore that the familiar form of the self-energy in 2D [$\varepsilon^2 \ln |\varepsilon|$), see Ref. [56]] is valid only on the Fermi surface ($\xi_k = 0$). The logarithmic singularity in $\mathrm{Im}\Sigma^R$ on the mass shell is eliminated by retaining the finite curvature of single-particle spectrum (which amounts to keeping the $q^2/2m$ term in $\xi_{\bar{k}+\bar{q}}$). This brings in a new scale ε^2/E_F. The emerging singularity in (2.3.2) is regularized at $|\varepsilon - \xi_k| \sim \varepsilon^2/E_F$ and the $\varepsilon^2 \ln |\varepsilon|$ behavior is restored. However, higher orders diverge as power-laws and finite curvature does not help to regularize them. This means that–in contrast to 3D–the perturbation theory must be re-summed even for an infinitesimally weak interaction. Once this is done, the singularities are removed. Re-summation also help to understand the reason for the problems in the perturbation theory. In fact, what we were trying to do was to take into account a non-perturbative effect–an interaction with the zero-sound mode–via a perturbation theory. Once all orders are re-summed, the zero-sound mode splits off the continuum boundary–now it is a propagating mode with velocity $c > v_F$. This splitting is what regularizes the divergences. The resulting state is essentially a FL: the leading term in Σ behaves as $\varepsilon^2 \ln |\varepsilon|$. However, some non-perturbative features remain: for example, the spectral function exhibits a second peak away from the mass shell corresponding to the emission of the zero-sound waves by fermions. A two-peak structure of the spectral function is reminiscent of the spin-charge separation, although we do not really have a spin-charge separation here: in contrast to the 1D case, the spin-density collective mode lies within the continuum and is damped by the particle-hole pairs.

3. Dzyaloshinskii-Larkin solution of the Tomonaga-Luttinger model

3.1. Hamiltonian, anomalous commutators, and conservation laws

In the Tomonaga-Luttinger model [57],[58] one considers a system of 1D spin-1/2 fermions with a linearized dispersion. Only forward scattering of left- and right-moving fermions is taken into account (g_2 and g_4- processes), whereas backscattering is neglected. This last assumption means that the interaction potential is of sufficiently long-range so that $U(2k_F) \ll U(0)$. [We will come back to this condition later.] Coupling between fermions of the same chirality (g_4) is assumed to be different from coupling between fermions of different chirality (g_2). If the original Hamiltonian contains only density-density interaction, then $g_2 = g_4$. A difference between g_2 and g_4 leads to an unphysical (within this model) current-current interaction. We will keep $g_2 \neq g_4$, however, at the intermediate steps of the calculations as it helps to elucidate certain points. At the end, one can make g_2 equal to g_4 without any penalty. In addition, in some physical situations, $g_2 \neq g_4$ [14]. In what follows I will follow the original paper by Dzyaloshinskii and Larkin (DL) [59] and a paper by Metzner and di Castro [60], where the Ward identity used by Dzyaloshinskii and Larkin is derived in a detailed way.

The Hamiltonian of the model is written as

$$
\begin{aligned}
H &= H_0 + H_{\text{int}}; \\
H_{\text{int}} &\equiv H_2 + H_4,
\end{aligned}
$$

---

[14]For example, Coulomb interaction between the electrons at the edges of a finite-width Hall bar (in the Integer Quantum Hall Effect regime) has this feature: electrons of the same chirality are situated on the same edge, whereas electrons of different chirality are on opposite edges; hence the matrix elements for the g_2- and g_4- interactions are different.

where

$$H_0 = v_F \sum_{k,\sigma} k \left(a_{+,\sigma}^{\dagger}(k) a_{+,\sigma}(k) - a_{-,\sigma}^{\dagger}(k) a_{-,\sigma}(k) \right)$$

is the Hamiltonian of free fermions ($\pm$ denote right/left moving fermions and σ is the spin projection) and

$$H_2 = \frac{g_2}{L} \sum_q \sum_{\sigma,\sigma'} \rho_{+,\sigma}(q) \rho_{-,\sigma'}(-q);$$

$$H_4 = \frac{g_4}{2L} \sum_q \sum_{\sigma,\sigma'} \rho_{+,\sigma}(q) \rho_{+,\sigma'}(-q) + \rho_{-,\sigma}(q) \rho_{-,\sigma'}(-q),$$

with

$$\rho_{\pm,\sigma} = \sum_k a_{\pm,\sigma}^{\dagger}(k+q) a_{\pm,\sigma}(k).$$

To avoid additional complications, I assume that the interaction is spin-independent. To simplify the notations and to emphasize the similarity with of our model QED, I will set v_F to unity in this section.

Introducing the chiral charge- and spin densities as

$$\rho_{\pm}^c = \rho_{\pm,\uparrow} + \rho_{\pm,\downarrow};$$
$$\rho_{\pm}^s = \rho_{\pm,\uparrow} - \rho_{\pm,\downarrow},$$

and total charge density and current as

$$\rho^c = \rho_+^c + \rho_-^c;$$
$$j^c = \rho_+^c - \rho_-^c,$$

the interaction part of the Hamiltonian can be represented as

$$H_{\text{int}} = \sum_q \frac{1}{2} (g_2 + g_4) \rho^c(q) \rho^c(-q) + \frac{1}{2} (g_4 - g_2) j^c(q) j^c(-q). \quad (3.1)$$

As we have already said, for $g_2 = g_4$, the interaction is of a pure density-density type. Notice also that the spin density and current drop out of the Hamiltonian– this is to be expected for a spin-invariant interaction. To make a link with QED, let us introduce Minkowski current j^{μ} with $\mu = 0, 1$ so that $j^0 = \rho^c (=j_0)$ and $j^1 = j^c (=-j_1)$. Then the interaction can be written as a 4-product of Minkowski currents in a Lorentz-invariant form

$$H_{\text{int}} = \sum_q g_{\mu\nu} j_{\nu} j^{\nu},$$

where

$$g_{00} = \frac{1}{2}(g_2 + g_4);$$

$$g_{11} = \frac{1}{2}(g_4 - g_2);$$

$$g_{01} = g_{10} = 0. \tag{3.2}$$

In what follows, we will need the following anomalous commutators

$$\left[\rho_{\pm,\sigma}(q), H_0\right] = \pm q\rho_{\pm,\sigma}(q);$$

$$\left[\rho_{\pm,\sigma}(q), H_2\right] = \pm \frac{g_2}{2\pi}q\rho_{\mp,\sigma}(q);$$

$$\left[\rho_{\pm,\sigma}(q), H_4\right] = \pm \frac{g_4}{2\pi}q\rho_{\pm,\sigma}(q).$$

The derivation of these commutation relations can be found in a number of standard sources [61, 10] and I will not present it here. Adding up the commutators, we get

$$\begin{aligned}
\left[\rho_{\pm,\sigma}, H\right] &= \left[\rho_{\pm,\sigma}, H_0 + H_2 + H_4\right] \\
&= \pm q\rho_{\pm,\sigma} \pm \frac{g_2}{2\pi}q\rho_{\mp}^c \pm \frac{g_4}{2\pi}q\rho_{\pm}^c
\end{aligned}$$

Adding up equations for spin-up and -down fermions, we obtain

$$\left[\rho_{\pm}^c, H\right] = \pm q\rho_{\pm}^c \pm \frac{g_2}{\pi}q\rho_{\mp}^c \pm \frac{g_4}{\pi}q\rho_{\pm}^c.$$

Finally, adding up the $\pm$ components yields

$$i\partial_t \rho^c = \left[\rho^c, H\right] = v_c q j^c \tag{3.3}$$

where

$$v_c \equiv 1 + \frac{g_4 - g_2}{\pi}$$

(recall that $v_F = 1$). Eq. (3.3) is a continuity equation reflecting charge conservation. As if we did not have enough new notations, here is another one

$$Q^\mu = (\omega, q)$$

and

$$\mathcal{Q}^\mu = (\omega, v_c q).$$

In these notations and after a Fourier transform in time, the continuity equations can be written as

$$\mathcal{Q}_\mu j^\mu = 0.$$

The same relation for free particles reads

$$Q_\mu j^\mu = 0.$$

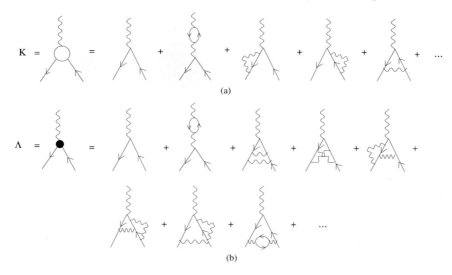

Fig. 9. a) Three-leg correlator K. b) Vertex part Λ.

3.2. Reducible and irreducible vertices

Now, construct a mixed (fermion-boson) correlator

$$K^\mu_{\pm,\sigma}\left(k, q|t, t_1, t_1'\right) = -\langle Tj^\mu\left(q, t\right) a_{\pm,\sigma}\left(k, t_1\right) a^\dagger_{\pm,\sigma}\left(k+q, t_1\right)\rangle, \qquad (3.4)$$

where $\mu = 0, 1$ and

$$\begin{aligned} j^0 &= \rho^c_+ + \rho^c_- \\ j^1 &= \rho^c_+ - \rho^c_- \end{aligned}$$

K^μ is an analog of the three-leg vertex in QED, except that in QED the "boson" is the $\mu-$ the component of the photon field

$$\text{QED} : K^\mu = -\langle T A^\mu a\bar{a}\rangle.$$

A diagrammatic representation of K^μ is a three-particle (one boson and two fermions) diagram (cf. Fig. 9a).

The diagrams with self-energy insertions to solid lines simply renormalize the Green's functions. Absorbing these renormalizations, we single out the vertex part, re-writing K^μ as

$$K^\mu = G^2\Lambda^\mu. \qquad (3.5)$$

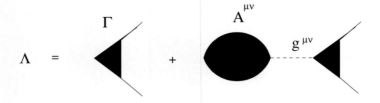

Fig. 10. Relation between vertices Λ and Γ.

Notice that there are as many vertex parts as there are bosonic degrees of freedom. In (3+1) QED, Λ^0 is a *scalar* vertex and $\Lambda^{\mu=1,2,3}$ are the components of the *vector* vertex. Diagrams representing Λ^μ are shown in Fig. 9b. These series can be re-arranged further by separating the *photon-irreducible* vertex part, Γ^μ. A photon-irreducible part is obtained by separating the corrections to the bosonic line, i.e., taking into account polarization. Vertices Λ^μ and Γ^μ are related via a kind of Dyson equation, which is simpler than the usual Dyson in a sense that there is no Λ^μ on the right-hand-side. Diagrammatically, this relation is represented by Fig.10 where a shaded bubble is an exact (renormalized) current-current correlation function

$$A^{\mu\nu}(q,t) = -\frac{i}{V}\langle j^\mu(q,t)\, j^\mu(-q,0)\rangle.$$

Algebraically, equation in Fig.10 says

$$\Lambda^\mu_{\pm,\sigma} = \Gamma^\mu_{\pm,\sigma} + A^{\mu\nu}g_{\mu\lambda}\Gamma^\lambda_{\pm,\sigma}. \tag{3.6}$$

(We remind the reader that indices $\pm, \sigma$ simply specify the fermionic flavor which is not mixed in our approximation of forward-scattering and spin-independent forces, so all relations are applicable to each individual flavor.) The coupling constants $g^{\mu\nu}$ relate currents to densities. According to Eqs. (3.1) and (3.2), densities couple to densities and currents to currents with no cross terms. Opening the matrix product in Eq. (3.6), we obtain

$$\Lambda^\mu_i = \Gamma^\mu_i + A^{\mu 0}g_{00}\Gamma^0 + A^{\mu 1}g_{11}\Gamma^1. \tag{3.7}$$

3.3. Ward identities

A Ward identity for vertex Λ^μ is obtained by applying $i\partial_t$ to K^μ in Eq. (3.4) and using the continuity equation (3.4) [15]. Performing this operations and Fourier

---

[15] When differentiating, recall that the $T-$ product can be represented by step-functions in time which, upon differentiating, yield delta-functions.

transforming in time, we obtain

$$\mathcal{Q}_\mu K_i^\mu (K, Q) = G_i (K) - G_i (K + Q),$$

where i denotes the branch

$$i \equiv \pm, \sigma.$$

Recalling Eq. (3.5), we see that the Ward identity takes a form

$$\mathcal{Q}_\mu \Lambda_i^\mu (K, Q) = G_i^{-1} (K + Q) - G_i^{-1} (K), \tag{3.8}$$

which is identical to a corresponding identity in QED. For those who like to see things not masked by fancy notations, here is Eq. (3.8) in an explicit form

$$\omega \Lambda_i^0 (\varepsilon, k; \omega, q) - v_c q \Lambda_i^1 (\varepsilon, k; \omega, q) = G_i^{-1} (\varepsilon + \omega, k + q) - G_i^{-1} (\varepsilon, k). \tag{3.9}$$

Notice that Eqs. (3.8,3.9) contain *renormalized* velocity v_c. In what follows, we will actually need a Ward identity not for Λ^μ but for the photon-irreducible vertex Γ^μ. This one is obtained by deriving the continuity equation for 4-current correlation function $A^{\mu\nu}$. To this end, one applies $i\partial_t$ to $A^{0\nu}$ and uses continuity equation (3.3), which yields [16]

$$\mathcal{Q}_\mu A^{\mu\nu} = \frac{2}{\pi} q \delta_{\nu,1}. \tag{3.11}$$

Now, we form a scalar product between $\mathcal{Q}_\mu$ and Eq. (3.7), using continuity equation for $A^{\mu\nu}$ (3.11). This brings us to

$$\mathcal{Q}_\mu \Lambda_i^\mu = \mathcal{Q}_\mu A^\mu,$$

---

[16]To get this result, recall the form of the anomalous density-density commutator

$$[j^\mu (q), j^\nu (-q)] = \epsilon^{\mu\nu} \frac{2}{\pi} q L,$$

where $\epsilon^{00} = \epsilon^{11} = 0$, $\epsilon^{01} = -\epsilon^{10} = 1$. Now open the $T-$ product in $A^{0\nu}$ and apply $i\partial_t$

$$
\begin{aligned}
i\partial_t A^{0\nu} (q, t) &= -\frac{i}{V} (i\partial_t) \langle \theta (t) j^0 (q, t) j^\nu (-q, 0) + \theta (-t) j^\nu (-q, 0) j^0 (q, t) \rangle \\
&= \frac{1}{V} \delta (t) [j^0 (q, 0), j^\nu (-q, 0)] + v_a q A^{1\nu} = 2 \frac{q}{\pi} \delta_{\nu,1} + v_a q A^{1\nu}. \tag{3.10}
\end{aligned}
$$

In 4-notations, (3.10) is equivalent to (3.11).

where

$$Q_\mu \Lambda_i^\mu = Q_\mu \left(\Gamma_i^\mu + A^{\mu 0} g_{00} \Gamma_i^0 + A^{\mu 1} g_{11} \Gamma_i^1 \right)$$

$$= Q_\mu \Gamma_i^\mu + \underbrace{Q_\mu A^{\mu 0}}_{=0} g_{00} \Gamma_i^0 + \underbrace{Q_\mu A^{\mu 1}}_{=2q/\pi} g_{11} \Gamma_i^1$$

$$= \omega \Gamma_i^0 - \underbrace{v_c}_{=1+(g_4-g_2)/\pi} q \Gamma_i^1 + \frac{2q}{\pi} \frac{1}{2} \frac{g_4 - g_2}{\pi} \Gamma_i^1$$

$$= \omega \Gamma_i^0 - q \Gamma_i^1 = Q_\mu \Gamma^\mu.$$

Finally, the Ward identity for photon-irreducible vertex is

$$Q_\mu \Gamma^\mu = G_i^{-1}(K + Q) - G_i^{-1}(K). \tag{3.12}$$

It is remarkable that the left-hand-side of Eq. (3.12) contains *bare* Fermi velocity
(= 1) instead of the renormalized one. This is true even if we allowed for spin-
dependent interaction in the Hamiltonian.

It seems that we have not achieved much, as the conservation law was simply
cast into a different form. However, in our 1D problem with a linearized spectrum a further progress can be made because the current and density (for given
chirality) are just the same quantity (up to an overall Fermi velocity which we
put to unity anyhow):

$$\Gamma_{\pm,\sigma}^1 = \pm \Gamma_{\pm,\sigma}^0$$

Therefore, we have a closed relation between just one vertex and Green's functions. Suppressing the 4-vector index μ, we get the Ward identity for the density
vertex

$$\Gamma_{\pm,\sigma}^0 (K, Q) = \frac{G_{\pm,\sigma}^{-1}(K + Q) - G_{\pm,\sigma}^{-1}(K)}{\omega \mp q}. \tag{3.13}$$

This is the identity that we need to proceed further with the Dzyaloshinskii-
Larkin solution of the Tomonaga-Luttinger problem. Notice that (3.13) contains
fully interacting Green's functions.

3.4. Effective interaction

Effective interaction is obtained by collecting polarization corrections to the bare
one. Diagrammatically, this procedure is described by the Dyson equation, represented in Fig.11. The interaction and polarization bubble are matrices with
components

$$\hat{V} = \begin{pmatrix} V_{++} & V_{+-} \\ V_{+-} & V_{++} \end{pmatrix}, \hat{V}_0 = \begin{pmatrix} g_4 & g_2 \\ g_2 & g_4 \end{pmatrix}, \hat{\Pi} = \begin{pmatrix} \Pi_+ & 0 \\ 0 & \Pi_- \end{pmatrix},$$

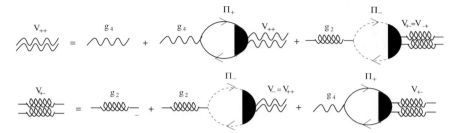

Fig. 11. Dyson equation for the effective interaction. Solid line: Green's function of a right-moving fermion. Dashed line: Green's function of a left-moving fermion. Single wavy line: bare interaction of fermions of the same chirality; spiral line: same for the fermions of opposite chirality. Double wavy and spiral lines represent the renormalized interactions.

where we used an obvious symmetry $V_{++} = V_{--}$, $V_{+-} = V_{-+}$. The Dyson equation in the matrix form reads

$$\hat{V} = \hat{V}_0 + \hat{V}_0 \hat{\Pi} \hat{V},$$

or, in components,

$$
\begin{aligned}
V_{++} &= g_4 + g_4 \Pi_+ V_{++} + g_2 \Pi_- V_{+-}; \\
V_{+-} &= g_2 + g_2 \Pi_- V_{++} + g_4 \Pi_- V_{+-}.
\end{aligned}
\tag{3.14}
$$

Bubbles in these equations are *fully renormalized* ones, *i.e.,* they are built on exact Green's functions and contain a vertex (hatched corner):

$$\Pi_\pm (\omega, q) = -2i \int \int \frac{dk\, d\varepsilon}{(2\pi)^2} G_\pm (\varepsilon + \omega, k + q)\, G_\pm (\varepsilon, k)\, \Gamma^0_\pm (\varepsilon, k; \omega, q).$$

Now we use the Ward identity for $\Gamma^0_\pm$ (3.13) to get [17]

$$\Pi_\pm (\omega, q) = -2i \frac{1}{\omega \mp q} \int \int \frac{dk\, d\varepsilon}{(2\pi)^2} \left[G_\pm (\varepsilon, k) - G_\pm (\varepsilon + \omega, k + q) \right].$$
$$\tag{3.15}$$

Eq. (3.15) looks exactly the same as a *free* bubble [cf. Eq. (B.1)] except that it contains exact rather than free Green's functions. Because we managed to

---

[17]I skipped over a subtlety related to the infinitesimal imaginary parts $i0^+$ in the denominator. Works the same way. If you are unhappy with this, imagine that we work with Matsubara frequencies. Then there are no $i0^+$s whatsoever.

transform the product of two Green's functions into a difference, frequency integration in Eq. (3.15) can be performed term by term yielding *exact* momentum distribution functions $n_{\pm}(k)$ and $n_{\pm}(k+q)$:

$$\Pi_{\pm}(\omega, q) = \frac{1}{\omega \mp q} \int \int \frac{dk}{\pi} \left[n_{\pm}(k) - n_{\pm}(k+q) \right]. \tag{3.16}$$

It seems that we have not achieved much so far. Indeed, we traded one unknown quantity ($\Pi_{\pm}$) for another ($n_{\pm}$). Both of them include the interaction to all orders and without any further simplification we are stuck. In fact, we have already made an important simplification: when specifying the model, we assumed only forward scattering. This means that the interaction is sufficiently long-range in real space so that backscattering can be neglected. Equivalently, in the momentum space it means that our interaction operates only in a narrow window of width q_0 near the Fermi points, $\pm k_F$. Thus the states far away from the Fermi points are not affected by the interaction. The momentum integral in (3.16) comes from regions far away from the Fermi surface where unknown functions $n_{\pm}$ can be approximated by free Fermi steps. This approximation is good as long as $q_0 \ll k_F$. The solution is going to be exact only in a sense that there will be no constraints on the amplitude of the interaction (parameters g_2 and g_4) but not its range [18]. Now we understand better why the title of the paper by Dzyaloshinskii and Larkin [59] is "Correlation functions for a one-dimensional Fermi system with *long-range* interaction (Tomonaga model)" [19].

With this simplification, the momentum integration proceeds in the same way as for free fermions (see Appendix B) with the result *that the fully interacting bubbles are the same as free ones*

$$\Pi_{\pm}(\omega, q) = \Pi_{\pm}^0(\omega, q) = \pm \frac{1}{\pi} \frac{q}{\omega - q + i0^+ \mathrm{sgn}\omega}. \tag{3.17}$$

This is a truly remarkable result which is a cornerstone for the DL solution [20].

---

[18] In higher dimensions, we have a familiar problem of the Coulomb potential. Because it's a power-law potential, one cannot separate it into "amplitude" and "range". There is in fact a single dimensionless parameter, r_s, which must be small for the perturbation theory–Random Phase Approximation–to work. Once $r_s \ll 1$, we have two things: the screened potential is simultaneously small *and* long-ranged. The Tomonaga-Luttinger model unties these two things: the interaction is assumed to be long-ranged but not necessarily small.

[19] What seemed to be just a matter of mathematical convenience in the 50 and 60s, turns out to be quite a realistic case these days. If a wire of width a is located at distance d to the metallic gate, the Coulomb potential between electrons in the wire is screened by their images in the gate. Typically, $d \gg a$. A simple exercise in electrostatics shows that in this case $U(0)$ is larger than $U(2k_F)$ by large factor $\ln(d/a)$ [62].

[20] In QED, this statement is known as Furry theorem (W. H. Furry, 1937).

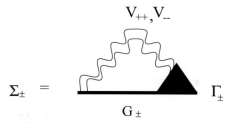

Fig. 12. Dyson equation for the self-energy.

Because our bubbles were effectively "liberated" from the interaction effects, system (3.14) is equivalent to what we would have obtained from the Random Phase Approximation (RPA). It turns out that RPA is *asymptotically* exact in 1D in the limit $q_0/k_F \to 0$. Solving the 2 by 2 system, we obtain for the effective interaction

$$V_{++}(\omega, q) = (\omega - q)\frac{g_4(\omega + q) + \left(g_4^2 - g_2^2\right)q/\pi}{\omega^2 - u^2 q^2 + i0^+},$$

where [21]

$$u = \sqrt{1 + \frac{2g_4}{\pi} + \frac{g_4^2 - g_2^2}{\pi}}.$$

For $g_4 = g_2 \equiv g$,

$$V_{++}(\omega, q) = g\frac{\omega^2 - q^2}{\omega^2 - u^2 q^2 + i0^+}. \tag{3.18}$$

3.5. Dyson equation for the Green's function

Dyson equation for right-moving fermions reads

$$\Sigma_+(P) = i \int \frac{d^2 Q}{(2\pi)^2} G_+(P - Q) V_{++}(Q) \Gamma_+^0(P, Q).$$

Diagrammatically, this equation is shown in Fig. 12. For linear dispersion,

$$\Sigma_\pm(\varepsilon, p) = \varepsilon \mp p - G_\pm^{-1}(\varepsilon, p)$$

---

[21] Notice that as long as $g_4 \neq g_2$, the left-right symmetry is broken, i.e., the potential is not symmetric with respect to $q \to -q$.

Substituting this relation back into the Dyson equations, we obtain

$$(\varepsilon - p)\, G_+ \,(\varepsilon, p) \;=\; 1 + i \int \int \frac{d\omega dq}{(2\pi)^2} G_+ \,(\varepsilon, p)\, G_+ \,(\varepsilon - \omega, p - q)$$
$$\times V_{++} \,(\omega, q)\, \Gamma_+^0 \,(\varepsilon, p; \omega, q)\,.$$

Using the Ward identity (3.13), we get

$$(\varepsilon - p + \Sigma_0)\, G_+ \,(\varepsilon, p) = 1 + i \int \int \frac{d\omega dq}{(2\pi)^2} G_+ \,(\varepsilon, p)\, G_+ \,(\varepsilon - \omega, p - q)$$
$$\times \frac{V_{++} \,(\omega, q)}{\omega - q} \left[G_+^{-1} \,(\varepsilon, p) - G_+^{-1} \,(\varepsilon - \omega, p - q) \right]$$
$$= 1 + i \int \int \frac{d\omega dq}{(2\pi)^2} G_+ \,(\varepsilon - \omega, p - q)\, \frac{V_{++} \,(\omega, q)}{\omega - q} + G_+ \,(\varepsilon, p) \times \text{const},$$

where

$$\text{const} = i \int \int \frac{d\omega dq}{(2\pi)^2} \frac{V_{++} \,(\omega, q)}{\omega - q}.$$

A constant term can always be absorbed into Σ, which simply results in a shift of the chemical potential. We are free to choose this shift in such a way that const=0, so that the Dyson equation reduces to

$$(\varepsilon - p)\, G_+ \,(\varepsilon, p) = 1 + i \int \int \frac{d\omega dq}{(2\pi)^2} G_+ \,(\varepsilon - \omega, p - q)\, \frac{V_{++} \,(\omega, q)}{\omega - q}. \quad (3.19)$$

Notice that Eq. (3.19) is an integral equation with a difference kernel, which can be reduced to a differential equation for G. Before we demonstrate how it is done, let's have a brief look at a case when there is no coupling between left- and right-moving fermions: $g_2 = 0$. In this case,

$$V_{++} \,(\omega, q) = \pi \frac{(w - 1)\,(\omega - q)}{\omega - wq + i0^+},$$

where

$$w = 1 + g_4/\pi.$$

Eq. (3.19) takes the form

$$(\varepsilon - p)\, G_+ \,(\varepsilon, p) = 1 + i\,(w - 1) \int \int \frac{d\omega dq}{4\pi} \frac{G_+ \,(\varepsilon - \omega, p - q)}{\omega - wq + i0^+}.$$

This equation is satisfied by the following function

$$G_+ \,(\varepsilon, p) = \frac{1}{\sqrt{\varepsilon - p + i0^+}\sqrt{\varepsilon - wp + i0^+}}. \quad (3.20)$$

This is an example of a non-Fermi-liquid behavior: the pole of a free G splits into the product of two branch cuts, one peaked on the mass shell of free fermions ($\varepsilon = p$) and another one at the renormalized mass shell ($\varepsilon = wp$). As left- and right movers are totally decoupled in this problem, the same result would have been obtained for two separate subsystems of left- and right movers. For example, Eq. (3.20) predicts that an edge state of an *integer* quantum Hall system is not a Fermi liquid, if spins are not yet polarized by the magnetic field [63]. The same procedure for a spinless system would give us a pole-like G with a renormalized Fermi velocity. The non-Fermi-liquid behavior described by Eq. (3.20) is rather subtle: it exists only if both ε and p are finite. In the limiting case of $p = 0$ (tunneling DoS) we are back to a free-fermion behavior $G(\varepsilon, 0) = \varepsilon^{-1}$. Also, the $\varepsilon-$ integral of Eq. (3.19) gives a step-like distribution function in momentum space. The spectral function, however, is characteristically non-FL-like: instead of delta-function peak we have a whole region $|p| < |\varepsilon| < w|p|$ in which ImG is finite. At the edges of this interval ImG has square-root singularities.

3.6. Solution for the case $g_2 = g_4$

Substituting the effective interaction (3.18) into Dyson equation (3.19), we obtain

$$(\varepsilon - p)\, G_+ (\varepsilon, p) = 1 + i \int \frac{d\omega dq}{4\pi^2} G(\varepsilon - \omega, p - q)\, g(q)\, \frac{\omega + q}{\omega^2 - u^2 q^2},$$

where

$$u = \sqrt{1 + 2g/\pi}.$$

Notice that the constant g is replaced by a momentum-dependent interaction, $g(q)$. The reason is that without such a replacement the integral diverges at the upper limit. Here, the assumption of a cut-off in the interaction becomes important again. Transforming back to real time and space

$$G(x, t) = \int \int \frac{d\varepsilon dp}{(2\pi)^2} G(\varepsilon, p)\, e^{i(px - \varepsilon t)},$$

we obtain the Dyson equation in a differential form

$$\left(\frac{\partial}{\partial t} + \frac{\partial}{\partial x} \right) G(x, t) = P(x, t)\, G(x, t) - i\delta(x)\delta(t), \tag{3.21}$$

where

$$P(x, t) = \frac{1}{4\pi^2} \int \int d\omega dq\, e^{i(qx - \omega t)} g(q)\, \frac{\omega + q}{\omega^2 - u^2 q^2 + i0^+}. \tag{3.22}$$

The integral for P diverges if g is constant. To ensure convergence, we will approximate $g(q) = ge^{-|q|/q_0}$. An actual form of the cut-off function is not important as long as we are interested in such times and spatial intervals such that $x, t \gg q_0^{-1}$. The integral over ω is solved by closing the contour around the poles of the denominator $\omega = \pm u|q|(1 + i0^+)$. For $t > 0$, we need to choose the one with $\text{Im}\,\omega < 0$. Doing so, we obtain

$$\int \frac{d\omega}{2\pi} \cdots = i\,\frac{\text{sgn}\,q + u}{2u}\,e^{i(qx - u|qt|)}.$$

Solving the remaining $q-$ integral, we obtain for $P(x, t)$

$$P(x, t) = \frac{g}{4\pi u}\left(\frac{u + 1}{x - ut + i/q_0} - \frac{u - 1}{x + ut + i/q_0}\right).$$

For $t < 0$, one needs to change $q_0 \to -q_0$ in the last formula.

The delta-function term can be viewed as a boundary condition

$$G(x, 0+) - G(x, 0-) = -i\delta(x). \tag{3.23}$$

Once the function $P(x, t)$ is known, Eq. (3.21) is trivially solved in terms of new variables $r = x - t$, $s = x + t$. For example, for $t > 0$

$$G_+(r, t > 0) = G_0(x, t)\,f_>(r)\exp\left[i\int_r^s ds'\,P(r, s')\right], \tag{3.24}$$

where function $f_>(r)$ is determined by the analytic properties of G as a function of ε. Substituting result for $P(x, t)$ into Eq. (3.24), we get

$$G_+(x, t > 0)$$
$$= \frac{1}{2\pi}G_0(x, t)\,f_>(x - t)\left(\frac{x - t + i/q_0}{x - ut + i/q_0}\right)^{\alpha + 1/2}\left(\frac{x - t - i/q_0}{x + ut - i/q_0}\right)^{\alpha}.$$

where

$$\alpha = \frac{(u - 1)^2}{8u}. \tag{3.25}$$

Formula for $t < 0$ is obtained by choosing another function $f_<$ and replacing $q_0 \to -q_0$. Functions $f_{>,<}$ are determined from the analytic properties. First of all, recall that

$$G_0(x, t) = \frac{1}{x - t + i\,\text{sgn}\,t0^+}.$$

From this form, we see that although G_0 is *not* an analytic function of t for any t, it is analytic for $\text{Re}\,t > 0$ in the right lower quadrant ($\text{Im}\,t < 0$) and for $\text{Re}\,t < 0$

in the upper left quadrant ($\text{Im}t > 0$). The interaction cannot change analytic properties of a Green's function hence we should expect the same properties to hold for full G [22].

From the boundary condition (3.23), it follows that

$$f_> (x) = f_< (x)$$

and $f (0) = 0$.

Analyzing different factors in the formula for G, we see that only the term $(x - t \mp i/q_0)^\alpha$ does not satisfy the required analyticity property. This term is eliminated by choosing function $f (x)$ as

$$f (x) = \left(q_0^2 x^2 + 1\right)^{-\alpha} .$$

Finally, the result for G takes the form

$$G_+ (x, t) = \frac{1}{2\pi} \frac{1}{x - t + i\,\text{sgn}t0^+} \left(\frac{x - t + i\gamma}{x - ut + i\gamma}\right)^{1/2}$$
$$\times \frac{1}{\left[q_0^2 (x - ut + i\gamma)(x + ut - i\gamma)\right]^\alpha},$$

where $\gamma = \text{sgn}t/q_0$. It seems somewhat redundant to keep two different damping terms ($i\,\text{sgn}t0^+$ and γ) in the same equation. However, these terms contain different physical scales. Indeed, $i\,\text{sgn}t0^+$ enters a free Green's function and 0^+ there has to be understood as the limit of the inverse system size. On the other hand, γ contains a cut-off of the interaction. Obviously, $|\gamma| \gg 1/L \to 0^+$ for a realistic situation. The difference between the two cutoffs becomes important for the momentum distribution function and tunneling DoS, discussed in the next Section.

---

[22] Indeed, this property follows immediately from the Lehmann representation for G

$$G (x, t) = -i \sum_\nu |M_{\nu 0}|^2 e^{ip_\nu x} e^{-i(E_\nu - E_0)t}, \text{ for } t > 0;$$
$$= i \sum_\nu |M_{\nu 0}|^2 e^{-ip_\nu x} e^{i(E_\nu - E_0)t}, \text{ for } t < 0,$$

where $M_{\nu 0}$ are the matrix elements between the ground state and state ν with energy $E_\nu > E_0$. The required property simply follows from the condition for convergence of the sum.

3.7. Physical properties

3.7.1. Momentum distribution

Having an exact form of the Green's function, we can now calculate the momentum distribution of, *e.g.,* right-moving fermions:

$$
\begin{aligned}
n_+(p) &= -i \int_{-\infty}^{\infty} dx\, e^{-ipx} G_+\left(x, t \to 0^+\right) \\
&= -\frac{i}{2\pi} \int_{-\infty}^{\infty} dx \frac{e^{-ipx}}{x + i0^+} \frac{1}{\left[q_0^2 x^2 + 1\right]^\alpha} \\
&= -\frac{i}{2\pi} \int_{-\infty}^{\infty} dx\, e^{-ipx} \left[\mathcal{P}\frac{1}{x} - i\pi\delta(x)\right] \frac{1}{\left[q_0^2 x^2 + 1\right]^\alpha} \\
&= \frac{1}{2} - \frac{1}{\pi} \mathrm{sgn}\, p \int_0^{\infty} dx \frac{\sin|p|\,x}{x} \frac{1}{\left[q_0^2 x^2 + 1\right]^\alpha},
\end{aligned}
$$

We are interested in the behavior at $p \to 0$ (which means $|p| \ll q_0$). The final result for $n_+(p)$ depends on whether α is larger or smaller than $1/2$ [59, 64].

• For $\alpha < 1/2$ ("weak interaction"), one cannot expand $\sin px$ in x because the resulting integral diverges at $x = \infty$. Instead, rescale $px \to y$

$$
n_+(p) = \frac{1}{2} - \frac{1}{\pi} \int_0^{\infty} dy \frac{\sin y}{y} \frac{1}{\left[(q_0/p)^2 y^2 + 1\right]^\alpha}
$$

and neglect 1 in the denominator. This gives

$$
n_+(p) = \frac{1}{2} + C_1 \left(\frac{|p|}{q_0}\right)^{2\alpha} \mathrm{sgn}\, p \tag{3.26}
$$

where

$$
C_1 = \frac{\sin \pi\alpha}{\pi} \Gamma(-2\alpha).
$$

Notice that $n_+(p)$ is finite ($= 1/2$) at $p = 0$, although its derivative is singular. We should be able to recover the Fermi-gas step at $p = 0$ by setting $\alpha = 0$ in (3.26). Indeed,

$$
\lim_{\alpha \to 0} C_1 = \alpha \frac{1}{-2\alpha} = -\frac{1}{2}
$$

and

$$
n(p) = \frac{1 - \mathrm{sgn}\, p}{2},
$$

which is just the Fermi-gas result. Notice also that [23] there is nothing special about the limit $\alpha \to 0$. Indeed, constant C_1 has a regular expansion in α

$$C_1 = -\frac{1}{2} - \gamma a + \ldots$$

and factor $(|p|/q_0)^{2\alpha}$ can be expanded for finite p and small α as

$$(|p|/q_0)^{2\alpha} = 1 + 2\alpha \ln |p|/q_0.$$

To leading order in α, we obtain

$$n_+(p) = \frac{1}{2} - \mathrm{sgn} p \frac{1}{2}[1 + 2\alpha \ln |p|/q_0] = n_0(p) - \alpha \, \mathrm{sgn} p \ln |p|/q_0,$$

which is a perfectly regular in α (but logarithmically divergent at $p \to 0$) behavior. Once again, it is not surprising: despite the fact that the results for a 1D system differ dramatically from that for the Fermi gas, they are still *perturbative, i.e., analytic,* in the coupling constant.

• For $\alpha > 1/2$ ("strong interaction"), it is safe to expand $\sin px$ and the result is

$$n_+(p) = \frac{1}{2} - C_2 p/q_0,$$

where

$$C_1 = \frac{1}{2\sqrt{\pi}} \frac{\Gamma(\alpha - 1/2)}{\Gamma(\alpha)}.$$

In this case, no remains of a jump at the Fermi point is present in $n_+(p)$ which is a regular, linear function near $p = 0$.

• Finally, $\alpha = 1/2$ is a special case, where expansion in p results in a log-divergent integral. To log-accuracy

$$n_+(p) = \frac{1}{2} - \frac{1}{\pi} \frac{p}{q_0} \ln \frac{q_0}{|p|}.$$

In general, $n(p)$ is some hypergeometric function of p/q_0 which decays rapidly for $p \gg q_0$ and approaches 1 for $p \ll -q_0$. A posteriori, this justifies the replacement of exact $n(p)$ by its free form in the Dyson equation.

---

[23] Contrary to some statements in the literature.

3.7.2. Tunneling density of states

Now we turn to the tunneling DoS

$$N(\varepsilon) = -\frac{1}{\pi} \text{Im} G^R(\varepsilon, x = 0).$$

Recalling that [23]

$$
\begin{aligned}
G^R(\varepsilon) &= G(\varepsilon), \text{ for } \varepsilon > 0; \\
&= G^*(\varepsilon), \text{ for } \varepsilon < 0,
\end{aligned}
$$

we see that

$$\text{Im} G^R(\varepsilon, 0) = \text{sgn}\varepsilon \, \text{Im} G(\varepsilon, 0)$$

and

$$
\begin{aligned}
N(\varepsilon) &= -\frac{1}{\pi} \text{sgn}\varepsilon \, \text{Im} G(0, \varepsilon) \\
&= -\frac{1}{\pi} \text{sgn}\varepsilon \left[\int dt e^{i\varepsilon t} G(0, t) - \int dt e^{-i\varepsilon t} G^*(0, t) \right] \\
&= -\frac{1}{\pi} \text{sgn}\varepsilon \left[\int dt e^{i\varepsilon t} \{ G(0, t) - G^*(0, -t) \} \right]
\end{aligned}
$$

For $t \to \infty$,

$$G(0, t) = \frac{\text{const}}{(-t)^{1+2\alpha}}$$

and

$$G(0, t) - G^*(-t)$$

is an odd function of t. Thus

$$
\begin{aligned}
N(\varepsilon) &= -\frac{1}{\pi} \text{sgn}\varepsilon \, \text{Im} G(0, \varepsilon) \\
&= -\frac{1}{\pi} \text{sgn}\varepsilon \frac{1}{2i} \left[\int dt e^{i\varepsilon t} G(0, t) - \int dt e^{-i\varepsilon t} G^*(0, t) \right] \\
&= -\frac{1}{\pi} \text{sgn}\varepsilon \left[\int_0^\infty dt \sin \varepsilon t \, \{ G(0, t) - G^*(0, -t) \} \right] \\
&\propto \text{sgn}\varepsilon \int_0^\infty dt \frac{\sin \varepsilon t}{t^{1+2\alpha}}.
\end{aligned}
$$

The integral is obviously convergent for $\alpha < 1/2$. In this case,

$$N_s(\varepsilon) \propto |\varepsilon|^{2\alpha}.$$

which means that the local tunneling DoS is suppressed at the Fermi level. Actually, the exponent for $\alpha > 1/2$ is the same, however, the prefactor is a different function of α [65].

The DoS in Eq. (3.7.2) with exponent 2α, where α is given by Eq. (3.25) corresponds to tunneling into the "bulk" of a 1D system, *i.e.*, when the tunneling contact–with a tip of an STM or another carbon nanotube crossing the first one– is far away from its ends. In the next Section, we will analyze tunneling into an edge of a 1D conductor, which is characterized by a different exponent, α'.

4. Renormalization group for interacting fermions

The Tomonaga-Luttinger model can be solved exactly as it was done in the previous Section– only in the absence of backscattering. Backscattering can be treated via the Renormalization Group (RG) procedure. This treatment is standard by now and discussed in a number of sources [1, 2, 3, 4, 5, 6, 7, 8, 9, 10]. For the sake of completeness, I present here a short derivation of the RG equations. A reader familiar with the procedure can skip this Section and go directly to Sec. 5, where these equations will be used in the context of a single-impurity problem.

An exact solution of the previous Section is parameterized by two coupling constants, g_2 and g_4, which are equal to their bare values. In the RG language, it means that these couplings do not flow. Let's see if this is indeed the case. In what follows, I will neglect the g_4- processes, as their effect on the flow of other couplings is trivial, and, for the sake of simplicity, consider a spin-independent interaction. To second order, the renormalization of the g_2- coupling is accounted for by two diagrams: diagrams a) and b) of Fig. 13.

Diagram a) is a correction to g_2 in the particle-particle channel. The correction to g_2 is given by

$$\left(g_2^{(2)}\right)_a = \frac{g_2^2}{(2\pi)^2} \int dq \int d\omega G_+ \left(i\varepsilon_1 + i\omega, k_1 + q\right) G_- \left(i\varepsilon_2 - i\omega, k_2 - q\right).$$

Without a loss of generality, one can choose all momenta to be on the Fermi "surface": $k_1 = k_2 = k_3 = k_4 = 0$. Choose $q > 0$ (the other choice $q < 0$ will simply double the result)

$$
\begin{aligned}
\left(g_2^{(2)}\right)_a &= \frac{g_2^2}{(2\pi)^2} \int dq \int d\omega G_+ \left(i\varepsilon_1 + i\omega, q\right) G_- \left(i\varepsilon_2 - i\omega, -q\right) \\
&= \frac{g_2^2}{(2\pi)^2} \int_0^{\Lambda/2} dq \int d\omega \frac{1}{i\left(\varepsilon_1 + \omega\right) - q} \frac{1}{i\left(\varepsilon_2 - \omega\right) - q} \\
&= \frac{2\pi i g_2^2}{(2\pi)^2} \int_0^{\Lambda/2} dq \frac{1}{\varepsilon_1 + \varepsilon_2 + \omega + 2iq} = \frac{g_2^2}{4\pi} \ln \frac{i\Lambda}{\varepsilon_1 + \varepsilon_2}
\end{aligned}
$$

Adding the result up with the (identical) $q < 0$ contribution, we find

$$\left(g_2^{(2)}\right)_a = \frac{g_2^2}{2\pi} \ln \frac{i\Lambda}{\varepsilon_1 + \varepsilon_2}.$$

Diagram b) is a correction to g_2 in the particle-hole channel:

$$\begin{aligned}
\left(g_2^{(2)}\right)_b &= \frac{g_2^2}{(2\pi)^2} \int dq \int d\omega\, G_+\, (i\varepsilon_1 + i\omega, q)\, G_-\, (i\varepsilon_4 + i\omega, q) \\
&= \frac{g_2^2}{(2\pi)^2} \int_0^{\Lambda/2} dq \int d\omega \frac{1}{i\,(\varepsilon_1 + \omega) - q} \frac{1}{i\,(\varepsilon_4 + \omega) + q} \\
&= -\frac{2\pi i}{(2\pi)^2} g_2^2 \int_0^{\Lambda/2d} q \frac{1}{\varepsilon_1 - \varepsilon_4 + \omega + 2iq} = -\frac{g_2^2}{4\pi} \ln \frac{i\Lambda}{\varepsilon_1 - \varepsilon_4}.
\end{aligned}$$

As in the previous case, the final result is:

$$\left(g_2^{(2)}\right)_b = -\frac{g_2^2}{2\pi} \ln \frac{i\Lambda}{\varepsilon_1 - \varepsilon_4}.$$

If we sum only the Cooper ladders, adding up more vertical interaction lines to diagram a), the full vertex becomes

$$\Gamma_{pp} = \frac{g_2}{1 + g_2 \ln \frac{i\Lambda}{\varepsilon_1 + \varepsilon_2}}.$$

(To keep track of the signs, one needs to recall that in Matsubara frequencies each interaction line comes with the minus sign from the expansion of the $S-$ matrix.) The resulting vertex blows up for attractive interaction ($g_2 < 0$) as $\varepsilon_1 + \varepsilon_2 \to 0$, which is nothing more than a Cooper instability.

Likewise, untwisting the crossed lines in diagram b) and adding more interaction lines, we get the particle-hole vertex

$$\Gamma_{ph} = \frac{g_2}{1 - g_2 \ln \frac{i\Lambda}{\varepsilon_1 - \varepsilon_4}}.$$

This vertex has an instability for repulsive interaction ($g_2 > 0$). In fact, none of these instabilities occur. To see this, add up the results of diagrams a) and b)

$$\left(g_2^{(2)}\right)_{a+b} = \frac{g_2^2}{2\pi} \left[\ln \frac{i\Lambda}{\varepsilon_1 + \varepsilon_2} - \ln \frac{i\Lambda}{\varepsilon_1 - \varepsilon_4} \right] = \frac{g_2^2}{2\pi} \ln \frac{\varepsilon_1 - \varepsilon_4}{\varepsilon_1 + \varepsilon_2}.$$

In the RG, one changes the cut-off and follow the corresponding evolution of the couplings. As the cut-off dependence cancelled out in the result for $(g_2^{(2)})_{a+b}$, coupling g_2 remains invariant under the RG flow.

Backscattering generates additional diagrams: diagrams c)-f) in Fig.13.

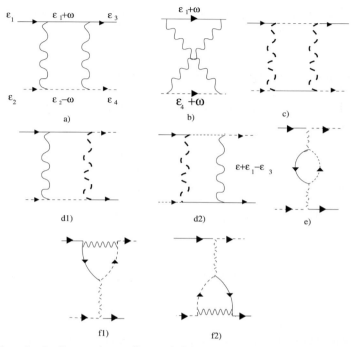

Fig. 13. Second order diagrams for couplings g_2 (solid wavy line) and g_1 (dashed wavy line). Straight solid and dashed lines correspond to Green's functions of right- and left moving fermions, correspondingly.

Diagram c) describes repeated backscattering in the particle-particle channel, which is equivalent to forward scattering. Therefore, this diagram gives a correction to g_2- coupling. Using the relation between $G_\pm$, *i.e.*, $G_\pm = -(G_\mp)^*$, we find

$$
\left(g_2^{(2)}\right)_c = \frac{g_1^2}{(2\pi)^2} \int dq \int d\omega G_- (i\varepsilon_1 + i\omega, q) \, G_+ (i\varepsilon_4 + i\omega, q)
$$

$$
= \frac{g_1^2}{(2\pi)^2} \left[\int dq \int d\omega G_+ (i\varepsilon_1 + i\omega, q) \, G_- (i\varepsilon_4 + i\omega, q) \right]^*.
$$

The last integral is the same as for $(g_2^{(2)})_a$. Thus,

$$
\left(g_2^{(2)}\right)_c = \frac{g_1^2}{g_2^2} \left[dg_2^{(1)} \right]^* = \frac{g_1^2}{2\pi} \ln \frac{-i\Lambda}{\varepsilon_1 + \varepsilon_2}.
$$

The rest of the diagrams provide corrections to g_1.

Diagram d1) is the same as diagram a) except for the prefactor being equal to $g_1 g_2$:

$$\left(g_1^{(2)}\right)_{d1} = \frac{g_1 g_2}{2\pi} \ln \frac{i\Lambda}{\varepsilon_1 + \varepsilon_2}.$$

Diagram d2) is a complex-conjugate of diagram d1). The sum of diagrams d1) and d2) is equal to

$$\left(g_1^{(2)}\right)_{d1+d2} = \frac{g_1 g_2}{2\pi} \ln \frac{i\Lambda}{\varepsilon_1 + \varepsilon_2} + \frac{g_1 g_2}{2\pi} \ln \frac{-i\Lambda}{\varepsilon_1 + \varepsilon_2}$$

$$= \frac{g_1 g_2}{\pi} \ln \frac{\Lambda}{\varepsilon_1 + \varepsilon_2}.$$

Diagrams e) is a polarization correction to the bare g_1−coupling:

$$\left(g_1^{(2)}\right)_e = \underbrace{-}_{\text{fermionic loop}} g_1^2 \Pi_{2k_F} \left(\omega = \varepsilon_1 - \varepsilon_2, q = 0\right).$$

Using Eq. (B.8), we obtain

$$\left(g_1^{(2)}\right)_e = \frac{N_s}{2\pi} g_1^2 \ln \frac{\Lambda}{|\varepsilon_1 - \varepsilon_2|},$$

where N_s is the degeneracy factor (=2 for spin 1/2 fermions, occupying a single valley in the momentum space).

Diagram f1) is the same as the bubble insertion, except for no - sign, no degeneracy factor (N_s) factor, and the overall coefficient is $g_1 g_2$:

$$\left(g_1^{(2)}\right)_{f1} = -\frac{1}{2\pi} g_1 g_2 \ln \frac{\Lambda}{|\varepsilon_1 - \varepsilon_2|}.$$

Diagram f2) is equal to f1). Their sum

$$\left(g_1^{(2)}\right)_{f1+f2} = -\frac{1}{\pi} g_1 g_2 \ln \frac{\Lambda}{|\varepsilon_1 - \varepsilon_2|}$$

Collecting all contributions together, we obtain

$$-\Gamma_2 = -g_2 + \left(g_2^{(2)}\right)_a + \left(g_2^{(2)}\right)_b + \left(g_2^{(2)}\right)_c;$$

$$\Gamma_2 = g_2 - \underbrace{\frac{g_2^2}{2\pi} \ln \frac{i\Lambda}{\varepsilon_1 + \varepsilon_2} + \frac{g_2^2}{2\pi} \ln \frac{i\Lambda}{\varepsilon_1 - \varepsilon_4}}_{\text{cancel out in the RG sense}} - \frac{g_1^2}{2\pi} \ln \frac{-i\Lambda}{\varepsilon_1 + \varepsilon_2};$$

$$-\Gamma_1 = -g_1 + \left(g_2^{(2)}\right)_d + \left(g_2^{(2)}\right)_e + \left(g_2^{(2)}\right)_f;$$

$$\Gamma_1 = g_1 - \frac{g_1 g_2}{\pi} \ln \frac{\Lambda}{\varepsilon_1 + \varepsilon_2} - \frac{N_s}{2\pi} g_1^2 \ln \frac{\Lambda}{|\varepsilon_1 - \varepsilon_2|} + \frac{1}{\pi} g_1 g_2 \ln \frac{\Lambda}{|\varepsilon_1 - \varepsilon_2|}.$$

Second and fourth terms in Γ_1 also cancel out in the RG sense. Changing the cut-off from Λ to $\Lambda + d\,\Lambda$, we obtain two differential equations

$$\frac{d\Gamma_2}{dl} = -\frac{\Gamma_1^2}{2\pi}; \tag{4.1}$$

$$\frac{d\Gamma_1}{dl} = -N_s \frac{\Gamma_1^2}{2\pi}, \tag{4.2}$$

where $l = \ln \Lambda$. We see that a quantity

$$\bar{\Gamma} = \Gamma_2 - \frac{1}{N_s}\Gamma_1 = \text{const} = g_2 - \frac{1}{N_s}g_1. \tag{4.3}$$

is invariant under RG flow, therefore its value can be obtained by substituting the bare values of the coupling constants (g_2 and g_1) into (4.3). The RG-invariant combination is then

$$\bar{\Gamma} = g_2 - \frac{1}{N_s}g_1. \tag{4.4}$$

For spinless electrons ($N_s = 1$),

$$\bar{\Gamma} = \Gamma_2 - \Gamma_1 = U(0) - U(2k_F).$$

This last result can be understood just in terms of the Pauli principle. Indeed, the anti-symmetrized vertex for spinless electrons is obtained by switching the outgoing legs of the diagram ($p_1, p_2 \to p_3, p_4$). To first order,

$$\Gamma(p_1, p_2; p_3, p_4) = U(p_1 - p_3) - U(p_1 - p_4).$$

Choosing $p_3 = p_1 - q$ and $p_4 = p_2 + q$, we obtain [recall that $U(q) = U(-q)$]

$$\Gamma(p_1, p_2|q) = U(q) - U(p_1 - p_2 - q).$$

One of the incoming fermions is a right mover ($p_1 = p_F$) and the other one is a left mover ($p_2 = -p_F$). As q is small compared to p_F, we obtain

$$\Gamma(p_1, p_2|q) = U(0) - U(2k_F).$$

In fact, for spinless electrons g_2 and g_1 processes are indistinguishable [24] as we do not know whether the right-moving electron in the final state is a right-mover

---

[24] That does not mean that backscattering is unimportant! It comes with a different scattering amplitude $U(2k_F)$. In fact, it is only backscattering which guarantees that the Pauli principle is satisfied, namely, for a contact interaction, when $U(0) = U(2k_F)$, we must get back to a Fermi gas as fermions are not allowed to occupy the same position in space and hence they cannot interact via contact forced. Our invariant combination $U(0) - U(2k_F)$ obviously satisfies this criterion. We will see that bosonization does have a problem with respecting the Pauli principle, and it takes some effort to recover it.

of the initial state, which experienced forward scattering, or the left-mover of the initial state, which experienced backscattering. A proper way to treat the case of spinless fermions is to include backscattering into Dzyaloshinskii-Larkin scheme from the very beginning, re-write the Hamiltonian in terms of *forward scattering* with invariant coupling $\bar{\Gamma}$, and proceed with the solution. All the results will then be expressed in terms of $\bar{\Gamma}$ rather than of g_2.

Solving the equation for Γ_1, gives on scale ε

$$\Gamma_1 = \frac{1}{(g_1)^{-1} + \frac{N_s}{2\pi} \ln \Lambda/\varepsilon}. \tag{4.5}$$

At low energies, Γ_1 renormalizes to zero ($\Gamma_1^* = \Gamma_1 (l = \infty) = 0$), if the interaction is repulsive, and blows up at $\varepsilon = \Lambda \exp(-1/|g_1|)$, if the interaction is attractive. Coupling Γ_2 also flows to a new value which can be read off from Eq. (4.4)

$$\Gamma_2^* = g_2 - \frac{1}{N_s} g_1.$$

Roughly speaking, g_1 is not important for repulsive interaction as the effective low-energy theory will look like a theory with forward scattering only. This does not really mean, however, that one can consider a fixed point as a new problem in which backscattering is absent, and apply our exact solution to this problem. Instead, one should calculate observables, derive the RG equations for flows, and use current values of coupling constants in these RG equations. An example of this procedure will be given in the next Section, where we will see that the flow of Γ_1 provides additional renormalization of the transmission coefficient in an interacting system.

Assigning different coupling constants to the interaction of fermions of parallel ($g_{1\parallel}$) and anti-parallel ($g_{1\perp}$) spins, one could see that the coupling which diverges for attractive interaction is in fact $g_{1\perp}$. This clarifies the nature of the gap that RG hints at (in fact, a perturbative RG can at most just give a hint): it is a spin gap. This becomes obvious in the bosonization technique, as the instability occurs in the spin-sector of the theory. An exact solution by Luther and Emery [66] for a special case of attractive interaction confirms this prediction.

5. Single impurity in a 1D system: scattering theory for interacting fermions

A single impurity or tunneling barrier placed in a 1D Fermi gas reduces the conductance from its universal value–e^2/h per spin orientation–to

$$\mathcal{G} = N_s \frac{e^2}{h} |t_0|^2, \tag{5.1}$$

where t_0 is the transmission amplitude. The interaction renormalizes the bare transmission amplitude, making the conductance depend on the characteristic energy scale (temperature or applied bias), which is observed as a zero-bias anomaly in tunneling. This effect is not really a unique property of 1D: in higher dimensions, zero-bias anomalies in both dirty and clean (ballistic) regimes [67, 12, 13, 14] as well as the interaction correction to the conductivity [67, 15], stem from the same physics, namely, scattering of electrons from Friedel oscillations produced by tunneling barriers or impurities. 1D is special in the magnitude of the effect: the conductance varies significantly already on the energy scale comparable to the Fermi energy, whereas in higher dimensions the effect of the interaction is either small at all energies or becomes significant only at low energies (below some scale which is much smaller than E_F as long as the parameter $k_F l$, where l is the elastic mean free path, is large). The 1D zero-bias anomaly is described quite simply in a bosonized language [68], which does not require the interaction to be weak. We will use this description in Sec.6. However, in this Section I will choose another description–via the scattering theory for fermions rather than bosons–developed by Matveev, Yue, and Glazman [11]. Although this approach is perturbative in the interaction, it elucidates the underlying mechanism of the zero-bias anomaly and allows for an extension to higher-dimensional case (which was done for the case of tunneling in Ref. [12] and transport in Ref. [15]).

5.1. First-order interaction correction to the transmission coefficient

In this section we consider a 1D system of *spinless* fermions with a tunneling barrier located at $x = 0$ [11]. For the sake of simplicity, I assume that the barrier is symmetric, so that transmission and reflection amplitude for the waves coming from the left and right are the same. Also, I assume that e-e interaction is present only to the right of the barrier, whereas to the left we have a Fermi gas. Such a situation models a setup when a tunneling contact separates a 1D interacting system (quantum wire or carbon nanotube) and a "good metal", in which interactions can be neglected. We also assume that the interaction potential $U(x)$ is sufficiently short-ranged, so that $U(0)$ is finite and one can neglect over-the-barrier interaction. However, $U(0) \neq U(2k_F)$ (otherwise, spinless electrons would not interact at all [25]).

---

[25] For a contact potential (which leads to $U(0) = U(2k_F)$), the four-fermion interaction for the spinless case reduces to $[\Psi^\dagger(0)]^2 \Psi^2(0)$. By Pauli principle, $[\Psi^\dagger(0)]^2 = \Psi^2(0) = 0$, so that the interaction is absent.

The wave function of the free problem for a right-moving state is:

$$\psi_k^0(x) = \frac{1}{\sqrt{L}}\left(e^{ikx} + r_0 e^{-ikx}\right), x < 0; \tag{5.2}$$

$$= \frac{1}{\sqrt{L}} t_0 e^{ikx}, x > 0.$$

For a left-moving state:

$$\psi_{-k}^0(x) = \frac{1}{\sqrt{L}}\left(e^{-ikx} + r_0 e^{ikx}\right), x > 0; \tag{5.3}$$

$$= \frac{1}{\sqrt{L}} t_0 e^{-ikx}, x < 0.$$

Here $k = \sqrt{2mE} > 0$. To begin with, we consider a high barrier: $|t_0| \ll 1$, $r_0 \approx -1$. Then the free wavefunction reduces to

$$\psi_k^0(x) = \frac{2i}{\sqrt{L}} \sin kx, x < 0 \quad \text{(incoming from the left+reflected);} \tag{5.4}$$

$$= \frac{1}{\sqrt{L}} t_0 e^{ikx}, x > 0 \quad \text{(transmitted left} \rightarrow \text{right);} \tag{5.5}$$

$$\psi_{-k}^0(x) = \frac{1}{\sqrt{L}} t_0 e^{-ikx}, x < 0 \quad \text{(transmitted right} \rightarrow \text{left);} \tag{5.6}$$

$$= -\frac{2i}{\sqrt{L}} \sin kx, x > 0 \quad \text{(incoming from the right+reflected).} \tag{5.7}$$

The barrier causes the Friedel oscillation in the electron density on both sides of the barrier. The interaction is treated perturbatively, via finding the corrections to the transmission coefficient due to additional scattering at the potential produced by the Friedel oscillation. Diagrammatically, the corrections to the Green's function are described by the diagrams in Fig.14, where a) represents the Hartree and b) the exchange (Fock) contributions, correspondingly. Compared to the textbook case, though, the solid lines in these diagrams are the Green's functions composed of the exact eigenstates in the presence of the barrier (but no interaction). Because the barrier breaks translational invariance, these Green's functions are not translationally invariant as well. I emphasized this fact by drawing the diagrams in real space, as opposed to the momentum -space representation. Notice also that the Hartree diagram is usually discarded in textbooks because a bubble there corresponds to the total charge density (density of electrons minus that of ions), which is equal to zero in a translationally invariant and neutral system.

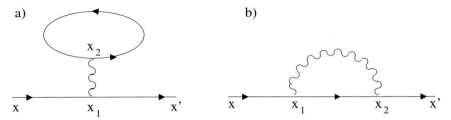

Fig. 14. Correction to the Green's function: exact with respect to the barrier and first order in the interaction

However, what we have in our case is the *local* density of electrons at some distance from the barrier. Friedel oscillation is a relatively short-range phenomenon (the period of the oscillation is comparable to the electron wavelength), and it is possible to violate the charge neutrality locally on such a scale. As a result, the Hartree correction is not zero.

To first-order in the interaction, an equivalent way of solving the problem is to find a correction to the wave-function, rather than the Green's function, in the Hartree-Fock method. The electron wave-function which includes both the barrier potential and the electron-electron interaction is

$$
\begin{aligned}
\psi_k(x) &= \psi_k^0(x) + \int_0^\infty dx' G_0^>(x,x',E) \\
&\times \int_0^\infty dx'' [V_H(x'')\delta(x'-x'') + V_{ex}(x',x'')]\psi_k^0(x''),
\end{aligned}
\tag{5.8}
$$

where $G_0^>$ is the Green's function of free electrons on the right semi-line, E is the full energy of an electron, V_H and V_{ex} are the Hartree and the exchange potentials. The Hartree potential is

$$
V_H(x) = \int dx' U(x-x')\delta n(x')
\tag{5.9}
$$

where $\delta n(x) = n(x) - n_0$ is the deviation of the electron density from its uniform value (in the absence of the potential) and $U(x)$ is the interaction potential. Hartree interaction is a direct interaction with the modulation of the electron density produced by the Friedel oscillation. For a high barrier, which is essentially equivalent to a hard-wall boundary condition, the electron density is

$$
n(x) = 4\int_0^{k_F} \frac{dk}{2\pi} \sin^2(kx) = n_0 \left(1 - \frac{\sin 2k_F x}{2k_F x}\right) \rightarrow
\tag{5.10}
$$

$$
\delta n(x) = -\sin(2k_F x)/2\pi x
\tag{5.11}
$$

where $n_0 = k_F/\pi$ is the density of electrons. Then,

$$V_H(x) = -\frac{1}{2\pi} \int_0^\infty dx' U(x - x') \frac{\sin 2k_F x'}{x'}. \tag{5.12}$$

Notice that although the bare interaction is short-range, the effective interaction has a slowly-decaying tail due to the Friedel oscillation. (The integral goes over only for positive values of x' because electrons interact only there.)

The exchange potential is equal to

$$
\begin{aligned}
V_{ex}(x, x) \;=\; & -U(x - x')\Big[\int_0^{k_F} \frac{dk}{2\pi} \big[\psi_k^0(x')\big]^* \psi_k^0(x) \\
& + \int_0^{k_F} \frac{dk}{2\pi} \big[\psi_{-k}^0(x')\big]^* \psi_{-k}^0(x)\Big].
\end{aligned}
\tag{5.13}
$$

Since we assumed that electrons interact only if they are located to the right of the barrier, the integral in (5.8) runs only over $x, x' > 0$ and the Green's function is a Green's function on a semi-line. The wave-function in (5.8) needs to be evaluated at $x \to \infty$, which means that we will only need an asymptotic form of the Green's function far away from the barrier. This form is constructed by the method of images

$$G_0^>(x, x', E) = G_0(x, x', E) - G_0(x, -x', E), \tag{5.14}$$

where

$$G_0(x, x', E) = \frac{1}{iv_k} e^{ik|x-x'|}$$

is the free Green's function on a line with $k = \sqrt{2mE}$ and $v_k = k/m$. Coordinate x' is confined to the barrier, whereas $x \to \infty$, thus $x > x'$ and

$$G_0^>(x, x', E) = -\frac{2}{v_k} \sin(kx') e^{ikx}.$$

5.1.1. Hartree interaction

Our goalis to present the correction to the wave-function for electrons incoming from $x < 0$ to $x > 0$ in the form

$$\psi_k - \psi_k^0 = \frac{1}{\sqrt{L}} \delta t e^{ikx}, \tag{5.15}$$

where δt is the interaction correction to the transmission coefficient. Substituting (5.15) into (5.8), we obtain for the Hartree contribution to t

$$\frac{\delta t^H}{t_0} = -\frac{2}{v_F} \int_0^\infty dx \sin kx\, e^{ikx} V_H(x),$$

where one can replace $v_k \rightarrow v_F$ in all non-oscillatory factors. For a delta-function potential, $U(x) = U\delta(x)$

$$V_H(x) = -\frac{U}{2\pi}\frac{\sin 2k_F x}{x}. \tag{5.16}$$

However, the $\delta-$ function potential is not good enough for us, because it will be cancelled by the exchange contribution. Friedel oscillation arises due to backscattering. With a little more effort, one can show that U in the last formula is replaced by $U(2k_F)$:[26]

$$V_H(x) = -\frac{U(2k_F)}{2\pi}\frac{\sin 2k_F x}{x}.$$

Substituting this into $\delta t/t$ yields

$$
\begin{aligned}
\frac{\delta t_H}{t_0} &= \frac{U(2k_F)}{\pi v_F}\int_0^\infty dx \sin(kx)\, e^{ikx}\frac{\sin 2k_F x}{x} \\
&= \frac{U(2k_F)}{\pi v_F}\int_0^\infty dx \frac{1}{2i}\left(e^{2ikx} - \underbrace{1}_{\text{regular correction to Imt}}\right) \\
&\times \frac{\sin 2k_F x}{x} = \frac{U(2k_F)}{\pi v_F}\int_0^\infty dx \frac{1}{2i}e^{2ikx}\frac{\sin 2k_F x}{x} = \frac{U(2k_F)}{2\pi v_F} \\
&\times \int_0^\infty dx\left(\sin 2kx + \underbrace{i^{-1}\cos 2kx}_{\text{yet another regular correction}}\right) \\
&\times \frac{\sin 2k_F x}{x} = \frac{U(2k_F)}{2\pi v_F}\int_0^\infty dx \sin 2kx\frac{\sin 2k_F x}{x} \\
&= \frac{U(2k_F)}{4\pi v_F}\ln\frac{k+k_F}{|k-k_F|} \approx \alpha'_{2k_F}\ln\frac{k_F}{|k-k_F|},
\end{aligned}
$$

where

$$\alpha'_{2k_F} = \frac{g_1}{4\pi v_F},$$

and $g_1 = U(2k_F)$. In deriving the final result, I discarded all terms that are regular in the limit $k \rightarrow k_F$.

---

[26]Notice that the sign of the Hartree interaction is attractive near the barrier (assuming the sign of the e-e interaction is repulsive at $2k_F$): for $x \rightarrow 0$, $V_H(x) \rightarrow -U(2k_F)k_F/\pi$. The reason is that the depletion of electron density near the barrier means that the positive background is uncompensated. As a result, electrons are *attracted* to the barrier and transmission is *enhanced* by the Hartree interaction.

5.1.2. Exchange

Now both x and $x' > 0$. We need to select the largest wave-function, *i.e.*, such that does not involve a small transmitted component. Obviously, this is only possible for $k < 0$ (second term in (5.13)) and ψ^0_{-k}, given by (5.7). Substituting the free wave-functions into the equation for the exchange interaction, we get

$$V_{ex}(x, x') = -U(x - x')\rho(x, x'), \tag{5.17}$$

where the 1D density-matrix is

$$\rho(x, x') = 4 \int_0^{k_F} \frac{dk}{2\pi} \sin(kx) \sin(kx') \tag{5.18}$$

$$= 2 \int_0^{k_F} \frac{dk_x}{2\pi} \left[\cos k(x - x') - \cos k(x + x') \right] \tag{5.19}$$

$$= \cdots - \frac{\sin k_F(x + x')}{\pi(x + x')}, \tag{5.20}$$

where $\ldots$ stand for the term which depends on $x - x'$. This term does not lead to the log-divergence in δt and will be dropped [27]. For $x = x'$, we get the correction to the density $\delta n(x)$, as we should.

Correction to the transmission coefficient

$$\delta t_{ex}/t_0 = -\frac{2}{\pi v_F} \int_0^\infty dx' \int_0^\infty dx'' U(x' - x'') \sin kx' e^{ikx''} \frac{\sin k_F(x' + x'')}{x' + x''}. \tag{5.21}$$

After a little manipulation with trigonometric functions, which involves dropping of the terms depending only on $x - x'$, we arrive at

$$\frac{\delta t_{ex}}{t_0} = -\frac{1}{4\pi^2 v_F} \int_0^{+\infty} \frac{dq}{q} U(q) \int_0^\infty \frac{dx_+}{x_+} \tag{5.22}$$

$$\times \{\sin 2(k - k_F + q)x_+ - \sin 2(k - k_F - q)x_+\}, \tag{5.23}$$

where

$$x_+ = \frac{x' + x''}{2}. \tag{5.24}$$

Integral over x_+ provides a lower cut-off for the $q-$ integral

$$\int_0^{+\infty} \frac{dx_+}{x_+} \{\sin 2(k - k_F + q)x_+ - \sin 2(k - k_F - q)x_+\} \tag{5.25}$$

$$= \frac{\pi}{2} \text{sgn}(q + k - k_F) + \frac{\pi}{2} \text{sgn}(q - k + k_F) \tag{5.26}$$

$$= \pi\theta(q - |k - k_F|). \tag{5.27}$$

---

[27] Notice that the important part of the exchange potential is *repulsive* near the barrier. This means that electrons are repelled from the barrier and transmission is suppressed.

Now

$$\frac{\delta t_{ex}}{t_0} = -\frac{1}{4\pi v_F} \int_{|k-k_F|}^{+\infty} \frac{dq}{q} U(q).$$ (5.28)

As $U(q)$ is finite at $q \to 0$ [28], one can take $U(q)$ out of the integral at $q = 0$ (denoting $U(0) = g_2$)

$$\frac{\delta t_{ex}}{t_0} \approx -\frac{1}{4\pi v_F} g_2 \int_{|k-k_F|}^{q_0} \frac{dq}{q} = -\alpha_0' \ln \frac{q_0}{|k - k_F|}.$$

Combining the exchange and Hartree corrections together (in doing so we choose the smallest upper cut-offs for the log which we assume to be inverse interaction range, q_0), we get

$$\delta t = -t_0 \alpha' \ln \frac{q_0}{|k - k_F|},$$ (5.29)

where

$$\alpha' = \alpha_0' - \alpha_{2k_F}' = \frac{g_2 - g_1}{4\pi v_F}; \text{ asymmetric geometry.}$$ (5.30)

It can be shown in a similar manner that if we had interacting regions on *both* sides of the barrier, the result for α' would be double of that in Eq. (5.30).

$$\alpha' = \alpha_0' - \alpha_{2k_F}' = \frac{g_2 - g_1}{2\pi v_F}; \text{ symmetric geometry.}$$ (5.31)

The sign of the correction to t depends on the sign of $g_2 - g_1 = U(0) - U(2k_F)$. Notice that transmission is *enhanced*, if $U(2k_F) > U(0)$. Usually, this behavior is associated with attraction. We see, however, that even if the interaction is repulsive at all q but $U(2k_F) > U(0)$, it works effectively as an attraction. The case $U(2k_F) > U(0)$ is not a very realistic one, at least not in a situation when electrons interact only among themselves. Other degrees of freedom, *e.g.*, phonons, must be involved to give a preference to $2k_F-$ scattering.

5.2. Renormalization group

It is tempting to think that the first-order in interaction correction to t_0 in Eq. (5.30) is just an expansion of the scaling form $t \propto |k - k_F|^{\alpha'}$. A poor-man RG indeed

---

[28] If $U(q)$ has a strong dependence on q for $q \to 0$ (which is the case for a bare Coulomb potential $U(q) \propto \ln q$), this dependence affects the resulting dependence of the transmission coefficient on energy $|k - k_F|$, *i.e.*, on the temperature and/or bias. Instead of a familiar power-law scaling of the tunneling conductance for the short-range interaction, the conductance falls off with energy faster than any power law for the bare Coulomb potential.

shows that this is the case. Near the Fermi level, $k - k_F = (E - E_F)/v_F = \varepsilon/v_F$ so that the first-order correction to t is

$$t_1 = t_0 \left(1 - \alpha' \ln \frac{W_0}{|\varepsilon|} \right),$$

where $W_0 = q_0 v_F$ is the effective bandwidth. The meaning of this bandwidth is that the states at $\pm W_0$ from the Fermi level (=0) are not affected by the inter-action. For $|\varepsilon| = W_0$, $t_1 = t_0$. Suppose that we want to reduce to bandwidth $W_0 \to W_1 < W_0$ and find t at $|\varepsilon| = W_1$

$$t_1 = t_0 \left(1 - \alpha' \ln \frac{W_0}{W_1} \right).$$

It is of crucial importance here that coefficient α' (which will become the tunneling exponent in the scaling form we are about to get) is proportional to the *RG-invariant* combination $U(0) - U(2k_F) = g_2 - g_1$ for spinless electrons. This means that α' is to be treated as a constant under the RG flow. Repeating this procedure using t, found at the previous stage instead of a bare t_0, n times, we get

$$t_{n+1} = t_n \left(1 - \alpha' \ln \frac{W_n}{W_{n+1}} \right).$$

The renormalization process is to be stopped when bandwidth coincides with the physical energy $|\varepsilon|$, at which t is measured. In the continuum limit ($t_{n+1} - t_n = dt$; $W_{n+1} = W_n - dW$), this equation reduces to a differential one

$$\frac{dt}{t} = \alpha' \frac{dW}{W}$$

$$\frac{dt}{d \ln W} = \alpha' t$$

Integrating from $t(\varepsilon)$ to t_0 (and, correspondingly, from $W = |\varepsilon|$ to $W = W_0$) we obtain

$$t(\varepsilon) = t_0 \left(|\varepsilon| / W_0 \right)^{\alpha'}.$$

5.3. Electrons with spins

Now let's introduce the spin. The effect will be more interesting than just multiplying the result for the tunneling conductance by a factor of two (which is all what happens for non-interacting electrons.) To keep things general, I will assume an arbitrary "spin" (which may involve other degrees of freedom) degeneracy N_s and put $N_s = 2$ at the end. In this section we will exploit the result of

Sec. 4 stating the backscattering amplitude flows under RG. This flow affects the renormalization of the transmission coefficient at low energies.

Repeating the steps for the first-order correction to t, including spin is easy– one just have to realize that the Hartree correction is multiplied by N_s (as the polarization bubble involves summation over all isospin components, it is simply multiplied by a factor of N_s). On the contrary, the exchange interaction is possible only between electrons of the same spin, so there are N_s identical exchange potentials for every spin component. I am going to discuss the strong barrier case first in the symmetric geometry. Then, taking into account what we have just said about the factor of N_s, we can replace the result for spinless electrons (5.30) by

$$\alpha' \to \alpha' = \alpha'_0 - N_s \alpha'_{2k_F} = \frac{g_2 - N_s g_1}{4\pi v_F}. \tag{5.32}$$

(and similarly for the symmetric geometry of the tunneling experiment). Correspondingly, the correction to the transmission coefficient (for a given spin projection) changes to

$$t_\sigma = t_0 \left(1 - \alpha' \ln W / |\varepsilon|\right).$$

The tunneling conductance is found from the Landauer formula

$$\mathcal{G} = \frac{e^2}{h} \sum_{\sigma=1}^{N_s} |t_\sigma|^2,$$

where, as the barrier is spin-invariant, the sum simply amounts to multiplying the result for a given spin component by N_s. Now, the result in Eq. (5.32) seems to be interesting, as the $2k_F$ contribution gets a boost. If $N_s U (2k_F) > U (0)$, we have in increase of the barrier transparency. It does not seem too hard to satisfy this condition. For example, it is satisfied already for the delta-function potential [29] and $N_s = 2$. However, as opposed to the spinless case, α' is *not* an RG-invariant but flows under renormalizations. Let's split α' into an RG-invariant part (4.4) and the rest

$$
\begin{aligned}
\alpha' &= \frac{1}{4\pi v_F} \left[U (0) - \frac{1}{N_s} U (2k_F)\right] - \frac{1}{4\pi v_F} \frac{N_s^2 - 1}{N_s} U (2k_F) \\
&= \alpha'_s - \frac{1}{4\pi v_F} \frac{N_s^2 - 1}{N_s} g_1,
\end{aligned}
$$

where

$$\alpha'_s = \frac{1}{4\pi v_F} \left(g_2 - \frac{1}{N_s} g_1\right) = \frac{\bar{\Gamma}}{4\pi v_F}. \tag{5.33}$$

---

[29] As now electrons have spins, they are allowed to be at the same point in space and interact.

The condition for the tunneling exponent to be negative is more restrictive that it seemed to be: $g_1 > N_s g_2$. It is not hard to see that the RG equation for t_σ now changes to

$$\frac{dt}{dl} = -t \left(\alpha_s' - \frac{1}{4\pi v_F} \frac{N_s^2 - 1}{N_s} \Gamma_1(l) \right), \tag{5.34}$$

where $\Gamma_1(l)$ is given by

$$\Gamma_1 = \frac{1}{(g_1)^{-1} + \frac{N_s}{2\pi} l}.$$

Integrating (5.34), we find

$$t_\sigma = t_0 \left(1 + \frac{N_s g_1}{4\pi v_F} \ln \frac{W}{|\varepsilon|} \right)^{\beta_s} (|\varepsilon| / W)^{\alpha_s'},$$

where

$$\beta_s = \frac{N_s^2 - 1}{N_s^2}.$$

In particular, for $N_s = 2$, we get

$$t_\sigma = t_0 \left(1 + \frac{g_1}{2\pi v_F} \ln \frac{W}{|\varepsilon|} \right)^{3/4} (|\varepsilon| / W)^{\alpha_s'}.$$

and conductance

$$\mathcal{G} = \mathcal{G}_0 \left(1 + \frac{g_1}{2\pi v_F} \ln \frac{W}{|\varepsilon|} \right)^{3/2} (|\varepsilon| / W)^{2\alpha_s'}, \tag{5.35}$$

where $\mathcal{G}_0$ is the conductance for the free case. Thus the flow of the backscattering amplitude results in a multiplicative log-renormalization of the transmission coefficient. One can check the first-order result is reproduced if expand the RG result to the first log.

An interesting feature of this result is that it predicts *three* possible types of behavior of the conductance as function of energy.

1. weak backscattering:
$$\alpha' > 0 \rightarrow g_1 < g_2/N_s.$$

In this regime already the first-order correction corresponds to suppression of the conductance, which decreases monotonically as the energy goes down.

2. intermediate backscattering:

$$\alpha' > 0 \text{ but } \alpha_s' < 0 \rightarrow g_2/N_s < g_1 < N_s g_2.$$

In this regime, the first-order correction enhances the transparency, but the RG result shows that the for $\varepsilon \to 0$, the transmission goes to zero. It means that at higher energies, when the RG has not set in yet, the conductance increases as the energy goes down, but at lower energies the conductance decreases. The dependence of $\mathcal{G}(\varepsilon)$ on ε is non-monotonic–there is a maximum at the intermediate energies.

3. strong backscattering:

$$\alpha'_s < 0 \to g_1 > N_s g_2.$$

In this regime, tunneling exponent α'_s is negative and the conductance increases as the energy goes down.

5.4. *Comparison of bulk and edge tunneling exponents*

Tunneling into the bulk of a 1D system is described by the density of states obtained, *e.g.*, in the DL solution of the Tomonaga-Luttinger model (no backscattering). The "bulk" tunneling exponent is equal to

$$2\alpha = \frac{(u-1)^2}{4u},$$

where

$$u = \sqrt{1 + 2g_2/\pi v_F}.$$

In this Section, we considered tunneling into the edge for weak interaction and found that the conductance scales with exponent $2\alpha'_s$ (5.33). To compare the two exponents, we need to expand the DL exponent for weak interaction

$$2\alpha = \frac{g_2^2}{4\pi^2 v_F}.$$

For $g_1 = 0$, the edge exponent is

$$2\alpha'_s = \frac{1}{2\pi v_F} g_2.$$

We see that for weak coupling tunneling into the edge is stronger affected by the interaction than tunneling into the bulk: the former effect starts at the first order in the interaction whereas the latter starts at the second order. This difference has a simple physical reason which is general for all dimensions. In a translationally invariant system, the shape of the Green's function is modified in a non-trivial way only starting at the second order. For example, the imaginary part of the self-energy (decay of quasi-particles) occur only at the second order. The first-order corrections lead only to a shift in the chemical potential and, if the potential is of a finite-range, to a renormalization of the effective mass. If the translational invariance is broken, non-trivial changes in the Green's function occur already at the first order in the interaction. That tunneling into the bulk and edge are characterized by different exponents is also true in the strong-coupling case (cf. Sec. 6). As the relation between the bulk and edge exponents is known for an arbitrary coupling, one can eliminate the unknown strength of interaction and express one exponent via the other. Knowing one exponent from the experiment, one can check if the observed value of the second exponent agrees with the data. This cross-check was cleverly used in the interpretation of the experiments on single-wall carbon nanotubes [70, 71].

6. Bosonization solution

Bosonization procedure in described in a number of books and reviews [1]-[10]. Without repeating all standard manipulations, I will only emphasize the main steps in this Section, focusing on a couple of subtle points not usually discussed in the literature. A reader familiar with bosonization may safely skip the first part of this Section, and go directly to Secs. 6.1.5 and 6.2.1, where tunneling exponents are calculated. Some technical details of the bosonization procedure are presented in Appendix C. As in Sec. 3, $v_F = 1$ in this Section, unless specified otherwise.

6.1. Spinless fermions

6.1.1. Bosonized Hamiltonian
We start from a Hamiltonian of interacting fermions without spin

$$H = \frac{1}{L} \sum_{p,k,q} \xi_k a_k^\dagger a_k + \frac{1}{2L^2} \sum_{p,k,q} V_q a_{p-q}^\dagger a_{k+q}^\dagger a_k a_p.$$

The interacting part of the Hamiltonian can be re-written using chiral densities

$$\rho_\pm(q) = \sum_{p \gtrless 0} a_{p-q/2}^\dagger a_{p+q/2}$$

as

$$H_{\text{int}} = g_2 \frac{1}{L} \sum_q \rho_+(q) \rho_-(-q) + \frac{g_4}{2} \frac{1}{L} \sum_q \rho_+(q) \rho_+(-q) + \rho_-(q) \rho_-(-q).$$

The interacting part is in the already bosonized form. For a linearized dispersion

$$\xi_k = |k| - k_F,$$

it is also possible to express the free part via densities. To check this, let's assume that H_0 can indeed be written as

$$H_0 = \frac{1}{L} \sum_q A_q \left[\rho_+(q) \rho_+(-q) + \rho_-(q) \rho_-(-q) \right],$$

where A_q is some unknown function. Commute ρ_+ with H_0, making use of the anomalous commutator $\left[\rho_+(q), \rho_+(-q) \right] = qL/2\pi$:

$$\left[\rho_+(q), H_0 \right] = \frac{1}{L} \sum_{q'} A_{q'} \left[\rho_+(q), \rho_+(q') \rho_+(-q') \right]$$

$$= A_q \rho_+(q) \frac{q}{\pi}.$$

On the other hand, the same commutator can be calculated directly in a model with linearized spectrum, using only fermionic anticommutation relations. This gives

$$\left[\rho_+(q), H_0 \right] = q\rho_+(q).$$

Comparing the two results, we see that

$$A_q = \pi$$

and thus

$$H_0 = \pi \frac{1}{L} \sum_q \rho_+(q) \rho_+(-q) + \rho_-(q) \rho_-(-q).$$

Combining the free and interacting parts of the Hamiltonian, we obtain

$$H = \pi \frac{1}{L} \left(1 + \frac{g_4}{2\pi} \right) \sum_q \{ \rho_+(q) \rho_+(-q) + \rho_-(q) \rho_-(-q) \}$$

$$+ \pi \frac{1}{L} g_2 \sum_q \rho_+(q) \rho_-(-q).$$

Notice that if only the g_4- interaction is present, the system remains free but the Fermi velocity changes.

It is convenient to expand the density operators over the normal modes

$$\rho_+(x) = \sum_{q>0} \sqrt{\frac{|q|}{2\pi L}} \left(b_q e^{iqx} + b_q^\dagger e^{-iqx} \right);$$

$$\rho_-(x) = \sum_{q<0} \sqrt{\frac{|q|}{2\pi L}} \left(b_q e^{iqx} + b_q^\dagger e^{-iqx} \right).$$

One can readily make sure that density operators defined in this way reproduce the correct commutation relations, given that $[b_q, b_{q'}^\dagger] = \delta_{q,q'}$. In terms of these operators, the Hamiltonian reduces to

$$H = \pi \left(1 + \frac{g_4}{2\pi}\right) \sum_{q>0} q \left\{b_q^\dagger b_q + b_{-q}^\dagger b_{-q}\right\} + \pi g_2 \sum_q q \left(b_q^\dagger b_{-q}^\dagger + b_q b_{-q}\right).$$

Introducing new bosons via a Bogoliubov transformation

$$
\begin{aligned}
c_q^\dagger &= \cosh\theta_q b_q + \sinh\theta_q b_{-q}^\dagger; \\
c_{-q}^\dagger &= \cosh\theta_q b_{-q}^\dagger + \sinh\theta_q b_q
\end{aligned}
$$

and choosing θ_q so that the Hamiltonian becomes diagonal, *i.e.*,

$$\tanh 2\theta_q = \frac{g_2/2\pi}{1 + g_4/2\pi},$$

we obtain

$$H = \frac{1}{L} \sum_q \omega_q c_q^\dagger c_q,$$

where

$$\omega_q = u\,|q|\,,$$

and [30]

$$u = \left[\left(1 + \frac{g_4}{2\pi}\right)^2 - \left(\frac{g_2}{2\pi}\right)^2\right]^{1/2}. \tag{6.2}$$

For the spinless case, backscattering can be absorbed into forward scattering. The resulting expression for the renormalized velocity for the case $g_2 = g_4 \neq g_1$ is (cf. Appendix C.3)

$$u = \sqrt{1 + \frac{g_2 - g_1}{2\pi}}.$$

---

[30]Let's now introduce backscattering. Since for spinless fermions backscattering is just an exchange process to forward scattering of fermions of opposite chirality (think of a diagram where you have right and left lines coming in and then interchange them at the exit), the only effect of backscattering is to replace $g_2 \to g_2 - g_1$. (cf. discussion in Sec. 4). Instead of (6.2), we then have

$$u = \left[\left(1 + \frac{g_4}{2\pi}\right)^2 - \left(\frac{g_2 - g_1}{2\pi}\right)^2\right]^{1/2}. \tag{6.1}$$

Now, consider a delta-function interaction, when $g_1 = g_2 = g_4$. Pauli principle says that we should get back to the Fermi gas in this case. However, u still differs from unity (Fermi velocity) and thus our result violates the Pauli principle. See Appendix C.3 for a resolution of the paradox.

6.1.2. Bosonization of fermionic operators

The $\Psi-$ operators of right/left movers can be represented as

$$\Psi_{\pm}(x) = \frac{1}{\sqrt{2\pi a}} e^{\pm 2\pi i \int_{-\infty}^{x} \rho_{\pm}(x') dx'}, \tag{6.3}$$

where a is the ultraviolet cut-off in real space. Using the commutation relations for $\rho_{\pm}$, one can show that the (anti) commutation relations

$$\left\{ \Psi_{\pm}(x), \Psi_{\pm}^{\dagger}(x') \right\} = \delta(x - x')$$

are satisfied.

The argument of the exponential can be re-written as two bosonic fields

$$\begin{aligned}
\Psi_{\pm}(x) &= \frac{1}{\sqrt{2\pi a}} e^{\pm i \sqrt{\pi} \varphi_{\pm}(x)}; \\
\varphi_{\pm}(x) &= \varphi(x) \mp \vartheta(x).
\end{aligned} \tag{6.4}$$

Equating exponents in (6.3) and (6.4), we obtain

$$\begin{aligned}
\sqrt{\pi} [\varphi(x) - \vartheta(x)] &= 2\pi \int_{-\infty}^{x} dx' \rho_{+}(x') \tag{6.5} \\
&= 2\pi \sum_{q>0} \sqrt{\frac{|q|}{2\pi L}} \frac{1}{iq} \left(b_q e^{iqx} - b_q^{\dagger} e^{-iqx} \right); \\
\sqrt{\pi} [\varphi(x) + \vartheta(x)] &= 2\pi \int_{-\infty}^{x} dx' \rho_{-}(x') \tag{6.6} \\
&= 2\pi \sum_{q<0} \sqrt{\frac{|q|}{2\pi L}} \frac{1}{iq} \left(b_q e^{iqx} - b_q^{\dagger} e^{-iqx} \right).
\end{aligned}$$

Solving for $\varphi(x)$ and $\vartheta(x)$ gives

$$\varphi(x) = -i \sum_{-\infty < q < \infty} \frac{1}{\sqrt{2|q|L}} \mathrm{sgn} q \left(b_q e^{iqx} - b_q^{\dagger} e^{-iqx} \right); \tag{6.7}$$

$$\vartheta(x) = i \sum_{-\infty < q < \infty} \frac{1}{\sqrt{2|q|L}} \left(e^{iqx} b_q - b_q^{\dagger} e^{-iqx} \right). \tag{6.8}$$

Using (6.7) and (6.8), one can prove that $\varphi(x)$ and $\partial_x \vartheta(x)$ satisfy canonical commutation relations between coordinate and momentum (cf. Appendix C.2.1)

$$\left[\varphi(x), \partial_{x'} \vartheta(x') \right] = i\delta(x - x'). \tag{6.9}$$

Using Eqs. (6.5) and (6.6), we obtain the density and current as the gradients of the bosonic fields

$$
\begin{aligned}
\varphi(x) &= \sqrt{\pi} \int_{-\infty}^{x} dx' \left(\rho_+(x') + \rho_-(x') \right) = \sqrt{\pi} \int_{-\infty}^{x} dx' \rho(x') \rightarrow \\
\rho(x) &= \frac{1}{\sqrt{\pi}} \partial_x \varphi; \\
\vartheta(x) &= -\sqrt{\pi} \int_{-\infty}^{x} dx' \left(\rho_+(x') + \rho_-(x') \right) = -\sqrt{\pi} \int_{-\infty}^{x} dx' j(x') \rightarrow \\
j(x) &= -\frac{1}{\sqrt{\pi}} \partial_x \vartheta(x).
\end{aligned}
$$

The continuity equation,

$$
\partial_t \rho + \partial_x j = 0
$$

relates the Heisenberg fields $\varphi(x,t)$ and $\vartheta(x,t)$

$$
\partial_t \varphi = \partial_x \vartheta.
$$

The current can be also found as

$$
j(x,t) = -\frac{1}{\sqrt{\pi}} \partial_t \varphi(x,t).
$$

We will use this relation later. Expressing H in the real space via $\rho_\pm$

$$
H = \pi \int dx \left[\rho_+^2 + \rho_-^2 \right] + \frac{g_4}{2} \int dx \left[\rho_+^2 + \rho_-^2 \right] + g_2 \int dx \rho_+ \rho_-
$$

and using the relations

$$
\rho_\pm = \frac{1}{2\sqrt{\pi}} \left(\partial_x \varphi \mp \partial_x \vartheta \right),
$$

we obtain a canonical form of H in terms of the bosonic fields

$$
\begin{aligned}
H &= \frac{1}{2} \int dx \left[(\partial_x \varphi)^2 + (\partial_x \vartheta)^2 \right] + \frac{g_2 + g_4}{4\pi} \int dx \, (\partial_x \varphi)^2 \\
&\quad + \frac{g_2 - g_4}{4\pi} \int dx \, (\partial_x \vartheta)^2 \\
&= \frac{1}{2} \int dx \left[\frac{u}{K} (\partial_x \varphi)^2 + uK (\partial_x \vartheta)^2 \right],
\end{aligned} \tag{6.10}
$$

where [31]

$$u = \sqrt{\left(1 + \frac{g_4}{2\pi}\right)^2 - \left(\frac{g_2}{2\pi}\right)^2}; \ K = \sqrt{\frac{1 + \frac{g_4 - g_2}{2\pi}}{1 + \frac{g_4 + g_2}{2\pi}}}. \tag{6.11}$$

For $g_4 = g_2 \equiv g$, we have

$$u = \sqrt{1 + g/\pi}; \ K = \frac{1}{\sqrt{1 + g/\pi}}. \tag{6.12}$$

Notice that in this case $uK = 1$. This is important: in the next Section, we will see that this product renormalizes the Drude weight (and the persistent current). Neither of these quantities are supposed to be affected by the interactions, as the Galilean invariance remains intact. We see that it is indeed the case in our model.

If backscattering is present (but $g_2 = g_4$), the parameters change to (cf. Appendix C.3)

$$u = \sqrt{1 + \frac{g_2 - g_1}{\pi}}; \tag{6.13}$$

$$K = \frac{1}{\sqrt{1 + (g_2 - g_1)/\pi}}. \tag{6.14}$$

Had we started with another microscopic model, *e.g.*, with fermions on a lattice but away from half-filling, the effective low-energy theory would have also been described by Hamiltonian (6.10) albeit with different–and, in general, unknown– parameters u and K. The term "Luttinger liquid" (LL) [69] refers to a universal Hamiltonian of type (6.10), which describes the low-energy properties of many seemingly different systems. In that sense, the LL is a 1D analog of higher-dimensional Fermi liquids, which also encapture the low-energy properties of a large class of fermionic systems, while encoding the quantitative differences in their high-energy properties by a relatively small set of parameters.

6.1.3. Attractive interaction

What happens for the case of an attractive interaction, $g < 0$? Formally, for $g < -\pi$ (or $g_2 - g_1 < \pi$), u^2 in Eqs. (6.12,6.13) is negative, which seems to suggest some kind of an instability. Actually, this is not the case [59], as a 1D system of spinless fermions does not have any phase transitions even at $T = 0$. All it means that the interacting system is a liquid rather than a gas, *i.e.,* it does

---

[31] As we have already seen in Sec. 3, a difference between g_2 and g_4 leads to the current-current interaction in the Hamiltonian. In the bosonized form, this interaction is the $(g_2 - g_4)(\partial_x \vartheta)^2$ term in the first line of Eq. (6.10).

not require external pressure to mantain its volume. An equilibrium value of the density is fixed by given ambient pressure. To see this, restore the Fermi velocity $v_F = \pi n / m$, where n is the density

$$u^2 = v_F^2 \left(1 + \frac{g}{\pi v_F} \right) = \left(\frac{\pi n}{m} \right)^2 + \frac{ng}{m} \tag{6.15}$$

and recall the thermodynamic relation

$$u^2 = m^{-1} \partial P / \partial n, \tag{6.16}$$

where P is the pressure. Integrating (6.16) with the boundary condition $P(n = 0) = 0$, we obtain the constituency relation

$$P = \left(\frac{\pi}{m} \right)^2 \frac{n^3}{3} + \frac{g}{2m} n^2.$$

For $g < 0$, there is a metastable region of negative pressure. This means that if the ambient pressure is equal to zero, the thermodynamically stable value of the density is given by the non-zero root of the equation $P(n) = 0$, by

$$n^* = \frac{3}{2\pi^2} |g| m.$$

The square of the sound velocity at this density is positive:

$$\left(u^* \right)^2 = \frac{3}{4\pi^2} g^2.$$

The Fermi velocity at $n = n^*$ is

$$v_F^* = \pi n^* / m = \frac{3}{2\pi^2} |g|$$

and parameter K

$$K = \frac{v_F^*}{u^*} = \frac{\sqrt{3}}{\pi}$$

is a universal number, independent of the interaction.

6.1.4. Lagrangian formulation

In what follows, it will be more convenient to work in the Lagrangian rather the Hamiltonian formulation (and also in complex time). A switch from the Hamiltonian to Lagrangian formulation is done via the usual canonical transformation

$$S = \int dx \int dt \left(\dot{q} p - \mathcal{H}(p, q) \right), \tag{6.17}$$

where $\mathcal{H}$ is the Hamiltonian density defined such that

$$H = \int dx\, \mathcal{H}$$

and q and p are the canonical coordinate and momentum, correspondingly. According to commutation relation (6.9),

$$q = \varphi \quad p = \partial_x \vartheta. \tag{6.18}$$

Performing a Wick rotation, $t \to -i\tau$, we reduce the quantum-mechanical problem into a statistical-mechanics one with the partition function

$$Z = \int D\varphi \int D\vartheta\, e^{-S_E},$$

where the Euclidian action

$$S_E = \int d\tau \int dx \left[\frac{1}{2}\frac{u}{K}(\partial_x \varphi)^2 + \frac{1}{2}uK(\partial_x \vartheta)^2 - i\partial_\tau \varphi \partial_x \vartheta \right].$$

In a Fourier-transformed form

$$S_E = \int d^2k \left[\frac{1}{2}\frac{u}{K}q^2 \varphi_{\vec{k}}\varphi_{-\vec{k}} + \frac{1}{2}uK \vartheta_{\vec{k}}\vartheta_{-\vec{k}} + iq\omega\varphi_{\vec{k}}\vartheta_{-\vec{k}} \right],$$

where $\vec{k} \equiv (q, \omega)$. If one needs only an average composed of fields of one type (φ or ϑ), then the other field can be integrated out. This leads to two equivalent forms of the action

$$
\begin{aligned}
S_\varphi &= \frac{1}{2K} \int d^2k \left[\frac{1}{u}\omega^2 + uq^2 \right] \varphi_{\vec{k}}\varphi_{-\vec{k}} & (6.19a) \\
&= \frac{1}{2K} \int dx \int d\tau \left[\frac{1}{u}(\partial_\tau \varphi)^2 + (\partial_x \varphi)^2 \right]; & (6.19b) \\
S_\vartheta &= \frac{K}{2} \int d^2k \left[\frac{1}{u}\omega^2 + uq^2 \right] \vartheta_{\vec{k}}\vartheta_{-\vec{k}} & (6.19c) \\
&= \frac{K}{2} \int dx \int d\tau \left[\frac{1}{u}(\partial_\tau \vartheta)^2 + (\partial_x \vartheta)^2 \right]. & (6.19d)
\end{aligned}
$$

In calculating certain correlation functions, *e.g.*, the fermionic Green's function, one also needs a cross-correlator $\langle \varphi\vartheta \rangle$. This one is computed by keeping both φ and ϑ in the action.

It is convenient to re-write the action in the matrix form

$$S_E = \frac{1}{2} \int d^2 k \hat{\eta}_{\bar{k}}^{\dagger} \hat{D}^{-1} \hat{\eta}_{\bar{k}},$$

where

$$\hat{\eta}_{\bar{k}} = \begin{pmatrix} \varphi_{\bar{k}} \\ \vartheta_{\bar{k}} \end{pmatrix}$$

and the inverse matrix of propagators

$$\hat{D}^{-1} = \begin{pmatrix} q^2 u K & iq\omega \\ iq\omega & q^2 \frac{u}{K} \end{pmatrix}.$$

Inverting the matrix, we obtain

$$\hat{D} = \frac{1}{u^2 q^2 + \omega^2} \begin{pmatrix} uK & -i\omega/q \\ -i\omega/q & \frac{u}{K} \end{pmatrix}.$$

The space-time propagators can be found by performing the Fourier transforms of $\hat{D}$. For diagonal terms, one really does not need to do it, as it is obvious from (6.19a) and (6.19c) that these propagators just coincide with that of a 2D Laplace's equations. Recalling that the potential of a line charge is a log-function of the distance, we obtain

$$\Phi(z) = \langle \varphi(z)\varphi(0) - \varphi^2(0) \rangle = \frac{K}{4\pi} \ln \frac{a^2}{x^2 + (u|\tau| + a)^2},$$

$$\Theta(z) = \langle \vartheta(z)\vartheta(0) - \vartheta^2(0) \rangle = \frac{1}{4\pi K} \ln \frac{a^2}{x^2 + (u|\tau| + a)^2}$$

where a is "lattice constant", $z \equiv (x, \tau)$, and $x^2 + u^2\tau^2 \gg a^2$ [32]. These are the two correlation functions we will need the most. In addition, there is also an off-diagonal propagator

$$\Xi(z) = \langle \varphi(z)\vartheta(0) - \varphi(0)\vartheta(0) \rangle = \langle \vartheta(z)\varphi(0) - \vartheta(0)\varphi(0) \rangle$$

$$= \int d^2 k \left(e^{i\bar{k}\cdot\bar{z}} - 1 \right) \langle \varphi_{\bar{k}}\vartheta_{-\bar{k}} \rangle = -i \int d^2 k \left(e^{i\bar{k}\cdot\bar{z}} - 1 \right) \frac{\omega/q}{u^2 q^2 + \omega^2}.$$

---

[32] A non-symmetric appearance of the cut-off with respect to time and space coordinates reflect an asymmetric way the sums over bosonic momenta and frequencies were cut. We adopted a standard procedure in which the sum of over q is regularized by $\exp(-|k|a)$, whereas the frequency sum is unlimited. Other choices of regularization are possible.

$\Xi(z)$ depends only on the ratio $x/u\tau$ and thus does not change the power-counting. To see this, introduce polar coordinates $q = k\cos\alpha/u$, $\omega = \sin\alpha$, $x = (z/u)\cos\beta$, and $\tau = z\sin\beta$. Then

$$
\begin{aligned}
\Xi(x,\tau) &= -i\int \frac{d^2k}{(2\pi)^2}\left(e^{i(qx-\omega\tau)}-1\right)\frac{\omega/q}{u^2q^2+\omega^2} \\
&= -i\frac{1}{(2\pi)^2}\int_0^\infty \frac{dk}{k}\int_0^{2\pi} d\alpha \left(e^{ik\cos(\alpha+\beta)}-1\right)\tan\alpha.
\end{aligned}
$$

The resulting integral is a function of only $\beta = \tan^{-1}(x/u\tau)$.

6.1.5. Correlation functions

Now we can calculate various correlation functions, including the Green's function.

Non-time-ordered Green's function for right movers:

$$
\begin{aligned}
G_+(x,\tau) &= -\langle T_\tau^B \psi_+(x,\tau)\psi_+^\dagger(0,0)\rangle && (6.20)\\
&= \frac{1}{2\pi a}\langle T_\tau^B e^{i\sqrt{\pi}(\varphi(1)-\vartheta(1))}e^{-i\sqrt{\pi}(\varphi(0)-\vartheta(0))}\rangle,
\end{aligned}
$$

where $(1) \equiv (x,\tau)$ and $(0) \equiv (x=0,\tau=0)$, and where T_τ^B is a bosonic time-ordering operator [33]. I will use the well-known result, valid for an average of the product of the exponentials of gaussian fields (see books by Tsvelik [6] or Giamarchi [10] for a derivation)

$$
\langle T_\tau \prod_j e^{iA_j\gamma(z_j)}\rangle = \delta_{\sum_j A_j,0} \times e^{-\sum_{k>j} A_jA_k\langle T_\tau\gamma(z_j)\gamma(z_k)\rangle}e^{-\frac{1}{2}\sum_k A_k^2\langle\gamma^2(z_j)\rangle}. \tag{6.21}
$$

[Eq. (6.21) is essentially a field-theoretical analog of the probability theory result for the average of $e^{iA\gamma}$, where γ is a Gaussian random variable.] For example, in the average

$$
Av(z) = \langle T_\tau e^{i\sqrt{\pi}\varphi(z)}e^{-i\sqrt{\pi}\varphi(0)}\rangle
$$

$A_1 = \sqrt{\pi}$, $A_2 = -\sqrt{\pi}$ and

$$
Av(z) = e^{\pi\langle T_\tau[\varphi(z)\gamma(0)-\varphi^2(0)]\rangle} = \left(\frac{a^2}{x^2+(u|\tau|+a)^2}\right)^{1/4K}.
$$

---

[33] Surely, it is not a conventional definition of the Green's function, but it is easier to work with this one for now, and restore the fermionic T_τ product at the end.

Similarly, with the help of (6.21), Eq. (6.20) reduces to

$$
\begin{aligned}
G_+ & (x, \tau) \\
&= \frac{1}{2\pi a} e^{\pi \langle \varphi(1)\varphi(0) - \varphi^2(0)\rangle_\tau} e^{\pi \langle \vartheta(1)\vartheta(0) - \vartheta^2(0)\rangle_\tau} e^{-2\langle \varphi(1)\vartheta(0) - \varphi(0)\vartheta(0)\rangle_\tau} \\
&= \frac{1}{2\pi a} e^{\pi \Phi(x,\tau)} e^{\pi \Theta(x,\tau)} e^{-2\Xi(x,\tau)} \\
&= \frac{1}{2\pi a} \left(\frac{a^2}{x^2 + (u\,|\tau| + a)^2} \right)^{\frac{K + K^{-1}}{4}} e^{if(x/u\tau)},
\end{aligned}
\tag{6.22}
$$

where $\langle \ldots \rangle_\tau$ stands for a time-ordered product and where I used that in a translationally invariant and equilibrium system $\langle \varphi^2(0) \rangle = \langle \varphi^2(1) \rangle$ (same for ϑ). Function $f(x/u\tau)$ is a phase factor which does not effect the power-counting.

Bulk tunneling DoS   For $x = 0$,

$$
G(0, \tau) \propto \tau^{-\frac{K + K^{-1}}{2}}.
$$

By power-counting,

$$
\nu(\varepsilon) \propto |\varepsilon|^{\frac{K + K^{-1}}{2} - 1} = |\varepsilon|^{\frac{(K-1)^2}{2K}}.
\tag{6.23}
$$

This is an analog of the DL result for the spinless case.

Edge tunneling DoS   In a tunneling experiment, one effectively measures the *local* DoS at the sample's surface. In a correlated electron system, the boundary condition affects the wavefunction over a long (exceeding the electron wavelength) distance from the surface. Therefore, the surface DoS differs significantly from the "bulk" one. If a tunneling barrier is high, then—to leading order in transmission— the DoS can be found via imposing a hard-wall boundary condition. The presence of the surface (boundary) can be taken into account by imposing the boundary conditions on the number current

$$
j(x = 0, \tau) = -\frac{1}{i\sqrt{\pi}} \partial_\tau \varphi = 0.
\tag{6.24}
$$

at $x = 0$. This means that φ is *pinned* at the boundary, *i.e.*, it takes some time-independent value. In the gradient-invariant theory, we can always choose this constant to be zero. Thus,

$$
\varphi(0, \tau) = 0.
$$

This suggests that the local correlator $\Phi(0, \tau) = 0$, and the long-time behavior of the Green's function in Eq. (6.22) is determined only by the correlator of the ϑ fields. If the boundary would not have affected this correlation, we would have arrived at

$$G\left(x = x' = 0, \tau\right) \propto \exp\left(\pi\Theta\left(x = 0, \tau\right)\right) \propto \frac{1}{|\tau|^{1/2K}}. \qquad \text{(wrong)}$$

But then we have a problem, as Eq. (wrong) does not reproduce the free-fermion behavior for $K = 1$. Consequently, the DoS at the edge $\nu_e(\varepsilon) \propto |\varepsilon|^{\frac{1}{2K}-1}$ would have not reproduced the free behavior either. What went wrong is that we pinned one field but forgot the other one is canonical conjugate to the first one. By the uncertainty principle, fixing the "coordinate" (φ) increases the uncertainty in the "momentum" (ϑ)–and vice versa. Thus, fluctuations of ϑ fields should increase. A rigorous solution to this problem is to change the fermionic basis from the plane waves to the solutions of the Schrodinger equation with the hard-wall boundary condition and to bosonize in this basis. This was done by Eggert and Affleck [74] and Fabrizio and Gogolin [75], [9]. Here I will give an heuristic argument based on a simple image construction, which leads to the same result.

Eq. (6.24) translates into the boundary conditions for the bosonic propagators:

$$\Phi_e(x, x', \tau) = 0; \quad \partial_{x,x'}\Theta_e(x, x', \omega) = 0, \qquad (6.25)$$

for $x, x' = 0$, where subindex e denotes the correlators in a semi-infinite system. Since Φ_e and Θ_e satisfy the Laplace's equation, we can view these propagators as potentials produced by some fictitious charges. Then, Φ_e and Θ_e can be constructed from the propagators of an infinite sample by the method of images:

$$\Phi_e(x, x', \tau) = \Phi(x - x', \tau) - \Phi(x + x', \tau);$$
$$\Theta_e(x, x', \tau) = \Theta(x - x', \tau) + \Theta(x + x', \tau).$$

For $x = x'$,

$$\Phi_e(0, 0, \tau) = 0; \quad \Theta_e(0, 0, \tau) = 2\Theta(0, \tau). \qquad (6.26)$$

Hence, pinning the φ field enhances the rms fluctuations of the ϑ field by a factor of two. This leads us to

$$\begin{aligned} G_+(0, 0, \tau) &\propto \exp\left(\pi\Phi_e(0, 0, \tau)\right)\exp\left(\pi\Theta_e(0, 0, \tau)\right) \\ &= \exp\left(2\pi\Theta_e(0, \tau)\right) \propto \exp\left(\frac{2\pi}{2\pi K}\ln\frac{a}{|\tau|}\right) \propto |\tau|^{-1/K}. \end{aligned}$$

Consequently, the DOS becomes

$$\nu_e(\varepsilon) \propto |\varepsilon|^{K^{-1}-1}. \qquad (6.27)$$

This result by Kane and Fisher [68] initiated the new (and still continuing) surge of interest to 1D systems (in terms of the impurity scattering time, this result was obtained earlier in Refs. [72, 73]). For tunneling from a contact with energy-independent DoS ("Fermi liquid") into a 1D system, the tunneling conductance scales as $\nu_e(\varepsilon)$

$$\mathcal{G}(\varepsilon) \propto \nu_e(\varepsilon) \propto |\varepsilon|^{K^{-1}-1}.$$

Now we see that the free-fermion behavior is correctly reproduced for $K = 1$.

Expanding the tunneling exponent $K^{-1} - 1$ with parameter K from Eq. (6.14) for the weak-coupling case gives

$$K^{-1} - 1 \approx \frac{g_2 - g_1}{2\pi v_F}.$$

This is the same result as the weak-coupling tunneling exponent (5.30) obtained in Sec. 5 via the scattering theory for interacting fermions.

Where do the "bulk" and "edge" forms of DoS match? Consider an object $G(x = x', \varepsilon)$. At the boundary, the DoS is of the "edge" form (6.27). Far away from the boundary, the Green's function does not depend on x and $\nu(\varepsilon)$ acquires a "bulk" form (6.23). As a function of x, $G(x = x', \varepsilon)$ varies on the scale $\simeq u/|\varepsilon|$ and the crossover between two limiting forms of ν occurs on this scale. Choosing the energy in a tunneling experiment, i.e., temperature or bias–whichever is larger, determines how far from the boundary one should go in order to see a change in the scaling behavior.

6.2. Fermions with spin

For fermions with spin, each component of the fermionic operator is bosonized separately

$$\psi_{\pm,\sigma} = \frac{1}{\sqrt{2\pi a}} \exp\left[\pm i\sqrt{\pi}\,(\varphi_\sigma \mp \vartheta_\sigma)\right].$$

Index σ of the bosonic field does not mean that bosons acquired spin. We simply have more bosonic fields. Charge and spin densities and currents are related to the derivatives of the bosonic fields

$$\rho_{\pm,\sigma} = \frac{1}{2\sqrt{\pi}} \left(\varphi'_\sigma \mp \vartheta'_\sigma \right) \quad \rho_\sigma = \rho_{+,\sigma} + \rho_{-,\sigma} = \frac{1}{\sqrt{\pi}} \varphi'_\sigma ;$$

$$j_\sigma = \rho_{+,\sigma} - \rho_{-,\sigma} = \frac{1}{\sqrt{\pi}} \vartheta'_\sigma ;$$

$$\rho_c = \rho_\uparrow + \rho_\downarrow = \frac{1}{\sqrt{\pi}} \left(\varphi'_\uparrow + \varphi'_\downarrow \right) = \sqrt{\frac{2}{\pi}} \varphi'_c$$

$$\rho_s = \rho_\uparrow - \rho_\downarrow = \frac{1}{\sqrt{\pi}} \left(\varphi'_\uparrow - \varphi'_\downarrow \right) = \sqrt{\frac{2}{\pi}} \varphi'_s ;$$

$$j_c = j_\uparrow + j_\downarrow = \frac{1}{\sqrt{\pi}} \left(\vartheta'_\uparrow + \vartheta'_\downarrow \right) = \sqrt{\frac{2}{\pi}} \vartheta'_c ;$$

$$j_s = j_\uparrow - j_\downarrow = \frac{1}{\sqrt{\pi}} \left(\vartheta'_\uparrow - \vartheta'_\downarrow \right) = \sqrt{\frac{2}{\pi}} \vartheta'_s ,$$

where $'$ denotes ∂_x and where the charge and spin bosons are defined as

$$\varphi_{c,s} = \frac{\varphi_\uparrow \pm \varphi_\downarrow}{\sqrt{2}}; \quad \vartheta_{c,s} = \frac{\vartheta_\uparrow \pm \vartheta_\downarrow}{\sqrt{2}}. \tag{6.28}$$

I assume that the interaction is spin-invariant, i.e., couplings of $\uparrow\uparrow$ and $\uparrow\downarrow$ fermions are the same. Substituting the relations between charge- and spin-densities into the Hamiltonian, one arrives at the familiar bosonized Hamiltonian which consists of totally independent charge and spin parts

$$H = H_c + H_s;$$

$$H_c = \frac{1}{2} \int dx \frac{u_c}{K_c} (\partial_x \phi_c)^2 + u_c K_c (\partial_x \theta_c)^2 ;$$

$$H_s = \frac{1}{2} \int dx \frac{u_s}{K_s} (\partial_x \phi_s)^2 + u_s K_s (\partial_x \theta_s)^2$$

$$+ \frac{2g_1}{(2\pi a)^2} \int dx \cos \left(\sqrt{8\pi} \phi_\sigma \right). \tag{6.29}$$

Parameters of the Gaussian parts are related to the microscopic parameters of the original Hamiltonian

$$u_c = \left(1 + \frac{g_1}{2\pi} \right)^{1/2} \left(1 + \frac{4g_2 - g_1}{2\pi} \right)^{1/2}; \quad K_c = \left(\frac{1 + g_1/2\pi}{1 + (4g_2 - g_1)/2\pi} \right)^{1/2};$$

$$u_s = \left(1 - \left(\frac{g_1}{2\pi} \right)^2 \right)^{1/2}; \quad K_s = \left(\frac{1 + g_1/2\pi}{1 - g_1/2\pi} \right)^{1/2}.$$

Notice that $K_c < 1$ for $g_1 < 2g_2$ ("repulsion") and $K_c > 1$ for $g_1 > 2g_2$ ("attraction"). The boundaries for "repulsive" and "attractive" behaviors coincide with those obtained when studying tunneling of interacting electrons. The velocity of the charge part for $g_1 = 0$ coincides with that found in the DL solution (Sec. 3)

$$u_c = \left(1 + \frac{2g_2}{\pi}\right)^{1/2}.$$

Scaling dimension of the backscattering term in the spin part can be read off from the correlation function

$$\frac{1}{a^4}\langle e^{i\sqrt{8\pi}\phi_\sigma(z)}e^{-i\sqrt{8\pi}\phi_s}\rangle$$

$$= \frac{1}{a^4}\exp\left(\frac{8\pi K_s}{4\pi}\ln\frac{a^2}{z^2}\right) = \frac{1}{a^4}\left(\frac{a}{|z|}\right)^{4K_s} \propto a^{4(K_s-1)}.$$

If we allowed for different coupling constants between electrons of different spin orientations, then the coefficient in front of the cos term would have been $g_{1\perp}$. For $K_s > 1$, the operator scales down to zero as $a \to 0$, whereas for $K_s < 1$, it blows up signaling an instability: a spin-gap phase.

The RG-flow of the spin-part is described by the Berezinskii-Kosterlitz-Thouless phase diagram. The fixed-point value of $g_1^* = 0$ for $K_s^* > 1$. In the weak coupling limit, the RG reduces to a single equation for g_1, which we have derived in the fermionic language in Sec. 4

$$\frac{dg_1}{dl} = -g_1^2 \to g_1 = \frac{1}{\left(g_1^0\right)^{-1} + l}, \tag{6.30}$$

6.2.1. Tunneling density of states
The procedure of finding the scaling behavior for the DoS reduces to a simple trick.

• Take the free Green's function and split it formally into spin and charge parts

$$G(x, t) = \frac{1}{x - t} = \frac{1}{(x - t)^{1/2}}\frac{1}{(x - t)^{1/2}}.$$

• In an interacting system, $1/2$ in the charge part is replaced by $(K_c + K_c^{-1})/4$ and in the spin-part by $(K_s + K_s^{-1})/4$. If the spin-rotational invariance is preserved, then the spin exponent remains equal to $1/2$.

• Take $x = 0$

$$G(t) \propto \frac{1}{t^{(K_c+K_c^{-1})/4+1/2}}.$$

and read off the scaling behavior of the DoS

$$\nu\left(\varepsilon\right) \propto |\varepsilon|^{(K_c + K_c^{-1})/4 - 1/2} = |\varepsilon|^{\frac{(K_c - 1)^2}{4K_c}} = |\varepsilon|^{\frac{(u_c - 1)^2}{4u_c}}.$$

Comparing this result for $g_1 = 0$ with that by DL (Sect. 3), we see that the bosonization solution gives the same result as the fermionic one.

• For tunneling into the edge, remove K_c, which comes from the correlator $\langle \varphi \varphi \rangle$ pinned by the boundary, and multiply K_c^{-1}, which comes from $\langle \vartheta \vartheta \rangle$, by a factor of 2. This gives

$$G_e\left(t\right) \propto \frac{1}{t^{K_c^{-1}/2 + 1/2}}$$

and

$$\mathcal{G} \propto \nu_e\left(\varepsilon\right) \propto |\varepsilon|^{\frac{1}{2}\left(K_c^{-1} - 1\right)}.$$

Expanding K_c back in the interaction

$$K_c = \left(\frac{1 + g_1/2\pi}{1 + (4g_2 - g_1)/2\pi}\right)^{1/2} \approx 1 - \frac{g_2 - (1/2)g_1}{\pi},$$

we obtain the weak-coupling limit for the tunneling exponent

$$(1/2)\left(1/K_c - 1 \approx \frac{g_2 - (1/2)g_1}{2\pi}\right).$$

This coincides with the result obtained in the fermionic language (Sec. 5). What was missed in a bosonization solution is a multiplicative log-renormalization, present in Eq. (5.35). This is because we evaluated G at the fixed point, where $g_1^* = 0$, rather then derived an independent RG equation for the flow of the conductance. This procedure should bring in the log-factors (cf. Ref. [73] where these factors were obtained for the impurity scattering time).

7. Transport in quantum wires

7.1. Conductivity and conductance

7.1.1. Galilean invariance
Interactions between electrons cannot change the response to an electric field in a Galilean-invariant system–the electric field couples only to the center-of-mass whose motion is not affected by the inter-electron interaction. This property is reproduced by the bosonized theory provided that the product $uK = 1 \ (= v_F$ in

dimensional form.) To see this, combine the Heisenberg equation of motion for density ρ (spinless fermions) with the continuity equation:

$$\partial_t \rho = i[H, \rho] = -\partial_x j. \tag{7.1}$$

Calculating the commutator in Eq. (7.1) with the help of Eq. (6.9), we identify the current operator as

$$
\begin{aligned}
\partial_t \rho &= \frac{uK}{\sqrt{\pi}} \partial_x^2 \vartheta = -\partial_x \left(-\frac{uK}{\sqrt{\pi}} \partial_x \vartheta \right) \rightarrow \\
j &= -\frac{uK}{\sqrt{\pi}} \partial_x \vartheta \ .
\end{aligned}
$$

The current is not affected by the interaction as long as $uK = 1$.

7.1.2. Kubo formula for conductivity
The Kubo formula relates the conductivity, a response function to an electric field at finite ω and q, to the current-current correlation function

$$\sigma(\omega, q) = \frac{1}{i\omega} \left[-\frac{e^2}{\pi} + \langle JJ \rangle_{q\omega}^R \right], \tag{7.2}$$

where I used $n = k_F / \pi$ and $k_F / m = v_F = 1$ in our units.

Electric current for electrons ($e > 0$)

$$J = -ej = \frac{e}{\sqrt{\pi}} \partial_x \vartheta.$$

In complex time,

$$
\begin{aligned}
\langle JJ \rangle_{x,\tau}^R &= \left(\frac{e}{\sqrt{\pi}} \right)^2 (-\partial_x^2) \langle \vartheta\vartheta \rangle_{x,\tau} \rightarrow \\
\langle JJ \rangle_{q,\omega_m}^R &= \frac{e^2}{\pi} q^2 \langle \vartheta\vartheta \rangle_{q,\omega_m} \\
&= \frac{e^2}{\pi} - \frac{e^2}{\pi} \frac{\omega_m^2}{\omega_m^2 + u^2 q^2} = \frac{e^2}{\pi} + \langle \tilde{J}\tilde{J} \rangle_{q,\omega_m}.
\end{aligned}
\tag{7.3}
$$

The first term in (7.3) cancels the diamagnetic response in (7.2). Continuing analytically to real frequencies, we find

$$
\begin{aligned}
\sigma(\omega, q) &= \frac{1}{i\omega} \langle \tilde{J}\tilde{J} \rangle_{q,\omega_m \rightarrow -i\omega+\delta} = -\frac{e^2}{\pi} \frac{1}{i\omega} \frac{-\omega^2}{-(\omega + i\delta)_m^2 + u^2 q^2} \\
&= i\frac{e^2}{\pi} \frac{\omega}{\omega^2 - u^2 q^2 + i \operatorname{sgn}\omega\delta}.
\end{aligned}
\tag{7.4}
$$

Consequently, the dissipative conductivity is equal to

$$
\begin{aligned}
\mathrm{Re}\sigma\,(\omega, q) &= -\frac{e^2}{\pi}\omega\mathrm{Im}\frac{1}{\omega^2 - u^2 q^2 + i\,\mathrm{sgn}\omega\delta} \\
&= \frac{e^2}{2}\left[\delta\,(\omega - uq) + \delta\,(\omega + uq)\right].
\end{aligned}
\tag{7.5}
$$

7.1.3. Drude conductivity

In a macroscopic system, one is accustomed to take the limit $q \to 0$ first: this corresponds to applying a spatially uniform but time-dependent electric field [61]. (For the lack of a better name, I will refer to the conductivity obtained in this way as to the *Drude conductivity*). The Drude conductivity in our case is the same as for the Fermi gas as the charge velocity drops of the result

$$
\mathrm{Re}\sigma\,(\omega, 0) = e^2\delta\,(\omega)
$$

or, restoring the units,

$$
\mathrm{Re}\sigma\,(\omega, 0) = \frac{e^2 v_F}{\hbar}\delta\,(\omega)\,.
$$

All it means that when a static electric field is applied to a continuous system of either free or interacting electrons, the center-of-mass moves with an acceleration and there is no linear response, as there is no "friction" that can balance the electric force.

For electrons with spins, the electrical current is related only to the charge component of the $\vartheta-$ field:

$$
J_c = -ej_c = e\sqrt{\frac{2}{\pi}}\partial_x\vartheta_c,
$$

where again $u_c K_c = 1$. Because of the $\sqrt{2}$ factor in the current, the conductivity is by a factor of two different from that in the spinless case

$$
\mathrm{Re}\sigma\,(\omega, 0) = \frac{2e^2 v_F}{\hbar}\delta\,(\omega)\,.
$$

(Notice, however, that at fixed density v_F is by a factor of 2 smaller.)

7.1.4. Landauer conductivity

Let's consider now the opposite order of limits, corresponding to a situation when a static electric field is applied over a part of the infinite wire. (Again, for the lack of a better name, I will refer to this conductivity as to *Landauer conductivity*.)

The electric field might as well be non-uniform; the only constraint we are going to use is that the integral

$$\int dx \, E(x),$$

equal to the applied voltage, is finite. The induced current (which in 1D coincides with the current density) is given by

$$
\begin{aligned}
J(t, x) &= \int dx' \int dt' \sigma \left(t - t'; x, x'\right) E(t', x') \\
&= \int dx' \int \frac{d\omega}{2\pi} e^{-i\omega t} \sigma(\omega; x, x') E(\omega, x').
\end{aligned}
$$

In linear response, the conductivity is defined in the absence of the field. As such, it is still a property of a translationally invariant system and depends thus only on $x - x'$. This allows one to switch to Fourier transforms

$$J(t, x) = \int dx' \int \frac{d\omega}{2\pi} \int \frac{dq}{2\pi} e^{iq(x-x')} e^{-i\omega t} \sigma(\omega, q) E(\omega, x'). \qquad (7.6)$$

Now use the fact that the applied field is static: $E(x, \omega) = 2\pi \delta(\omega) E_0(x)$ (upon which the t-dependence of the current disappears, as it should be in the steady state)

$$J(x) = \int dx' \int \frac{dq}{2\pi} e^{iq(x-x')} \sigma(0, q) E_0(x'). \qquad (7.7)$$

From (7.5),

$$\sigma(0, q) = \frac{1}{u} e^2 \delta(q) = K e^2 \delta(q), \qquad (7.8)$$

where $uK = 1$ was used again. Substituting (7.8) into (7.7), we see that the $x-$ dependence of the current also disappears

$$J = \frac{K e^2}{2\pi} \int dx' E_0(x') = \frac{K e^2}{2\pi} V.$$

Conductance $\mathcal{G} = J/V$ is given by

$$\mathcal{G} = \frac{K e^2}{2\pi},$$

or, restoring the units,

$$\mathcal{G} = K \frac{e^2}{h}. \qquad (7.9)$$

For electrons with spin, a similar consideration gives

$$\mathcal{G} = K_c \frac{2e^2}{h}. \tag{7.10}$$

We see that the conductance is renormalized by the interactions from it universal value given by the Landauer formula for an ideal wire [68].

7.2. Dissipation in a contactless measurement

What kind of an experiment Eqs. (7.9) and (7.10) correspond to?

Suppose that we connect a wire of length L to an external resistor and place the whole circuit into a resonator [78]. Now, we apply an *ac* electric field $E(x, t)$ of frequency ω_0 and parallel to a segment of the wire of length $L_E \ll L$, and measure the losses in the resonator. The external resistor takes care of energy dissipation: as the wire is ballistic (also in a sense that electrons travel through the wire without emitting phonons), the Joule heat can be generated only outside the wire. Dissipated energy, averaged over many periods of the field, is given by

$$\dot{Q} = -\int dx \langle J(x, t) E(x, t) \rangle.$$

For a monochromatic field, $E(x, t) = E_0(x) \cos \omega_0 t$ and after averaging over many periods of oscillations, we obtain

$$\dot{Q} = -\int dx \int dx' \mathrm{Re}\sigma(\omega_0; x, x') E_0(x) E_0(x').$$

Now, choose the frequency in such a way that

$$L_E \ll \frac{u}{\omega_0} \ll L, \tag{7.11}$$

where u is the velocity of the charge mode in the wire. Because the wavelength of the charge excitations at frequency ω_0–acoustic plasmons– is much shorter than the distance to contacts (L), the conductivity is essentially the same as for an infinite wire and depends only on $x - x'$. Performing partial Fourier transform in Eq. (7.5), we find

$$\mathrm{Re}\sigma(\omega, x) = \frac{e^2}{2\pi u} \cos(\omega x / u) = \frac{e^2}{2\pi} K \cos(\omega x / u), \tag{7.12}$$

so that

$$\dot{Q} = -\frac{1}{2} \frac{e^2}{2\pi} K \int dx \int dx' \cos[\omega_0(x - x')/u] E_0(x) E_0(x').$$

On the other hand, because $|x, x'| \leq L_E \ll u/\omega_0$, the cosine can be replaced by unity, and

$$\dot{Q} = -\frac{e^2}{2\pi} K V^2 \equiv -\mathcal{G} V^2,$$

or

$$\mathcal{G} = \frac{e^2}{2\pi} K.$$

Therefore, dissipation in a contactless measurements under the conditions specified by Eq. (7.11) corresponds to a renormalized conductance. To the best of my knowledge, this experiment has not been performed. A typical (two-probe) transport measurement is done by applying the current and measuring the voltage drop between the reservoirs. In this case, the measured conductance does *not* correspond to Eqs. (7.9,7.10) but is rather given simply by e^2/h per spin projection–*regardless of the interaction in the wire* [79],[80],[81]. This result is discussed in the next Section.

7.3. Conductance of a wire attached to reservoirs

The reason why the two-terminal conductance is not renormalized by the interactions within the wire is very simple. For the Fermi-gas case, the conductance of e^2/h per channel is actually not the conductance of wire itself–a disorder-free wire by itself does not provide any resistance to the current. In a four-probe measurement, when the voltage and current are applied to and measured in different contacts, the conductance of a disorder-free wire is, in fact, infinite. However, in a two-probe measurement, the voltage and current contacts are the same. Finite resistance comes only from scattering of electrons from the boundary regions, connecting wide reservoirs to the narrow wire [82, 83], as shown in Fig.15a). The universal value of e^2/h is approached in the limit of an adiabatic (smooth on the scale of the electron wavelength) connection between the reservoirs and the wire [84] [34]. As the resistance comes from the regions *exterior* to the wire, the interaction *within* the wire is not going to modify the $e^2/h-$ result. Another way to think about it is to notice that the renormalized conductance (7.9,7.10) can be interpreted as a manifestation of a fractional charge $e^* = \sqrt{K}(\sqrt{K_c})$, associated with the excitations in a 1D system. However, the current coming from, *e.g.,* the left reservoir is carried by integer charges, and as all these charges get eventually transmitted through the wire, the current collected in the right reservoir is carried again by integer charges. Fractional charges is a transient phenomena which, in

---

[34] Accidentally, the actual constraint on the adiabaticity of the connection is rather soft–it is enough to require the radius of curvature of the transition region be just comparable to, rather than much larger than, the electron wavelength [84].

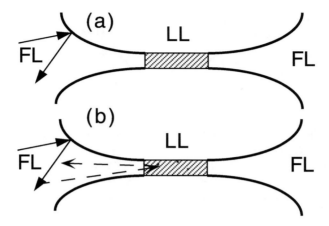

Fig. 15. a) A Luttinger-liquid (LL) wire attached to Fermi-liquid (FL) reservoirs. b) same for a single impurity within the wire.

principle, can be observed in an *ac* conductance or noise measurements but not in a *dc* experiment. In the rest of this Section, these arguments will be substantiated with some simple calculations.

7.3.1. Inhomogeneous Luttinger-liquid model

An actual system consists of two Fermi-liquid reservoirs connected via a Luttinger-liquid (LL) wire and, due to the presence of the reservoirs, is not one-dimensional. In the *inhomogeneous Luttinger-liquid model,* the actual system is replaced by an effective 1D system, which is an infinite LL with inhomogeneous interaction parameter $K(x)$ (cf. Fig. 16). The actual reservoirs are higher ($D = 2$ or 3) systems, where the effect of the interaction can be disregarded. Consequently, the reservoirs are modeled but one-dimensional free conductors with $K_L = 1$. In between, $K(x)$ goes through some variation. Similarly, the charge velocity is equal to the Fermi one in the reservoirs and varies with x in the middle of the system. The potential difference applied to the system produces some distribution of the electric field in the wire whose particular shape is irrelevant in the *dc* linear-response limit.

7.3.2. Elastic-string analogy

The (real-time) bosonic action for a spinless LL is

$$S = \frac{1}{2} \int d^2x \, \frac{1}{K(x)} \left\{ u(x)(\partial_x \varphi)^2 - \frac{1}{u(x)} (\partial_t \varphi)^2 \right\}. \tag{7.13}$$

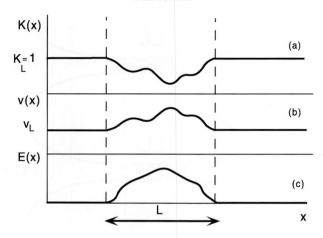

Fig. 16. Inhomogeneous Luttinger-liquid model.

The density of the electrons (minus the background density) and the (number) current are given by

$$\rho = \partial_x \varphi / \sqrt{\pi}, \qquad j = -\partial_t \varphi / \sqrt{\pi}. \tag{7.14}$$

The interaction with an external electromagnetic field A_μ is described by

$$S_{int} = \frac{e}{2\sqrt{\pi}} \int d^2x \left\{ A_0 \partial_x \varphi - A_1 \partial_t \varphi \right\}, \tag{7.15}$$

so that the equation of motion for φ is

$$\partial_t \left(\frac{1}{Ku} \partial_t \varphi \right) - \partial_x \left(\frac{u}{K} \partial_x \varphi \right) = \frac{e}{\sqrt{\pi}\hbar} E(x, t), \tag{7.16}$$

where $E = -\partial_x A_0 + \partial_t A_1$ is the electric field. We assume that the electric field is switched on at $t = 0$, so that $E(x, t) = 0$ for $t < 0$ and $E(x, t) = E(x)$ for $t \geq 0$. The problem reduces now to determining the profile of an infinite elastic string under the external force. In this language, $\varphi(x, t)$ is the transverse displacement of the string at point x and at time t, while the number current $j = -\partial_t \varphi / \sqrt{\pi}$ is proportional to the transverse velocity of the string.

To develop some intuition into the solution of Eq. (7.16), we first solve it in the homogeneous case, when $K =$ const, $u =$ const, and $E(x) =$ const for $|x| \leq L/2$, and is equal to zero otherwise. In this case, the solution of Eq. (7.16) for $t > L/u$,

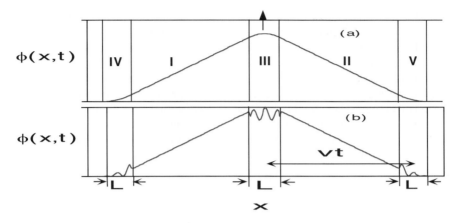

Fig. 17. a) Solution of the wave equation in the homogeneous case for $t = 5L/u$. b) Schematic solution in the inhomogeneous case for $t \gg L/u$.

is

$$\varphi(x, t) = \frac{KeV}{2\sqrt{\pi}} \times \begin{cases} t - \frac{x^2 + L^2/4}{Lu} \text{ for } |x| \le L/2; \\ t - |x|/u \text{ for } L/2 \le |x| \le ut - L/2 \\ \frac{u}{2L}\left(t - \frac{|x|-L/2}{u}\right)^2 \text{ for } ut - L/2 \le |x| \le ut + L/2 \\ 0, \text{ for } |x| \ge ut + L/2, \end{cases}$$

where $V = EL$ is the total voltage drop. This solution is depicted in Fig. 17a.

The profile of the string consists of two segments (I and II in Fig. 17a) whose widths, equal to $(ut - L)$, grow with time, and of three segments (III, IV and V in Fig. 17a) whose widths are constant in time and equal to L. In segments I and II, the profile of the string $\varphi(x, t)$ is linear in x, and therefore, being the solution of the wave equation, also in t; in segments III-V, the profile is parabolic. Outside segments IV and V, the string is not perturbed yet, and $\varphi(x, t) = 0$. As time goes on, the larger and larger part of the profile becomes linear. For late times, the pulse produced by the force spreads outwards with velocity u, involving the yet unperturbed parts of the string in motion; simultaneously, in all but narrow segments in the middle and at the leading edges of the pulse, the string moves upwards with the t- and x-independent "velocity" $\partial_t \varphi = KeV/2\sqrt{\pi}$. In terms of the original transport problem, it means that the charge current $J = -ej$ is constant outside the wire (but not too close to the edges of the regions of where the electron density is not yet perturbed by the electric field) and given by $J = Ke^2V/h$. Therefore, the conductance (per spin orientation) is $\mathcal{G} = Ke^2/h$.

We now turn to the inhomogeneous case. As in the previous case, the profile consists of several characteristic segments (cf. Fig. 17b). In segments III-V, the profile is affected by the inhomogeneities in $K(x)$, $u(x)$, and $E(x)$ and depends on the particular choice of the x-dependences in all these quantities. In segments I and II however, the profile, being the solution of the free wave equation, is again linear in x (and in t). Requiring the slopes of the string be equal and opposite in segments I and II (which is consistent with the condition of the current conservation), the solution in these regions can be written as $\varphi(x,t) = A(t - |x|/u_L)$. The constant A can be found by integrating Eq. (7.16) between two symmetric points $\pm a$, chosen outside the wire

$$-\int_{-a}^{+a} dx \, \partial_x \left(\frac{u}{K} \partial_x \varphi \right) = \frac{e}{\sqrt{\pi}} \int_{-a}^{+a} dx \, E(x) = \frac{eV}{\sqrt{\pi}}. \tag{7.17}$$

Outside the wire, $K(x) = K_L$ and $u(x) = u_L$, thus $A = K_L eV/2\sqrt{\pi}\hbar$. Calculating the current, we get $\mathcal{G} = K_L e^2/h$ and, recalling that $K_L = 1$, we finally arrive at $\mathcal{G} = e^2/h$. Thus, *the conductance is not renormalized by the interactions in the wire.*

7.3.3. Kubo formula for a wire attached to reservoirs

The Kubo formula for a translationally non-invariant system can be written as

$$\sigma(\omega; x, x')$$
$$= -\frac{e^2}{i\pi\omega} \delta(x - x')$$
$$+ \frac{e^2}{i\pi\omega} \left\{ \int d(\tau - \tau') e^{i\omega_m \tau} \langle T_\tau \partial_\tau \varphi(x, \tau) \partial_{\tau'} \varphi(x', \tau') \rangle \right\} \Big|_{i\omega_m \to \omega + i\delta}.$$

The diamagnetic contribution is cancelled by a delta-function term, which is obtained when integrating by parts in the time-ordered product [85, 10]. Having this in mind, I will re-write the conductivity via the Fourier transform of the $\varphi\varphi$-correlator without the $T-$ product

$$\sigma(\omega; x, x') = i\frac{e^2}{\pi\omega} \omega_m^2 \Phi_{\omega_m}(x, x') \Big|_{i\omega_m \to \omega + i\delta}.$$

For a translationally invariant case, this reduces back to Eq. (7.4). Now, $K(x)$ and $u(x)$ depend on position. The propagator of the φ fields satisfy the wave equation (or a Laplace's equation as we are dealing with the imaginary time)

$$\left[\frac{\omega_m^2}{u(x)K(x)} - \partial_x \left(\frac{u(x)}{K(x)} \partial_x \right) \right] \Phi_{\omega_m}(x, x') = \delta(x - x'). \tag{7.18}$$

In a model of step-like variation of $K(x)$ and $u(x)$ ($K = K_W$ and $u = u_W$ within the wire and $K = K_L = 1$ and $u = u_L = 1$ outside the wire), Eq. (7.18) is complemented by the following boundary conditions: 1) $\Phi_{\omega_m}(x, x')$ is continuous at $x = \pm L/2$, 2) $u(x) K(x) \partial_x \Phi_{\omega_m}(x, x')$ is continuous at $x = \pm L/2$; but 3) undergoes a jump of unit height at $x = x'$. Solution of this problem is totally equivalent to finding a potential of a point charge located somewhere in a sandwich-like system, consisting of three insulators with different dielectric constants. Two of these layers are semi-infinite and the third one (in the middle) is of finite thickness. "Potential" $\Phi_{\omega_m}(x, x')$ can be found in a general form for arbitrary x, x'. The expression for the current

$$ J(t, x) = \int dx' \int \frac{d\omega}{2\pi} \sigma(\omega; x, x') E(\omega, x'), $$

x' is within the wire; hence we need to know $\Phi_{\omega_m}(x, x')$ only for $-L/2 \le x' \le L/2$. In a steady-state regime, one is free to measure the current through any cross-section; let's choose x also within the wire. As we are interested in the limit $\omega \to 0$, when the plasmon wavelength is larger than the wire length, we can put $x = x'$. In the interval $-L/2 < x = x' < L/2$ the solution of the Laplace's equation is

$$ \Phi_{\omega_m}(x, x) = \frac{K_W}{2 |\omega_m|} + \frac{K_W}{2 |\omega_m|} \frac{\kappa_-^2 e^{-L/L_\omega} + \kappa_+ \kappa_- \cosh(2x/L_\omega)}{e^{L/L_\omega} \kappa_+^2 - e^{-L/L_\omega} \kappa_-^2}, \qquad (7.19) $$

where $L_\omega = u_W / |\omega_m|$, u_W is the charge velocity within the wire, and

$$ \kappa_\pm = K_W^{-1} - K_L^{-1}. $$

Letting ω_m in (7.19) to zero (and thus L_ω to ∞), we find that

$$ \Phi_\omega(x, x) = \frac{K_L}{2i\omega} = \frac{1}{2i\omega}, $$

as by our assumption $K_L = 1$. This result is true for any x, x' within the wire for $\omega \to 0$

$$ \Phi_\omega(x, x') = \frac{1}{2i\omega}, \text{ for } -L/2 < x, x' < L/2. $$

The Luttinger-liquid parameters of the wire drop out from the answer. The *dc* conductivity reduces to its free value

$$ \sigma(\omega \to 0; x, x') = \frac{e^2}{2\pi}, $$

and, consequently, the conductance

$$\mathcal{G} = \frac{e^2}{h}$$

is not renormalized by the interaction. The same consideration for electrons with spins gives

$$\mathcal{G} = \frac{2e^2}{h} \qquad (7.20)$$

7.3.4. Experiment

Most of the experiments on quantum wires indeed show that the conductance is quantizated in units of $2e^2/h$ at relatively high temperatures [35,36]. At lower temperatures, the conductance decreases beyond the universal value and also plateaux exhibit some structure as a function of the gate voltage [89]. This can be interpreted as the effect of residual disorder: as was discussed in Secs. 5,6 transmission decreases at lower energy scales. An effect of single impurity in a quantum wire will be largely insensitive to the presence of reservoirs: as long as the transmission coefficient for an impurity is much smaller than one, the largest voltage drops occurs near the impurity rather than at the contacts to the wire. One can show that the scaling of the conductance with energy is determined by the interaction parameter K *inside* the wire [90], in a contrast to the disorder-free case when only K outside the wire matters. Also, the mesoscopic conductance fluctuations increase as the temperature goes down (the theory predicts that this effect is enhanced by the interaction [91]). As one is dealing here with a crossover regime from scattering at a single impurity to that at many impurities, a quantitative analysis of the temperature dependences is difficult; another complication arises from the finite-length of the wire which cuts-off the scaling with temperature and voltage. In addition, at higher temperatures the first (and sometimes the second) quantization plateau exhibits a well-defined step at about $0.7 \times 2e^2/h$ [92, 93, 94, 95, 96, 97]. This "0.7" feature is not likely to result from spurious impurity scattering but rather reveals some interesting physics beyond what has been discussed so far in this review. Although the "0.7" feature deserves a review on its own, I will come back to this subject briefly in the next Section.

---

[35] However, it has been observed recently that the conductance of carbon nanotubes is quantized in units of e^2/h –as opposed to $4e^2/h$, predicted by the non-interacting theory for this case [86].

[36] A special case of a non-universal conductance quantization is very long wires grown by cleaved-edge overgrowth technique [87] can be attributed to a non-trivial coupling between the wire and 2D reservoirs [88], characteristic for these systems.

7.4. Spin component of the conductance

As we have shown in the previous sections, the Luttinger-liquid models predicts that conductance of a disorder-free wire is given by e^2/h per channel at any temperature. Also, thanks to spin-charge separation, spin degrees of freedom do not play an essential role in charge transport except for giving an overall factor of two to the conductance. These two results hold as long as the Luttinger-liquid model is a good description for interacting electrons in the wire. When does this model break down? If the interaction is strong, electrons form almost a periodic 1D structure: quasi-Wigner crystal. The exchange energy of almost localized electrons is exponentially small and, correspondingly, the spin velocity is small too: $u_s \ll u_c$. The Luttinger-liquid model should work for energies (temperatures) much smaller than the smallest of the two (spin and charge) bandwidth $T \ll u_s k_F \ll u_c k_F$, when both spin- and charge degrees of freedom are coherent. The *spin-incoherent regime*, i.e., $u_s k_F \ll T \ll u_c k_F$, has attracted considerable interest recently [19, 20, 21], and was shown to spoil the conductance quantization in integer multiples of $2e^2/h$ [19] at temperatures larger than the spin bandwidth ($u_s k_F$). In what follows, I present a short summary of Ref. [19].

In a quasi-Wigner-crystal regime, a reasonable starting point for describing the spin sector is the Heisenberg model

$$H_s = J_{\text{ex}} \sum_l \vec{S}_l \cdot \vec{S}_{l+1},$$

where spins are localized at "lattice sites" corresponding to positions of electrons. Because the Lieb-Mattis theorem [46] forbids ferromagnetic ordering in 1D, the sign of the exchange interaction must be antiferromagnetic: $J_{\text{ex}} > 0$. Assuming that electrons are well localized at distances $a = 1/n$ from each other, J_{ex} can be estimated in the WKB approximation: $J_{\text{ex}} \sim E_F \exp\left(-c/\sqrt{a_B n}\right)$, where $c \sim 1$ and a_B is the Bohr radius. A spin-1/2 chain is then mapped onto a Hubbard 1/2-filled model of spinless fermions via the Jordan-Wigner transormation

$$S_z(l) = a_l^\dagger a_l - 1/2;$$

$$S_x(l) + i S_y(l) = a_l^\dagger \exp\left(i\pi \sum_{J_{\text{ex}}=1}^{l-1} a_{J_{\text{ex}}}^\dagger a_{J_{\text{ex}}}\right)$$

with the result

$$H_s = -\frac{J_{\text{ex}}}{2} \sum \left[c_{l+1}^\dagger c_l + H.c.\right] + J_{\text{ex}} \sum : c_l^\dagger c_l - 1/2 :: c_{l+1}^\dagger c_l - 1/2 : .$$

The spinless Hubbard model can be bosonized

$$H_s = \frac{1}{2} \int dx \, \frac{u}{K} \, (\partial_x \phi)^2 + uK \, (\partial_x \theta)^2 + \frac{a J_{ex}}{(2\pi a)^2} \cos\left(\sqrt{16\pi}\phi\right). \qquad (7.21)$$

A comparison to the Bethe Ansatz solution of the Hubbard model at half-filling enables one to identify the parameters of the spinless LL with the microscopic parameters of the spin-1/2 chain. In particular, for an isotropic spin chain ($J_x = J_y = J_z$)

$$u = \frac{\pi}{2} J_{ex} a; \quad K = 1/2. \qquad (7.22)$$

Comparing the spin-part of the Hamiltonian of the original LL model, Eq. (6.29) with that of the spinless LL, Eq. (7.21), one notices that they the same upon the following mapping

$$\phi = \frac{1}{\sqrt{2}} \phi_s; \quad \theta = \sqrt{2}\theta_s; \quad \frac{u_s}{K_s} = \frac{u}{2K}; \quad u_s K_s = 2uK, \qquad (7.23)$$

or

$$u_s = u = \frac{\pi}{2} J_{ex} a; \quad K_s = 2K = 1. \qquad (7.24)$$

As J_{ex} is exponentially small, so is the spin bandwidth. Therefore, the Luttinger-liquid description is valid only at very low temperatures.

A translationally invariant LL still possesses spin-charge separation. However, this is no longer true for a wire connected to non-interacting leads. To understand this point, let's come back to the inhomogeneous LL model (cf. Sec. 7.3.1), where the electron density changes from a higher value in the leads to a lower value within the wire. Because J_{ex} depends on the local density, it is modulated along the wire, and its minimum value is at the middle of the wire. In the leads, we have a non-interacting system, where $J_{ex} \sim E_F \gg T$. However, in the middle of the wire spins are incoherent, if $J_{ex}^{min} \ll T$. Thus a spin part of the electron incoming from the lead at energy T above the Fermi level cannot propagate freely through the wire because the spin band narrows down: it works as if there is a barrier for spin excitations in the wire. Although charge plasmons propagate freely, backscattering of spin plasmons leads to additional dissipation, and thus to additional resistance. The total resistance of the wire consists now of two parts

$$R = R_c + R_s.$$

The charge part, R_c, is due to propagation of charge plasmons. Since the charge part is still described by the LL model, our previous result for universal conductance, Eq. (7.20), still holds and $R_c = \mathcal{G}^{-1} = h/2e^2$. For $T \ll J_{ex}^{min}$, only athermal spin plasmons, with energies exceeding the width of the spin band, contribute to R_s. The number of such plasmons is exponentially small, hence

$$R_s \propto \exp\left(-J_{ex}^{min}/T\right),$$

and total conductance $\mathcal{G} = (R_c + R_s)^{-1}$ is exponentially close to $2e^2/h$. At high temperatures $\left(T \gg J_{ex}^{min}\right)$, almost all spin plasmons are reflected by the wire. Then $R_s \sim R_c \sim h/e^2$, and the conductance differ substantially from its universal value. This qualitative picture can be confirmed in a particular simple case of the $XY-$ model for the spin-chain. In this case, R_s can be calculated explicitly [19] with the result

$$R_s = \frac{h}{2e^2} \frac{1}{\exp\left(J_{ex}^{min}/T\right) + 1}$$

and, consequently, the conductance is equal to

$$\mathcal{G} = \frac{2e^2/h}{1 + \left[\exp\left(J_{ex}^{min}/T\right) + 1\right]^{-1}}.$$

For $T \to 0$, $\mathcal{G}$ approaches the universal value of $2e^2/h$. For $T \gg J_{ex}^{min}$, $\mathcal{G}$ approaches another T -independent limit, equal to $(2/3)2e^2/h$. The actual number in the high-temperature limit of the conductance is model-dependent (it is different, for example, for an isotropic spin-chain), but the main result, *i.e.*, the non-universality of conductance quantization at higher temperatures, survives.

As it was mentioned in Sec. 7.3.4, the experiment shows that there is a shoulder in the conductance preceding the first quantization plateau at a fractional value of about $0.7 \times 2e^2/h$. Surprisingly, this "0.7 feature" is more pronounced at *higher* temperatures, and the T- dependence of this feature was reported to be of an activated type [94]. The magnetic field transforms the "0.7 feature" into a fully developed quantization plateau at e^2/h, which is to be expected in a fully polarized, and thus spinless, regime. The sensitivity to the magnetic field hints at the spin origin of the effect, and a significant theoretical effort was invested in understanding how spins can explain the observed phenomena. Although the effect, described in this Section, does have all qualitative characteristics of the observed "0.7 feature", it is not clear at the moment whether this feature indeed corresponds to the spin-incoherent regime. Other explanations of the "0.7 feature" have been suggested (most prominently, the Kondo physics is believed to be involved [98, 97]), but a further discussion of this point goes beyond the scope of these notes.

7.5. Thermal conductance: Fabry-Perrot resonances of plasmons

There is an important difference between charge and thermal (electronic) conductances [22]. As we have just shown, the charge conductance is equal to e^2/h regardless of interaction in the wire. This means that the transmission coefficient of electrons is equal to unity. The effect of the temperature on the charge conductance is the same as for a non-interacting, perfectly transmitting wire: at finite temperature, not only the lowest but also higher subbands of transverse quantization are populated, and the quantization plateaux are smeared. However, this effect is exponentially small for temperatures smaller than either the Fermi energy, E_F, or the difference between the Fermi energy and the threshold of the next subband of transverse quantization, Δ; whichever is smaller.

Thermal current is carried not by electrons but bosonic excitations: acoustic plasmons. In contrast to electrons, plasmons get reflected at the boundary between the reservoirs and the wire due to the mismatch of charge velocities (this reflection happens even for an adiabatically smooth transition). From the plasmon's point-of-view, a wire coupled to reservoirs represents a Fabry-Perrot interferometer. Interference of plasmon waves scattered from the opposite ends of the wire result in an oscillatory dependence of the transmission coefficient on the frequency with a period, given by the travel time of plasmons through the wire

$$2\pi\omega_L = L/u_W.$$

As long as $\lambda_F \ll L$, this period is long: $\omega_L \ll E_F$. The difference between charge and heat transport is that the chemical potential of plasmons is equal to zero, and thus the characteristic scale for frequency is set by T. Therefore, the thermal conductance varies with the temperature on a scale $T \simeq \omega_L$.

Suppose that a small temperature difference δT is maintained between the reservoirs, connected by a quantum wire. As the Hamiltonian of an interacting system is diagonalized of terms of plasmons, plasmons modes are decoupled and contribute to the energy flux independently. Then the thermally averaged energy current, i.e., the thermal current can be found via a Landauer-like argument

$$J_T = \int_0^\infty \frac{d\omega}{2\pi} \omega \, |t\,(\omega)|^2 \left(n_L \left(\frac{\omega}{T + \delta T} \right) - n_R \left(\frac{\omega}{T} \right) \right),$$

where $n_{L,R}\,(\omega/T)$ are the Bose distribution functions in the reservoirs. Expanding in small δT, we obtain the thermal conductance

$$\mathcal{G}_T = \frac{J_T}{\delta T} = \frac{1}{8\pi T^2} \int_0^\infty d\omega \frac{\omega^2}{\sinh^2 \omega/2T} \, |t\,(\omega)|^2. \qquad (7.25)$$

For a free system, $|t(\omega)|^2 = 1$ and $\mathcal{G}_T = \pi T/6$. The charge and thermal conductances of a free system obey the Wiedemann-Franz law

$$\frac{\mathcal{G}_T}{T\mathcal{G}} = L_0 = \frac{\pi^2}{3e^2}, \tag{7.26}$$

where L_0 is the Lorentz number. This means that charge and energy are carried by the same excitations. This is not so in a Luttinger liquid.

For an interacting system, relation (7.26) holds in the limit of $T \to 0$. The characteristic scale for frequencies in integral (7.25) is determined by T. For $T \ll \omega_L$, one can substitute $\omega = 0$ into $|t(\omega)|^2$. Regardless of the interaction strength, $|t(0)|^2 = 1$: a Fabry-Perot interferometer becomes transparent in the long wavelength limit. For $T \gtrsim \omega_L$, the result for $\mathcal{G}_T$ depends on how the charge velocity varies along the wire, and is thus non-universal. On the other hand, the charge conductance is universal. Therefore, their ratio is non-universal and the Wiedemann-Franz law is violated.

In a step-like model of Sec. 7.3.1, the transmission coefficient of plasmons is equal to

$$|t(\omega)|^2 = \frac{1}{1 + \frac{(K^2-1)^2}{4K^2}\sin^2\frac{\omega}{\omega_L}}.$$

Obviously, $|t(0)|^2 = 1$ regardless of K, an agreement with what was said above. For $T \ll \omega_L$, the Lorentz number is close to the universal value of $\pi^2/3e^2$. For $T \gg \omega_L$, the oscillations of $|t(\omega)|^2$ become very fast, so that $|t(\omega)|^2$ can be replaced by its averaged value

$$\langle|t(\omega)|^2\rangle = \int_0^{\omega_L} \frac{d\omega}{\omega_L} |t(\omega)|^2 = \frac{2K}{K^2+1}.$$

The thermal conductance increases linearly with T, so that L_0 approaches a constant but a non-universal value

$$L|_{T\gg\omega_L} = \frac{2K}{K^2+1}L_0 < L_0. \tag{7.27}$$

As the Lorentz number varies with temperature in between two limits specified by Eqs. (7.26) and (7.27), the Wiedemann-Franz law is violated.

Acknowledgments

Late R. Landauer and H. J. Schulz helped me–either directly or indirectly, through their writings–to form my way of thinking about interactions and transport, and my gratitude goes to their memories. I benefited tremendously from discussions with B. L. Altshuler, C. Biagini, C. Pèpin, A. V. Chubukov, V. V. Cheianov, E. Fradkin, F. Essler, S. Gangadharaiah, Y. Gefen, L. I. Glazman, P. M. Goldbart, I. V. Gornyi, Y. B. Levinson, D. Loss, K. A. Matveev, B. N. Narozhny, A. A. Nersesyan, D. G. Polyakov, C. Pépin, M. Yu. Reizer, I. Safi, R. Saha, B. I. Shklovskii, M. Stone, O. A. Starykh, S. Tarucha, S.-W. Tsai, A. Yacoby, O. M. Yevtushenko, and M. B. Zvonarev on various subjects discussed in these notes. I am also grateful to the organizers of the 2004 Summer School in Les Houches–H. Bouchiat, Y. Gefen, S. Guéron, and G. Montambaux–for assembling a very interesting program and providing a stimulating environment, as well as to the all participants of the School for their questions, attention, and patience. S. Gangadharaiah, L. Merrill, and R. Saha kindly helped me with producing the images. S. Gangadharaiah, P. Kumar, R. Saha, and S.-W. Tsai proofread parts of the manuscript (which does not make them responsible for the typos). I acknowledge the hospitality of the Abdus Salam International Centre for Theoretical Physics (Trieste, Italy), where part of this manuscript was written. Financial support from NSF DMR-0308377 is greatly appreciated.

Appendix A. Polarization bubble for small q in arbitrary dimensionality

The polarization bubble in Matsubara frequencies and at $T = 0$ is given by

$$
\begin{aligned}
\Pi\left(i\omega, q\right) &= \frac{N_s}{(2\pi)^{D+1}} \int \int d^D p \, d\varepsilon \, G\left(i\varepsilon + i\omega, \vec{p} + \vec{q}\right) G\left(i\varepsilon, p\right) \\
&= \frac{N_s}{(2\pi)^{D+1}} \int \int d^D p \, d\varepsilon \, \frac{1}{i\omega - \xi_{\vec{p}+\vec{q}} + \xi_{\vec{p}}} \\
&\quad \times \left[G\left(i\varepsilon + i\omega, \vec{p} + \vec{q}\right) - G\left(i\varepsilon, p\right) \right] \\
&= \frac{N_s}{(2\pi)^{D+1}} \int d^D p \, \frac{f\left(|\vec{p} + \vec{q}|\right) - f\left(p\right)}{i\omega - \xi_{\vec{p}+\vec{q}} + \xi_{\vec{p}}},
\end{aligned}
$$

where f is the Fermi function. Expanding in $\vec{q}$, and switching from the integration over $d^D p$ to $d\xi$, we obtain

$$
\Pi\left(i\omega, q\right) = -N_s \nu_D \left(1 - \int \frac{d\Omega}{\Omega_D} \frac{i\omega}{i\omega - v_F q \cos\theta} \right).
$$

where $\Omega_D = 4\pi$ (3D), $= 2\pi$ (2D), $= 2$ (1D) and ν_D is the DoS in D dimensions per one of the N_s isospin components. For D=1, the integral over Ω is understood as a sum of terms with $\cos\theta = \pm 1$. It is obvious already this form that the small $q-$form of the bubble depends on the combination $\omega/v_F q$ for any D. The final result depends on the dimensionality. Performing analytic continuation to real frequencies $i\omega \to \omega + i0^+$, we obtain

$$\Pi^R(\omega, q) = -N_s \nu_D \left(1 - \int \frac{d\Omega}{\Omega_D} \frac{\omega}{\omega - v_F q \cos\theta + i0}\right).$$

Taking the imaginary part

$$\mathrm{Im}\Pi^R(\omega, q) = -\pi N_s \nu_D \omega \int \frac{d\Omega}{\Omega_D} \delta(\omega - v_F q \cos\theta). \qquad (\mathrm{A.1})$$

From here

$$\cos\theta = \omega/v_F q$$

which means that $\theta \approx \pi/2$ for $\omega \ll v_F q$. Thus, the fermionic momentum $\vec{p}$ is almost perpendicular to the bosonic one, $\vec{q}$, in this limit.

Appendix B. Polarization bubble in 1D

Appendix B.1. Small q

Free time-ordered (causal) Green's function in 1D is equal to

$$G_\pm^0(\varepsilon, k) = \frac{1}{\varepsilon - \xi_k^\pm + i0^+ \mathrm{sgn}\xi_k^\pm},$$

where

$$\xi_k^\pm = \pm v_F(k \mp k_F),$$

and $\pm$ signs correspond to right/left moving fermions. We will be measuring the momenta from the corresponding Fermi points. For +branch: $k - k_F \to k$ and for -branch: $k + k_F \to k$. Consequently,

$$G_\pm^0(\varepsilon, k) = \frac{1}{\varepsilon \mp v_F k + i0^+ \mathrm{sgn}k}.$$

I assume N_s— fold degeneracy ($N_s = 2$ for electrons with spin, $N_s = 1$ for spinless electrons), so that

$$\Pi_\pm(\omega, q) = -\frac{i}{(2\pi)^2} N_s \int d\varepsilon \int dk G_\pm^0(\varepsilon + \omega, k + q) G_\pm^0(\varepsilon, k).$$

Calculate, *e.g.*, Π_+:

$$\Pi_+ (\omega, q)$$

$$= -\frac{i}{(2\pi)^2} N_s \int d\varepsilon \int dk \frac{1}{\varepsilon + \omega - v_F(k+q) + i0^+\mathrm{sgn}(k+q)}$$

$$\times \frac{1}{\varepsilon - v_F k + i0^+\mathrm{sgn}k}$$

$$= -\frac{i}{(2\pi)^2} N_s \int d\varepsilon \int dk \frac{1}{\omega - v_F q + i0^+\mathrm{sgn}(k+q) - i0^+\mathrm{sgn}k} \quad \text{(B.1)}$$

$$\times \left[G^0_+ (\varepsilon, k) - G^0_+ (\varepsilon + \omega, k+q) \right]. \quad \text{(B.2)}$$

The integral of the Green's over the frequency gives a Fermi distribution function [23]

$$n_+ (k) = -i \int \frac{d\varepsilon}{2\pi} G^0_+ (\varepsilon, k) .$$

For free fermions,

$$n_+ (k) = \theta (-k)$$

Now,

$$\Pi^0_+ (\omega, q) = \frac{N_s}{2\pi} \int dk \frac{1}{\omega - v_F q + i0^+\mathrm{sgn}(k+q) - i0^+\mathrm{sgn}k}$$
$$\times [\theta (-k) - \theta (-k-q)].$$

The integral is not equal to zero only if the arguments of the θ-functions are of the opposite signs. Consider different situations.

1) $k > 0; k + q < 0 \rightarrow 0 < k < -q \rightarrow q < 0$. In this case,

$$\Pi^0_+ (\omega, q) = \frac{N_s}{2\pi} \frac{q}{\omega - v_F q - i0^+}, \quad q < 0;$$

2) $k < 0, k + q > 0 \rightarrow -q < k < 0 \rightarrow q > 0$

$$\Pi^0_+ (\omega, q) = \frac{N_s}{2\pi} \frac{q}{\omega - v_F q + i0^+}, q > 0.$$

Combining the results for $q > 0$ and $q < 0$ together,

$$\Pi^0_+ (\omega, q) = \frac{N_s}{2\pi} \frac{q}{\omega - v_F q + i0^+\mathrm{sgn}q}. \quad \text{(B.3)}$$

Similarly,

$$\Pi^0_- (\omega, q) = -\frac{N_s}{2\pi} \frac{q}{\omega - v_F q + i0^+\mathrm{sgn}\omega}. \quad \text{(B.4)}$$

The total bubble

$$\Pi^0(\omega, q) = \Pi^0_+(\omega, q) + \Pi^0_-(\omega, q) = \frac{N_s}{\pi} \frac{v_F q^2}{\omega^2 - v_F^2 q^2 + i0^+}. \tag{B.5}$$

In what follows, we will also need the retarded and advanced form of the bubble. These forms can easily be obtained by repeating the calculation above in Matsubara frequencies and analytically continuing $i\omega_m \to \omega + i0$. Even simpler, one can use the general relation between time-ordered and retarded propagators [23] (which works equally well for fermionic and bosonic quantities)

$$
\begin{aligned}
\Pi^R_\pm(\omega, q) &= \Pi_\pm(\omega, q), \quad \text{for } \omega > 0 \\
&= \Pi^*_\pm(\omega, q), \quad \text{for } \omega < 0.
\end{aligned}
$$

Using Eqs. (B.3) and (B.4) we obtain

$$\Pi^R_\pm(\omega, q) = \pm \frac{N_s}{2\pi} \frac{q}{\omega - v_F q + i0^+} \tag{B.6}$$

and

$$
\begin{aligned}
\Pi^R(\omega, q) &= \frac{N_s}{\pi} \frac{v_F q^2}{\omega^2 - v_F^2 q^2 + i0^+ \text{sgn} \omega} \\
&= \frac{N_s}{\pi} \frac{v_F q^2}{(\omega + i0^+)^2 - v_F^2 q^2}. \tag{B.7}
\end{aligned}
$$

Appendix B.2. q near $2k_F$

We will also need the $2k_F$ bubble. This time, I choose to do the calculation in Matsubara frequencies:

$$
\begin{aligned}
\Pi_{2k_F}(i\omega, q) &= \frac{N_s}{(2\pi)^2} \int dk \int d\varepsilon\, G_+(i\varepsilon + i\omega, k + q) G_-(i\varepsilon, k) \\
&= -\frac{N_s}{(2\pi)^2} \int dk \int d\varepsilon \frac{1}{\varepsilon + \omega + i(k + q)} \frac{1}{\varepsilon - ik}.
\end{aligned}
$$

Poles in $\varepsilon_1 = ik$ and $\varepsilon_2 = -i(k + q) - \omega$ have to be on different sides of the real axis, otherwise the integral is equal to zero. Choose $q > 0$. Then this condition is satisfied in two intervals of k: $k > 0$ and $-\Lambda/2 < k < -q$, where Λ is the ultraviolet cut-off

$$
\begin{aligned}
\Pi_{2k_F} &= -\frac{iN_s}{2\pi}\left[\int_0^{\Lambda/2}dk - \int_{-\Lambda/2}^{-q}dk\right]\frac{1}{\omega + 2iv_Fk + iv_Fq}\\
&= -\frac{N_s}{4\pi}\left[\ln\frac{i\Lambda v_F}{\omega + iv_Fq} - \ln\frac{\omega - iv_Fq}{-i\Lambda v_F}\right]\\
&= -\frac{N_s}{4\pi}\ln\frac{\Lambda^2 v_F^2}{\omega^2 + v_F q^2}.
\end{aligned}
\tag{B.8}
$$

Because the result depends on q^2 there is no need for a separate calculation for the case $q < 0$.

Appendix C. Some details of bosonization procedure

Appendix C.1. Anomalous commutators

$$
\begin{aligned}
\rho(q) &= \sum_p a^\dagger_{p-q/2}a_{p+q/2} = \rho_+ + \rho_-;\\
\rho_\pm &= \sum_{p>0(p<0)} a^\dagger_{p-q/2}a_{p+q/2}.
\end{aligned}
$$

The operators of full density commute. The operators of left-right densities have non-trivial commutators. For example, let us calculate $[\rho_+(q), \rho_+(q')]$

$$
C_{++}(q,q') = [\rho_+(q), \rho_+(q')] = \sum_{p>0,k>0}\left[a^\dagger_{p-q/2}a_{p+q/2}, a^\dagger_{k-q'/2}a_{k+q'/2}\right]
$$

$$
= \sum_{p>0,k>0}\begin{pmatrix} a^\dagger_{p-q/2} & \underbrace{a_{p+q/2}a^\dagger_{k-q'/2}}_{=\delta_{p+q/2,k-q'/2}-a^\dagger_{k-q'/2}a_{p+q/2}} & a_{k+q'/2}\\[2em] -a^\dagger_{k-q'/2} & \underbrace{a_{k+q'/2}a^\dagger_{p-q/2}}_{=\delta_{k+q'/2,p-q/2}-a^\dagger_{p-q/2}a_{k+q'/2}} & a_{p+q/2}\end{pmatrix}.
$$

The first $\delta-$ function means that $k = p + q/2 + q'/2 > 0$ and the second one that $k = p - q/2 - q'/2$.

$$
\begin{aligned}
C_{++}(q,q') &= \sum_{p>0}a^\dagger_{p-q/2}a_{p+q/2+q'}\vartheta\left(p + q/2 + q'/2\right)\\
&\quad - a^\dagger_{p-q/2-q'}a_{p+q/2}\theta\left(p - q/2 - q'/2\right)\\
&\quad - [f(q,q') - f(q',q)],
\end{aligned}
$$

where

$$
\begin{aligned}
f(q, q') &= \sum_{p,k>0} a^\dagger_{p-q/2} a^\dagger_{k-q'/2} a_{p+q/2} a_{k+q'/2} \\
&= \sum_{p,k>0} a^\dagger_p a^\dagger_k a_{p+q} a_{k+q'}
\end{aligned}
$$

It is easy to show that $f(q, q') = f(q', q)$. Indeed,

$$
\begin{aligned}
f(q', q) &= \sum_{p,k>0} a^\dagger_p a^\dagger_k a_{p-q'} a_{k+q} \\
&= \text{re} - \text{labelling } k \longleftrightarrow p = \sum_{p,k>0} a^\dagger_k a^\dagger_p a_{k-q'} a_{p+q} \\
&= \text{anticommuting} = \sum_{p,k>0} a^\dagger_p a^\dagger_k a_{p+q} a_{k+q'} = f(q, q').
\end{aligned}
$$

Thus

$$
\begin{aligned}
C_{++}(q, q') &= \sum_{p>0} a^\dagger_{p-q/2} a_{p+q/2+q'} \theta \left(p + q/2 + q'/2 \right) \\
&\quad - a^\dagger_{p-q/2-q'} a_{p+q/2} \theta \left(p - q/2 - q'/2 \right).
\end{aligned}
$$

Introduce a new momentum

$$
Q = \frac{q + q'}{2}.
$$

In the first sum, shift $p + q'/2 \to p$ and in the second sum shift $p - q'/2 \to p$. Then

$$
\begin{aligned}
C_{++}(q, 2Q - q) &= \sum_{p>0} a^\dagger_{p-Q} a_{p+Q} \left[\theta\left(p + q/2 \right) - \theta\left(p - q/2 \right) \right] \\
&= \sum_{p>-q/2} a^\dagger_{p-Q} a_{p+Q} - \sum_{p>q/2} a^\dagger_{p-Q} a_{p+Q}
\end{aligned}
$$

If the main contribution to the sum is given by the states which lie either deep below or far above the Fermi levels, then the quantum fluctuations in the occupancy of these states are small, and the operators $a^\dagger_{p-Q} a_{p+Q}$ can be replaced by their expectation values $\langle a^\dagger_{\pi-Q} a_{p+Q} \rangle = \delta_{Q,0} n_p = \delta_{Q,0} \theta \left(p_F - p \right)$. Doing this, we find

$$
\begin{aligned}
C_{++}(q, 2Q - q) &= \delta_{Q,0} \left(\sum_{p>-q/2}^{p_F} - \sum_{p>q/2}^{p_F} \right) \\
&= \delta_{Q,0} \frac{L}{2\pi} \left(\int_{-q/2}^{p_F} dp - \int_{q/2}^{p_F} dp \right) = \delta_{Q,0} \frac{qL}{2\pi}.
\end{aligned}
$$

Therefore,

$$
[\rho_+(q), \rho_+(-q)] = \frac{qL}{2\pi}, \text{ spinless.} \tag{C.1}
$$

The same procedure for fermions with spin gives

$$
[\rho_{+,\sigma}(q), \rho_{+,\sigma'}(-q)] = \delta_{\sigma\sigma'} \frac{qL}{2\pi}, \text{ with spin.}
$$

Similarly,

$$
[\rho_{-,\sigma}(q), \rho_{-,\sigma'}(-q)] = -\delta_{\sigma\sigma'} \frac{qL}{2\pi}, \text{ with spin.}
$$

and

$$
[\rho_{+,\sigma}(q), \rho_{-,\sigma}(-q)] = 0.
$$

Combining these results together

$$
\left[\rho_{\alpha,\sigma}(q), \rho_{\alpha',\sigma'}(-q) \right] = \alpha \delta_{\alpha,\alpha'} \delta_{\sigma,\sigma'} \frac{qL}{2\pi},
$$

where $\alpha = \pm$ is the chirality index. For full charge density and current, it means that

$$
\begin{aligned}
\left[\rho^c(q), \rho^c(-q) \right] &= \left[\rho_+^c(q) + \rho_-^c(q), \rho_+^c(-q) + \rho_-^c(-q) \right] \\
&= \frac{qV}{2\pi} + \frac{qV}{2\pi} - \frac{qV}{2\pi} - \frac{qV}{2\pi} = 0.
\end{aligned}
$$

Similarly,

$$
[j^c(q), j^c(-q)] = 0,
$$

whereas

$$
[\rho^c(q), j^c(-q)] = \frac{qV}{2\pi} + \frac{qV}{2\pi} + \frac{qV}{2\pi} + \frac{qV}{2\pi} = \frac{2}{\pi} qL.
$$

In 4-notations,

$$
[j^\mu(q), j^\nu(-q)] = \epsilon^{\mu\nu} \frac{2}{\pi} qL,
$$

where $\epsilon^{00} = \epsilon^{11} = 0$, $\epsilon^{01} = -\epsilon^{10} = 1$.

Appendix C.2. Bosonic operators

Let's check that the representation of density operators via standard bosonic operators does reproduce commutation relation for density. Expand the density operators over the normal modes

$$
\rho_+ (x) \;=\; \frac{1}{L} \sum_{q>0} A_q \left(b_q e^{iqx} + b_q^\dagger e^{-iqx} \right) ;
$$

$$
\rho_- (x) \;=\; \frac{1}{L} \sum_{q<0} A_q \left(b_q e^{iqx} + b_q^\dagger e^{-iqx} \right) ,
$$

where, without a loss of generality, A_q can be chosen real and even function of q. Fourier transforming $\rho_+ (x)$

$$
\begin{aligned}
\rho_+ (q) \;&=\; \int_{-\infty}^{\infty} dx\, e^{-iqx} \frac{1}{L} \sum_{q'>0} A_{q'} \left(b_{q'} e^{iq'x} + b_{q'}^\dagger e^{-iq'x} \right) , &\text{(C.2)}\\
\;&=\; A_q \left(\theta (q)\, b_q + \theta (-q)\, b_{-q}^\dagger \right) .
\end{aligned}
$$

Choose $q > 0$ and substitute (C.2) into the commutation relation

$$
\begin{aligned}
\left[\rho_+ (q) , \rho_+ (-q) \right] \;&=\; A_q^2 \left[b_q , b_q^\dagger \right] = A_q^2 = \frac{qL}{2\pi} \;\rightarrow\\
A_q \;&=\; \sqrt{\frac{qL}{2\pi}} .
\end{aligned}
$$

Appendix C.2.1. Commutation relations for bosonic fields φ and ϑ

Using

$$
\varphi (x) \;=\; -i \sum_{-\infty<q<\infty} \frac{1}{\sqrt{2\,|q|\,L}} \mathrm{sgn} q \left(b_q e^{iqx} - b_q^\dagger e^{-iqx} \right) ;
$$

$$
\vartheta (x) \;=\; i \sum_{-\infty<q<\infty} \frac{1}{\sqrt{2\,|q|\,L}} \left(e^{iqx} b_q - b_q^\dagger e^{-iqx} \right) ;
$$

$$
\vartheta' (x) \;=\; - \sum_{-\infty<q<\infty} \frac{1}{\sqrt{2\,|q|\,L}} q \left(e^{iqx} b_q + b_q^\dagger e^{-iqx} \right) ,
$$

we find

$$
\begin{aligned}
\left[\varphi\left(x\right),\vartheta'\left(x\right)\right] &= i\sum_{q,q'}\frac{1}{\sqrt{2\,|q|\,L}}\frac{1}{\sqrt{2\,|q|'\,L}}\,|q'| \\
&\quad\times\underbrace{\left[b_q e^{iqx}-b_q^\dagger e^{-iqx},\,b_{q'}e^{iq'x'}-b_{q'}^\dagger e^{iq'x'}\right]}_{=2\delta_{q,-q'}} \\
&= i\frac{1}{L}\sum_q e^{iq(x-x')}=i\delta(x-x').
\end{aligned}
$$

Appendix C.3. Problem with backscattering

As it was pointed out in the main text, straightforward bosonization of the Hamiltonian for the spinless case encounters a problem if one tries to account for backscattering. As backscattering (g_1) is just an exchange process to forward scattering of fermions of opposite chiralities (g_2), the Luttinger liquid parameters with $g_1\neq0$ should be obtained from those with $g_1=0$ by a simple replacement: $g_2\rightarrow g_2-g_1$. However, if we do this, we cannot satisfy the Pauli principle which says that for a contact interaction, when $g_2=g_4=g_1$, all the interaction effects should disappear. Indeed, Eqs. (6.2) and (6.11) for u and K, correspondingly, change to

$$
\begin{aligned}
u^2 &= \left(1+\frac{g_4}{2\pi}\right)^2-\left(\frac{g_2-g_1}{2\pi}\right)^2; \\
K^2 &= \frac{1+\frac{g_4-g_2+g_1}{2\pi}}{1+\frac{g_4+g_2-g_1}{2\pi}}.
\end{aligned}
$$

For contact interaction, when $g_2=g_4=g_1$, we get

$$
\begin{aligned}
u^2 &= \left(1+\frac{g}{2\pi}\right)^2\neq1 \\
K &= 1.
\end{aligned}
$$

The charge velocity is different from 1. In addition, the product uK is renormalized from unity–this is also a problem, as it means that the current operator is renormalized by the interactions. How to fix this problem? Ref. [76] shows how to arrive at the expressions for u and K which satisfy all necessary constraints just on the basis on Galilean invariance and dimensional analysis. Ref. [77] arrives at the same result by using a careful point-splitting of the operators. Here, I present the method of Ref. [77].

Recall that the density operator, represented in terms of bosonic fields, contains not only the lowest harmonic ($q \to 0$), corresponding to long-wavelength excitations, but also harmonics oscillating at $q = 2k_F$, $4k_F$, etc. Indeed, taking into account only the $2k_F-$ oscillations, we have

$$
\begin{aligned}
\rho(x) &= \left(\psi_+^\dagger(x) e^{-ik_Fx} + \psi_-^\dagger(x) e^{ik_Fx}\right)\left(\psi_+(x) e^{ik_Fx} + \psi_-(x) e^{-ik_Fx}\right) \\
&= \psi_+^\dagger(x)\psi_+(x) + \psi_-^\dagger(x)\psi_-(x) + e^{-2ik_Fx}\psi_+^\dagger(x)\psi_-(x) + H.c.
\end{aligned}
$$

The first term in this equation has to be treated using the point-splitting procedure, because it involves two fermionic operators at the same point. The result is an infinite constant, ρ_0, which is just a uniform density, plus the gradient term. The $2k_F$-component can be bosonized without a problem, as it involves products of different fermions. The result is

$$
\rho(x) - \rho_0 = \frac{1}{\sqrt{\pi}}\partial_x\varphi + \frac{1}{2\pi\alpha}\exp\left[2\sqrt{\pi}\varphi + 2k_Fx\right] + H.c.
$$

Using this expression for the interaction part of H, we have

$$
\begin{aligned}
H_{\text{int}} &= \frac{1}{2}\int dx \int dx' V(x-x')\left[\rho(x) - \rho_0\right]\left[\rho(x') - \rho_0\right] \\
&= H_F + H_B,
\end{aligned}
$$

where the forward and backscattering parts of the Hamiltonian are given by

$$
\begin{aligned}
H_F &= \frac{1}{2\pi}\int dx \int dx' V(x-x')\partial_x\varphi\partial_{x'}\varphi; \\
H_B &= \frac{1}{2}\left(\frac{1}{2\pi a}\right)^2 \int dx \int dx' V(x-x') \\
&\quad \times \left\{\exp\left[2i\sqrt{\pi}\varphi(x)\right]\exp\left[-2i\sqrt{\pi}\varphi(x')\right]e^{2ik_F(x-x')} + H.c\right\}.
\end{aligned}
$$

In H_B, we neglected the terms that oscillate with x, x', and $x+x'$,, and kept only those terms that oscillate with $x-x'$. As our potential is sufficiently short-ranged, the oscillations of the first group of terms will average out, whereas the second group will survive. Introducing new coordinates

$$
\begin{aligned}
R &\equiv \frac{x+x'}{2}; \\
r &\equiv x - x',
\end{aligned}
$$

and assuming that $|R| \gg |r|$, the forward-scattering part of the Hamiltonian reduces to

$$H_F = \frac{1}{2\pi} \int dR \, (\partial_R \varphi)^2 \int dr V(r) = \frac{V(0)}{2\pi} \int dR \, (\partial_R \varphi)^2 .$$

The product of the two exponentials needs to be evaluated with care. Applying the Baker-Hausdorff identity

$$e^A e^B =: e^{A+B} : e^{\langle AB - \frac{1}{2} A^2 - \frac{1}{2} B^2 \rangle},$$

we get

$$\exp \left[2i \sqrt{\pi} \varphi(x) \right] \exp \left[-2i \sqrt{\pi} \varphi(x') \right]$$
$$= \exp \left[2i \sqrt{\pi} \left(\varphi(x) - \varphi(x') \right) \right] \exp[4\pi \langle \varphi(x - x') \varphi(0) - \varphi^2(0)],$$

Using the expression for the free bosonic propagator

$$\langle \varphi(x - x') \varphi(0) - \varphi^2(0)] = \frac{1}{4\pi} \ln \frac{a^2}{(x - x')^2},$$

and expanding in $r = x - x'$ under the normal-ordering sign, we obtain

$$\exp \left[2i \sqrt{\pi} \varphi(x) \right] \exp \left[-2i \sqrt{\pi} \varphi(x') \right] = -\frac{1}{2} 4\pi \, (\partial_R \varphi)^2 \, r^2 \frac{a^2}{r^2} = -2\pi \, (\partial_R \varphi)^2 \, a^2.$$

(While expanding, we neglected the first derivative of φ which can be always eliminated by choosing appropriate boundary condition.) H_B reduces to

$$\begin{aligned} H_B &= -\frac{1}{2} \left(\frac{1}{2\pi a} \right)^2 2\pi a^2 \int dR \, (\partial_R \varphi)^2 \int dr V(r) \, 2 \cos 2k_F r \\ &= -\frac{1}{2\pi} \int dR \, (\partial_R \varphi)^2 \int dr V(r) \cos 2k_F r \\ &= -\frac{V(2k_F)}{2\pi} \int dR \, (\partial_R \varphi)^2 . \end{aligned}$$

Therefore, the bosonized form of the total Hamiltonian

$$H_{int} = \frac{V(0) - V(2k_F)}{2\pi} \int dR \, (\partial_R \varphi)^2$$

manifestly obeys the Pauli principle. The Luttinger-liquid parameters are now given by

$$u = \sqrt{1 + \frac{V(0) - V(2k_F)}{2\pi}}; \quad K = \frac{1}{\sqrt{1 + \frac{V(0) - V(2k_F)}{2\pi}}}.$$

References

[1] J. Solóyom, Adv. Phys. **28**, 209 (1979).

[2] V. J. Emery, in *Highly Conducting One-Dimensional Solids*, eds. J. T. Devreese, R. E. Evrard, and V. E. van Doren, (Plenum Press, New York, 1979), p. 247.

[3] H. J. Schulz, in *Mesoscopic Quantum Physics*, Les Houches XXI (eds. E. Akkermans, G. Montambaux, J. L. Pichard, and J. Zinn-Justin), (Elsevier, Amsterdam, 1995), p. 533.

[4] R. Shankar, Rev. Mod. Phys. **66**, 129 (1994).

[5] J. Voit, Rep. Prog. Phys. **58**, 977 (1995).

[6] A. M. Tsvelik, *Quantum Field Theory in Condensed Matter Physics*, (Cambridge, 1995).

[7] M. P. A. Fisher and L. I. Glazman, in *Mesoscopic Electron Transport*, ed. L. Kouwenhoven et al. (Kluwer, Dordrecht); cond-mat/9610037.

[8] J. von Delft, H. Schoeller, Ann. Phys. **7**, 225 (1998).

[9] A. O. Gogolin, A. A. Nersesyan, ad A. M. Tsvelik, *Bozonization and Strongly Correlated Systems* (Cambridge, 1998).

[10] T. Giamarchi, *Quantum Physics in One Dimension* (Oxford, 2004).

[11] K. A. Matveev, D. Yue, and L. I. Glazman, Phys. Rev. Lett. **71**, 3351-3354 (1993); D. Yue, L. I. Glazman and K. A. Matveev, Phys. Rev. B **49**, 1966 (1994).

[12] A. M. Rudin, I. L. Aleiner, and L. I. Glazman, Phys. Rev. B **55**, 9322 (1997).

[13] D. V. Khveshchenko and M. Yu. Reizer, Phys. Rev. B **57**, 4245 (1998).

[14] E. G. Mishchenko and A. V. Andreev, Phys. Rev. B **65**, 235310 (2002).

[15] G. Zala, B. N. Narozhny, and I. L. Aleiner, Phys. Rev. B **65**, 020201 (2002).

[16] I. L. Aleiner, B. L. Altshuler, and L. I. Glazman (unpublished).

[17] M. Pustilnik, E. G. Mishchenko, L. I. Glazman, and A. V. Andreev, Phys. Rev. Lett. **91**, 126805 (2003).

[18] I. V. Gornyi, A. D. Mirlin, D. G. Polyakov, cond-mat/0407305.

[19] K. A. Matveev, Phys. Rev. Lett. **92**, 106801 (2004).

[20] V. Cheianov and M. B. Zvonarev, Phys. Rev. Lett. **92**, 176401 (2004).

[21] G. A. Fiete and L. Balents, Phys. Rev. Lett. **93**, 226401 (2004).

[22] R. Fazio, F. Hekking, and D. E. Khmelnitskii, Phys. Rev. Lett. **80**, 5611 (1998).

[23] A. A. Abrikosov, L. P. Gorkov, and I. E. Dzyaloshinskii, *Methods of quantum field theory in statistical physics*, (Dover Publications, New York, 1963).

[24] E. M. Lifshitz and L. P. Pitaevski, *Statistical Physics*, (Pergamon Press, 1980).

[25] D. Pines and P. Nozières, *The Theory of Quantum Liquids* (Addison-Wesley, Redwood City, 1966).

[26] G. M. Eliashberg, Sov. Phys. JETP **16**, 780 (1963).

[27] S. Doniach and S. Engelsberg, Phys. Rev. Lett. **17** , 750 (1966).

[28] W. F. Brinkman and S. Engelsberg, Phys. Rev. **169**, 417 (1968).

[29] D. J. Amit, J. W. Kane, and H. Wagner, a) Phys. Rev. **175**, 313 (1968); b) *ibid.* **175**, 326 (1968).

[30] D. Coffey and K. S. Bedell, Phys. Rev. Lett. **71**, 1043 (1993).

[31] D. Belitz, T. R. Kirkpatrick, and T. Vojta, Phys. Rev. B **55**, 9452 (1997).

[32] M. A. Baranov, M. Yu. Kagan, and M. S. Mar'enko, JETP Lett. **58**, 709 (1993).

[33] G. Y. Chitov and A. J. Millis, Phys. Rev. Lett. **86**, 5337 (2001); Phys. rev. B **64**, 0544414 (2001).

[34] a) A. V. Chubukov and D. L. Maslov, Phys. Rev. B **68**, 155113 (2003); b) *ibid.* **69**, 121102 (2004).

[35] A. V. Chubukov, C. Pépin, and J. Rech, Phys. Rev. Lett. **92**, 147003 (2004).

[36] S. Das Sarma, V. M. Galitski, and Y. Zhang, Phys. Rev. B **69**, 125334 (2004); V. M. Galitski and S. Das Sarma, *ibid.*, **70**, 035111 (2004).

[37] A. V. Chubukov, D. L. Maslov, S. Gangadharaiah, and L. I. Glazman, cond-mat/0412283; S. Gangadharaiah, D. L. Maslov, A. V. Chubukov, and L. I. Glazman, cond-mat/0501013.

[38] V. M. Galitski, A. V. Chubukov, and S. Das Sarma, cond-mat/0501132.

[39] D. Belitz, T. D. Kirkpatrick, and T. Vojta, cond-mat/0403182 (to appear in Rev. Mod. Phys.).

[40] see D. S. Greywall, Phys. Rev. B **27**, 2747 (1983) and references therein.

[41] G. R. Stewart, Rev. Mod. Phys. **86**, 755 (1984).

[42] G. Catelani and I. L. Aleiner, JETP **100**, 2005, 331(2005).

[43] A. Casey, H. Patel, J. Nyeki, B. P. Cowan, and J. Saunders, Phys. Rev. Lett. **90**, 115301 (2003).

[44] T. Holstein, R. E. Norman, and P. Pincus, Phys. Rev. B **8**, 2649 (1973); M. Reizer, *ibid.* **40**, 11571 (1989).

[45] P. A. Lee, Phys. Rev. Lett. **63**, 680 (1989).

[46] E. Lieb and D. Mattis, Phys. Rev. **125,** 164 (1962).

[47] I. E. Dzyaloshinskii and A. I. Larkin, Zh. Sov. Phys. JETP **34**, 202 (1972).

[48] G. I. Japaridze and A. A. Nersesyan, Phys. Lett. **94** A, 224 (1983).

[49] R. Saha and D. L. Maslov (unpublished).

[50] A. V. Chubukov, D. L. Maslov, and A. Millis (unpublished).

[51] C. M. Varma, P. B. Littlewood, S. Schmitt-Rink, E. Abrahams, and A. E. Ruckenstein, Phys. Rev. Lett. **63**, 1996 (1989).

[52] Yu. A. Bychkov, L. P. Gor'kov, and I. E. Dzyaloshisnkii, Sov. Phys. JETP **23**, 489 (1966).

[53] C. Castellani, C. Di Castro, and W. Metzner, Phys. Rev. Lett. **72**, 316 (1994).

[54] H. Fukuyama and M. Ogata, J. Phys. Soc. Jpn. **63**, 3923 (1995).

[55] C. Halboth and W. Metzner, Phys. Rev. B **57**, 8873 (1998).

[56] A. V. Chaplik, Sov. Phys. JETP **33**, 997 (1971); C. Hodges, H. Smith and J. W. Wilkins, Phys. Rev. B **4**, 302 (1971); P. Bloom, Phys. Rev. B **12**, 125 (1975).

[57] S. Tomonaga, Prog. Theor. phys. **5**, 544 (1950).

[58] J. M. Luttinger, J. Math. Phys. **4**, 1154 (1963).

[59] I. E. Dzyaloshinskii and A. I. Larkin, JETP, **38**, 202 (1974).

[60] W. Metzner and C. di Castro, Phys. Rev. B **47**, 16107 (1993).

[61] G. D. Mahan, *Many-Particle Physics* (Plenum Press, New York, 1990).

[62] L. I. Glazman, I. M. Ruzin, B. I. Shklovskii, Phys. Rev. B **45**, 8454 (1992).

[63] A. M. Finkel'stein and A. I. Larkin, Phys. Rev. B **47**, 10461 (1993).

[64] H. Gutfreund and M. Schick, Phys. Rev. **168,** 418 (1968).

[65] J. Voit, J. Phys. Condensed Matter **5**, 8355 (1993).

[66] A. Luther and V. J. Emery, Phys. Rev. Lett. **33**, 589 (1974).

[67] See B. L. Al'tshuler and A. G. Aronov, in *Electron-electron interactions in disordered conductors*, edited by A. L. Efros and M. Pollak (Elsevier, 1985), p. 1. and references therein.

[68] C. Kane and M. P. A. Fisher, Phys. Rev. Lett. **68**, 1220; Phys. Rev. B **45**, 15233.

[69] F. D. M. Haldane, J. Phys. C **14**, 2585 (1981), Phys. Rev. Lett. **47**, 1840 (1981).

[70] M. Bockrath, D. H. Cobden, J. Lu, A. G. Rinzler, R. E. Smalley, L. Balents, and P. L. McEuen, Nature **397**, 598 (1999).

[71] Z. Yao, H. W. Postma, C. Balents, and C. Dekker, Nature **402**, 273 (1999).

[72] D. C. Mattis, J. Math. Phys. **15**, 609 (1974).

[73] L. P. Gor'kov and I. E. Dzyaloshinskii, JETP Lett. **18**, 401 (1973).

[74] S. Eggert and I. Affleck, Phys. Rev. B **46**, 10866 (1992).

[75] M. Fabrizio and A. O. Gogolin, Phys. Rev. B **51**, 17827 (1995).

[76] O. A. Starykh, D. L. Maslov, W. Häusler, and L. I. Glazman, in *Interactions and Transport Properties of Lower Dimensional Systems*, Lecture Notes in Physics, eds. B. Kramer and T. Brandeis, (Springer 2000) p. 37; cond-mat/9911286.

[77] S. Capponi, D. Poilblanc, T. Giamarchi, Phys. Rev. B 61, 13410 (2000).

[78] The idea of this *gedanken* experiment was suggested by Y. B. Levinson.

[79] D. L. Maslov and M. Stone, Phys. Rev. **52**, R5539 (1995).

[80] V. V. Ponomarenko, Phys. Rev. B **52**, R8666 (1995).

[81] I. Safi and H. J. Schulz, Phys. Rev. B **52**, 17040 (1995).

[82] Y. Imry, in *Directions in Condensed Matter Physics*, edited by G. Grinstein and G. Mazenko (World Scientific, Singapore, 1986), Vol. 1, p. 101.

[83] R. Landauer, Z. Phys. B **68**, 217 (1987).

[84] L. I. Glazman, G. B. Lesovik, D. E. Khmel'nitskii and R. I. Shekhter, JETP. Lett. **48**, 238 (1988].

[85] R. Shankar, Int. J. of Mod. Phys. B **4**, 2371 (1990).

[86] M. Biercuk, N. Mason, J. Martin, A. Yacoby, and C. M. Marcus, Phys. Rev. Lett. **94**, 026801 (2004).

[87] A. Yacoby, H. L. Stormer, N. S. Wingreen, L. N. Pfeiffer, K. W. Baldwin, and K. W. West, Phys. Rev. Lett. **77**, 4612 (1996).

[88] R. de Picciotto, H. L. Stormer, A. Yacoby, L. N. Pfeiffer, K. W. Baldwin, and K. W. West, Phys. Rev. Lett. **85**, 1730 (2000).

[89] S. Tarucha, T. Honda, T. Saku, Sol. State Commun. **94**, 413 (1995).

[90] D. L. Maslov, Phys. Rev. B **52**, R14368 (1995).

[91] S. R. Renn and D. P. Arovas, Phys. Rev. Lett. **78**, 4091 (1997).

[92] K. J. Thomas, J. T. Nicholls, M. Y. Simmons, M. Pepper, D. R. Mace, and D. A. Ritchie, Phys. Rev. Lett. **77**, 135 (1996).

[93] K. J. Thomas, J. T. Nicholls, N. J. Appleyard, M. Y. Simmons, M. Pepper, D. R. Mace, W. R. Tribe, and D. A. Ritchie, Phys. Rev. B **58**, 4846 (1998).

[94] A. Kristensen, H. Bruus, A. E. Hansen, J. B. Jensen, P. E. Lindelof, C. J. Marckmann, J. Nygard, and C. B. Sorensen, Phys. Rev. B **62**, 10950 (2000).

[95] B. E. Kane, G. R. Facer, A. S. Dzurak, N. E. Lumpkin, R. G. Clark, L. N. Pfeiffer and K. W. West, Appl. Phys. Lett. **72**, 10950 (1998).

[96] D. J. Reilly, G. R. Facer, A. S. Dzurak, B. E. Kane, R. G. Clark, P. J. Stiles, R. G. Clark, A. R. Hamilton, J. L. O'Brien, N. E. Lumpkin, L. N. Pfeiffer, and K. W. West, Phys. Rev. B 63, 121311 (2001).

[97] S. M. Cronenwett, H. J. Lynch, D. Goldhaber-Gordon, L. P. Kouwenhoven, C. M. Marcus, K. Hirose, N. S. Wingreen, and V. Umansky, Phys. Rev. Lett. **88**, 226805 (2002).

[98] Y. Meir, K. Hirose, and N. S. Wingreen Phys. Rev. Lett. **89**, 196802 (2002).

Seminar 1

IMPURITY IN THE TOMONAGA-LUTTINGER MODEL: A FUNCTIONAL INTEGRAL APPROACH

I.V. Lerner and I.V. Yurkevich

School of Physics and Astronomy, University of Birmingham, B15 2TT, United Kingdom

H. Bouchiat, Y. Gefen, S. Guéron, G. Montambaux and J. Dalibard, eds.
Les Houches, Session LXXXI, 2004
Nanophysics: Coherence and Transport

Contents

111

1. Introduction

The Tomonaga-Luttinger model [1–3] of one-dimensional (1D) strongly correlated electrons gives a striking example of non-Fermi-liquid behaviour [4–6].
Since the seminal paper by Haldane [6], where the notion of the Luttinger liquid (LL) was coined and fundamentals of a modern bosonization technique were
formulated, this model and its various modifications remain at the focus of interest in research in strongly correlated systems. Without diminishing the role
of standard theoretical methods, such as diagrammatic techniques or the renormalization group approach, one can say that the bosonization methods constitute
the most powerful theoretical tool for strongly correlated 1D systems. All these
methods, and in particular the 'canonical' operator bosonization, where the creation and annihilation operators of electrons are explicitly represented in terms
of Bose operators and a 4-fermionic Hamiltonian is eventually diagonalized in
the bosonic representation, are described by Dmitrii Maslov in lecture notes published in this volume.

The standard operator bosonization is one of the most elegant methods developed in theoretical physics. However, by its very formulation it seems both
limited to and specific for one-dimensional physics. A subject of this seminar
is to demonstrate the existence and usefulness of an alternative way to bosonize
1D interacting electrons, called the 'functional bosonization'. It is based on the
Hubbard-Stratonovich decoupling of the four-fermion interaction – a typical way
to "bosonize" a fermionic system in higher-dimensional problems. This method
was elaborated in different ways in a set of papers [7–11]. In this seminar, we
will describe such a functional method in the form similar to that developed earlier [10, 11] for the treatment of the pure LL as well as a single-impurity problem
in the Luttinger model. However, we will employ here the Keldysh technique
(see for reviews [12]) rather than the Matsubara one used in [10, 11].

The essence of the method is to eliminate a mixed fermion-boson term in the
action (resulted from the Hubbard-Stratonovich decoupling) by a gauge transformation. Such a procedure is exact for the pure 1D Luttinger model and gives a
convenient starting point for including a single backscattering impurity.

The problem of a single impurity in the LL has been actively investigated by
many authors. [13–20] One of the main results of these considerations [13, 14,
18–20] was the suppression at low temperatures of the local density of states

(TDoS) *at the impurity site* and at a distance from the impurity [11] and the related suppression of the conductance [13, 14] and the X ray edge singularity [18–20]. Another prominent result was the dependence of the Friedel oscillations [9, 15, 16] on the distance from the impurity.

In this seminar, we will only briefly outline only the most important results for the one-impurity problem in the Tomonaga-Luttinger model, while giving a slightly more detailed description of the functional bosonization in the Keldysh technique that was not previously published. For simplicity, we will only address a single-mode Tomonaga-Luttinger model, with one species of right- and left-moving electrons, thus omitting spin indices and considering eventually the simplest linearized model of a single-valley parabolic electron band. We will also skip over a physical introduction, referring the reader to lecture notes by Dmitrii Maslov published in this volume.

2. Functional integral representation

Within the limitations outlined above, the most generic Hamiltonian of interacting 1D electrons in the presence of an external scattering potential $v(x)$ can be written as

$$
\hat{H} = \int dx \hat{\psi}^\dagger \left[-\frac{\partial_x^2}{2m} - \varepsilon_F + v(x) \right] \hat{\psi}(x)
$$
$$
+ \frac{1}{2} \int dx dx' \hat{\psi}^\dagger(x) \hat{\psi}^\dagger(x') V_0(x - x') \hat{\psi}(x') \hat{\psi}(x), \qquad (2.1)
$$

where $V_0(x-x')$ is a bare electron-electron interaction. The observable quantities to be calculated with this Hamiltonian are the tunneling density of states (TDoS) and the current as a linear response to an applied field characterised by the vector potential $A(x, t)$. We will give the results for the TDoS at the end of this presentation, but will mainly describe techniques for calculating the current $j(x, t)$. It can be written with the help of the Kubo formula in terms of the current-current correlation function thermally averaged $(\langle \ldots \rangle_0)$ in the equilibrium Gibbs ensemble:

$$
j(x, t) = i \int_{-\infty}^{t} dt' \int dx' \left\langle \left[\hat{j}(x, t), \hat{j}(x', t') \right] \right\rangle_0 A(x', t') - \frac{ne^2}{m} A(x, t), \quad (2.2)
$$

where the current operator is defined in the standard way (in the units $\hbar = 1$ here and elsewhere):

$$
\hat{j}(x, t) = -\frac{ie}{2m} \hat{\psi}^\dagger(x, t) \partial_x \hat{\psi}(x, t) + \text{h.c.} \qquad (2.3)
$$

The Kubo formula can be rewritten in Keldysh techniques by defining the contour C_K which runs from $-\infty$ to the observation time t along the upper bank of the cut along the real time axis in the complex time plane and then returns to $-\infty$ along the lower cut:

$$j(x, t) = i \int_{C_K} dt' \int dx' \left\langle T_K \hat{j}(x, t) \hat{j}(x', t') \right\rangle_0 A(x', t') - \frac{ne^2}{m} A(x, t). \quad (2.4)$$

Here we have introduced the chronological operator T_K which time-orders the operators in a descending order along the contour C_K (with all the times on the lower cut considered as later times as compared to those on the upper cut). Although the current operators are bilinear in Fermi-operators, for future usage we assume in the standard way that the Fermi-operators anticommute under T_K, assuming additionally that at equal times $\hat{\psi}^\dagger(t)$ is taken at an infinitesimally later moment than $\hat{\psi}(t)$. Such an agreement is consistent with the definition of the current (Eq. 2.3). In what follows, we will choose a more general contour (the Keldysh contour), running from minus to plus infinity above the cut and returning below the cut. In the operator language, such an extension of the contour corresponds to the insertion of extra evolution operators: having not been coupled to the observables (the current operator in our case), such evolution operators above and below the cut simply cancel each other.

Now, any expression written as the chronological operator average can be straightforwardly represented as a functional integral over the fields defined on the contour of the time ordering. We introduce the action,

$$S[\bar{\psi}, \psi] = \int_{C_K} dt \left\{ i \bar{\psi} \partial_t \psi - H[\bar{\psi}, \psi] \right\}, \quad (2.5)$$

where the last term is the normal ordered Hamiltonian with $\hat{\psi}^\dagger \to \bar{\psi}$ and $\hat{\psi} \to \psi$, where $\bar{\psi}$ and ψ are the anticommuting Fermi fields. Then, the linear response current of Eq. (2.4) takes the form

$$j(x, t) = i \int_{C_K} dt' \int dx' \left\langle \tilde{j}(x, t) \tilde{j}(x', t') \right\rangle A(x', t') - \frac{ne^2}{m} A(x, t), \quad (2.6)$$

where the brackets stand for the functional integral

$$\langle \ldots \rangle = \int \mathcal{D}\bar{\psi} \, \mathcal{D}\psi \, (\ldots) e^{i S[\bar{\psi}, \psi]}, \quad (2.7)$$

and the current field is defined by

$$\tilde{j}(x, t) = \frac{e}{2mi} \bar{\psi}(x, t) \partial_x \psi(x, t) + \text{c.c.}$$

3. The effective action for the Tomonaga-Luttinger Model

For free electrons, the plane waves basis is natural. Assuming that the scattering by impurities and electron-electron interaction involve energy scales much smaller than the Fermi energy, the plane wave basis presents a natural starting point. Therefore we may assume that the main contribution to the functional integral above comes from the fields representing separately right-moving and left-moving electrons

$$\psi(x,t) \approx \psi_R(x,t)\,e^{ip_Fx} + \psi_L(x,t)\,e^{-ip_Fx} \tag{3.1}$$

where $\psi_{R,L}$ are smooth on the p_F^{-1} scale. Such a separation is the essence of the Tomonaga-Luttinger model, and corresponds to the linearization of the initially parabolic electron band. The right- and left-moving electrons can be transformed into another due to backscattering processes. We shall neglect such processes due to the electron-electron interaction, thus keeping only the small momentum transfer part of interaction, $V(q \ll 2p_F)$. This part of interaction is non-trivial by itself as it breaks the Fermi Liquid theory in 1D. On the other hand we will keep only backscattering in the impurity potential $v(x)$ since small-momentum elastic scattering does not result in any qualitative change in the Luttinger Liquid behavior.

Now we make the substitution (3.1) neglecting higher order derivatives of smooth functions and discarding integrals over fast oscillating terms. After some straightforward manipulations we come to the action for the Tomonaga-Luttinger model:

$$S_{TL} = S_0 + S_{\text{int}}. \tag{3.2}$$

The first term describes free electrons in the presence of the external scattering potential (which can also be changing in time):

$$S_0 = \int dx\,dt\,\Psi^\dagger(x,t) \begin{pmatrix} i\partial_R & v(x,t) \\ \bar{v}(x,t) & i\partial_L \end{pmatrix} \Psi(x,t). \tag{3.3}$$

We have assumed (here and below) that all the time integrations are performed along the Keldysh contour, and introduced the following notations:

$$\Psi = \begin{pmatrix} \psi_R \\ \psi_L \end{pmatrix}, \qquad \Psi^\dagger = \begin{pmatrix} \bar{\psi}^R & \bar{\psi}^L \end{pmatrix}, \qquad \partial_{R/L} \equiv \partial_t \pm v_F \partial_x. \tag{3.4}$$

The second term in Eq. (3.2) gives the interaction part of the action

$$S_{\text{int}} = \frac{i}{2} \int dx\,dx'\,dt\, n(x,t)\,V_0(x-x')n(x',t), \qquad \left(n \equiv \Psi^\dagger\Psi\right). \tag{3.5}$$

4. The bosonized action for free electrons

Before dealing with interacting electrons, we convert the action (3.3) for free electron in the presence of the impurity potential into the action in terms of bosonic fields.

Expanding in the impurity strength, we integrate over the fermion fields using the Wick theorem:

$$Z_0 = \sum_{n=0}^{\infty} \frac{(-1)^n}{(n!)^2} \int d^n z d^n z' \left[\prod_{k=1}^{n} v(z_k) \bar{v}(z_k') \right] \det g_L(z_i, z_j') \det g_R(z_i', z_j).$$

(4.1)

We have introduced the notation $z = (x, t)$ and defined the Green functions of left- and right-moving electrons as follows:

$$\partial_\eta g_\eta(z, z') = \delta(x - x')\delta(t - t'), \qquad\qquad (\eta \equiv R, L). \qquad (4.2)$$

Their explicit form is

$$g_{R/L}(z, z') = -\frac{T}{2v_F} \frac{1}{\sinh \pi T \left[(t - t') \mp \frac{x - x'}{v_F} \right]}. \qquad (4.3)$$

These Green functions are defined on the Keldysh contour, so that each of them is the matrix

$$g \equiv -i \langle \psi(t)\bar{\psi}(t') \rangle = \begin{pmatrix} g^{++} & g^{<} \\ g^{>} & g^{--} \end{pmatrix}, \qquad (4.4)$$

where the time arguments of both the fields in g^{++} and g^{--} are, respectively, on the upper or lower parts of the contour, while in $g^{<(>)}$ the first (second) argument is on the upper part and the second (first) is on the lower. We should recall at this point that the functional average (2.7), invoked in the definition (4.4), automatically arranges for the time ordering along the Keldysh contour. The components of g are not independent, obeying the usual relation $g^{++} + g^{--} = g^{<} + g^{>}$. In what follows, we will perform the standard Keldysh rotation, reducing all the appropriate matrices to the triangular form with the Keldysh component ($g^K = g^{<} - g^{>}$) in the upper right corner, and the retarded and advanced components ($g^r = (g^a)^* = g^{++} - g^{<}$) on the main diagonal (see, e.g., ref. [12]). Then, without writing this explicitly, we shall assume that the time arguments belonging to the upper and lower branch of the contour have, respectively, positive and negative infinitesimal shift into the complex plane. Note finally that in the equilibrium case presented here the (Fourier transform of the) Keldysh component is related to the (Fourier transforms of the) retarded and advanced ones via the Fermi distribution function $f_\varepsilon(T)$ as follows:

$$g^K(\varepsilon) = (1 - 2f_\varepsilon(T))(g^r(\varepsilon) - g^a(\varepsilon)).$$ (4.5)

The same relations are valid for other Green's functions to be considered so that, wherever this does not involve an ambiguity, we will give only the retarded/advanced components in the explicit form assuming that the expression like Eq. (4.5) is valid for the Keldysh component.

To calculate the partition function in Eq. (4.1) we use the Cauchy identity [21]:

$$\det \frac{1}{\sinh(z_i - z_j')} = (-1)^{n+1} \frac{\prod_{i<j} \sinh(z_i - z_j) \sinh(z_i - z_j)}{\prod_{i,j} \sinh(z_i - z_j')}.$$ (4.6)

It reduces Eq. (4.1) for the partition function to the following one:

$$Z_0 = \sum_{n=0}^{\infty} \frac{1}{(n!)^2} \left(\frac{T}{2v_F}\right)^{2n} \int d^n z \, d^n z' \left[\prod_{k=1}^{n} v(z_k)\bar{v}(z_k')\right] \frac{\prod_{i<j} s(z_i - z_j)s(z_i' - z_j')}{\prod_{i,j} s(z_i - z_j')}$$ (4.7)

with

$$s(z - z') \equiv \sinh \pi T\left(t - t' - \frac{x - x'}{v_F}\right) \sinh \pi T\left(t - t' + \frac{x - x'}{v_F}\right)$$ (4.8)

Introducing

$$i G_0(z - z') = -\ln s(z - z')$$ (4.9)

one can write

$$Z_0 = \sum_{n=0}^{\infty} \frac{1}{(n!)^2} \left(\frac{T}{2v_F}\right)^{2n} \int d^n z \, d^n z' \left\{\prod_{k=1}^{n} v(z_k)\bar{v}(z_k')\right.$$

$$\left. \times \exp\left[-i\sum_{i<j}\left[G_0(z_i - z_j) + G_0(z_i' - z_j')\right] + i\sum_{i,j} G_0(z_i - z_j')\right]\right\}$$ (4.10)

This is a partition function of the Coulomb gas with the logarithmic interaction and it can be represented as the functional integral over the bosonic field $\varphi(x, t)$:

$$Z_0 = \int \mathcal{D}\varphi \, e^{i S_0[\varphi]} \exp\left\{\alpha \int dz \left[v e^{-i\varphi} + \text{c.c.}\right]\right\},$$ (4.11)

Here the free bosonic action $S_0[\varphi]$ is defined in terms of the Green's function G_0 of Eq. (4.9) as follows

$$S_0[\varphi] = \frac{1}{2} \int dz dz' \varphi(z) G_0^{-1}(z - z') \varphi(z'), \qquad (4.12)$$

while the Fourier transform of the retarded/advanced components of $G_0(z)$ is given by (with $\omega_\pm \equiv \omega \pm i0$)

$$G_0^{r/a}(q, \omega) = \frac{4\pi v_F}{\omega_\pm^2 - v_F^2 q^2}. \qquad (4.13)$$

The constant α in Eq. (4.11) absorbs an ill defined value of $G_0(x, t)$ at $x = 0$, $t = 0$. We use the ultra-violet cutoff which corresponds to the scale of order ε_F:

$$\alpha = \frac{T}{2v_F} e^{\frac{i}{2}G_0(0)} \simeq \frac{\varepsilon_F}{2\pi v_F}.$$

Thus, we have cast the original free fermion problem into that of interacting bosons, represented by the partition function (4.11). The interaction between bosons, i.e. the second term in the exponent in Eq. (4.11) comes from the backscattering impurity term in the original fermionic problem. The Gaussian action for noninteracting bosons, Eq. (4.12), can be explicitly written in the x - t representation in the standard form

$$S_0[\varphi] = \frac{1}{8\pi v_F} \int dz \left[(\partial_t \varphi)^2 - v_F^2 (\partial_x \varphi)^2 \right], \qquad (4.14)$$

although Eq. (4.12) is no less convenient, especially for the generalisation for the interaction. The main advantage of the bosonization, either in standard or functional form, is that including the quadric electron-electron interaction does not substantially change the free action.

5. Gauging out the interaction

The first step in dealing with the interaction term (3.5) in the action is to perform the Hubbard-Stratonovich transformation which can be symbolically written as

$$\exp\left\{ i S_{\text{int}}[\bar\psi, \psi] \right\} = \int \mathcal{D}\phi \exp\left\{ \frac{i}{2} \phi V_0^{-1} \phi + \phi \Psi^\dagger \Psi \right\}. \qquad (5.1)$$

Note that the auxiliary 'Hubbard-Stratonovich' (HS) bosonic field ϕ here is different from the field φ in Eqs. (4.14) and (4.11). Substituting this representation into the full action (3.2), we bring the partition function to the following form:

$$Z = \int \mathcal{D}\phi \, \mathcal{D}\Psi^{\dagger} \mathcal{D}\Psi \, e^{iS_0[\phi] + iS[\Psi,\phi]}. \tag{5.2}$$

The action $S[\Psi, \phi]$ for fermions interacting with the HS field is given by

$$S[\Psi, \phi] = \int dz \, \Psi^{\dagger}(z) \begin{pmatrix} i\partial_R + \phi & v(z) \\ \bar{v}(z) & i\partial_L + \phi \end{pmatrix} \Psi(x, t). \tag{5.3}$$

To cast this integral into a form identical to that of the previous section we apply the local gauge transformation,

$$\psi_{\eta}(z) \equiv \psi_{\eta}(x, t) \rightarrow \psi_{\eta}(x, t) \, e^{i\theta_{\eta}(x,t)} \quad \text{with} \quad \partial_{\eta}\theta_{\eta}(x, t) = \phi(x, t), \tag{5.4}$$

which removes the bosonic field ϕ from the diagonal part of the action (5.3) but at a cost: the off-diagonal terms are rotated with the factors $e^{\pm i\theta}$, and the Jacobian of the transformation J changes the quadratic in ϕ part of the action. Let us now deal with this bosonic part of the action.

It is shown in Appendix A that the Jacobian J of the gauge transformation (5.4) can be represented as

$$\ln J[\phi] = \frac{i}{2} \int dz dz' \phi(z) \, \Pi(z, z') \, \phi(z'). \tag{5.5}$$

The polarization operator Π is given in the random phase approximation (RPA) by

$$\Pi = \sum_{\eta=R,L} \Pi_{\eta}, \qquad \Pi_{\eta}(z - z') = ig_{\eta}(z - z')g_{\eta}(z' - z), \tag{5.6}$$

where $g_{\eta}(x - x', t - t')$ is the free electron Green function given by Eq. (4.3). It is well known that the RPA is exact for the Luttinger Liquid [4]. Note that we give in Appendix A a very simple and straightforward proof of this.

The Jacobian gives an additional quadratic in ϕ contribution to the action that should be added to the quadratic term in Eq. (5.1). This results in the free bosonic action with the kernel corresponding to the screened interaction:

$$S[\phi] = \frac{1}{2} \int dz dz' \phi(z) V^{-1}(z - z') \phi(z'), \qquad V^{-1} = V_0^{-1} + \Pi. \tag{5.7}$$

The above expression should be understood in the operator sense: V and Π are the operators whose kernels are defined with the proper time and spatial dependence on the Keldysh contour.

Using the expressions for the retarded and advanced components of g_η (with $\varepsilon_\pm = \varepsilon \pm i0$),

$$g_R^{r/a}(q, \varepsilon) = i\,(\varepsilon_\pm - v_F q)^{-1}\,, \qquad g_L^{r/a}(q, \varepsilon) = i\,(\varepsilon_\pm + v_F q)^{-1}\,, \qquad (5.8)$$

one finds the appropriate components of the polarisation operator,

$$\Pi_R^{r/a}(q, \omega) = -\frac{1}{2\pi}\frac{v_F q}{\omega_\pm - v_F q}\,, \qquad \Pi_L^{r/a}(q, \omega) = \frac{1}{2\pi}\frac{v_F q}{\omega_\pm + v_F q}\,,$$

and thus the total polarisation operator as

$$\Pi^{r/a}(q, \omega) = -\frac{1}{\pi}\frac{v_F q^2}{(\omega_\pm)^2 - v_F^2 q^2}\,. \qquad (5.9)$$

Assuming that the Fourier transform of the forward-scattering pair interaction only weakly depends on momentum, i.e. $V_0(q \ll 2p_F) \approx \text{const} \equiv \bar{V}$, and substituting Eq. (5.9) into the free bosonic action (5.7), one finds the components of the free HS bosonic propagator as follows:

$$V^{r/a}(q, \omega) = \frac{\omega_\pm^2 - v_F^2 q^2}{\omega_\pm^2 - v^2 q^2}\,\bar{V}\,,$$

$$V^K(q, \omega) = \tanh\left(\frac{\omega}{2T}\right)\left[V^R(q, \omega) - V^A(q, \omega)\right]. \qquad (5.10)$$

Here we introduced the renormalized velocity v which defines the effective coupling constant g:

$$v^2 \equiv v_F^2 + \frac{v_F \bar{V}}{\pi}\,, \qquad g \equiv \frac{v_F}{v}\,. \qquad (5.11)$$

Therefore, the gauge transformation (5.4) reduces the action in (5.2) to

$$S = S[\theta] + S[\Psi^\dagger, \Psi; \theta]. \qquad (5.12)$$

Its fermionic part is given by

$$S[\Psi^\dagger, \Psi; \theta] = \int dz \Psi^\dagger(z)\left(\begin{array}{cc} i\partial_R & v\,e^{-i\theta} \\ v\,e^{i\theta} & i\partial_L \end{array}\right)\Psi(z) \qquad (5.13)$$

with $\theta = \theta_R - \theta_L$, while its bosonic part $S[\theta]$ is defined via the field ϕ by Eqs. (5.7), (5.10) and (5.11). It is convenient to write it explicitly as an integral over the field θ, which is straightforward since θ is linearly related to the field

ϕ as in Eq. (5.4). Thus we arrive at the following explicit expression for $S[\theta]$ in Eq. (5.12):

$$S[\theta] = \frac{1}{2} \int dz\, dz'\, \theta(z) G_B^{-1}(z - z')\theta(z').$$

(5.14)

The Gaussian kernel of this interaction, G_B, can be represented as

$$G_B = G - G_0,$$

(5.15)

where G_0 is defined by Eqs. (4.9) and (4.13), while G has the standard triangular matrix structure in Keldysh space, with the Fourier transform of its retarded/advanced component given by

$$G^{r/a}(q, \omega) = \frac{4\pi v_f}{\omega_{\pm}^2 - v^2 q^2},$$

(5.16)

i.e. it differs from G_0 only by substituting v for v_F in the denominator.

The effective action (5.12) is quadratic in fermionic fields which can now be easily integrated out. Before doing so, let us stress that the representation of Eqs. (5.12)–(5.14) seems to be more convenient for some problems than the fully bosonized action.

To perform the fermionic integration, we note that the fermionic part of the action, Eq. (5.13), differs from that of the free electrons, Eq. (3.3), only by the substitution $v \to v\, e^{-i\theta}$. Therefore, repeating the same procedure as in the previous section, we represent this part of the action with the help of the bosonic field φ so that the full action in Eq. (5.12) goes (in symbolical notations) to

$$S[\varphi, \theta] = \frac{1}{2}\theta G_B^{-1}\theta + \frac{1}{2}\varphi G_0^{-1}\varphi + \alpha \left[v e^{-i(\varphi+\theta)} + \text{c.c.} \right]$$

Introducing the new bosonic field, $\Theta \equiv \theta + \varphi$, and noting again the relation (5.15) we arrive at the standard fully bosonized action:

$$S[\Theta] = \frac{1}{2} \int dz\, dz'\, \Theta(z) G^{-1}(z - z')\Theta(z') + \alpha \int dz \left[v(z)\, e^{-i\Theta(z)} + \text{c.c.} \right].$$

(5.17)

6. Tunnelling density of states near a single impurity

As an application of the formalism developed above we consider a single impurity in the Luttinger Liquid characterised by the following local time-independent potential:

$$v(x) = \lambda v_F \delta(x)$$

(6.1)

where λ is the dimensionless impurity strength. The potential (6.1) should be substituted into the action (5.17). Then one can integrate out the fields with $x \neq 0$ which results in the local action in terms of $\Theta(t) \equiv \Theta(x=0, t)$:

$$S_{\text{imp}} = \frac{1}{2} \int_{C_K} dt\, dt'\, \Theta(t) G_{\text{imp}}^{-1}(t - t') \Theta(t') + 2i\alpha\lambda v_F \int_{C_K} dt \cos\Theta(t) \qquad (6.2)$$

where the Fourier transform of the retarded/advanced components of the Gaussian kernel are given by

$$G_{\text{imp}}^{r/a}(\omega) = \int \frac{dq}{2\pi} \frac{4\pi v_F}{\omega_{\pm}^2 - v^2 q^2} = -ig \frac{2\pi}{\omega \pm i0} \qquad (6.3)$$

We now employ the self-consistent harmonic approximation (see, e.g., [22, 23]), i.e. substitute the impurity $\cos\Theta$ term with the quadratic one:

$$i2\alpha v_F \lambda \int dt \cos\Theta(t) \rightarrow -\frac{i}{2}\Lambda \int dt\, \Theta^2(t)$$

The coefficient Λ is to be found from the condition that this substitution is optimal,

$$\frac{\partial}{\partial\Lambda}\left[2\alpha v_F \lambda \langle\cos\Theta\rangle - \frac{1}{2}\Lambda \langle\Theta^2\rangle\right] = 0, \qquad (6.4)$$

where the averages are taken with the effective action symbolically represented as

$$S_{\text{eff}} = \frac{1}{2}\Theta\left(G_{\text{imp}}^{-1} + \Lambda\right)\Theta. \qquad (6.5)$$

Solving self-consistently Eq. (6.4) with the action (6.5) (which involves preserving the proper analytical structure, causing Λ to be of the standard matrix structure in the Keldysh space) one finds with logarithmic accuracy:

$$\Lambda^{r/a} = \pm\varepsilon_F \left(\frac{a v_F}{\varepsilon_F}\lambda\right)^{\frac{1}{1-g}} \simeq \pm\varepsilon_F \lambda^{\frac{1}{1-g}}. \qquad (6.6)$$

Until now we have omitted any source terms while making transformations between different representations of the partition function. As we explained in Section 2, using the functional representation of the current (and thus conductance) as an example, in order to generate observable quantities (e.g., TDoS or conductance) all the transformations should be done with the source fields. Including the source terms does not bring any principal difficulties but makes the

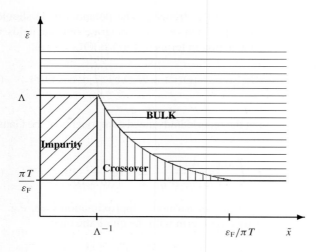

transformations more cumbersome. Therefore, we do not describe such transformations in the framework of this seminar presentation. Instead, we simply present the results for the tunnelling density of states [following from the long time asymptotics of the full electron Green function calculated in the SCHA with the bosonized action (5.12)] as a function of the distance from the impurity. We refer the reader interested in detail to our previous publication [11] (albeit for the Matsubara rather than the Keldysh functional integrals).

We present the explicit expressions for the TDoS smoothed over a length scale much larger than p_F^{-1} in three different regions:

$$\nu(x, \varepsilon) \sim \begin{cases} \tilde{\varepsilon}^{\frac{1}{g}-1} \Lambda^{-\frac{1}{2}\left(\frac{1}{g}-g\right)}, & \tilde{x} \ll \Lambda^{-1} \ll \tilde{\varepsilon}^{-1} \quad (a) \\[2ex] \tilde{\varepsilon}^{\frac{1}{g}-1} \tilde{x}^{\frac{1}{2}\left(\frac{1}{g}-g\right)}, & \Lambda^{-1} \ll \tilde{x} \ll \tilde{\varepsilon}^{-1} \quad (b) \\[2ex] \tilde{\varepsilon}^{\frac{1}{2}\left(\frac{1}{g}+g\right)-1}, & \min(\tilde{x}, \Lambda^{-1}) \gg \tilde{\varepsilon}^{-1} \quad (c) \end{cases} \qquad (6.7)$$

The regions of different behavior of the tunneling density of states are sketched in the figure. There and in Eq. (6.7), $\tilde{x} \equiv g p_F |x|$, $\tilde{\varepsilon} \equiv \varepsilon/\varepsilon_F$, and the renormalized impurity strength Λ is given by Eq. (6.6). Equation (6.7a) describes the TDoS in the vicinity of the impurity, in full correspondence with the original results of Kane and Fisher [13] obtained for the TDoS at $x = 0$, i.e. exactly at the impurity. In addition, we have established here the TDoS dependence on the impurity strength $\Lambda \equiv \lambda^{\frac{1}{1-g}}$. The region of applicability of Eq. (6.7a) corresponds to the diagonally hatched region in Fig. 1. Equation (6.7c) gives the TDoS at very large distances from the impurity. As expected, it coincides with a well-known result for the TDoS in the homogenous Luttinger liquid. Its region of applicability is horizontally hatched in Fig. 1. In the intermediate region, vertically hatched in Fig. 1, the TDoS depends both on the energy and the distance from the impurity. This analytic dependence given by Eq. (6.7b) describes the crossover from the impurity-induced dip in the TDoS to the bulk behavior. Finally, the unhatched region for $\tilde{\varepsilon} < \alpha$ corresponds to small energies, $\varepsilon < T$, where the energy dependence saturates (by $\varepsilon \to T$) in all the three lines of Eq. (6.7).

In conclusion, we have demonstrated how to develop the formalism of bosonization based on the functional integral representation of observable quantities within the Keldysh formalism. We have derived in this way the fully bosonized action for the interacting electrons in the presence of the scattering potential given by Eq. (5.17), and illustrated its usage on the example of the TDoS on a single impurity, Eq. (6.7). Let us stress finally that the intermediate representation of Eqs. (5.12)–(5.14), which still contains the part quadratic in fermion fields appears to be more convenient for some problems than the fully bosonized action.

Acknowledgements

We gratefully acknowledge support by the EPSRC grant GR/R95432.

Appendix A.   Jacobian of the gauge transformation

The Jacobian of the gauge transformation (5.4) can be defined as $J = J_R J_L$ with

$$J_\eta[\phi] = \frac{\int \mathcal{D}\psi \, e^{-\int dz \, \bar{\psi}[\partial_\eta - i\phi]\psi}}{\int \mathcal{D}\psi \, e^{-\int dz \, \bar{\psi}\partial_\eta \psi}} = e^{\operatorname{Tr}\ln[1 - ig_\eta \phi]}, \tag{A.1}$$

where the Green functions of non-interacting right- or left-moving electrons, obeying $\partial_\eta g_\eta = \hat{I}$, are given by Eq. (4.3). Note that in the matrix components $g^<(t)$ and $g^>(t)$ the time argument should be understood, respectively, as $t \pm i0$. The exponent in Eq. (A.1) can be represented as infinite series in the HS-field ϕ:

$$\ln J_\eta[\phi] = -\sum_{n=1}^{\infty} \frac{i^n}{n!} \operatorname{Tr}\left[g_\eta \phi\right]^n \tag{A.2}$$

The n-th order term in ϕ is proportional to the loop Γ_n^η with n external lines corresponding to ϕ's, each loop being built of the n Green functions g_η:

$$\text{Tr}\left[g_\eta\phi\right]^n \propto \int\left[\prod_{i=1}^{n}\mathrm{d}z_i\,\phi(z_i)\right]\Gamma_n\left(z_1;\ldots;z_n\right), \qquad (A.3)$$

where the n-th order vertex is given by

$$\Gamma_n\left(z_1;\ldots;z_n\right) = \prod_{i=1}^{n}g_\eta\left(z_i;z_{i+1}\right), \qquad \left(z_{n+1} = z_1\right).$$

Introducing the new variables $\xi = e^{2\pi T\left[\frac{x}{v_F}-t\right]}$, one represents the Green functions of Eq. (4.3) as

$$g_R(\xi_1,\xi_2) \propto \frac{\sqrt{\xi_1\xi_2}}{\xi_1 - \xi_2}$$

so that the vertex becomes

$$\Gamma_n\left(\xi_1;\ldots;\xi_n\right) \propto \gamma_n\left(\xi_1;\ldots;\xi_n\right)\prod_{i=1}^{n}\xi_i, \qquad \gamma_n = \prod_{i=1}^{n}\frac{1}{\xi_i - \xi_{i+1}}, \qquad (A.4)$$

with the boundary condition $\xi_{n+1} = \xi_1$. Since only the symmetric part of the vertex contributes to the integral (A.3), we may symmetrize γ_n:

$$\gamma_n \longmapsto \frac{\mathcal{A}_N\left(\xi_1;\ldots;\xi_n\right)}{\prod_{i<j}\left(\xi_i - \xi_j\right)}, \qquad (A.5)$$

where $\mathcal{A}_N$ is an absolutely antisymmetric polynomial of the N-th order which depends on n variables. A simple power counting gives $N = n(n-3)/2$ while the possible minimal order of such a polynomial is $N_{min} = n(n+1)/2$. This is not self-contradictory for $n = 1$ and $n = 2$ only. All other terms must therefore be equal to zero. The term with $n = 1$ is cancelled due to electroneutrality. The only non-vanishing vertex then is the one with two legs:

$$\ln J_\eta\left[\phi\right] = \frac{i}{2}\int\mathrm{d}z\mathrm{d}z'\,\phi(z)\Pi_\eta(z;z')\phi(z'), \qquad (A.6)$$

where

$$\Pi_\eta(z;z') = ig_\eta(z-z')g_\eta(z'-z). \qquad (A.7)$$

References

[1] S. Tomonaga, Prog. Theor. Phys. **5**, 544 (1950).

[2] J. M. Luttinger, J. Math. Phys. **4**, 1154 (1963).

[3] D. C. Mattis and E. H. Lieb, J. Math. Phys. **6**, 304 (1965).

[4] I. E. Dzyaloshinskii and K. B. Larkin, Sov. Phys. JETP **38**, 202 (1973).

[5] K. B. Efetov and A. I. Larkin, Sov. Phys. JETP **42**, 390 (1976).

[6] F. D. M. Haldane, *J. Phys. C.* **14**, 2585 (1981).

[7] H. C. Fogedby, J. Phys. C **9**, 3757 (1976).

[8] D. K. Lee and Y. Chen, J. Phys. A **21**, 4155 (1988).

[9] V. I. Fernández, K. Li, and C. Naon, Phys. Lett. B, **452**, 98 (1999); V. I. Fernandez and C. M. Naon, Phys. Rev. B. **64**, 033402 (2001).

[10] I. V. Yurkevich, *Bosonisation as the Hubbard-Stratonovich Transformation*, in "Strongly Correlated fermions and bosons in Low-Dimensional Disordered Systems", ed. by I. V. Lerner et al., Kluwer Academic Publishers, 69 (2002).

[11] A. Grishin, I. V. Yurkevich and I. V. Lerner, Phys. Rev. B. **69**, 165108 (2004).

[12] J. Rammer and H. Smith, Rev. Mod. Phys. **58**, 323 (1986).

[13] C. L. Kane and M. P. A. Fisher, Phys. Rev. Lett. **68**, 1220 (1992); Phys. Rev. B. **46**, 15233 (1992).

[14] K. A. Matveev, D. Yue, and L. I. Glazman, Phys. Rev. Lett. **71**, 3351 (1993).

[15] R. Egger and H. Grabert, Phys. Rev. Lett. **75**, 3505 (1995).

[16] A. Leclair, F. Lesage, and H. Saleur, Phys. Rev. B. **54**, 13597 (1996).

[17] Y. Oreg and A. M. Finkel'stein, Phys. Rev. Lett. **76**, 4230 (1996).

[18] A. Furusaki, Phys. Rev. B. **56**, 9352 (1997).

[19] M. Fabrizio and A. O. Gogolin, Phys. Rev. Lett. **78**, 4527 (1997).

[20] J. von Delft and H. Schoeller, Ann. Phys. -Berlin **7**, 225 (1998).

[21] J. Zinn-Justin, *Quantum Field Theory and Critical Phenomena* (Oxford Science Publications, 1996).

[22] Y. Saito, Z. Phys. B. **32**, 75 (1978).

[23] A. O. Gogolin, A. A. Nersesyan, and A. M. Tsvelik, *Bosonization in Strongly Correlated Systems*, University Press, Cambridge 1998.

Course 2

NOVEL PHENOMENA IN DOUBLE LAYER
TWO-DIMENSIONAL ELECTRON SYSTEMS

J.P. Eisenstein

*Dept. of Physics, California Institute of Technology,
Pasadena, CA, 91125 USA*

H. Bouchiat, Y. Gefen, S. Guéron, G. Montambaux and J. Dalibard, eds.
Les Houches, Session LXXXI, 2004
Nanophysics: Coherence and Transport

129

Contents

1. Introduction

This manuscript is the written companion to lectures presented at Les Houches in the summer of 2004. The intention is to introduce young scientists to some of the fascinating physics of two-dimensional electrons at high magnetic fields. With two Nobel prizes to its credit already, this topic surely rates as one of the most important in contemporary physics. The field is, of course, huge. I cannot hope, nor did I attempt, to do justice to more than a tiny fraction of it. My choice of topics on which to concentrate is highly personal and should not be taken as a judgment on what is most interesting or important in the field. They are simply the things that I like, sort of understand, and have worked very hard on over the last several years.

Section 2 is an overview of the field. It is divided into two main parts. The first aims to give a pedestrian description of those aspects of two-dimensional electron systems at high field that we understand pretty well. The basic ideas behind integer and fractional quantization of the Hall effect in single layer 2D systems is covered here. So is an experimentalist's view of the so-called composite fermion model of electron correlation in the lowest Landau level. The second part of the section introduces the novel new physics that emerges from double layer 2D electron systems. Bilayer systems are among my favorites and have yielded many important results which could not have been obtained using conventional single layer samples. Weakly coupled bilayers with large inter-layer separation have provided direct access to otherwise hidden aspects of single layer 2D electron systems. Strongly coupled bilayers with small layer separation support exotic many-body states that simply do not exist in single layer systems. Bilayer 2D electron systems were the main subject of my lectures and the remaining section of this manuscript are devoted to their description.

Section 3 deals with the novel technique called Coulomb drag. This is a method whereby the Coulomb interactions between two parallel 2D electron gases can be directly observed in a straightforward resistance measurement. The section begins with a discussion of the basic principle of Coulomb drag, how double layer 2D systems are fabricated, and how we are able to make the separate electrical connections to the individual layers essential to the Coulomb drag technique. The physics of drag at zero magnetic field is then discussed in some detail, although the focus is on the simplest cases.

Section 4 is about tunneling between parallel 2D systems. Once again, the section begins with a discussion of the zero magnetic field case. Even here there is much interesting physics, which may well surprise the reader. I go into some detail here because a thorough understanding of the zero field case is a prerequisite for understanding the much more subtle phenomena observed at high magnetic field, which are discussed at the end of the section. As with the discussion of Coulomb drag, the focus here is on weakly-coupled layers.

Section 5 deals with strongly coupled bilayer systems. The great majority of the discussion deals with what I think is one of the most fascinating of all low dimensional electronic systems: A strongly-coupled bilayer electron gas at unit total Landau level filling factor. This state, which may be equivalently viewed as a pseudo-spin ferromagnet or a Bose condensate of excitons, beautifully connects the fields of the quantum Hall effect, Bose condensation, and superconductivity.

2. Overview of physics in the quantum hall regime

2.1. Basics

Consider a collection of electrons confined to move in a two-dimensional plane. In the presence of a uniform magnetic field $\mathbf{B}$ the Hamiltonian is

$$H = \frac{(\mathbf{p} + e\mathbf{A})^2}{2m} + \frac{1}{2}g\mu_B\sigma \cdot \mathbf{B}. \tag{2.1}$$

The eigenvalues of this Hamitonian are particularly simple:

$$\epsilon_N = (N + \frac{1}{2})\hbar\omega_c \pm g\mu_B B/2. \tag{2.2}$$

Thus, the continuous energy spectrum present at zero magnetic field is resolved into a ladder of discrete Landau levels $N = 0, 1, \ldots$, each of which is further resolved into two spin sublevels. The cyclotron frequency $\omega_c = eB_\perp/m_b$ depends on the band mass m_b of the electrons and the magnetic field component $B_\perp$ perpendicular to the 2D plane. The spin Zeeman term depends on the total magnetic field B, the electron g-factor, and the Bohr magneton $\mu_B = e\hbar/2m_0$ (with m_0 the bare electron mass). For 2D electrons near the Γ-point of the conduction band in GaAs one has $m_b \approx 0.067m_0$ and $g \approx -0.44$. The ratio of the bare spin splitting $g\mu_B B$ to the Landau level separation $\hbar\omega_c$ in a perpendicular magnetic field is thus $gm_b/2m_0 \approx 1/70$.

Landau levels are highly degenerate. Indeed, the number of states within each spin-resolved sublevel is $eB_\perp S/h$, where S is the area of the 2D sample. At $B_\perp$ = 1 T, the density of electrons required to fill one spin sublevel is $N_0 = 2.41 \times$

10^{10}cm^{-2}, independent of the electron mass or any other sample parameters. Note that the degeneracy of the Landau levels is equivalent to the number n_ϕ of magnetic flux quanta $\phi_0 = h/e$ threading the sample.

It is useful to examine the wavefunctions associated with the Landau level states. These depend, of course, on the gauge chosen for the magnetic field. In Landau gauge a magnetic field $B_\perp$ along the $\hat{z}$ direction is represented by a undirectional vector potential, e.g. $\mathbf{A} = -B_\perp y\hat{x}$. In this case the problem reduces to a modified one-dimensional simple harmonic oscillator and the (unnormalized) wavefunctions are

$$\psi_{N,k,\sigma} = e^{ikx}\phi_N(y - y_k)\chi_\sigma. \tag{2.3}$$

In this equation k is a "momentum" index, $y_k = k\ell^2$ is the guiding center of the state (with $\ell = (\hbar/eB_\perp)^{1/2}$ being the so-called magnetic length), and χ_σ is the spin wavefunction. The function ϕ_N is describes the confinement of the state about the guiding center y_k and is given by

$$\phi_N(\xi) = e^{-\xi^2/2\ell^2} H_N(\xi/\ell). \tag{2.4}$$

with H_N being the Hermite polynomial. In this gauge the Landau level states are plane waves in one direction and harmonic oscillator states in the orthogonal direction. Applying periodic boundary conditions in the $\hat{x}$-direction over the length L_x and restricting the guiding center to lie within $0 < y_k < L_y$ restricts the allowed values of k to $2\pi j/L_x$ with $j = 1, 2, \ldots, L_x L_y/2\pi\ell^2 = eB_\perp S/h$. This explicitly demonstrates that the areal degeneracy of each Landau spin sublevel is $eB_\perp/h$. Note that the separation between the guiding centers y_k of adjacent k-states (i.e. $\Delta k = 2\pi/L_x$) is $\Delta y_k = 2\pi\ell^2/L_x \ll \ell$. Adjacent k-states thus overlap strongly in the $\hat{y}$-direction. However, since the states are extended in the $\hat{x}$-direction over the entire length L_x, the net *area* they occupy is, as expected, $2\pi\ell^2 = h/eB_\perp$. Since the energies of the states are independent of the momentum index k, they carry no current along the $\hat{x}$-direction.

Finally, it is also useful to write down the wavefunctions in the symmetric gauge $\mathbf{A} = \frac{1}{2}B_\perp(x\hat{y} - y\hat{x})$. In this case the wavefunctions in the lowest ($N = 0$) Landau level may be written as

$$\psi_{0,j,\sigma}(z) = z^j e^{-|z|^2/4}\chi_\sigma \tag{2.5}$$

where $z = (x + iy)/\ell$ is the position in the 2D plane, rendered as a complex number. The index j is an angular momentum and is restricted by the size of the sample (here assumed a circle of radius R): $0 \le j < R^2/2\ell^2$.

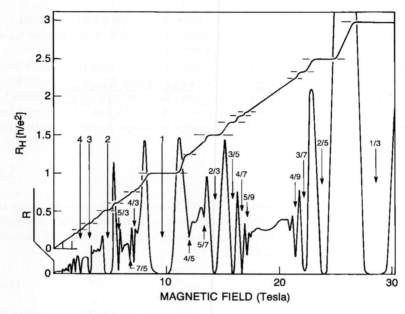

Fig. 1. The quantized Hall effects. Each plateau in the Hall resistance R_H is accompanied by a deep minimum in the longitudinal resistance R. The numbers indicate the filling factor of the Landau level spectrum. After Ref. [3].

2.2. Quantized hall effects

Figure 1 illustrates the integer [1] and fractional [2] quantized Hall effects (QHE) as typically observed in a good quality 2D electron gas at low temperature [3]. Both the Hall R_H and longitudinal resistance R are shown. The Hall resistance (diagonal trace) rises linearly with magnetic field at low field but then develops a complex series of plateaus at higher field. These plateaus, which are precisely flat, occur when $R_H = (h/e^2)/(p/q)$ with p and q integers. Meanwhile, the longitudinal resistance R exhibits regular oscillations at low field. At higher fields these oscillations become more intricate and in many cases R appears to vanish altogether. As the figure makes clear, there is a one-to-one correspondence between deep minima, or zeroes, in R and plateaus in R_H. Note that the broader the plateau in R_H, the deeper the minimum in R.

The carrier concentration N_s of the 2D electron gas in Fig. 1 may be determined either from the slope of the Hall resistance R_H at low field or from the Shubnikov-de Haas oscillations in the longitudinal resistance R. In the latter case, the formation of Landau levels modulates the density of states and essentially every observable in the 2D gas oscillates periodically in $1/B$. In Fig. 1 $N_s \approx 2.3 \times 10^{11} \text{cm}^{-2}$.

2.2.1. Integer QHE

The ratio of the electron density N_s to the Landau level degeneracy N_0 is known as the *filling factor*: $\nu = N_s/N_0 = hN_s/eB$. The filling factor provides a measure of how many Landau levels are occupied at low temperatures. In Fig. 1, the Hall plateau near $B \approx 10$T, where $R_H = h/e^2$, corresponds to complete filling of just one spin-resolved Landau level, i.e. $\nu \approx 1$. Similarly, near $B = 5$ T there is a plateau at $R_H = h/2e^2$. In this case two Landau levels (both spin subbands of the lowest $N = 0$ orbital level) are filled and $\nu \approx 2$. And so on. Integer quantized Hall plateaus are thus associated with the filling of an integer number of Landau levels.

The deep minimum in the longitudinal resistance R associated with each plateau in R_H implies the existence of an energy gap Δ in the system. At low temperatures dissipation, and thus longitudinal resistance, is suppressed if $k_B T << \Delta$. In the case of the *even* integer QHE states this gap is dominated by the splitting $\hbar\omega_c$ between Landau levels with orbital indices N and $N + 1$. This splitting can be quite large, around 20 K at $B = 1$ T. This is generally consistent with estimates of the splitting based upon measurements of the temperature dependence of the longitudinal resistance, R. For the *odd* integer QHE states the situation is more interesting. The relevant single particle gap is now the Zeeman splitting $g\mu_B B$ between spin sublevels of a given orbital Landau level. This is a very small energy, about 0.3K at $B = 1$ T. Remarkably, however, the temperature dependent resistance for the $R_H = h/e^2$ QHE again suggests an energy gap in the 10-20 K range [4]. The reason for this is that the true cost of flipping a spin in the 2D electron gas at high magnetic fields is dominated by electron-electron interaction effects, not the simple Zeeman energy. In fact, the nature of the charged excitations of the system at $\nu = 1$ are strange topological objects known as skyrmions.

If the magnetic field is adjusted so that the filling factor is an exact integer $\nu = j$, we can expect the classical Hall resistance to assume the magic value: $R_H = B/eN_s = B/e\nu N_0 = h/je^2$. While interesting, this is clearly *not* the entire explanation for the QHE: any deviation of the magnetic field from the value producing $\nu = j$ will change the Hall resistance from the special h/je^2 value. Additional physics must be responsible for the pinning of R_H to the quantized value over a finite range of magnetic fields. This new physics involves disorder in the 2D system and can be most easily understood in terms of the so-called edge state picture.

In the bulk of a 2D electron system the Landau energy levels ϵ_N are independent of position. Near the boundary, however, the forces which keep electrons

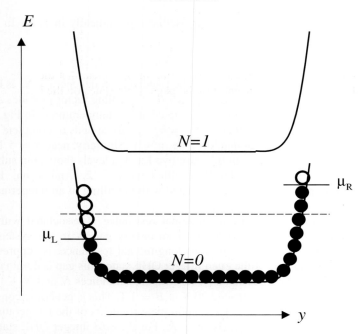

Fig. 2. Schematic illustration of Landau levels in a finite 2D electron system. Solid dots indicate occupied states in the lowest Landau level. Current is carried only by those states near the edge of the sample. A *net* current flows only when the chemical potentials at the two edges are different.

in the sample require the Landau levels to move up in energy. This situation is depicted in Fig. 2 where the two lowest Landau levels (ignoring spin) are displayed as a function of position y across a bar-shaped sample. At each side of the sample the Landau bands move up in energy. If we choose the Landau gauge used in Eq. 2.3 then the lowest Landau level eigenstates are plane waves in the $\hat{\mathbf{x}}$-direction (perpendicular to the page) and gaussians in $\hat{\mathbf{y}}$-direction centered at y_k. Of course, near the boundary this description is not exact, but so long as the potential confining the electrons in the sample is slowly-varying on the scale of the magnetic length ℓ, it is accurate enough. In Fig. 2 the magnetic field is high enough that in the bulk the $N = 0$ LL is fully occupied and the $N = 1$ LL is empty. The solid dots are meant to indicate the guiding centers y_k of those states which are occupied. The open dots represent empty states. With no current flowing through the sample all states below a uniform chemical potential (dashed line) would be filled.

As already mentioned, if the Landau level is independent of position (and thus of wavevector k), the eigenstates carry no current. This is obvious since $\partial \epsilon_N / \partial k = 0$. Hence the states in the bulk of the sample are carrying no current.

Near the edges of the sample the Landau level is not independent of position and $\partial\epsilon_N/\partial k \neq 0$. States in these regions do carry current. In fact, the electrical current carried in the $\hat{x}$-direction by a single k-state in any Landau level is just $i_k = (e/\hbar L_x)\partial\epsilon_N/\partial k = (e\ell^2/\hbar L_x)\partial\epsilon_N/\partial y_k = [\partial\epsilon_N/\partial y_k]/BL_x$, with L_x the length of the sample in the $\hat{x}$-direction. From this it is clear that the states on the right side of Fig. 2 carry current into the page while states on the left side carry current out of the page. It is also clear that in equilibrium, when all k-states up to a common chemical potential are occupied, there is no *net* current. This is true regardless of the precise shape of the edges or whether disorder makes the Landau level fluctuate in the bulk. To achieve a net current a non-equilibrium situation must be established. In the figure there is a different chemical potential μ on the right and left side of the sample, with $\mu_R > \mu_L$. The net current is just

$$I = \sum_{occ.} i_k = \int \frac{L_x dy_k}{2\pi\ell^2} \frac{\partial\epsilon_N/\partial y_k}{BL_x} = \frac{e}{h}(\mu_R - \mu_L) = \frac{e^2}{h}V_H. \tag{2.6}$$

In arriving at this expression we have used the fact that the spatial separation between the guiding centers of adjacent k-states is $\Delta y_k = 2\pi\ell^2/L_x$ and that the Hall voltage $V_H = (\mu_R - \mu_L)/e$. Thus, we find that the Hall resistance of the sample under the conditions shown in Fig. 2 is precisely h/e^2. (This is the expected value, since we have ignored the spin of the electron in this derivation.)

But does Eq. 2.6 really imply quantization of the Hall resistance? Suppose the magnetic field were changed by a small amount, δB. The degeneracy of the Landau levels changes and the Fermi level will move. In the situation depicted in Fig. 2 the edges of the sample create a small but finite density of states at the Fermi level. Thus, if δB is small enough, the picture shown in the figure will not change qualitatively and the Hall resistance will remain at $R_H = h/e^2$. This establishes the quantization of the Hall effect.

In reality, of course, it is not the tiny density of states at the edge of the sample which produces the broad Hall plateaus seen in real samples. Crudely speaking, the edge regions are only about one magnetic length wide. In a macroscopic Hall bar of width $L_y = 1$ mm this would lead to very narrow plateaus indeed: $\delta B/B \sim \ell/L_y \sim 10^{-5}$. Disorder in the 2D system (hills and valleys in the potential landscape in which the electron move) are much more effective in producing a large density of states in the gaps between Landau levels. If the potential is smooth enough these hills and valleys are surrounded by edge channels as well, but as long as they do not provide a "connection" from one side of the sample to the other they do not destroy the Hall quantization. That disorder is key to the width of the Hall plateaus is clearly proven by experiments: As samples have become cleaner and cleaner, the widths of quantized Hall plateaus have steadily shrunk.

2.2.2. Fractional QHE

Two ingredients are necessary for Hall quantization and vanishing longitudinal resistance, an energy gap and disorder. The gaps between Landau levels (and their spin sublevels) exhaust all possible *integer* QHE states. So then where does a plateau at $R_H = 3h/e^2$, which appears when the lowest Landau level is $\nu = 1/3$ filled, come from? The only possibility is electron-electron interactions.

To understand what is going on at $\nu = 1/3$ let us first return to $\nu = 1$ where all states in the lowest spin sublevel of the lowest Landau level are occupied. Assuming that the gap to the upper spin branch is sufficiently large, the wavefunction of the many-electron state is simply a Slater determinant constructed out of all possible states in the lowest Landau level. In the symmetric gauge this is

$$\Psi(z_1, z_2, \ldots, z_n) = \begin{vmatrix} 1 & 1 & \cdots & 1 \\ z_1 & z_2 & \cdots & z_n \\ z_1^2 & z_2^2 & \cdots & z_n^2 \\ \vdots & \vdots & \ddots & \vdots \\ z_1^{n-1} & z_2^{n-1} & \cdots & z_n^{n-1} \end{vmatrix} \times exp\left[-\sum_{j=1}^{n} \frac{|z_j|^2}{4} \right] \quad (2.7)$$

where n is the number of electrons in the system. At $\nu = 1$, of course, $n = n_\phi$, the number of magnetic flux quanta present. In Eq. 2.7 the spin wavefunction has been omitted for simplicity; all spins are assumed to point along the magnetic field direction. The determinant in Eq. 2.7 is a van der Monde determinant, thus allowing Ψ to be re-written as

$$\Psi(z_1, z_2, \ldots, z_n) = \prod_{i<j}(z_i - z_j) \, exp\left[-\sum_{k=1}^{n} \frac{|z_j|^2}{4} \right]. \quad (2.8)$$

Written in this fashion it is obvious that Ψ is odd under the interchange of any two electrons as it must be. If we imagine holding fixed $n-1$ electrons and then examine the dependence of Ψ on the coordinate of the one remaining electron, we see that Ψ vanishes whenever the remaining electron is at the same position as one of the fixed electrons. This is required by the Pauli principle, but it is obviously also sensible from the perspective of lowering the net Coulomb energy of the system.

Now consider $\nu = 1/3$. In this case $n_\phi = 3n$, and there are three times as many states in the lowest Landau level as electrons. It is no longer possible to construct a unique Slater determinant. Let us again focus on just one of the many electrons. This electron lies in the lowest Landau level, so taken alone the most general possible wavefunction it could have would be of the form

$$\psi(z_1) = \sum_{j=0}^{n_\phi-1} a_j z_1^j \; exp\left(\frac{-|z_1|^2}{4}\right) \tag{2.9}$$

with the a_j some unknown coefficients. The sum in Eq. 2.8 is a complex polynomial in $z = x + iy$ of order $n_\phi - 1$, and therefore has $n_\phi - 1$ zeroes in the complex plane. The Pauli principle requires that $n - 1$ of these zeroes coincide with the positions of the remaining electrons in the system. Hence ψ may be factored

$$\psi(z_1) = (z_1 - z_2)(z_1 - z_3)\ldots(z_1 - z_n)P(z_1) \; exp\left(\frac{-|z_1|^2}{4}\right) \tag{2.10}$$

where $P(z_1)$ is a polynomial of order $n_\phi - n$. The question is now what to do with the zeroes of $P(z_1)$. Having satisfied the Pauli principle, the only remaining consideration is to minimize the Coulomb repulsion between electrons by keeping them as far apart as possible. Laughlin's great insight was to realize how to do this: The "unused" zeroes of $P(z_1)$ should be positioned on top of all the remaining electrons. There are just enough (in the thermodynamic limit) to add two more zeroes per particle. In other words, $P(z_1)$ is chosen so that

$$\psi(z_1) = (z_1 - z_2)^3 (z_1 - z_3)^3 \ldots (z_1 - z_n)^3 \; exp\left(\frac{-|z_1|^2}{4}\right). \tag{2.11}$$

With this choice $\psi(z_1)$ vanishes as the *cube* of the separation between between electrons, thereby greatly reducing the Coulomb energy. From this point it easy to construct the Laughlin wavefunction for all the electrons in the system

$$\Psi(z_1, z_2, \ldots, z_n) = \prod_{i<j}(z_i - z_j)^3 \; exp\left[-\sum_{j=1}^{n} \frac{|z_j|^2}{4}\right]. \tag{2.12}$$

This wavefunction is odd under exchange of any pair of electrons and corresponds to filling factor $\nu = 1/3$. Notice that the argument used to construct Ψ *would not have worked* at $\nu = 1/2$, for it would have led to a wavefunction even under particle exchange. This is comforting because there is no fractional QHE state at $\nu = 1/2$ in a single layer 2DES.

If the filling factor is not precisely $\nu = 1/3$ the commensurability between flux quanta and particles is lost and there are either too many or too few zeroes around. If, for example, $n_\phi = 3n + 1$ there is a single unattached zero in Ψ. All electrons will avoid this spot. It is, in fact, a *quasi-hole* and costs a finite amount of energy to produce. Similarly, if $n_\phi = 3n - 1$ a *quasi-electron* is present, again costing a finite energy to produce. These quasiparticles are the lowest lying excitations of the $v = 1/3$ Laughlin liquid and they are separated by an energy gap from the ground state. The cost to produce a single quasi-electron/quasi-hole pair is close to $\Delta = 0.1e^2/\epsilon\ell$. This energy gap is essential for the existence of a quantized Hall effect.

The quasi-particles of the Laughlin liquid have charge $\pm e/3$. This is at least plausible on the basis of flux counting. If each electron has three flux quanta attached to it in the ground state, then the charge associated with a single unattached flux line ought to be $+e/3$.

2.2.3. Composite fermions

Laughlin's argument is readily generalized to fractional QHE states at $v = 1/m$, with m odd. In fact, Laughlin predicted the existence of the $v = 1/5$ prior to its experimental discovery. If we assume that particle-hole symmetry exists within each spin-split Landau level then FQHE states at $v = 1 - 1/m$ (e.g. $v = 2/3$) can be understood as $v = 1/m$ states of holes. But what about the $v = 2/5$ state which is plainly evident in Fig. 1? Indeed, the sequences of states at $v = 2/5$, $3/7$, $4/9$, ... and their particle-hole conjugates at $v = 3/5$, $4/7$, $5/9$, ... all need explanation.

Jain has argued that these states, and much of the phenomenology of the FQHE are best understood in terms of a so-called composite fermion model. The idea is based upon the application of a Chern-Simons singular gauge transformation. The roots of this approach go back a long way and many investigators have played significant roles. I direct the interested reader to the excellent review chapters by Jain and Halperin in Ref. [5] and the references cited therein.

In the simplest case, composite fermions (CF's) are constructed by attaching two fictitious flux quanta to each electron in the lowest Landau level. This flux attachment is done via a gauge transformation and thus has no effect on the physics of the system. It is not at all clear *a priori* that the transformation is a useful one. The many-body problem is no more soluble than it was to begin with, but the hope is that the lowest-order approximate solution to the transformed problem is both easy to construct and subsumes much of the complexity due to the strong and non-perturbative Coulomb interactions between electrons. A good analogy is to Landau Fermi liquid theory of the interacting electron gas in a metal. In that case the lowest order approximation is a non-interacting gas of quasiparticles. The quasiparticles are in 1:1 correspondence with the orginal electrons but have modified physical attributes such as a different effective mass and g-factor. Interactions between the quasiparticles are included at higher order and are hopefully relatively weak. Composite fermion theory performs a similar trick for 2D electrons in the lowest Landau level.

The fictitious flux quanta are attached with their fictitious magnetic field directed opposite to the applied external magnetic field. At the mean-field level the effective magnetic field B^* experienced by each composite fermion is the external field B plus the fictitious field of all the other composite fermions. Hence $B^* = B - 2\phi_0 N_s$, where $\phi_0 = h/e$ is the flux quantum and N_s is the areal density of electrons. The Landau level filling factor for the original electrons is $\nu = n/n_\phi = N_s/N_0$. For the CF's the situation is different. The Landau level degeneracy for CF's is $N_0^{CF} = eB^*/h$ and so their filling factor is $\nu^{CF} = N_s/N_0^{CF}$. Comparing these we get

$$\frac{1}{\nu^{CF}} = \frac{1}{\nu} - 2 \tag{2.13}$$

Suppose now that the magnetic field is adjusted so that the CF's fill up j CF Landau levels, *i.e.* $\nu^{CF} = j$. According to Eq. 2.13 this occurs when $\nu = j/(2j+1)$, or the sequence $\nu = 1/3, 2/5, 3/7$, etc. This is the main sequence of observed FQHE states, at least for $\nu < 1/2$. Allowing ν^{CF} to be $-j$ (corresponding to negative B^*) we find that the same j values occur when $\nu = 2/3, 3/5, 4/7$, etc., giving us the main sequence of FQHE states for $1 > \nu > 1/2$. The fractional QHE of electrons seems to correspond to the *integer* QHE of composite fermions.

The situation of greatest interest in these lectures occurs when $B^* = 0$, i.e. when the physical electrons fill up one-half of the states in the lowest Landau level. If the mean field picture is at all close to the truth, this situation corresponds to a strange new kind of Fermi sea. This is quite subtle, for remember that the first thing a big magnetic field does to the Fermi sea of real electrons is to destroy the Fermi surface and create Landau levels. CF theory amounts to the reconstruction, via electron-electron interactions, of a Fermi surface at $\nu = 1/2$. There will once again be a Fermi wavevector ($\sqrt{2}$ larger owing to spin polarization), and an effective mass, determined not by band structure but by Coulomb interactions alone.

Although the discussion of CF's that I have given here amounts to little more than numerology, the theory has experienced several spectacular successes in relation to experiment. The most dramatic of these concerns the $\nu = 1/2$ state. Surface acoustic wave studies by Willett, *et al.* [6] and transport experiments in anti-dot arrays by Kang, *et al.* [7] have very clearly demonstrated, among other things, that in the vicinity of $\nu = 1/2$ the quasiparticles in the system (the CF's) move in semi-classical cyclotron orbits whose radii diverge as $\nu \to 1/2$. At half-filling composite fermions move in straight lines, in spite of the very large magnetic field present.

2.3. Double layer systems

The preceding section dealt with the simplest and best understood aspects of quantum Hall physics in single layer 2D electron systems. We now turn to the

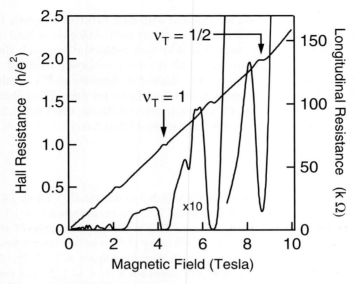

Fig. 3. Quantized Hall effect in a double layer 2D electron system. The states at $v_T = 1$ and $v_T = 1/2$ both disappear if the layers are too far apart.

main focus of these lectures, double layer 2D systems. Obviously, if the separation between the layers in a bilayer system is large, the two layers are independent and exhibit the same physics as single layers. Such weakly coupled bilayer systems are interesting nonetheless as they provide unique access to various physical parameters of the individual layers which are difficult if not impossible to reach experimentally using conventional single layer systems. In spite of this, it is strongly coupled bilayer systems which capture the most attention since they have been found to exhibit collective phases which do not exist at all in single layer 2D electron gases.

Figure 3 illustrates two intrinsically bilayer fractional quantized Hall phases as observed in 1992 using a sample consisting of two nearly identical 2D electron gases residing in 18 nm GaAs quantum wells separated by a 3 nm AlAs barrier layer [8]. Narrow plateaus in the Hall resistance are seen at $R_H = h/e^2$ and $R_H = 2h/e^2$ and deep minima in the longitudinal resistance accompany them. These QHE features are labeled by the *total* Landau level filling factors $v_T = 1$ and $v_T = 1/2$, respectively. The total filling factor is determined by the total electron density N_T in the double layer system. Hence $v_T = v_1 + v_2$, with v_1 and v_2 being the individual layer filling factors. For the balanced bilayer sample used in the figure $v_1 = v_2 = v_T/2$. Therefore, the Hall plateau at $R_H = h/e^2$ occurs when each individual layer is at $\nu_1 = \nu_2 = 1/2$ and the plateau at $R_H = 2h/e^2$ occurs at $\nu_1 = \nu_2 = 1/4$. As Fig. 1 clearly shows, no quantized Hall effect is observed at $\nu = 1/2$ in a single layer. Though not apparent in Fig. 1, no single layer QHE is observed at $\nu = 1/4$ either [9]. The QHE states at $\nu_T = 1$ and 1/2 shown in Fig. 3 must therefore exist by virtue of couplings between the two layers in the sample. In the case of the $\nu_T = 1/2$ state the ground state of the bilayer 2D system is believed to be well described with a generalized Laughin wavefunction first introduced by Halperin [10]

$$\Psi \sim \prod_{i<j}(z_i - z_j)^3 \prod_{k<l}(w_k - w_l)^3 \prod_{m<n}(z_m - w_n) \qquad (2.14)$$

where the ubiquitous gaussian factors have been omitted for simplicity. In Eq. 2.14 z and w represent the coordinates of electrons in the two layers. The first and second factors in this wavefunction are reminiscent of those in the Laughlin wavefunction for the $\nu = 1/3$ state given in Eq. 2.12. These factors embody the intra-layer correlations in the system and ensure that electrons avoid their neighbors in the same layer. The final factor, $(z_m - w_n)$, is new. This factor imposes *inter*-layer correlations and keeps electrons in opposite layers away from one another. Notice, however, that the smaller exponent on this last factor means that inter-layer repulsion is less important than intra-layer repulsion in bilayers at $\nu_T = 1/2$. Without the last factor in Eq. 2.14 each layer would be at $\nu = 1/3$. The last factor introduces quasi-holes into each layer. The number of quasi-holes in either layer is just the number $n = n_T/2$ of electrons in the other layer. The quasi-holes in either layer are, in effect, bound to the electrons in the opposite layer.

As interesting as the $\nu_T = 1/2$ FQHE in bilayers is, it pales in comparison to the $\nu_T = 1$ state. Here the wavefunction may be written as

$$\Psi \sim \prod_{i<j}(z_i - z_j) \prod_{k<l}(w_k - w_l) \prod_{m<n}(z_m - w_n). \qquad (2.15)$$

This state possesses a remarkable broken symmetry, spontaneous inter-layer phase coherence, which is responsible for several truly dramatic effects. Perhaps surprisingly, the collective electronic state represented by Eq. 2.15 may be viewed as a Bose-Einstein condensate of inter-layer excitons. The final chapter of this manuscript is devoted to the description of the still unfolding story of this new phase of matter.

3. Coulomb drag between parallel 2D electron gases

3.1. Basic concept

A current flowing in one low-dimensional electron system can exert a frictional force on a similar conducting system which is electrically isolated from it, but relatively nearby. This drag force is due to an exchange of electronic momentum between the two systems. In the simplest case, known as Coulomb drag, momentum exchange results from direct Coulomb scattering of the electrons in one conductor off of those in the other. Other processes, such as virtual phonon mediated electron-electron scattering can also contribute.

Normally this effect is unobservably small. For example, two thin metal films, of aluminum say, can not be placed close enough together for the Coulomb drag force to be detected. The reason for this is simple: bulk three-dimensional metals have an extremely small screening length, of order 0.05 nm. If the two films are separated by more than this distance, their charge distributions appear to be completely smooth and inter-film electron-electron scattering events are heavily suppressed.

The situation is quite different in low-dimensional electron systems in semiconductor heterostructures. In GaAs the screening length for conduction band electrons in two dimensions is 5 nm, a relatively large number. It is easy to create a double quantum well structure in which two two-dimensional electron gases are separated by a distance on this order. Provided that independent electrical contacts can be established to each of the 2D systems, Coulomb drag can be detected as a voltage V_D appearing in one of the layers when current I is driven through the other. This effect, predicted by Pogrebinsky [11] in 1978 and first observed by Gramila, *et al.* [12] in 1991, has proven to be extremely useful for the study of two-dimensional electron systems.

The usefulness of the Coulomb drag technique stems from the fact that it allows a direct determination of an electron-electron scattering rate via a simple resistance measurement. A simple one-dimensional force-balance argument, within the Drude model of conduction, is sufficient to see how this works. The forces F_1 and F_2 on electrons in layers 1 and 2 (here assumed to contain identical two-dimensional electron gases of density N_s) are:

$$F_1 = -eE_1 - mv_1\left(\frac{1}{\tau_t} + \frac{1}{\tau_D}\right) + mv_2/\tau_D$$

$$F_2 = -eE_2 - mv_2\left(\frac{1}{\tau_t} + \frac{1}{\tau_D}\right) + mv_1/\tau_D. \tag{3.1}$$

In these equations τ_t is the transport scattering time, including all momentum relaxation processes *except* those which transfer momentum between the layers.

Such inter-layer processes are included in the force balance equations via the drag relaxation time τ_D. In a steady state where a current I_1 flows in layer 1 but no current flows in layer 2, a voltage develops in layer 2 solely because of these inter-layer momentum transfer processes. Along a bar of width W and length L, the drag voltage V_D in layer 2 is:

$$V_D = E_2 L = -\frac{L}{W}\frac{m}{N_s e^2 \tau_D} I_1 = -\frac{L}{W}\rho_D I_1 = -R_D I_1. \tag{3.2}$$

This is a simple yet remarkable result. It demonstrates that the drag resistance R_D (and resistivity ρ_D) depends *only* on the inter-layer momentum relaxation rate, τ_D^{-1}. Momentum loss to phonons or impurities does not enter. Normally, of course, electron-electron scattering processes are not detectable (at the Drude level) in the ordinary resistance of a metal because they conserve the net momentum of the electron gas. In fact, however, when two conductors are in close proximity to one another, their individual resistances do depend upon the electron-electron processes which exchange momentum between them. This is usually a very tiny effect and is swamped by the loss of momentum to phonons and impurities. In contrast, the above equations demonstrate that the transresistance R_D between the two conductors depends *only* on the inter-layer momentum relaxation time.

Note the sign of the drag effect. In a double layer electron system, the drag voltage is opposite in sign to the resistive voltage drop (proportional to E_1) in the current-carrying layer. In a bilayer electron-hole system [13] the sign is reversed. In both cases the sign of the drag reflects the fact that carriers in the drag layer are swept along in the same direction as the carriers in the drive layer are moving. In steady state the drag electric field E_2 cancels this input of momentum.

The drag scattering time τ_D reflects all processes in which momentum lost by one layer is exactly picked up by the other. Direct Coulomb scattering of electrons in opposite layers is the simplest such process. Electron-phonon scattering can also contribute, provided the phonon emitted by one layer is absorbed by the other. But even without phonons, Coulomb drag itself contains both simple electron-electron scattering and more complex plasmon-assisted scattering processes. Although there have been interesting experimental and theoretical studies of phonon and plasmon-assisted drag, we will focus here on simple electron-electron Coulomb drag only. Not surprisingly, improvements to the Drude picture of conduction in low dimensional conductors have consequences for Coulomb drag. These higher-order effects, including weak-localization and diffusion corrections to the electron-electron interaction are also beyond our present scope.

3.2. Experimental

Modern crystal growth techniques, such as molecular beam epitaxy, allow the creation of bilayer 2D electron systems in which the two layers are separated by ~ 10 nm. For example, much of the data which we will describe below was obtained using a double layer system consisting of two 18 nm GaAs quantum wells separated by a 10 nm $Ga_{0.1}Al_{0.9}As$ barrier layer. Doping via Si δ-layers ~ 200 nm above and below the double quantum well populates the ground electric subband of each well with a 2DES having a density around $N_s = 5.5 \times 10^{10}$ cm^{-2}. At low temperatures the mobility of each 2DES is approximately 10^6 cm^2/Vs, corresponding to a transport scattering time τ_t of about 40 ps.

In order to make a drag measurement, separate electrical contacts to the individual 2D layers must be established. We solved this problem in 1990 using a localized selective depletion technique [14]. The basic idea is simple: an electrostatic gate is used to cut through one, but not both, of the 2D layers in the vicinity of a conventional ohmic contact which itself connects to both 2D layers. The gate, of course, is nothing more than a thin metal film (we use aluminum) evaporated on the surface of the sample. If the gate "surrounds" the contact, all current entering or leaving the contact must pass the gate. Applying a voltage (relative to the 2DES) to a gate deposited on the sample "front" surface can be used to deplete the top 2DES in the region under the gate. In this situation the ohmic contact "sees" the rest of the sample only via the lower 2D layer. Similarly, a gate on the back surface of the sample can be used to deplete the bottom 2DES and thus provide independent connection to the top 2DES. In order to attain reasonable lateral definition of the back gate depletion fields, we thin our samples from the back side to a total thickness of about $\sim 50\mu$m. Still better lateral definition can be achieved via the so-called EBASE technique [15].

Coulomb drag measurements are usually made with the 2DES layers confined laterally within a bar-shaped mesa. Figure 5 contains a schematic diagram of a generic drag sample, showing both the mesa and the various gates and ohmic contacts. Five ohmic contacts are shown, as are some of the relevant gates. Two of the contacts (1 and 2) are used to drive an ac current through the bottom 2D layer, while two more (3 and 4) are used to detect the drag voltage drop in the top 2D layer. The drive current is supplied via an isolation transformer and the fifth ohmic contact is often used to define ground potential in the drive layer. The drag voltage itself is detected with a low noise, high input impedance preamplifier. The drag layer can also be referenced to ground via a large resistance. Although not shown in the figure, top and backside gates covering the central portion of the mesa are usually present. These gates allow for measuring drag as a function of density in the two layers.

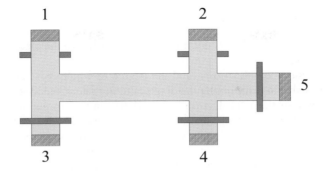

Fig. 4. Typical geometry for Coulomb drag measurement. Contacts 1-4 are used to inject current into one layer and measure the voltage across the other. Contact 5 can be used to locally ground one 2D layer. Gates used for establishing independent layer contact are shown.

Coulomb drag resistances are often quite small; Gramila, *et al.* worked in the mΩ regime [12]. Numerous spurious effects easily spoil the measurement of such a small effect. For example, inter-layer tunneling must be small enough that direct leakage of current from the drive to the drag layer is unimportant. In the sample described above, the 10 nm $Ga_{0.1}Al_{0.9}As$ barrier layer presents a tunnel resistance of typically 30 MΩ in a $40 \times 400 \mu m$ mesa. This is more than adequate at zero magnetic field. Another important spurious signal derives from the inter-layer capacitance, which is on the order of 100 pF in typical devices. Although capacitive coupling occurs 90° out of phase with the true drag voltage, it is generally difficult to set the phase of the lock-in measurement accurately enough to completely eliminate this problem. For this reason, low frequencies (2 - 13 Hz) are employed.

3.3. Elementary theory of Coulomb drag

MacDonald [12], and subsequently Jauho and Smith [16], developed the first theories of Coulomb drag directly appropriate to bilayer 2D electron systems in double quantum well heterostructures. In their very similar approaches, a linearized Boltzmann equation theory results in the following expression for the drag resistivity ρ_D between two identical 2D electron systems:

$$\rho_D = \frac{\hbar^2}{2\pi^2 e^2 N_s^2 k_B T} \int_0^\infty q dq \int_0^\infty d\omega \, q^2 |e\phi(q)|^2 \frac{[\text{Im}\chi(q,\omega)]^2}{\sinh^2(\hbar\omega/k_B T)}. \qquad (3.3)$$

In this formula N_s is the density of the individual 2D layers, T the temperature, q the momentum transfer, $\phi(q)$ the Fourier transformed screened inter-layer Coulomb interaction, ω the energy transfer, and $\text{Im}\chi(q, \omega)$ the electron gas susceptibility.

The interaction potential $\phi(q)$, and the degree to which it is screened, is of crucial importance in determining the drag scattering rate. Consider a test charge Q in one (infinitely thin) 2D electron layer and the Coulomb potential it produces in a second, parallel layer a distance d away. In the absence of screening this potential is just $\phi(q) = Qe^{-qd}/2\epsilon q$, with $\epsilon = \kappa\epsilon_0$ the dielectric constant of the host material. The large wave-vector components of $\phi(q)$ are thus exponentially suppressed by the finite layer separation. This implies that drag will be dominated by scattering events with momentum transfers $q < 1/d$. Now, if we allow screening (at the Thomas-Fermi level) of the test charge, but only by charges in the *same* layer, we get $\phi(q) = Qe^{-qd}/2\epsilon(q + q_{TF})$ where $q_{TF} = m^*e^2/2\pi\epsilon\hbar^2$ is the Thomas-Fermi screening wave-vector. Screening, therefore, reduces the impact of small q processes and so suppresses drag. However, we must also allow for screening of our test charge by electrons in the adjacent layer. We now get the more complicated result:

$$\phi(q) = \frac{Q}{2\epsilon} \frac{qe^{-qd}}{(q + q_{TF})^2 - q_{TF}^2 e^{-2qd}}. \tag{3.4}$$

For a rough approximation we assume $qd \ll 1$ and expand the exponentials to obtain $\phi(q) \approx Q/2\epsilon(q + q_{sc})$, with $q_{sc} = 2q_{TF}(1 + q_{TF}d) > 2q_{TF}$. Thus, the second layer more than doubles the screening-induced suppression of $\phi(q)$. Since $1/q_{TF} \approx 5$ nm in GaAs, and the layer separation d in typical samples is 30 nm, $q_{sc} \approx 14q_{TF}$. As $\phi(q)$ enters squared in the expression for ρ_D, the additional screening by the second layer suppresses the drag by roughly two orders of magnitude.

The susceptibility function $\text{Im}\chi(q, \omega)$ is, in effect, the phase space available for a scattering process of energy $\hbar\omega$ and momentum transfer $\hbar q$. At zero temperature and for $\hbar\omega$ much less than the Fermi energy E_F of the 2D gases, $\text{Im}\chi(q, \omega)$ scales as ω/q for small q and as $\omega/\sqrt{q - 2k_F}$ for q approaching $2k_F$. These divergences in q lead to well-known logarithmic divergences in the temperature dependence of the quasiparticle lifetime in two dimensions [17]. In Coulomb drag, however, the $q = 0$ divergence is eliminated by the q^2 factor in the integrand, this factor arising from the $1 - \cos\theta$ weighting over scattering angle θ essential in evaluating the momentum relaxation time. The backward scattering divergence at $q = 2k_F$ remains, but so long as $k_F d \gg 1$ its importance is slight owing to the exponential fall-off of the potential $\phi(q)$ for $q > 1/d$.

In general, Eq. 3.3 must be solved numerically. However, for two identical infinitely thin 2D electron systems at very low temperatures ($T \ll T_F$) and large inter-layer separation (so that both $k_F d$ and $q_{TF} d$ are $>> 1$), an analytic result can be obtained [16]:

$$\rho_D = \frac{h}{e^2} \frac{\zeta(3)\pi}{32} \frac{k_B^2 T^2}{E_F^2 k_F^2 q_{TF}^2 d^4}. \tag{3.5}$$

This equation has several notable features. First, the drag resistivity is proportional to T^2 at low temperatures. This is not unexpected; the joint phase space for inelastic inter-layer scattering processes is dominated by the $k_B T$ thermal smearing of both Fermi surfaces. Second, the drag scales with layer separation as d^{-4}; two of these powers of d arise from the enhanced screening in double layer systems and two more come from the restriction of momentum transfers to $q < 1/d$ and the $q^2 \sim 1 - \cos\theta$ weighting over scattering angle. Third, the predicted ρ_D varies with density as N_s^{-3} in Eq. 3.5. Finally, observe that ρ_D does not depend upon effective mass; the linear dependence of q_{TF} on m^* is cancelled by the $1/m^*$ scaling of E_F at constant density. Thus, at this level of approximation bilayer electron and bilayer hole systems at the same temperature, density, and layer separation ought to exhibit the same drag resistance.

3.4. Comparison between theory and experiment

Early Coulomb drag studies in bilayer 2D electron systems [12] were not performed in a regime where direct Coulomb scattering adequately described the momentum relaxation rate. Momentum transfer via virtual phonon exchange was substantial in the early experiments [12, 18] and by now a well-developed theory of this contribution to drag exists [19]. More recently, Kellogg, et al. [20] have measured drag in a bilayer electron sample in which, owing to its small layer separation and low density, direct Coulomb scattering dominates the drag at low temperatures. Figure 5 shows the temperature dependence of ρ_D in a sample consisting of two 18 nm GaAs quantum wells separated by a 10 nm $Al_{0.9}Ga_{0.1}As$ barrier. For the data shown the bilayer is balanced, i.e. has equal electron density N_s in the two layers.

While Fig. 5 shows that ρ_D is roughly quadratic in temperature, closer inspection reveals slight deviations: The dotted curve is a simple least-squares fit of the $N_s = 3.1 \times 10^{10} cm^{-2}$ data to $\rho_D = AT^2$. More importantly, the dashed curve in Fig. 5 gives the prediction of Eq. 3.5 for the same density data set; the theory falls short of the data by a factor of about 6 at this density. We find that this large discrepancy remains even after direct numerical integration of Eq. 3.3 is employed to account for finite layer thickness, finite T/T_F, and the fact that $k_F d$ is not large compared to unity [20, 21]. Our calculations, however, assume only the simplest static Thomas-Fermi screening of the inter-layer Coulomb interaction.

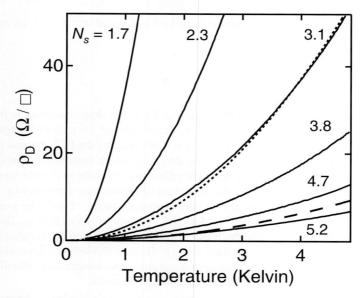

Fig. 5. Drag resistivity *vs.* temperature for six different layer densities N_s, labeled in units of 10^{10}cm^{-2}. The dotted line is a least-squares fit of the $N_s = 3.1 \times 10^{10}\text{cm}^{-2}$ data to $\rho_D = AT^2$. The dashed line is the drag resistance predicted by Eq. 3.5 for the same density. After Ref. [20].

The discrepancy between the static screening theory and the experimentally measured drag resistance grows steadily as the density is reduced. Figure 6 shows the density dependence of the drag at $T = 1$ K and 4 K. The solid lines, which qualitatively capture the density dependence, are proportional to N_s^{-4}. The dashed lines represent Eq. 3.5 and its N_s^{-3} density dependence.

More sophisticated theoretical calculations [22, 23] go beyond the random phase approximation to include higher order many-body effects on Coulomb drag. Such corrections generally become more important as the density is reduced and the dimensionless parameter $r_s = a_0^{-1}(\pi N_s)^{-1/2}$ grows. Giving the interparticle separation in units of the semiconductor Bohr radius a_0, the parameter r_s also gives the ratio of the mean Coulomb interaction energy to the Fermi kinetic energy. For the data in Fig. 6, r_s ranges from about 1.8 up to 4.2 at the lowest densities. It is thus not surprising that higher-order many-body effects are quantitatively important in these data. Indeed, Yurtserver, et al. [22] find quite good agreement with the data in Fig. 6.

Finally, we remark that for the data in Figs. 5 and 6, the parameter $k_F d$ ranges from about 2.2 down to 0.9. This means that backward scattering cannot be ignored. As mention above, the singularity in $\text{Im}\chi(q, \omega)$ at $q = 2k_F$ leads to logarithmic corrections to the temperature dependence of the drag. In fact, we find

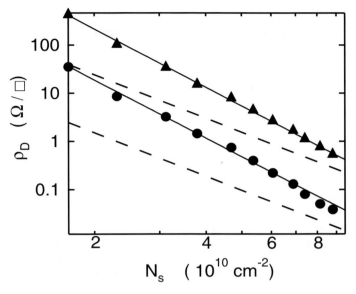

Fig. 6. Drag resistivity *vs.* density at $T = 4$ K (triangles) and $T = 1$ K (solid circles). The upper and lower dashed lines represent Eq. 3.5 at $T = 4$ K and 1 K, respectively. Adapted from Ref. [20].

that the data in Fig. 5 can be very well fit by a function of the form $AT^2 ln(T/T_0)$. While suggestive, however, such subtle temperature dependences are a poor way to prove the importance of backward scattering processes. Nonetheless, Kellogg, et al. demonstrated convincingly that such events do indeed dominate at low temperatures. Their method involved a careful study of the dependence of the drag on density *difference* between the two layers. The interested reader is referred to the original publication [20] for the details.

4. Tunneling between parallel two-dimensional electron gases

We now turn to the study of inter-layer tunneling. As in the previous section, we will here confine our attention to the case in which the two layers are only very weakly coupled by Coulomb interactions. Drag and tunneling in strongly coupled bilayers will be discussed in the final section of these notes.

Tunneling between two ordinary metals is quite simple. If a voltage V is imposed between the metals, the tunneling current will be:

$$I \sim \int dE\, P(E) N_L(E - eV) N_R(E)[f_L(E - eV) - f_R(E)] \qquad (4.1)$$

where $P(E)$ is the probability of tunneling through the barrier separating the metals, N_L and N_R are the densities of states of the metals on the left and right, respectively, and f_L and f_R are the associated Fermi functions. For small voltages (so that $P(E)$ may be taken as a constant) one has:

$$I \sim P N_L(E_{FL}) N_R(E_{FR}) eV = \text{const.} \times V. \tag{4.2}$$

In other words, a normal metal tunnel junction behaves ohmically; current is proportional to voltage.

4.1. Ideal 2D-2D tunneling

Tunneling between 2D electron gases confined to quantum wells separated by a barrier layer is very different than 3D-3D tunneling. The energy of an electron in either well will be of the form $\epsilon = E_{0,(L,R)} + \hbar^2 k^2/2m$, where $E_{0,(L,R)}$ is the bound state energy in the left or right quantum well. If the barrier is smooth, the in-plane momentum $\hbar k$ will be conserved upon tunneling. Energy conservation then implies that $E_{0,L} = E_{0,R}$ in any tunneling event [24]. Hence, 2D-2D tunneling is sharply resonant: An electron can tunnel only when the initial and final subband energy levels in the two quantum wells line up precisely. This remarkable fact is independent of temperature and the Fermi energy (*i.e.* density) of the electrons in either well. Figure 7 gives a typical example of the resonant nature of 2D-2D tunneling. The width of the observed resonance is about 0.3 meV, which is less than 10% of the Fermi energy of the 2D electron gases in the quantum wells. In a conventional 3D-3D tunnel junction the conductance dI/dV is independent of voltage over a wide range.

A straight-forward application of Fermi's Golden Rule for the tunneling current flowing between two weakly-coupled ideal 2D systems gives

$$I = \frac{2e\pi}{\hbar} \sum_{k,\sigma} \sum_{k',\sigma'} |t|^2 \delta_{k,k'} \delta_{\sigma,\sigma'} \delta(\epsilon_{R,k} - \epsilon_{L,k'}) (f_{R,k} - f_{L,k'}) \tag{4.3}$$

where the hopping matrix element $|t|$ has been assumed to be energy, momentum, and spin independent. Measuring the energies $\epsilon_{R,k}$ and $\epsilon_{L,k'}$ relative to single reference point (for example, the bottom of the right quantum well) allows us to re-write the delta function on energy as $\delta(\epsilon_{R,k} - \epsilon_{L,k'}) = \delta(\xi_k - \xi_{k'} + E_{0,R} - E_{0,L}) = \delta(\xi_k - \xi_{k'} + E_{F,L} - E_{F,R} + eV)$, with $\xi_k \equiv \hbar^2 k^2/2m$ being the kinetic energy and $E_{F,L}$ and $E_{F,R}$ the Fermi energies of the 2DES's in the left and right quantum wells. Summing over k', σ' and σ, and replacing the remaining sum on k by an integral over ξ yields:

$$I = \frac{2e\pi}{\hbar} |t|^2 \rho S \delta(E_{F,L} - E_{F,R} + eV) \int_0^\infty (f_{R,k} - f_{L,k}) d\xi \tag{4.4}$$

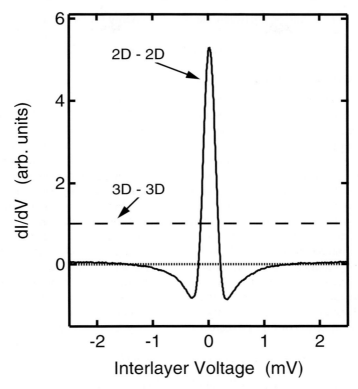

Fig. 7. Typical 2D-2D tunnel resonance at zero magnetic field. Tunneling conductance dI/dV is plotted against inter-layer voltage V at $T = 1.75$ K. Sample consists of two 20 nm quantum wells, each containing a 2DES with density $N_S \approx 1.5 \times 10^{11} \text{cm}^{-2}$, separated by a 17.5 nm $Al_{0.3}Ga_{0.7}As$ barrier layer. Dashed line represents 3D - 3D result.

where S is the area of the sample and $\rho = m/\pi\hbar^2$ is the density of states of the 2D systems, including both spin states. So long as the temperature is much less than the Fermi temperatures of the two 2D systems, the integral is just $E_{F,R} - E_{F,L}$. By virtue of the delta function, then, we have:

$$I = 2\pi |t|^2 \frac{e^2}{\hbar} S\rho V \delta(E_{F,L} - E_{F,R} + eV) \qquad (4.5)$$

Not surprisingly, the tunnel current is proportional to $|t|^2$ and S, the area of the sample. It is also makes sense that the current is proportional to ρV because this is number of states available to tunnel [25]. The delta function, however, is striking and implies that 2D-2D tunneling is singular. Note that the argument $E_{F,L} - E_{F,R} + eV$ of the delta function is, at *any* temperature, exactly equal to $E_{0,R} - E_{0,L}$, the gap between the subband energy levels in the two quantum wells. Hence, elastic, momentum-conserving 2D-2D tunneling can only occur between two precisely aligned quantum well energy levels [26].

4.2. Lifetime broadening

The resonant character of 2D-2D tunneling results from two-dimensionality and the conservation of momentum parallel to the 2D planes. Figure 7 reveals that although sharp, real 2D-2D tunnel resonances have a finite width. In many cases this width is a result of scattering. Put another way, the momentum eigenstates used above to determine the tunnel current do not possess an infinite lifetime in real samples. In order to incorporate this fact we introduce the electron spectral function, $A(E, \mathbf{k})$. This function gives the probability density for a momentum state $\mathbf{k}$ to have energy E. In the treatment given above, the implicit assumption was that $A(E, \mathbf{k}) = \delta(E - \xi_{\mathbf{k}})$. In general, $A(E, \mathbf{k})$ is quite complex, containing features due to independent quasiparticle states as well as structure associated with the interacting many-body aspects (e.g. collective modes) of the electron gas. For simplicity, we shall assume that $A(E, \mathbf{k})$ depends on E and $\mathbf{k}$ only through the combination $E - \xi_{\mathbf{k}} + E_F$, where E now is measured relative to the Fermi level. This assumption amounts, in effect, to assuming that all the spectral weight lies in the vicinity of the non-interacting single electron state. The generalization of Eq. 4.4 reads

$$I = 2\pi |t|^2 \frac{e}{\hbar} S\rho \int_0^\infty d\xi \int_{-\infty}^\infty dE \, A_R(E - \xi + E_{F,R})$$
$$\times A_L(E - \xi + E_{F,L} + eV)(f_R(E) - f_L(E + eV)) \tag{4.6}$$

Assuming the spectral functions consist of a single peak centered at $E = \xi - E_F$ with width γ much less than E_F, we can extend the lower integration limit on ξ to $-\infty$. Introducing the variable $x = E - \xi + E_{F,R}$ we get

$$I = 2\pi |t|^2 \frac{e^2}{\hbar} S\rho V \int_{-\infty}^\infty dx \, A_R(x) A_L(x + E_{F,L} - E_{F,R} + eV) \tag{4.7}$$

where, once again, the integral over the Fermi functions yielded a factor eV. This equation shows that the current-voltage characteristic of a 2D-2D tunnel junction directly measures the convolution of the electronic spectral functions in the two layers.

The spectral functions $A_{L,R}$ are often assumed to be lorentzian: $A(x) = (\gamma/2\pi)(x^2 + \gamma^2/4)$, with γ representing the lifetime broadening $\gamma = \hbar/\tau$ of the quasiparticles. In this case Eq. 4.7 becomes

$$I = 2|t|^2 \frac{e^2}{\hbar} S\rho V \frac{\Gamma}{(V - V_0)^2 + \Gamma^2} \tag{4.8}$$

with $\Gamma = (\gamma_R + \gamma_L)/2e$ and $eV_0 = E_{F,R} - E_{F,L}$. Hence, the ratio $F(V) \equiv I/V$ (not to be confused with the differential conductance dI/dV plotted in Fig. 7) is also a lorentzian, with half-width at half-maximum Γ equal to the average lifetime broadening $\hbar/\tau$ in the two layers.

Tunneling spectra like that in Fig. 7 have been used to quantitatively determine the lifetime of quasiparticle states [27] at zero magnetic field. At low temperature the tunneling linewidth is dominated by the scattering of electrons off of the static disorder potential in the sample, but at higher temperatures electron-electron scattering dominates. Complementary to the Coulomb drag measurement discussed previously, the tunneling experiment measures the *intra*-layer scattering rate rather than the inter-layer one. The experiments [27] revealed that early calculations of the $e - e$ rate [28] severely underestimated the rate. More recent calculations have corrected errors in the earlier ones and have achieved good agreement with experiment, especially when higher order many-body effects are included [29, 30].

4.3. 2D-2D tunneling in a perpendicular magnetic field

When a large magnetic field is applied perpendicular to the 2D plane the inter-layer tunneling conductance is drastically modified. Figure 8 shows data from a weakly-coupled bilayer sample structurally identical to the one used in Fig. 7. The sharp tunnel resonance observed at $B = 0$ is replaced, at high magnetic field, by a very broad spectrum. Perhaps the most notable feature present in the high field data is the wide region of heavily suppressed tunneling conductance surrounding zero inter-layer voltage [31].

The high field data shown in Fig. 8 suggest the presence of a gap in the density of states, centered at zero energy. Gaps are of course present in the 2DES energy spectrum at high fields when the system is in an integer or fractional quantized Hall state. Remarkably, however, the $B = 13$ T data shown correspond, roughly, to Landau level filling $\nu = 1/2$ in the individual 2D layers. At this filling no quantized Hall state exists (the layer separation is much too large to support the strongly correlated $\nu_T = 1/2 + 1/2 = 1$ excitonic QHE state (mentioned in section 2.3 and the subject of section 5) and the electronic energy spectrum is gapless.

In the absence of electron-electron interactions, Landau levels are sharp. Tunneling between two disorder-free non-interacting 2D electron gases at high field ought to be sharply resonant, just like at $B = 0$. The very broad spectrum shown in Fig. 8 therefore suggests that either disorder or interactions are very important at high field. Disorder can be discounted since the low temperature $B = 0$ tunnel resonance, which *is* dominated by disorder, is so much narrower than the high field spectrum. Thus, electron-electron interactions must somehow be re-

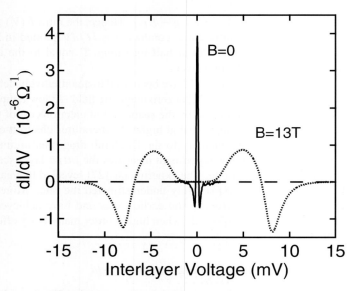

Fig. 8. Low temperature tunneling conductance at zero magnetic field and $B = 13$ T in a weakly-coupled double layer 2DES sample. After Ref. [31].

sponsible for the high field behavior. In particular, the strong suppression of the tunneling conductance around zero bias must be an interaction effect.

The suppression of the zero-bias tunneling conductance shown in Fig. 8 is a generic phenomenon at high magnetic fields. It occurs over wide swaths of magnetic field, and is largely indifferent to the presence or absence of quantized Hall states. The tunneling conductance at zero bias vanishes as the temperature is reduced, in an essentially thermally activated manner [31].

The main features of 2D-2D tunneling at high perpendicular magnetic field are by now fairly well understood [32, 33, 34, 35]. The strong suppression of the zero bias conductance reflects the presence of a *pseudo*-gap pinned to the Fermi level. Tunneling injects, essentially instantaneously, an electron into the 2DES. Even if the system is thermodynamically gapless, the $N + 1$ particle state so suddenly created will be a highly excited one. Low energy $N + 1$ particle states may exist, but tunneling cannot connect to them in the time available. The strong correlations in the electron gas mean that many electrons must adjust their positions to relax the charge defect created by tunneling. The time required to do this is vastly enhanced by the magnetic field which effectively bottles up the charge defect, converting its natural radial relaxation into a vortex-like motion by virtue of the Lorentz force. Crudely speaking, the minimum energy required

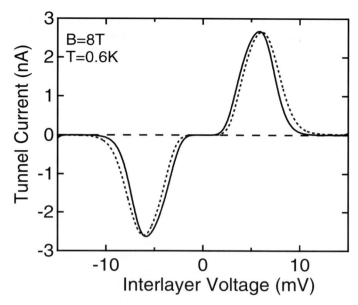

Fig. 9. Tunneling current-voltage characteristics in two double layer 2D electron gas samples whose only structural difference is the thickness of the tunnel barrier; 17.5 nm for the solid trace and 34 nm for the dashed trace. The dashed curve has been vertically scaled by a factor of 3.6. After Ref. [36].

to tunnel into a 2DES of density N_s at high field is the mean Coulomb energy between electrons: $e^2 N_s^{1/2}/\epsilon \sim e^2/\epsilon\ell$, with $\ell = (\hbar/eB)^{1/2}$ the magnetic length. This is quite consistent with experiment [31].

The suppression of the zero bias 2D-2D tunneling conductance at high magnetic field represents the convolution of two single layer effects. The Coulombic energy cost associated with the injection of an electron into one layer adds to the similar penalty incurred when that electron is first extracted from the other layer. Unless the two layers are quite close together, inter-layer Coulomb interactions are of only limited importance. To a rough approximation, the main effect of inter-layer interactions is to reduce the net Coulomb barrier to tunneling. This reduction may be thought of as an excitonic interaction, whereby an electron which has just tunneled into one layer is attracted to the hole left behind in the other layer. The very slow relaxation of charge defects at high magnetic fields render this final-state exciton much more readily observable than the analogous effect in ordinary tunnel junctions at zero magnetic field. Figure 9 shows one experimental manifestation of this effect. The figure shows current-voltage characteristics (I vs. V, not dI/dV vs. V) at $B = 8$ T for two samples with the same electron density ($N_s = 1.5 \times 10^{11} \text{cm}^{-2}$) and quantum well widths

(20 nm), but different tunnel barrier thicknesses: 17.5 nm for the solid curve, 34 nm for the dashed curve. The data suggest that the Coulomb pseudo-gap is larger in the sample with the thicker barrier. This is the expected result since the energy gain due to the formation of an inter-layer exciton ought to scale inversely with the separation between the layers. A more quantitative study of this effect, based on careful analysis of both the barrier thickness and electron density dependence of the tunnel spectrum, has confirmed this excitonic picture [36]. For the sample in Fig. 9, the effective excitonic "binding" energy is about $V_{ex} \approx -1.4$ meV.

5. Strongly-coupled bilayer 2D electron systems and excitonic superfluidity

As explained in section 2.3, double layer 2D electron systems with small interlayer separation support novel collective phases which do not exist in the individual layers. The prototypical examples are the quantized Hall effect states occurring at total Landau level filling fractions $\nu_T = 1$ and $\nu_T = 1/2$. For balanced (i.e. equal density) bilayer systems these two states correspond to individual layer filling fractions $\nu_1 = \nu_2 = 1/2$ and $1/4$, respectively. No quantized Hall states have ever been observed at these fillings in single layer 2D systems.

For the remainder of these lectures we shall concentrate on the case $\nu_T = 1$. The bilayer QHE phase which develops at this filling is particularly interesting. It possesses a remarkable broken symmetry known as spontaneous interlayer phase coherence and may be viewed equivalently as a pseudospin ferromagnet or a Bose condensate of interlayer excitons. Exciton condensation has been sought for over 40 years, largely in optically-generated electron-hole plasmas. Oddly enough, the $\nu_T = 1$ quantum Hall state in bilayer 2D electron systems turns out to be the first realization of this long-sought phenomenon.

5.1. Introduction

Equation 2.15 illustrates a specific Laughlin-like wavefunction which embodies the electron correlations present in the QHE phase at $\nu_T = 1$. In this wavefunction the z and w variables represent the (complex) coordinates of electrons in two distinguishable layers. In fact, Eq. 2.15 represents a quantum state in which the numbers, n_1 and n_2, of electrons in each 2D layer, are good quantum numbers. Indeed, Eq. 2.15 is incompatible with any tunneling between the layers for such tunneling will always render the number of electrons in each layer uncertain (of course the total number of electrons in the bilayer remains fixed). While this line of argument may lead one to think that tunneling will destroy the $\nu_T = 1$ QHE, quite the opposite is true.

If electron-electron interactions can be ignored, tunneling alone will establish a QHE at $\nu_T = 1$. Tunneling hybridizes the degenerate individual quantum well states into symmetric and antisymmetric combinations which are split in energy by an amount Δ_{SAS}. If the temperature is sufficiently low, and disorder is not too severe, then Δ_{SAS} provides the energy gap needed for a QHE. In this case the ground state of the bilayer electron system is simply a single filled Landau level of symmetric state electrons [37]. If this was the only mechanism for producing a bilayer QHE at $\nu_T = 1$, it would not garner much attention.

Coulomb interactions, both intra-layer and inter-layer, fundamentally change the picture. It is now known, in both experiment and theory, that the $\nu_T = 1$ QHE survives in the limit of *zero* tunneling. The situation is crudely analogous to the $\nu = 1$ QHE in single layers, where an (exchange-induced) energy gap persists even in the limit of zero Zeeman energy. There is, however, a crucial difference between the two cases. In single layers all pairs of electrons, regardless of their spin, experience the same Coulomb repulsive force. In the $\nu_T = 1$ bilayer however, electrons in opposite layers repel one another less strongly (for a given in-plane separation) than do electrons in the same layer. This reduced symmetry of the Coulomb interaction in bilayers has fundamental consequences. Most obvious among these is the existence of a quantum phase transition whereby the QHE energy gaps vanishes above a critical layer separation.

The collapse of odd-integer QHE states in bilayer 2D electron systems at large layer separation was first observed by Boebinger, *et al.* [38] in strongly tunneling samples. Subsequently, Murphy, *et al* [39] demonstrated that the $\nu_T = 1$ QHE persists in the zero tunneling limit if the layer separation is small enough. A very crude, yet informative, model for this behavior is as follows. Consider the energetics of just two electrons in a double layer system. Let the layer separation be d and the in-plane separation between the electrons be R. Consider two situations, one where both electrons are in identical symmetric linear combinations of individual layer eigenstates and one in which the two electrons are localized in opposite layers. The former case is representative of the QHE phase in which the total system consists of one filled Landau level of symmetric electrons, while the latter case represents two independent half-filled Landau levels with no QHE. In the first case we may think of each electron as consisting of two particles, each of charge $e/2$, connected by a rigid rod of length d. Roughly speaking, the energy of a pair of such electrons is

$$U_1 = \frac{e^2}{2}\left(\frac{1}{R} + \frac{1}{\sqrt{R^2 + d^2}}\right) - \Delta_{SAS} + U_{ex} \tag{5.1}$$

where U_{ex} is an intra-layer exchange energy. Being negative, and of order $-e^2/\epsilon R$, U_{ex} enhances the stability of this configuration. In contrast, the energy of the second configuration has no tunneling term and no intra-layer exchange:

$$U_2 = \frac{e^2}{\sqrt{R^2 + d^2}}. \tag{5.2}$$

The difference in energy between the two configurations is thus

$$U_2 - U_1 = \frac{e^2}{2}\left(\frac{1}{\sqrt{R^2 + d^2}} - \frac{1}{R}\right) + \Delta_{SAS} - U_{ex}. \tag{5.3}$$

The first term on the right of this equation is always negative. Thus, in the absence of tunneling and exchange the non-QHE phase is always favored in this model. With tunneling (but not exchange) the situation changes: The QHE phase will be favored, provided Δ_{SAS} is large enough for a given d. Note however, that in the absence of exchange the critical d vanishes as $\Delta_{SAS} \to 0$. Exchange qualitatively changes this, with the QHE phase stabilized even at zero tunneling, provided d/R is small enough [40].

Figure 10 presents the empirical phase diagram for the $\nu_T = 1$ QHE in bilayer 2D electron systems [39]. Each symbol represents a different sample whose tunneling strength Δ_{SAS} (in units of the Coulomb energy $e^2/\epsilon\ell$) and layer separation d (in units of the magnetic length ℓ) serve as coordinates. Solid symbols represent samples which do exihibit a QHE at $\nu_T = 1$, while open symbols represent samples which do not. The solid line is guide to the eye, approximating the location of the phase boundary. Most importantly, the phase boundary intersects the $\Delta_{SAS} = 0$ axis at a non-zero value of d/ℓ.

5.2. Quantum hall ferromagnetism

The layer degree of freedom of electrons in double layer structures is conveniently encoded using a pseudospin observable, τ. Electrons definitely in one layer are pseudospin "up" while electrons definitely in the other layer are pseudospin "down". The algebra of pseudospin is identical to that of ordinary spin. The up and down pseudospin states are taken as eigenstates of τ_z and are written as $|\uparrow\rangle$ and $|\downarrow\rangle$, respectively. Using this notation we may re-write the wavefunction for the bilayer QHE state at $\nu_T = 1$ given in Eq. 2.15:

$$\Psi(z_1, z_2, \ldots, z_n) = \prod_{i<j}(z_i - z_j) \, S|\uparrow\uparrow \ldots \uparrow, \downarrow\downarrow \ldots \downarrow\rangle. \tag{5.4}$$

In this equation z_i covers the coordinates of all electrons in the system, irrespective of which layer they are in. The pseudospin ket vector represents n_1 electrons definitely in one layer and n_2 electrons definitely in the other layer, with $n = n_1 + n_2$. S is the symmetrization operator, needed here because the spatial

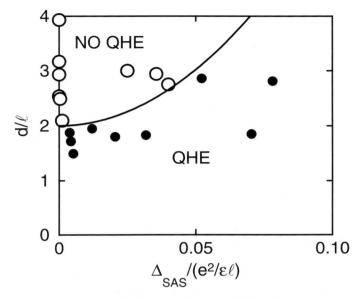

Fig. 10. Empirically determined phase diagram for the $\nu_T = 1$ QHE in bilayer 2D electron systems. Solid dots represent samples which exhibit a QHE at $\nu_T = 1$ while open dots represent samples which do not. Solid line is a guide to the eye. After Ref. [39].

part of the wavefunction is completely antisymmetric. The ubiquitous gaussian factors have been omitted for clarity, as has the real spin part of the wavefunction [37].

The wavefunction in Eq. 5.4 is an eigenstate of both total pseudospin T^2 and its z-component $T_z = \sum_i \tau_{i,z}$ with eigenvalue $(n_1 - n_2)/2$ (which vanishes in a balanced bilayer 2DES). With a definite number of particles in each layer, this wavefunction is not a good choice when interlayer tunneling is present, for then the layer index of each electron is uncertain. Tunneling is adds a term $H_t = -(\Delta_{SAS}/2)T_x$ to the Hamiltonian, explicitly breaking its otherwise perfect xy-symmetry. The tunneling energy is minimized when electrons are in symmetric bilayer states, $(|\uparrow\rangle + |\downarrow\rangle)/\sqrt{2}$. In this case Eq. 5.4 is replaced by

$$\Psi(z_1, z_2, \ldots, z_n) = \prod_{i<j}(z_i - z_j)\prod_i \frac{1}{\sqrt{2}}(|\uparrow\rangle + |\downarrow\rangle)_i. \qquad (5.5)$$

Eq. 5.5 is also a maximally polarized eigenstate of the total pseudospin T^2. However, unlike Eq. 5.4, Eq. 5.5 is not an eigenstate of T_z, but is instead an eigenstate of T_x with eiqenvalue $n/2$. In the absence of tunneling both states have the same energy, being members of a large multiplet of degenerate states which, in a balanced bilayer system, have zero expectation value for T_z. Indeed, this multiplet of states is encompassed by a generalization of Eq. 5.5:

$$\Psi(z_1, z_2, \ldots, z_n) = \prod_{i<j}(z_i - z_j) \prod_i \frac{1}{\sqrt{2}}(|\uparrow\rangle + e^{i\phi}|\downarrow\rangle)_i. \qquad (5.6)$$

In this equation every electron has its pseudospin lying in the xy-plane, inclined by the same angle ϕ with respect to the x-axis.

The analogy to ordinary ferromagnetism is now clear. The bilayer electron system, at small layer separation, is in a state in which all electrons are polarized in pseudospin space. As in an ordinary ferromagnet, this polarization is driven by the exchange energy. In the absence of tunneling the polarization is sponta-neous. Crudely speaking, the energy gap to the lowest-lying (charged) excitation in the system is equal to the exchange energy cost of flipping an electron from a symmetric to an antisymmetric pseudospin state. Tunneling explicitly breaks the xy-symmetry of the system and favors the pseudospin lying along the x-axis. This increases the symmetric/antisymmetric energy gap and thereby strengthens the quantized Hall effect in the system.

The above wavefunctions are all eigenstates of total pseudospin. This is in fact appropriate only in the limit of vanishing layer separation where the Coulomb interaction is pseudospin invariant. At finite layer separation total pseudospin is not a good quantum number and is subject to quantum fluctuations. If the layer separation is sufficiently small, these fluctuations are weak and do not de-stroy the energy gap necessary for the QHE. In this case the main effect of the layer separation is to keep the pseudospin close to the xy-plane, thereby reduc-ing the capacitive energy penalty of having more electrons in one layer than in the other [41]. As the layer separation increases, the quantum fluctuations of the pseudospin become more severe, eventually leading to the destruction of the QHE. This is the phase transition qualitatively discussed in the preceding section.

The elementary excitations of the $v_T = 1$ bilayer QHE state may be intuitively understood in the ferromagnetism picture. In the absence of tunneling, the Gold-stone theorem implies the existence of a collective mode whose energy vanishes at zero wavevector. This neutral mode, which involves spatio-temporal oscilla-tions of the phase ϕ, is analogous to spin waves in an ordinary ferromagnet. Here the finite layer separation leads to a linear dispersion $\omega \propto q$, of the pseudospin waves at small q [42]. At larger wavevector, the mode is expected to exhibit a roton minimum in its dispersion.

The charged excitations in the system are expected to be quite intricate. Vortex-like structures in the ϕ field, known as merons and anti-merons, carry both topo-logical and real charge ($\pm e/2$). Meron/anti-meron pairs with total charge e are expected to dominate dissipative electrical transport at low temperatures. Indeed, in the absence of tunneling, these vortex pairs are expected to unbind in a Kosterlitz-Thouless phase transition. Although not yet clearly observed in any experiment, such a finite temperature transition is unique among quantum Hall systems.

An early experiment by Murphy, *et al.* [39] provided impetus for the development of the ferromagnetism model. Murphy observed a dramatic anomaly in the measured charge gap of the $v_T = 1$ QHE when a small in-plane magnetic field component $B_{||}$ was applied. The gap was found to drop rapidly for very small $B_{||}$ and then level off. This was very surprising and suggested the existence of a hitherto unknown phase transition. Almost immediately, Yang, *et al.* [43] developed an explanation for this behavior based upon a commensurate-incommensurate textural phase transition in the pseudospin field ϕ. In Yang's model, the $v_T = 1$ QHE state takes on spiral pseudospin texture at small $B_{||}$ in order to maintain the energetic advantage of interlayer tunneling. The spiraling, however, costs exchange energy owing to the gradient in ϕ. At a critical $B_{||}$ this cost becomes too large and a phase transition to a new QHE phase occurs where the pseudospin is once again uniform. The predicted values of the critical $B_{||}$ were found to be in reasonable agreement with experiment. The initial drop in the energy gap at small $B_{||}$ suggests that the pseudospin field is coherent over distances approaching 1 μm, much larger than the mean separation between electrons. Details of both Murphy's experiment, Yang's model, and the quantum Hall ferromagnetism picture in general can be found in chapters 2 and 5 of Ref. [5].

5.3. *Tunneling and interlayer phase coherence at $v_T = 1$*

Beyond the basic observation that a QHE develops at $v_T = 1$ in closely-spaced bilayers and the discovery of a curious dependence of the energy gap on in-plane magnetic field, early experiments offered little direct evidence for the spontaneous interlayer phase coherence embodied in Eq. 5.6. According to theory, Coulomb interactions alone are sufficient to force the electrons in the system to assume identical quantum states possessing maximally uncertain layer index, even when the bare tunneling matrix elements are arbitrarily small. In effect, these interactions renormalize the single-particle tunnel splitting from essentially zero to a value comparable to the mean Coulomb energy in the system, thereby spontaneously imposing interlayer quantum phase coherence. From this perspective, it is clear that interlayer tunneling measurements of the type described in section 4 should be very interesting.

Figure 11 displays a series of low temperature tunneling conductance characteristics in a balanced bilayer 2DES at $v_T = 1$. The sample once again consists of two 18 nm GaAs quantum wells separated by a 10 nm $Al_{0.9}Ga_{0.1}As$ barrier layer. In each panel the total electron density N_T is different (having been set by gates),

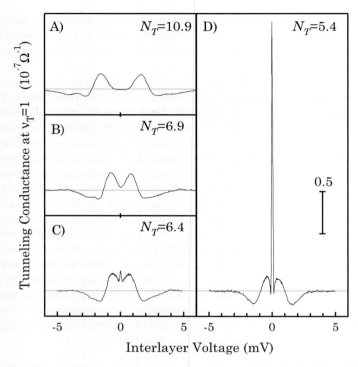

Fig. 11. Tunneling conductance characteristics at $\nu_T = 1$ in a variable density bilayer 2D electron system. For the total densities shown, the effective layer separations are $d/\ell = 2.3$, 1.84, 1.78, and 1.63, in panels A through D, respectively. After Ref. [44].

but the magnetic field is adjusted to maintain $\nu_T = 1$. Thus, the different densities correspond to different effective layer separations, ranging from $d/\ell = 2.3$ at $N_T = 10.9 \times 10^{10} cm^{-2}$ down to $d/\ell = 1.63$ at $N_T = 5.4 \times 10^{10} cm^{-2}$. The single-particle tunnel splitting Δ_{SAS} in this sample is estimated to be only about $100 \, \mu K$, roughly one million times smaller than the mean Coulomb energy $e^2/\epsilon\ell$ at $\nu_T = 1$.

In panel A of Fig. 11 the density is so high that the bilayer system is not in the QHE phase and the layers are behaving essentially independently. The tunnel spectrum displays the familiar conductance suppression around zero interlayer bias that is discussed in section 4.3. In panel B the density is lower and the zero bias suppression is weakened. This weakening is a result of both the reduced Coulomb repulsion between electrons in the individual layers and the stronger excitonic attraction between a tunneled electron and the hole it leaves behind.

Qualitatively, however, the spectra in panels A and B are similar. The data in panel C are different; a small peak in the tunneling conductance has appeared at zero bias. Reducing the density (and thus d/ℓ) further causes this peak to grow dramatically, soon becoming the dominant feature in the spectrum.

Detailed studies of the zero bias peak in the tunneling conductance reveal several important facts. First, the peak is clearly a $\nu_T = 1$ effect. Small changes of the magnetic field destroy it and restore the suppression effect. Furthermore, the conductance peak and the QHE both seem to develop at the same critical layer separation, near $d/\ell \approx 1.83$ in this sample, strongly suggesting that both phenomena reflect the same quantum phase transition. Although the $\nu_T = 1$ tunneling peak qualitatively resembles the simple tunnel resonances seen at zero magnetic field, several features sharply distinguish the two effects. For example, while small antisymmetric changes in the density of the two layers ($N_1 \rightarrow (N_T + \delta N)/2$, $N_2 \rightarrow (N_T - \delta N)/2$) shift the voltage location of the resonances seen at $B = 0$, the $\nu_T = 1$ peak remains locked at $V = 0$ [45]. In sharp contrast to zero field tunnel resonances, both the width Γ and height $G(0)$ of the $\nu_T = 1$ peak exhibit a strong temperature dependence down to the lowest temperatures. Recent experiments have revealed peak widths below $\Gamma = 2 \, \mu\text{V}$ at $T \sim 25$ mK, roughly 40 times smaller than that seen at $B = 0$ in the same sample. Similarly, the height $G(0)$ of the $\nu_T = 1$ peak exceeds that seen at $B = 0$ by more than two orders of magnitude. These results demonstrate tunneling peak at $\nu_T = 1$ is not governed by the simple single-particle effects operative at $B = 0$, but instead reflects a deeply collective phenomenon.

What is the physical origin of the giant tunneling peak at $\nu_T = 1$? What occurs at small layer separation that overcomes the near-universal tendency for interlayer tunneling at high magnetic field to be heavily suppressed? Although a quantitative answer to these questions has proven elusive [46, 62, 48], the qualitative explanation is clear: the enhanced zero bias tunneling is a direct manifestation of spontaneous interlayer phase coherence which lies at root of the strongly correlated $\nu_T = 1$ quantum fluid. The suppression effect ordinarily observed in high field tunneling results from the fact that an electron attempting to tunnel in arrives wholly uncorrelated with the strongly correlated electrons already present in the layer. There is insufficient time available for the necessarily large number of electrons to reconfigure themselves in order to allow the newcomer to enter at low energy. Instead, a electron can be rapidly injected only if its energy, supplied by the the interlayer voltage V, is large enough to simply shoulder aside nearby electrons. At $\nu_T = 1$ this is not the case. Electrons in the system are strongly correlated with their neighbors in *both* layers. An electron attempting to tunnel is guaranteed that no disturbance will be created by its entry. Indeed, as Eqs. 5.5 and 5.6 make clear, electrons in the condensed phase are uncertain as to which layer they are in in the first place. These wavefunctions make clear that if an electron is found definitely in one layer it is necessarily true that a hole exists in the opposite layer at the same location.

The zero bias tunneling peak observed at $\nu_T = 1$ is a direct indication that a low energy collective mode exists in the system. This mode is the Goldstone mode of the pseudo-ferromagnetic ground state. In the presence of a small interlayer tunneling matrix element (always present in real samples) the Goldstone mode actually acquires a tiny gap at zero wavevector q. This gap allows the mode to transfer charge between the two layers at $q = 0$ and thereby influence the conductance. In effect, the collective mode ensures that may electrons tunnel simultaneously. Indeed, a major challenge for theory is to understand why the tunneling conductance at zero bias is not *infinite* and the bilayer semiconductor device actually a Josephson junction [46, 62, 48].

The expected linear dispersion of the Goldstone mode has in fact been observed in experiment. By applying a small in-plane magnetic field component $B_{||}$, interlayer tunneling characteristics become sensitive to spectral features at the non-zero wavevector $q = eB_{||}d/\hbar$. If there is a collective mode which disperses linearly with wavevector, $\omega = cq$, then some feature should be detected in the tunnel spectrum at voltage $eV = \hbar cq = ceB_{||}d$. Spielman, *et al.* [49] observed weak satellites (see Fig. 12) in the tunneling conductance spectrum and from them determined the collective mode velocity to be about $c \approx 14$ km/sec, in good agreement with theory. The weakness of the features, and the persistence of a substantial zero bias peak out to $B_{||} \sim 0.6$ T are puzzling facts and suggest that disorder plays an important role in limiting the pseudospin coherence length in the bilayer system.

5.4. Excitonic superfluidity at $\nu_T = 1$

Perhaps the most dramatic prediction about the coherent $\nu_T = 1$ bilayer QHE phase is that of counter-flow superfluidity [50, 51]. While this new dissipationless mode of transport may be understood completely within the ferromagnetism picture (for the supercurrents then correspond to dissipationless pseudo-spin currents), a more intuitive understanding can be achieved via the exciton condensation point of view [52]. In second-quantized notation the ground state given in Eq. 5.6 is written

$$\Psi = \prod_k \frac{1}{\sqrt{2}}(c_{k,T}^{\dagger} + e^{i\phi}c_{k,B}^{\dagger})|0\rangle \tag{5.7}$$

where the product is over all single-particle momentum states k in the lowest Landau level and $c_{k,T}^{\dagger}$ and $c_{k,B}^{\dagger}$ create electrons with momentum k in the top and bottom layers, respectively. The ket $|0\rangle$ is the vacuum state, here representing empty conduction band Landau levels in each layer. If, however, one starts with

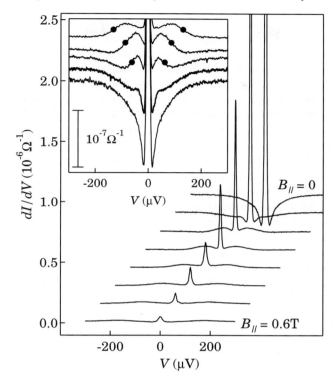

Fig. 12. Tunnel spectra at $\nu_T = 1$ and $T = 25$ mK in the strongly coupled phase at $d/\ell = 1.61$, as a function of in-plane magnetic field. Inset shows expanded views of low energy region at $B_\| = 0.07$, 0.11, 0.15, 0.24, and 0.35T (bottom to top). Dots indication positions of satellite resonances. After Ref. [49].

a vacuum state $|0'\rangle$ consisting of no electrons in the bottom layer and a single filled Landau level in the top layer then we can re-write Eq. 5.7 as

$$\Psi = \prod_k \frac{1}{\sqrt{2}}(1 + e^{i\phi}c^\dagger_{k,B}c_{k,T})|0'\rangle. \qquad (5.8)$$

In this equation the operator $c^\dagger_{k,B}c_{k,T}$ creates an electron-hole pair, with the electron in the bottom layer and the hole in the top layer. The wavefunction is very similar to the standard BCS wavefunction describing conventional superconductors, only the "Cooper pairs" in the present case are charge neutral excitons. The "coherence factors" u_k and v_k essential to BCS theory are here equal to $1/\sqrt{2}$, independent of k. In an ordinary superconductor pairing occurs only in the vicinity of the Fermi surface; in our case the non-interacting system consists of dispersionless Landau levels and thus all k states are treated on the same footing.

It is apparent from Eq. 5.8 that the ground state of a closely-spaced bilayer 2D electron system may be viewed as a BCS-like condensate of interlayer excitons. As in a superconductor, the language of Bose condensation may also be used, bearing in mind that the heavily overlapping fermionic pairs (here an electron and hole) are far from being point-like bosons [53].

Exciton condensation in semiconductors has been part of the literature for 40 years, the first concrete suggestions appearing shortly after the advent of BCS theory [54]. Since that time herculean efforts have been made in order to find such a novel state of matter, almost entirely using optically-generated electron-hole pairs. Recent research has concentrated on optically-generated indirect excitons in coupled quantum wells, and several interesting observations have been made [55, 56]. Nevertheless, clear signatures of excitonic Bose condensation in such systems have remained elusive.

In our case the valence band is, in effect, replaced by a single filled Landau level in one of the layers and the conduction band is the empty lowest Landau level in the other layer. Optical excitation is not needed since these two bands are degenerate to begin with. The excitons here do not optically recombine, thus eliminating the most serious obstacle to establishing a cold gas of excitons in equilibrium.

The collective mode spectrum studied experimentally by tunneling is rooted in the stiffness of the system against spatial variations in the phase ϕ. This stiffness is due to the exchange energy cost of having nearby pseudospins not exactly parallel. While these collective modes transport no *net* electrical charge through the system, they do involve the motion of excitons. A uniform phase gradient, $\nabla\phi = $ const., corresponds to a uniform flow of excitons with velocity $v_{ex} = \rho_s \nabla\phi/(\hbar N_T)$, where $\rho_s \sim 0.5K$ is the pseudospin stiffness [43]. In an ideal sample at temperatures below the Kosterlitz-Thouless transition, the lack of unpaired vortices makes it very difficult to relax such currents and superfluidity should result [57].

5.5. *Detecting excitonic superfluidity*

Excitonic currents, whether superfluid or not, present an interesting experimental challenge. Lacking charge, how does one excite and detect such currents? The answer to this is simple in principle: A uniform flow of excitons in a bilayer systems implies the existence of *counterflowing* electrical currents in the individual layers. Thus, if equal but opposite currents are driven through the bilayer at $v_T = 1$, and the layer separation is sufficiently small, these currents will be carried by excitons in the condensate rather than by independent charged excitations

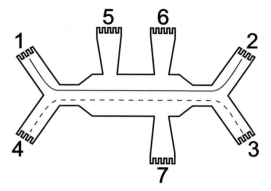

Fig. 13. Schematic diagram of sample geometry for counterflow transport measurements. Solid and dashed lines indicate current flow in different layers. After Ref. [58].

in the individual layers. If this is the case, such an experiment ought to reveal clear signs of the expected superfluidity.

The ability to make separate electrical connections to the individual layers has allowed us to make exactly this measurement. Fig. 13 depicts the experimental arrangement. Current is driven into the sample at contact 1 and is restricted to just one of the layers using appropriate gates (not shown). This current, indicated by the solid line, flows through the body of the device and exits via contact 2. The current is then sent back through the device in the other 2D layer, using contacts 3 and 4, as indicated by the dashed line [59]. Obviously, this current can be made to run in the same direction as that in the first layer or, for counterflow, in the opposite direction. Contacts 5,6 and 7 allow for measuring the longitudinal and Hall voltages that develop under counterflow or parallel flow conditions. Most importantly, these voltage probes are connected to one layer or the other, *not both*.

The most important results of these experiments are shown in Fig. 14. The solid and dotted lines are the Hall (R_{xy}) and longitudinal (R_{xx}) resistances (voltages divided by current in one layer), respectively, in the counterflow configuration. Over most of the magnetic field range these resistances reveal the ordinary signatures of the quantum Hall effect: plateaus in R_{xy} and deep minima or zeroes in R_{xx}. At these magnetic fields the two layers are effectively independent and the observed resistances reflect the familiar quantum Hall physics of the layer to which the voltage probes are connected; the current flowing in the adjacent layer having no effect.

For the sample used in Fig. 14 $\nu_T = 1$ occurs at $B \approx 2.0$ T. Given that the center-to-center quantum well separation in the sample is 28 nm, the effective

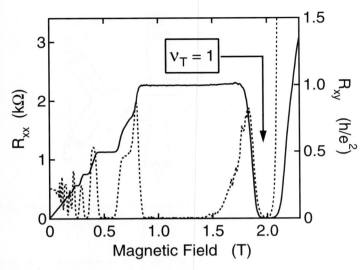

Fig. 14. Vanishing Hall (solid) and longitudinal (dotted) resistances at $\nu_T = 1$ at $T = 30$ mK and $d/\ell = 1.54$ in a bilayer 2D electron system.

layer separation at this magnetic field is $d/\ell = 1.54$. At this d/ℓ the bilayer 2DES is well within the excitonic (or pseudo-ferromagnetic) phase and the two layers are strongly coupled. Counterflowing electrical currents in the system ought to be transported by excitons. The data in Fig. 14 strongly support this: The Hall resistance R_{xy} measured in one layer (it does not matter which one) has fallen to nearly zero. At the same time, the longitudinal resistance R_{xx} is also essentially zero. This is a remarkable result! A semiconducting system containing a low density of electrons displays no Hall effect in spite of the presence of a large magnetic field. Exciton transport naturally explains this; the net Lorentz force on these neutral particles is zero, the force on the electron exactly cancelling that on the hole.

The vanishing of all components of the resistivity tensor measured in counterflow strongly suggests a type of superfluidity, or counter-current superconductivity, is present. To study this more carefully, the full temperature dependence of both R_{xy} and R_{xx} were measured. From these the longitudinal conductivity σ_{xx} could be determined. Figure 15 displays these results, for both counterflow and conventional parallel flow through the two layers [58].

For parallel flow, Fig. 15 reveals the expected behavior. R_{xy} is essentially constant and equal to $2h/e^2$ [60]. R_{xx} vanishes as $T \to 0$ owing to the presence of an energy gap for charged excitations. Parallel currents, of course, cannot be carried

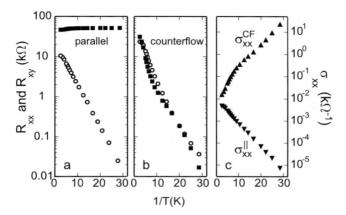

Fig. 15. Temperature dependence of longitudinal and Hall resistances in parallel and counterflow at $\nu_T = 1$. Panel c shows the deduced conductivities $\sigma_{xx}^{\parallel}$ and σ_{xx}^{CF}. After Ref. [58].

by the excitonic condensate, but must instead be transported (in edge channels) by these excitations. Since R_{xy} remains constant, the longitudinal conductivity $\sigma_{xx}^{\parallel}$ vanishes in the low temperature limit. This is the ordinary QHE result.

In counterflow *both* R_{xx} and R_{xy} appear to vanish at low temperature. The computed conductivity σ_{xx}^{CF} *rises* steadily as the temperature falls. Indeed, at the lowest temperatures the counterflow conductivity has significantly exceeded the very high conductivity the 2D electron systems in this sample exhibit at zero magnetic field!

The vanishing Hall resistance in counterflow is compelling evidence that charge neutral excitons transport the current. This is a key observation. The fact that the conductivity of these excitons appears to diverge as $T \to 0$, rather than saturating as the conductivity of a normal metal would, is a strong indicator of collective behavior of the exciton gas. On the other hand, the conductivity in counterflow is neither infinite nor non-linear in the current. This is not the expected Kosterlitz-Thouless behavior. The origin of the excess dissipation in the counterflow channel is a major outstanding problem for the theory of the excitonic phase. Conventional wisdom suggests that there must be unpaired vortices in the system or, equivalently, that the effective Kosterlitz-Thouless transition temperature is zero or at least very small. The same conventional wisdom places the blame for this slight blemish in an otherwise textbook demonstration of excitonic superfluidity on the inevitable disorder in the sample. Recent theoretical work has found strong numerical evidence for such unpaired vortices, and there are even predictions of intriguing "Bose glass" phases in which true superfluidity only exists at $T = 0$ [61, 62, 63].

6. Conclusions

These notes have highlighted but a few of the many interesting aspects of double layer 2D electron systems. It is my feeling that the field is as yet young and many challenges lie ahead. On the immediate theoretical side it should be now clear to the reader that while a good qualitative picture of the excitonic phase at $\nu_T = 1$ exists, the quantitative situation is poor. Excess dissipation in the counterflow channel, the persistence of the Josephson-like tunneling peak to significant in-plane magnetic field strengths, are only the most obvious outstanding questions. Beyond $\nu_T = 1$, there are several virtually unexplored directions to pursue. At $\nu_T = 2$ an intricate array of collective phases which mix the true spin and pseudospin degrees of freedom are expected. Very few experiments have addressed this system. At $\nu_T = 1/3$ there is the likelihood that two distinct QHE phases exist, one possessing interlayer phase coherence and the other not, in addition to a compressible phase at large layer separation. At $\nu_T = 1/2$ Coulomb drag experiments might be able to cleanly distinguish between the Halperin state given in Eq. 2.14 and other possible ground state wavefunctions. And the list goes on. Hopefully some of the students who attended the 2004 Les Houches summer school, or otherwise came across these notes, will take up the challenge.

Acknowledgements

It is a pleasure to acknowledge the essential help of my students Ian Spielman and Melinda Kellogg, who performed a number of the experiments discussed here. As always, my collaborators Loren Pfeiffer and Ken West, who designed and grew the many heterostructure samples needed for this research, cannot be thanked enough. I am also indebted to the National Science Foundation, the Department of Energy, and Caltech, for their generous support of this work.

References

[1] K. von Klitzing, G. Dorda, and M. Pepper, Phys. Rev. Lett. **45**, 494 (1980).

[2] D.C. Tsui, H.L. Stormer, and A.C. Gossard, Phys. Rev. Lett. **48**, 1559 (1982).

[3] J.P. Eisenstein and H.L. Stormer, Science **248**, 1510 (1990).

[4] A. Schmeller, J.P. Eisenstein, L.N. Pfeiffer, and K.W. West, Phys. Rev. Lett. **75**, 4290 (1995).

[5] *Perspectives in Quantum Hall Effects*, edited by A. Pinczuk and S. Das Sarma, (John Wiley, New York, 1997).

[6] R.L. Willett, R.R. Ruel, K.W. West, and L.N. Pfeiffer, Phys. Rev. Lett. **71**, 3846 (1993).

[7] W. Kang, H.L. Stormer, L.N. Pfeiffer, K.W. Baldwin, and K.W. West, Phys. Rev. Lett. **71**, 3850 (1993).

[8] J.P. Eisenstein, G.S. Boebinger, L.N. Pfeiffer, K.W. West, and Song He, Phys. Rev. Lett. **68**, 1383 (1992).

[9] Within the $N = 0$ lowest Landau level no fractional QHE states at $R_H = (h/e^2)/(p/q)$ with q even have ever been observed. In the $N = 1$ Landau level fractional QHE states at $\nu = 5/2$ and $7/2$ are observed. At the time of this writing the prevailing opinion seems to be that these are examples of the exotic Moore-Read state which possesses non-abelian quasi-particles.

[10] B.I. Halperin, Helv. Phys. Acta, **56**, 75 (1983).

[11] M.B. Pogrebinskii, Fiz. Tekh. Poluprovod. **11**, 637 (1977) [Sov. Phys. Semicond. **11**, 372 (1977)].

[12] T.J. Gramila, J.P. Eisenstein, A.H. MacDonald, L.N. Pfeiffer, and K.W. West, Phys. Rev. Lett. **66**, 1216 (1991).

[13] U. Sivan, P.M. Solomon, and H. Shtrikman, Phys. Rev. Lett. **68**, 1196 (1992).

[14] J.P. Eisenstein, L.N. Pfeiffer and K.W. West, Appl. Phys. Lett. **57**, 2324 (1990).

[15] M.V. Weckwerth, *et al.*, Superlattices and Microstructures **20**, 561 (1996).

[16] A.P. Jauho and H. Smith, Phys. Rev. B **47**, 4420 (1993).

[17] C. Hodges, H. Smith and J.W. Wilkins, Phys. Rev. B **4**, 302 (1971).

[18] T.J. Gramila, J.P. Eisenstein, A.H. MacDonald, L.N. Pfeiffer and K.W. West, Phys. Rev. B **47**, 12957 (1993).

[19] M.C. Bonsager, K. Flensberg, B.Y.-K. Hu, and A.H. MacDonald, Phys. Rev. B **57**, 7085 (1998). For an earlier approach see H. Tso, P. Vasilopolous, and F.M. Peeters, Phys. Rev. Lett. **68**, 2516 (1992).

[20] M. Kellogg, J.P. Eisenstein, L.N. Pfeiffer and K.W. West, Solid State Commun. **123**, 515 (2002).

[21] In evaluating Eq. 3.5 we set $d = 28$ nm, the center-to-center separation between the quantum wells.

[22] A. Yurtserver, V. Moldoveanu, and B. Tanatar, Solid State Commun. **125**, 575 (2003).

[23] S. Das Sarma, private communication.

[24] We are assuming inelastic tunneling processes (involving a phonon, for example) do not occur and that the effective mass of electrons in the two wells are the same.

[25] Unlike Eq. 4.2 the density of states ρ here enters only linearly. This is because of our assumption of complete momentum conservation.

[26] Strictly speaking, of course, this is only true in the limit of vanishing $|t|$. In reality, the energy levels are always hybridized. As it turns out, this is not important for us here since the level broadening due to scattering always greatly exceeds the hybridization gap.

[27] S.Q. Murphy, J.P. Eisenstein, L.N. Pfeiffer and K.W. West, Phys. Rev. B **52**, 14825 (1995).

[28] G.F. Giuliani and J.J. Quinn, Phys. Rev. B **26**, 4421 (1982).

[29] T. Jungwirth and A.H. MacDonald, Phys. Rev. B **53**, 7403 (1996).

[30] L. Zheng and S. Das Sarma, Phys. Rev. B **53**, 9964 (1996).

[31] J.P. Eisenstein, L.N. Pfeiffer, and K.W. West, Phys. Rev. Lett. **69**, 3804 (1992).

[32] Y. Hatsugai, P.-A. Bares, and X.G. Wen, Phys. Rev. Lett. **71**, 424 (1993).

[33] Song He, P.M. Platzman, and B.I. Halperin, Phys. Rev. Lett. **71**, 777 (1993).

[34] Peter Johannson and Jari M. Kinaret, Phys. Rev. Lett. **71**, 1435 (1993) and Phys. Rev. B **50**, 4651 (1994).

[35] A.L. Efros and F.G. Pikus, Phys. Rev. B **48**, 14694 (1993).

[36] J.P. Eisenstein, L.N. Pfeiffer, and K.W. West, Phys. Rev. Lett. **74**, 1419 (1995).

[37] We shall assume throughout that the spin Zeeman energy is so large that all electron spins are polarized by the magnetic field. This is an approximation; see cond-mat/0410092.

[38] G. S. Boebinger, H. W. Jiang, L. N. Pfeiffer, and K. W. West, Phys. Rev. Lett. **64**, 1793 (1990).

[39] S. Q. Murphy, J. P. Eisenstein, G. S. Boebinger, L. N. Pfeiffer, and K. W. West, Phys. Rev. Lett. **72**, 728 (1994).

[40] A.H. MacDonald, P.M. Platzman, and G.S. Boebinger, Phys. Rev. Lett. **65**, 775 (1990).

[41] This is the case for balanced bilayers. If the system is imbalanced, then the pseudospin must lie near the surface of a cone with the mean value of T_z equal to $(n_1 - n_2)/2$.

[42] H.A. Fertig, Phys. Rev. B **40**, 1087 (1989).

[43] Kun Yang, *et al.*, Phys. Rev. Lett. **72**, 732 (1994).

[44] I.B. Spielman, J.P. Eisenstein, L.N. Pfeiffer, and K.W. West, Phys. Rev. Lett. **84**, 5808 (2000).

[45] At $B = 0$ the voltage location of tunnel resonances is proportional to the difference in density (i.e. Fermi energy) of the two layers. See sections 4.1 and 4.2. At $\nu_T = 1$ small antisymmetric density changes do not shift the peak location, but they do affect its height. See I.B. Spielman, M. Kellogg, J.P. Eisenstein, L.N. Pfeiffer, and K.W. West, Phys. Rev. B **70**, 081303 (2004).

[46] L. Balents and L. Radzihovsky, Phys. Rev. Lett. **86**, 1825 (2001).

[47] A. Stern, S.M. Girvin, A.H. MacDonald, and N. Ma, Phys. Rev. Lett. **86**, 1829 (2001).

[48] M. Fogler and F. Wilczek, Phys. Rev. Lett. **86**, 1833 (2001).

[49] I.B. Spielman, J.P. Eisenstein, L.N. Pfeiffer and K.W. West, Phys. Rev. Lett. **87**, 036803 (2001).

[50] X.G. Wen and A.Zee, Phys. Rev. Lett. **69**, 1811 (1992).

[51] A.H. MacDonald, Physica (Amsterdam) **298B**, 129 (2001).

[52] For an accessible account, see J.P. Eisenstein and A.H. MacDonald, Nature **432**, 691 (2004), and the references cited therein.

[53] In the present case the "size" of an electron-hole pair is about the same as the distance between pairs. This is much more "point-like" than the usual Cooper pairs (in, say, aluminum) where the pair size can be orders of magnitude larger than the inter-electron spacing.

[54] L.V. Keldysh and Y.V. Kopaev, Fiz. Tverd. Tela **6**, 2791 (1964). See also Yu.E. Lozovik and V.I. Yudson, Sov. Phys. JETP **44**, 389 (1976) and S.I. Shevchenko, Sov. J. Low-Temp. Phys. **2**, 251 (1976) for early theoretical work on layered exciton condensates.

[55] D.Snoke, Science **298**, 1368 (2001).

[56] L. Butov, Solid State Commun. **127**, 89 (2003).

[57] In 2D this is not strictly true. A finite current creates a Magnus force which can lead to vortex pair unbinding. The more correct statement is that ideal excitonic currents exhibit non-linear current-drive characteristics and are dissipationless only in the limit of zero current.

[58] M. Kellogg, J.P. Eisenstein, L.N. Pfeiffer, and K.W. West.

[59] The current is measured before entering the first layer and just before redirection into the second. This allows for determination of how much current has "leaked" from one layer to the other. Typically this is less than 1% of the total.

[60] Here we have defined the resistances as the measured voltages divided by the current flowing in one layer. For parallel flow this is one-half the total current. Hence R_{xy} is quantized at $2h/e^2$ rather than h/e^2.

[61] H.A. Fertig and J.P. Straley, Phys. Rev. Lett. **91**, 046806 (2003).

[62] A. Stern, S. Das Sarma, M.P.A. Fisher, and S.M. Girvin, Phys. Rev. Lett. **84**, 139 (2000).

[63] D.N. Sheng, L. Balents, and Ziqiang Wang, Phys. Rev. Lett. **91**, 116802 (2003).

Course 3

MANY–BODY THEORY OF NON–EQUILIBRIUM SYSTEMS

Alex Kamenev

Department of Physics, University of Minnesota,
Minneapolis, MN 55455, USA

H. Bouchiat, Y. Gefen, S. Guéron, G. Montambaux and J. Dalibard, eds.
Les Houches, Session LXXXI, 2004
Nanophysics: Coherence and Transport

Contents

1. Introduction

1.1. Motivation and outline

These lectures are devoted to the Keldysh formalism for the treatment of out–of–equilibrium many–body systems. The name of the technique takes its origin from the 1964 paper of L. V. Keldysh [1]. Among the earlier approaches that are closely related to the Keldysh technique, one should mention J. Schwinger [2] and R. P. Feynman and F. L. Vernon [3]. The classical counterparts of the Keldysh technique are extremely useful and interesting on their own. Among them the Wyld diagrammatic technique [4] and the Martin–Siggia–Rose method [5] for stochastic systems.

There are a number of pedagogical presentations of the method [6–8]. The emphasis of these notes is on the functional integration approach. It makes the structure of the theory clearer and more transparent. The notes also cover modern applications such as the Usadel equation and the nonlinear σ–model. Great attention is devoted to exposing connections to other techniques such as the equilibrium Matsubara method and the classical Langevin and Fokker-Planck equations, as well as the Martin–Siggia–Rose technique.

The Keldysh formulation of the many–body theory is useful for the following tasks:
- Treatment of systems that are not in thermal equilibrium (either due to the presence of external fields, or in a transient regime) [1, 6, 8].
- Calculation of the full counting statistics of a quantum mechanical observable (as opposed to an average value or correlators) [9, 10].
- As an alternative to the replica and the supersymmetry methods in the theory of disordered and glassy systems [11–15].
- Treatment of equilibrium problems, in which the Matsubara analytical continuation may prove to be cumbersome.

The outline of these lectures is as follows. The technique is introduced and explained for the simplest possible system, that of a single bosonic state (harmonic oscillator), which is later generalized to real (phonons), or complex (atoms) bosonic fields. Their action and Green functions are introduced in Chapter 2. Boson interactions, the diagrammatic technique and the quantum kinetic equation are treated in Chapter 3. Chapter 4 is devoted to a bosonic particle in contact with a dissipative environment (bath). This example is used to establish connections with the classical methods (Langevin and Fokker–Planck) as well as with the quantum equilibrium technique (Matsubara). Fermions and fermion–boson systems are treated in Chapter 5. Covered topics include the random phase approximation and the quantum kinetic equation. Non–interacting fermions in the presence of quenched disorder are treated in Chapter 6 with the help of the Keldysh non-linear σ-model.

1.2. Closed time contour

The standard construction of the zero temperature (or equilibrium) many–body theory (see e.g. [7, 16]) involves the adiabatic switching "on" of interactions at a distant past, and "off" at a distant future. A typical correlation function has the form of a time ordered product of operators in the Heisenberg representation: $C(t, t') \equiv \langle 0|T\hat{A}(t)\hat{B}(t')|0\rangle$, where $|0\rangle$ is the ground-state (or thermal equilibrium state) of the *interacting* Hamiltonian, $\hat{H}$. This state is supposed to be given by $|0\rangle = \hat{S}(0, -\infty)|\rangle_0$, where $|\rangle_0$ is the (known) ground-state of the *non–interacting* Hamiltonian, $\hat{H}_0$, at $t = -\infty$. The $\hat{S}$–matrix operator $\hat{S}(t, t') = e^{i\hat{H}_0 t}e^{-i\hat{H}(t-t')}e^{-i\hat{H}_0 t'}$ describes the evolution due to the interaction Hamiltonian, $\hat{H} - \hat{H}_0$, and is thus responsible for the adiabatic switching "on" of the interactions. An operator in the Heisenberg representation is given by $\hat{A}(t) = [\hat{S}(t, 0)]^\dagger \hat{A}(t)\hat{S}(t, 0) = \hat{S}(0, t)\hat{A}(t)\hat{S}(t, 0)$, where $\hat{A}(t)$ is the operator in the interaction representation. As a result, the correlation function takes the form:

$$
\begin{aligned}
C(t, t') &= {}_0\langle| T\hat{S}(-\infty, 0)\hat{S}(0, t)\hat{A}(t)\hat{S}(t, t')\hat{B}(t')\hat{S}(t', 0)\hat{S}(0, -\infty)|\rangle_0 \\
&= \frac{{}_0\langle| T\hat{A}(t)\hat{B}(t')\hat{S}(\infty, -\infty)|\rangle_0}{{}_0\langle| \hat{S}(\infty, -\infty)|\rangle_0},
\end{aligned} \tag{1.1}
$$

where one employed: ${}_0\langle| \hat{S}(-\infty, 0) = e^{-iL} {}_0\langle| \hat{S}(\infty, -\infty)\hat{S}(-\infty, 0)$, and interchanged the order of operators, which is always allowed under the T–operation (time ordering). The idea is that, starting at $t = -\infty$ at the ground (or equilibrium) state, $|\rangle_0$, of the non–interacting system and then adiabatically switching interactions "on" and "off", one arrives at $t = +\infty$ at the state $|\infty\rangle$. The crucial *assumption* is that this state is unique, independent of the details of the switching procedure and is again the ground–state, up to a possible phase factor: $e^{iL} = {}_0\langle| |\infty\rangle = {}_0\langle| \hat{S}(\infty, -\infty)|\rangle_0$.

Clearly this is *not* the case out of equilibrium. Starting from some arbitrary non–equilibrium state and then switching interactions "on" and "off", the system

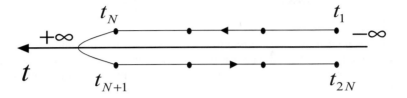

Fig. 1. The closed time contour $\mathcal{C}$. Dots on the forward and the backward branches of the contour denote discrete time points.

evolves to some unpredictable state. The latter depends, in general, on the peculiarities of the switching procedure. The entire construction sketched above fails since we have no knowledge of the final state.

One would like, thus, to build a theory that avoids references to the state at $t = +\infty$. Since traces are calculated, one still needs to know the final state. Schwinger's suggestion is to take the final state to be exactly the same as the initial one. The central idea is to let the quantum system evolve first in the forward direction in time and then to "unwind" its evolution backwards, playing the "movie" in the backward direction. One ends up, thus, with the need to construct a theory with the time evolution along the two–branch contour, $\mathcal{C}$, depicted on Fig. 1. Then, no matter what the state at $t = +\infty$ is, after the backward evolution the system returns back to the known initial state. As a result, the unitary evolution operator, $\hat{U}_{t,t'} \equiv e^{-iH(t-t')}$, along such a closed time contour is always a unit operator:

$$\hat{U}_\mathcal{C} \equiv 1. \tag{1.2}$$

In this construction there is no switching of interactions in the future. Both switchings "on" and "off" take place in the past: "on" – at the forward branch of the contour and "off" – at the backward one. This way the absence of information about the $t = +\infty$ state is bypassed. There is a price to pay for such luxury: a doubling of degrees of freedom. Indeed at every moment of time one needs to specify a field residing on the forward branch as well as on the backward branch of the contour. As a result, the algebraic structure of the theory is more complicated. The difficulties may be minimized, however, by a proper choice of variables based on the internal symmetries of the theory.

2. Free boson systems

2.1. Partition function

Let us consider the simplest possible many–body system: bosonic particles occupying a single quantum state with an energy ω_0. It is completely equivalent, of course, to a harmonic oscillator. The secondary quantized Hamiltonian has the form:

$$\hat{H} = \omega_0\, a^\dagger a, \tag{2.1}$$

where $a^\dagger$ and a are bosonic creation and annihilation operators with the commutation relation $[a, a^\dagger] = 1$. Let us define the "partition function" as:

$$Z = \frac{\text{Tr}\{\hat{U}_{\mathcal{C}}\hat{\rho}\}}{\text{Tr}\{\hat{\rho}\}}. \tag{2.2}$$

If one assumes that all external fields are exactly the same on the forward and backward branches of the contour, then $\hat{U}_{\mathcal{C}} = 1$ and therefore $Z = 1$.

The initial density matrix $\hat{\rho} = \hat{\rho}(\hat{H})$ is some operator–valued function of the Hamiltonian. To simplify the derivations one may choose it to be the equilibrium density matrix, $\hat{\rho}_0 = \exp\{-\beta(\hat{H}-\mu\hat{N})\} = \exp\{-\beta(\omega_0-\mu)a^\dagger a\}$. Since arbitrary external perturbations may be switched on (and off) at a later time, the choice of the equilibrium initial density matrix does not prevent one from treating non–equilibrium dynamics. For the equilibrium initial density matrix:

$$\text{Tr}\{\hat{\rho}_0\} = \sum_{n=0}^{\infty} e^{-\beta(\omega_0-\mu)n} = [1 - \rho(\omega_0)]^{-1}, \tag{2.3}$$

where $\rho(\omega_0) = e^{-\beta(\omega_0-\mu)}$. An important point is that, in general, $\text{Tr}\{\hat{\rho}\}$ is an interaction– and disorder–*independent* constant. Indeed, both interactions and disorder are supposed to be switched on (and off) on the forward (backward)

Reminder: The bosonic coherent state $|\phi\rangle$ ($\langle\phi|$), parametrized by a complex number ϕ, is defined as a right (left) eigenstate of the annihilation (creation) operator:
$a|\phi\rangle = \phi|\phi\rangle$ ($\langle\phi|a^\dagger = \langle\phi|\bar{\phi}$).
The matrix elements of a *normally ordered* operator, such as the Hamiltonian, take the form $\langle\phi|\hat{H}(a^\dagger, a)|\phi'\rangle = H(\bar{\phi}, \phi')\langle\phi|\phi'\rangle$.
The overlap between two coherent states is [17] $\langle\phi|\phi'\rangle = \exp\{\bar{\phi}\phi'\}$.
Since the coherent state basis is overcomplete, the trace of an operator, $\hat{A}$, is calculated with the weight: $\text{Tr}\{\hat{A}\} = \pi^{-1}\iint d(\Re\phi)\,d(\Im\phi)\,e^{-|\phi|^2}\,\langle\phi|\hat{A}|\phi\rangle$.

parts of the contour some time after (before) $t = -\infty$. This constant is, therefore, frequently omitted – it never causes a confusion.

The next step is to divide the C contour into $(2N - 2)$ time steps of length δ_t, such that $t_1 = t_{2N} = -\infty$ and $t_N = t_{N+1} = +\infty$ as shown in Fig. 1. One then inserts the resolution of unity in the coherent state overcomplete basis [17]

$$1 = \iint \frac{d(\Re \phi_j) \, d(\Im \phi_j)}{\pi} \, e^{-|\phi_j|^2} \, |\phi_j\rangle\langle\phi_j| \qquad (2.4)$$

at each point $j = 1, 2, \ldots, 2N$ along the contour. For example, for $N = 3$ one obtains the following sequence in the expression for $\mathrm{Tr}\{\hat{U}_C \hat{\rho}_0\}$ (read from right to left):

$$\langle\phi_6|\hat{U}_{-\delta_t}|\phi_5\rangle\langle\phi_5|\hat{U}_{-\delta_t}|\phi_4\rangle\langle\phi_4|\hat{1}|\phi_3\rangle\langle\phi_3|\hat{U}_{+\delta_t}|\phi_2\rangle\langle\phi_2|\hat{U}_{+\delta_t}|\phi_1\rangle\langle\phi_1|\hat{\rho}_0|\phi_6\rangle, \quad (2.5)$$

where $\hat{U}_{\pm\delta_t}$ is the evolution operator during the time interval δ_t in the positive (negative) time direction. Its matrix elements are given by:

$$\langle\phi_{j+1}|\hat{U}_{\pm\delta_t}|\phi_j\rangle \equiv \langle\phi_{j+1}|e^{\mp i\hat{H}(a^\dagger, a)\delta_t}|\phi_j\rangle \approx \langle\phi_{j+1}|\phi_j\rangle \, e^{\mp iH(\bar{\phi}_{j+1}, \phi_j)\delta_t}, \qquad (2.6)$$

where the last approximate equality is valid up to the linear order in δ_t. Obviously this result is not restricted to the toy example, Eq (2.1), but holds for any *normally–ordered* Hamiltonian. Notice that there is no evolution operator inserted between t_N and t_{N+1}. Indeed, these two points are physically indistinguishable and thus the system does not evolve during this time interval.

Exercise: Show that $\langle\phi|e^{\kappa a^\dagger a}|\phi'\rangle = \exp\left\{\bar{\phi}\phi' e^\kappa\right\}$.
Putting $\kappa = -\beta(\omega_0 - \mu)$, one finds $\langle\phi_1|\hat{\rho}_0|\phi_{2N}\rangle = \exp\left\{\bar{\phi}_1\phi_{2N}\rho(\omega_0)\right\}$.

Combining all such matrix elements along the contour together with the exponential factors from the resolutions of unity, Eq. (2.4), one finds for the partition function (2.2):

$$Z = \frac{1}{\mathrm{Tr}\{\rho_0\}} \iint \prod_{j=1}^{2N} \left[\frac{d(\Re \phi_j) \, d(\Im \phi_j)}{\pi}\right] e^{i \sum_{j,j'=1}^{2N} \bar{\phi}_j G_{jj'}^{-1} \phi_{j'}}, \qquad (2.7)$$

where the $2N \times 2N$ matrix $i G_{jj'}^{-1}$ stands for:

$$
i\,G_{jj'}^{-1} \equiv
\left[
\begin{array}{cccc|cccc}
-1 & & & & & & & \rho(\omega_0) \\
1-h & -1 & & & & & & \\
 & 1-h & -1 & & & & & \\
\hline
 & & 1 & -1 & & & \\
 & & & 1+h & -1 & & \\
 & & & & 1+h & -1 \\
\end{array}
\right],
\tag{2.8}
$$

and $h \equiv i\omega_0\delta_t$. It is straightforward to evaluate the determinant of such a matrix

$$
\det\!\big[i G^{-1}\big] = 1 - \rho(\omega_0)(1-h^2)^{N-1} \approx 1 - \rho(\omega_0)\,e^{(\omega_0\delta_t)^2(N-1)} \;\to\; 1 - \rho(\omega_0),
\tag{2.9}
$$

where one has used that $\delta_t^2 N \to 0$ if $N \to \infty$ (indeed, the assumption was $\delta_t N \to$ const). Employing Eqs. (A.1) and (2.3), one finds:

$$
Z = \frac{\det^{-1}\!\big[i G^{-1}\big]}{\mathrm{Tr}\{\rho_0\}} = 1,
\tag{2.10}
$$

as it should be, of course. Notice that keeping the upper–right element of the discrete matrix, Eq. (2.8), is crucial to maintain this normalization identity.

One may now take the limit $N \to \infty$ and formally write the partition function in the continuous notations, $\phi_j \to \phi(t)$:

$$
Z = \int \mathcal{D}\bar{\phi}\phi \; e^{\,i S[\bar{\phi},\phi]} = \int \mathcal{D}\bar{\phi}\phi \, \exp\left\{ i \int_C \big[\bar{\phi}(t)\,G^{-1}\phi(t)\big]\,dt \right\},
\tag{2.11}
$$

where according to Eqs. (2.7) and (2.8) the action is given by

$$
S[\bar{\phi},\phi] = \sum_{j=2}^{2N}\left[i\bar{\phi}_j\,\frac{\phi_j - \phi_{j-1}}{\delta t_j} - \omega_0\bar{\phi}_j\phi_{j-1}\right]\delta t_j + i\,\bar{\phi}_1\Big(\phi_1 - \rho(\omega_0)\phi_{2N}\Big),
\tag{2.12}
$$

where $\delta t_j \equiv t_j - t_{j-1} = \pm\delta_t$. Thus a continuous form of the operator G^{-1} is:

$$
G^{-1} = i\partial_t - \omega_0.
\tag{2.13}
$$

It is important to remember that this continuous notation is only an abbreviation that represents the large discrete matrix, Eq. (2.8). In particular, the upper–right element of the matrix (the last term in Eq. (2.12)), that contains the information about the distribution function, is seemingly absent in the continuous notations.

To avoid integration along the closed time contour, it is convenient to split the bosonic field $\phi(t)$ into the two components $\phi_+(t)$ and $\phi_-(t)$ that reside on the forward and the backward parts of the time contour correspondingly. The continuous action may then be rewritten as

$$S = \int\limits_{-\infty}^{\infty} dt \, \bar{\phi}_+(t)[i\partial_t - \omega_0]\phi_+(t) - \int\limits_{-\infty}^{\infty} dt \, \bar{\phi}_-(t)[i\partial_t - \omega_0]\phi_-(t), \qquad (2.14)$$

where the relative minus sign comes from the reverse direction of the time integration on the backward part of the contour. Once again, the continuous notations are somewhat misleading. Indeed, they create an undue impression that the $\phi_+(t)$ and $\phi_-(t)$ fields are completely independent from each other. In fact, they are connected due to the presence of the non–zero off–diagonal blocks in the discrete matrix, Eq. (2.8).

2.2. Green functions

One would like to define the Green functions as:

$$G(t, t') = -i\int \mathcal{D}\bar{\phi}\phi \, e^{i S[\bar{\phi}, \phi]} \, \phi(t)\bar{\phi}(t') \equiv -i \langle \phi(t)\bar{\phi}(t') \rangle, \qquad (2.15)$$

where both time arguments reside somewhere on the Keldysh contour. Notice the absence of the factor Z^{-1} in comparison with the analogous definition in the equilibrium theory [17]. Indeed in the present construction $Z = 1$. This seemingly minor difference turns out to be the major issue in the theory of disordered systems, Chapter 6.

According to the general property of Gaussian integrals (see Appendix A), the Green function is the inverse of the correlator matrix G^{-1}, Eq. (2.8), standing in the quadratic action. Thus, one faces the unpleasant task of inverting the large $2N \times 2N$ matrix, Eq. (2.8). It may seem more attractive to invert the differential operator, Eq. (2.13). Such an inversion, however, is undefined due to the presence of the zero mode ($\sim e^{-i\omega_0 t}$). The necessary regularization is provided by the off–diagonal blocks of the discrete matrix. The goal is to develop a formalism that avoids dealing with the large discrete matrices and refers to the continuous notations only.

The easiest way to proceed is to recall [17] that the Green functions are traces of *time–ordered* products of the field operators (in the Heisenberg representation), where the ordering is done along the contour $\mathcal{C}$. Recalling also that the time arguments on the backward branch are always *after* those on the forward, one finds:

$$\langle \phi_+(t)\bar{\phi}_-(t') \rangle \equiv iG^<(t,t') = \frac{\mathrm{Tr}\{a^\dagger(t')a(t)\hat{\rho}_0\}}{\mathrm{Tr}\{\hat{\rho}_0\}}$$

$$= \frac{\mathrm{Tr}\{e^{i\hat{H}t'}a^\dagger e^{i\hat{H}(t-t')}a\,e^{-i\hat{H}t}\hat{\rho}_0\}}{\mathrm{Tr}\{\hat{\rho}_0\}}$$

$$= \frac{e^{-i\omega_0(t-t')}}{\mathrm{Tr}\{\hat{\rho}_0\}} \sum_{m=0}^{\infty} m[\rho(\omega_0)]^m = ne^{-i\omega_0(t-t')};$$

$$\langle \phi_-(t)\bar{\phi}_+(t') \rangle \equiv iG^>(t,t') = \frac{\mathrm{Tr}\{a(t)a^\dagger(t')\hat{\rho}_0\}}{\mathrm{Tr}\{\hat{\rho}_0\}}$$

$$= \frac{\mathrm{Tr}\{e^{i\hat{H}t}a e^{i\hat{H}(t'-t)}a^\dagger\,e^{-i\hat{H}t'}\hat{\rho}_0\}}{\mathrm{Tr}\{\hat{\rho}_0\}}$$

$$= \frac{e^{i\omega_0(t'-t)}}{\mathrm{Tr}\{\hat{\rho}_0\}} \sum_{m=0}^{\infty} (m+1)[\rho(\omega_0)]^m = (n+1)e^{-i\omega_0(t-t')};$$

$$\langle \phi_+(t)\bar{\phi}_+(t') \rangle \equiv iG^T(t,t') = \frac{\mathrm{Tr}\{T[a(t)a^\dagger(t')]\hat{\rho}_0\}}{\mathrm{Tr}\{\hat{\rho}_0\}} \qquad (2.16)$$

$$= \theta(t-t')iG^>(t,t') + \theta(t'-t)iG^<(t,t');$$

$$\langle \phi_-(t)\bar{\phi}_-(t') \rangle \equiv iG^{\tilde{T}}(t,t') = \frac{\mathrm{Tr}\{\tilde{T}[a(t)a^\dagger(t')]\hat{\rho}_0\}}{\mathrm{Tr}\{\hat{\rho}_0\}}$$

$$= \theta(t'-t)iG^>(t,t') + \theta(t-t')iG^<(t,t');$$

where the symbols T and $\tilde{T}$ denote time–ordering and anti–time–ordering correspondingly. Hereafter the time arguments reside on the open time axis $t \in]-\infty, \infty[$. The Planck occupation number n stands for $n(\omega_0) \equiv \rho(\omega_0)/(1-\rho(\omega_0))$.

Notice the presence of non–zero off–diagonal Green functions $\langle \phi_-\bar{\phi}_+ \rangle$ and $\langle \phi_+\bar{\phi}_- \rangle$. This is seemingly inconsistent with the continuous action, Eq. (2.14). This is due to the presence of the off–diagonal blocks in the discrete matrix, that are lost in the continuous notations. The existence of the off–diagonal Green functions does not contradict to continuous notations. Indeed, $[i\partial_t - \omega_0]G^{>,<} = 0$, while $[i\partial_t - \omega_0]G^{T,\tilde{T}} = \pm\delta(t-t')$. Therefore in the obvious 2×2 matrix notations $G^{-1} \circ G = 1$, as it should be. The point is that the inverse of the operator $[i\partial_t - \omega_0]$ is not well-defined (due to the presence of the eigenmode ($\sim \exp\{-i\omega_0t\}$) with zero eigenvalue). A regularization must be specified and the off–diagonal blocks of the discrete matrix do exactly this.

The θ–function in Eq. (2.16) is the usual Heaviside step function. There is an uncertainty, however, regarding its value at coinciding time arguments. To resolve it, one needs to refer to the discrete representation one last time. Since the fields $\bar{\phi}$ always appear one time step δ_t *after* the fields ϕ on the Keldysh contour, cf. Eq. (2.6), the proper convention is:

$$G^T(t, t) = G^{\tilde{T}}(t, t) = G^<(t, t) = n. \tag{2.17}$$

Obviously not all four Green functions defined above are independent. Indeed, a direct inspection shows that for $t \neq t'$:

$$G^T + G^{\tilde{T}} = G^> + G^<. \tag{2.18}$$

One would like therefore to perform a linear transformation of the fields to benefit explicitly from this relation. This is achieved by the Keldysh rotation.

2.3. Keldysh rotation

Define new fields as

$$\phi_{cl}(t) = \frac{1}{\sqrt{2}}\big(\phi_+(t) + \phi_-(t)\big); \qquad \phi_q(t) = \frac{1}{\sqrt{2}}\big(\phi_+(t) - \phi_-(t)\big) \tag{2.19}$$

with the analogous transformation for the conjugated fields. The subscripts "*cl*" and "*q*" stand for the *classical* and the *quantum* components of the fields correspondingly. The rationale for these notations will become clear shortly. First, a simple algebraic manipulation of Eq. (2.16) shows that

$$-i\langle \phi_\alpha(t)\bar{\phi}_\beta(t')\rangle \equiv \hat{G}^{\alpha\beta} = \begin{pmatrix} G^K(t, t') & G^R(t, t') \\ G^A(t, t') & 0 \end{pmatrix}, \tag{2.20}$$

where hereafter $\alpha, \beta = (cl, q)$. The cancellation of the (q, q) element of this matrix is a manifestation of identity (2.18). Superscripts R, A and K stand for the *retarded, advanced* and *Keldysh* components of the Green function correspondingly. These three Green functions are the fundamental objects of the Keldysh technique. They are defined as

$$
\begin{aligned}
G^R(t, t') &= \frac{1}{2}\left(G^T - G^{\tilde{T}} - G^< + G^>\right) = \theta(t - t')(G^> - G^<); \\
G^A(t, t') &= \frac{1}{2}\left(G^T - G^{\tilde{T}} + G^< - G^>\right) = \theta(t' - t)(G^< - G^>); \quad (2.21) \\
G^K(t, t') &= \frac{1}{2}\left(G^T + G^{\tilde{T}} + G^> + G^<\right) = G^< + G^>.
\end{aligned}
$$

In the discrete representation each of these three Green functions is represented by an $N \times N$ matrix. Since both $G^<$ and $G^>$ are, by definition, anti-Hermitian (cf. Eq. (2.16)), Eq. (2.21) implies:

$$G^A = \left[G^R\right]^\dagger \qquad\qquad G^K = -[G^K]^\dagger, \tag{2.22}$$

A. Kamenev

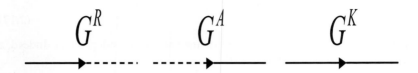

Fig. 2. Graphic representation of G^R, G^A, and G^K correspondingly. The full line represents the classical field component, ϕ_{cl}, while the dashed line – the quantum component, ϕ_q.

where the Hermitian conjugation includes complex conjugation, as well as transposition of the time arguments. The retarded (advanced) Green function is a lower (upper) triangular matrix. Due to the algebra of triangular matrices, a product of any number of upper (lower) triangular matrices is again an upper (lower) triangular matrix. This leads to the simple rule:

$$
\begin{aligned}
G_1^R \circ G_2^R \circ \ldots \circ G_l^R &= G^R ; \\
G_1^A \circ G_2^A \circ \ldots \circ G_l^A &= G^A ,
\end{aligned}
\tag{2.23}
$$

where the circular multiplication signs are understood as integrations over intermediate times (discrete matrix multiplication). At coinciding time arguments, one finds (cf. Eqs. (2.17) and (2.21)):

$$
G^R(t, t) + G^A(t, t) = 0.
\tag{2.24}
$$

Although in the discrete representation both G^R and G^A *do* contain non–zero (pure imaginary, due to Eqs. (2.22), (2.24)) main diagonals (otherwise the matrix $\hat{G}$ is not invertible), the proper continuous convention is: $\theta(0) = 0$. The point is that in any diagrammatic calculation, $G^R(t, t)$ and $G^A(t, t)$ always come in symmetric combinations and cancel each other due to Eq. (2.24). It is thus a convenient and noncontradictory agreement to take $\theta(0) = 0$.

It is useful to introduce graphic representations for the three Green functions. To this end, let us denote the classical component of the field by a full line and the quantum component by a dashed line. Then the retarded Green function is represented by a full-arrow-dashed line, the advanced by a dashed-arrow-full line and the Keldysh by full-arrow-full line, see Fig. 2. Notice that the dashed-arrow-dashed line that would represent the $\langle \phi_q \bar{\phi}_q \rangle$ Green function is identically zero due to identity (2.18). The arrow shows the direction from ϕ_α towards $\bar{\phi}_\beta$.

For the single bosonic state (cf. Eq. (2.16)): $G^> = -i(n + 1)e^{-i\omega_0(t-t')}$ and $G^< = -ine^{-i\omega_0(t-t')}$, where $n = n(\omega_0) = \rho(\omega_0)/(1 - \rho(\omega_0))$ is the Planck occupation number (since the system is non–interacting the initial distribution function does not evolve). Therefore:

$$
\begin{aligned}
G^R(t, t') &= -i\theta(t - t')\, e^{-i\omega_0(t-t')}; \\
G^A(t, t') &= i\theta(t' - t)\, e^{-i\omega_0(t-t')}; \\
G^K(t, t') &= -i(1 + 2n(\omega_0))\, e^{-i\omega_0(t-t')}.
\end{aligned}
\tag{2.25}
$$

Notice that the retarded and advanced components contain information only about the spectrum and are independent of the occupation number, whereas the Keldysh component does depend on it. Such a separation is common for systems that are not too far from thermal equilibrium. Fourier transforming with respect to $(t - t')$ to the energy representation, one finds:

$$
\begin{aligned}
G^{R(A)}(\epsilon) &= (\epsilon - \omega_0 \pm i0)^{-1}; \\
G^K(\epsilon) &= (1 + 2n(\omega_0))(-2\pi i)\delta(\epsilon - \omega_0) = (1 + 2n(\epsilon))(-2\pi i)\delta(\epsilon - \omega_0).
\end{aligned}
\tag{2.26}
$$

Therefore for the case of thermal equilibrium one notices that

$$
G^K(\epsilon) = \coth \frac{\epsilon}{2T} \left(G^R(\epsilon) - G^A(\epsilon) \right).
\tag{2.27}
$$

The last equation constitutes the statement of the *fluctuation–dissipation theorem* (FDT). As is shown below, the FDT is a general property of thermal equilibrium that is not restricted to the toy example considered here. It implies a rigid relation between the response and correlation functions.

In general, it is convenient to parameterize the anti-Hermitian (see Eq. (2.22)) Keldysh Green function by a Hermitian matrix $F = F^\dagger$, as:

$$
G^K = G^R \circ F - F \circ G^A,
\tag{2.28}
$$

where $F = F(t', t'')$ and the circular multiplication sign implies integration over the intermediate time (matrix multiplication). The Wigner transform (see below), $f(\tau, \epsilon)$, of the matrix F is referred to as the distribution function. In thermal equilibrium: $f(\epsilon) = \coth(\epsilon/2T)$.

2.4. Keldysh action and causality

The Keldysh rotation from the (ϕ_+, ϕ_-) field components to (ϕ_{cl}, ϕ_q) considerably simplifies the structure of the Green functions (cf. Eqs. (2.16) and (2.20)). It is convenient, therefore, to write the action in terms of ϕ_{cl}, ϕ_q as well. A simple way of doing this is to apply the Keldysh rotation, Eq. (2.19), to the continuous action, Eq. (2.14), written in terms of ϕ_+, ϕ_-. However, as was discussed above, the continuous action, Eq. (2.14), loses the crucial information about the off–diagonal blocks of the discrete matrix, Eq. (2.8). To keep this information, one may invert the matrix of Green functions, Eq. (2.20), and use the result as the correlator in the quadratic action. The inversion is straightforward:

$$\hat{G}^{-1} = \begin{pmatrix} G^K & G^R \\ G^A & 0 \end{pmatrix}^{-1} = \begin{pmatrix} 0 & [G^{-1}]^A \\ [G^{-1}]^R & [G^{-1}]^K \end{pmatrix}, \qquad (2.29)$$

where the three components of the inverted Green function, labelled in advance as A, R and K, satisfy:

$$[G^{-1}]^{R(A)} = [G^{R(A)}]^{-1} = (i\partial_t - \omega_0 \pm i0)\delta(t - t'); \qquad (2.30)$$

$$[G^{-1}]^K = -[G^R]^{-1} \circ G^K \circ [G^A]^{-1} = [G^R]^{-1} \circ F - F \circ [G^A]^{-1},$$

where the parameterization (2.28) was employed in the last line. It is easy to see that $[G^R]^{-1}$ and $[G^A]^{-1}$ are lower and upper triangular matrices correspondingly, thus justifying their superscripts. The continuous notations may create the impression that $[G^{-1}]^K = (2i0)F$ and thus may be omitted. One should remember, however, that this component is non–zero in the discrete form and therefore it is important to acknowledge its existence (even if it is not written explicitly).

Once the correlator Eqs. (2.29), (2.30) is established, one may immediately write down the corresponding action:

$$S[\phi_{cl}, \phi_q] = \iint\limits_{-\infty}^{\infty} dt\,dt' \, (\bar{\phi}_{cl}, \bar{\phi}_q)_t \begin{pmatrix} 0 & [G^A]^{-1} \\ [G^R]^{-1} & [G^{-1}]^K \end{pmatrix}_{t,t'} \begin{pmatrix} \phi_{cl} \\ \phi_q \end{pmatrix}_{t'}, \qquad (2.31)$$

where it is acknowledged that the correlator is, in general, a non–local function of time. The Green functions, Eq. (2.20), follow from the Gaussian integral with this action, by construction. Notice that the presence of $[G^{-1}]^K = (2i0)F$ (with a positive imaginary part) is absolutely necessary for the convergence of the corresponding functional integral.

The structure of the Gaussian action given by Eq. (2.31) is very general and encodes regularization of the functional integral. Since the Keldysh component carries the information about the density matrix, there is no further need to recall the discrete representation. The main features of this structure are:

• The $cl - cl$ component is zero.

This zero may be traced back to identity (2.18). It has, however, a much simpler interpretation. It reflects the fact that for a pure classical field configuration ($\phi_q = 0$) the action is zero. Indeed, in this case $\phi_+ = \phi_-$ and the action on the forward part of the contour is cancelled exactly by that on the backward part. The very general statement is, therefore, that

$$S[\phi_{cl}, \phi_q = 0] = 0. \qquad (2.32)$$

Obviously Eq. (2.32) is not restricted to a Gaussian action.

• The $cl - q$ and $q - cl$ components are mutually Hermitian conjugated upper and lower (advanced and retarded) triangular matrices in the time representation. This property is responsible for the causality of the response functions as well as for protecting the $cl - cl$ component from a perturbative renormalization (see below).

• The $q - q$ component is an anti-Hermitian matrix (cf. Eq. (2.22)) with a positive–definite imaginary spectrum. It is responsible for the convergence of the functional integral. It also keeps the information about the distribution function.

As was already mentioned, these three items are generic and reproduce themselves in every order of perturbation theory. For the lack of a better terminology, we'll refer to them as the *causality structure*.

2.5. Free bosonic fields

It is a straightforward matter to generalize the entire construction to bosonic systems with more than one state. Suppose the states are labelled by an index k, that may be, e.g., a momentum vector. Their energies are given by a function ω_k, for example $\omega_k = k^2/(2m)$, where m is the mass of the bosonic atoms. One introduces then a doublet of complex fields (classical and quantum) for every state k: $(\phi_{cl}(t; k), \phi_q(t; k))$ and writes down the action in the form of Eq. (2.31) including a summation over the index k. Away from equilibrium, the Keldysh component may be non–diagonal in the index k: $F = F(t, t'; k, k')$. The retarded (advanced) component, on the other hand, has the simple form $[G^{R(A)}]^{-1} = i\partial_t - \omega_k$.

If k is momentum, it is more instructive to Fourier transform to real space and to deal with $(\phi_{cl}(t; r), \phi_q(t; r))$. Introducing a combined time–space index $x = (t; r)$, one may write down for the action of the free complex bosonic field (atoms):

$$S_0 = \iint dx \, dx' \, (\bar{\phi}_{cl}, \bar{\phi}_q)_x \begin{pmatrix} 0 & [G^A]^{-1} \\ [G^R]^{-1} & [G^{-1}]^K \end{pmatrix}_{x,x'} \begin{pmatrix} \phi_{cl} \\ \phi_q \end{pmatrix}_{x'}, \tag{2.33}$$

where in the continuous notations

$$[G^{R(A)}]^{-1}(x, x') = \delta(x - x') \left(i\partial_{t'} + \frac{1}{2m} \nabla_{r'}^2 \right), \tag{2.34}$$

while in the discrete form it is a lower (upper) triangular matrix in time (not in space). The $[G^{-1}]^K$ component for the free field is only the regularization factor, originating from the (time) boundary terms. It is in general non–local in x and x', however, being a pure boundary term it is frequently omitted. It is kept here as a reminder that the inversion, $\hat{G}$, of the correlator matrix must possess the causality structure, Eq. (2.20).

In an analogous way, the action of free real bosons (phonons) is (cf. Eq. (B.9)):

$$S_0 = \iint dx\, dx'\, (\varphi_{cl}, \varphi_q)_x \begin{pmatrix} 0 & [D^A]^{-1} \\ [D^R]^{-1} & [D^{-1}]^K \end{pmatrix}_{x,x'} \begin{pmatrix} \varphi_{cl} \\ \varphi_q \end{pmatrix}_{x'}, \qquad (2.35)$$

where

$$[D^{R(A)}]^{-1}(x, x') = \delta(x - x')\left(-\partial_{t'}^2 + v_s^2 \nabla_{r'}^2\right), \qquad (2.36)$$

in the continuous notations. In the discrete representations $[D^{R(A)}]^{-1}$ are again the lower (upper) triangular matrices. Here too the Keldysh component $[D^{-1}]^K$ is just a regularization, originating from the (time) boundary terms. It is kept in Eq. (2.35) to emphasize the casuality structure of the real boson Green function $\hat{D}(x, x')$, analogous to Eq. (2.20):

$$\hat{D}(x, x') = \begin{pmatrix} D^K & D^R \\ D^A & 0 \end{pmatrix}; \qquad \begin{array}{l} D^{R(A)}(\epsilon, k) = ((\epsilon \pm i0)^2 - v_s^2 k^2)^{-1}; \\ D^K = D^R \circ F - F \circ D^A, \end{array} \qquad (2.37)$$

where $F = F(t, t'; r, r')$ is a symmetric distribution function matrix.

3. Collisions and kinetic equation

3.1. Interactions

The short range two–body collisions of bosonic atoms are described by the local "four–boson" Hamiltonian: $H_{int} = \lambda \sum_r a_r^\dagger a_r^\dagger a_r a_r$, where index r "numerates" spatial locations. The interaction constant, λ, is related to a commonly used s–wave scattering length, a_s, as $\lambda = 4\pi a_s / m$ [18]. The corresponding term in the continuous Keldysh action takes the form:

$$S_{int} = -\lambda \int dr \int_C dt\, (\bar{\phi}\phi)^2 = -\lambda \int dr \int_{-\infty}^{\infty} dt\, [(\bar{\phi}_+\phi_+)^2 - (\bar{\phi}_-\phi_-)^2]. \qquad (3.1)$$

It is important to remember that there are no interactions in the distant past $t = -\infty$ (while they are present in the future, $t = +\infty$). The interactions are supposed to be adiabatically switched on and off on the forward and backward branches correspondingly. That guarantees that the off–diagonal blocks of the matrix, Eq. (2.8), remain intact. Interactions modify only those matrix elements of the evolution operator, Eq. (2.6), that are away from $t = -\infty$. It is also

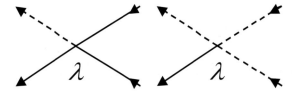

Fig. 3. Graphic representation of the two interaction vertexes of the $|\phi|^4$ theory. There are also two conjugated vertexes with a reversed direction of all arrows.

worth remembering that in the discrete time form the $\bar{\phi}$ fields are taken one time step δ_t *after* the ϕ fields along the Keldysh contour $\mathcal{C}$. Therefore the two terms on the r.h.s. of the last equation should be understood as $(\bar{\phi}_+(t + \delta_t)\phi_+(t))^2$ and $(\bar{\phi}_-(t)\phi_-(t+\delta_t))^2$ correspondingly. Performing the Keldysh rotation, Eq. (2.19), one finds

$$S_{int}[\phi_{cl}, \phi_q] = -\lambda \int_{-\infty}^{\infty} dt \left[\bar{\phi}_q \bar{\phi}_{cl} (\phi_{cl}^2 + \phi_q^2) + c.c. \right], \qquad (3.2)$$

where $c.c.$ stands for the complex conjugate of the first term. The collision action, Eq. (3.2), obviously satisfies the causality condition, Eq. (2.32). Diagrammatically the action (3.2) generates two types of vertexes depicted in Fig. 3 (as well as two complex conjugated vertexes, obtained by reversing the direction of the arrows): one with three classical fields (full lines) and one quantum field (dashed line) and the other with one classical field and three quantum fields.

Let us demonstrate that adding the collision term to the action does not violate the fundamental normalization $Z = 1$. To this end one may expand $e^{iS_{int}}$ in powers of λ and then average term by term with the Gaussian action, Eq. (2.33). To show that the normalization $Z = 1$ is not altered by the collisions, one needs to show that $\langle S_{int} \rangle = \langle S_{int}^2 \rangle = \ldots = 0$. Applying the Wick theorem, one finds for the terms that are linear order in λ: $\langle \bar{\phi}_q \bar{\phi}_{cl} \phi_{cl}^2 + c.c. \rangle \sim [G^R(t, t) + G^A(t, t)]G^K(t, t) = 0$, and $\langle \bar{\phi}_q \bar{\phi}_{cl} \phi_q^2 + c.c \rangle = 0$. The first term vanishes due to identity (2.24), while the second one vanishes because $\langle \phi_q \bar{\phi}_q \rangle = 0$. There are two families of terms that are second order in λ. The first one is $\langle \bar{\phi}_q \bar{\phi}_{cl} \phi_{cl}^2 \phi_q' \phi_{cl}' (\bar{\phi}_{cl}')^2 \rangle \sim G^R(t', t)G^A(t', t)[G^K(t, t')]^2$, while the second is $\langle \bar{\phi}_q \bar{\phi}_{cl} \phi_{cl}^2 \phi_q' \phi_{cl}' (\bar{\phi}_q')^2 \rangle \sim [G^R(t, t')]^2 G^R(t', t)G^A(t', t)$, where $\phi_\alpha' \equiv \phi_\alpha(t')$. Both of these terms are zero, because $G^R(t', t) \sim \theta(t' - t)$, while $G^A(t', t) \sim$

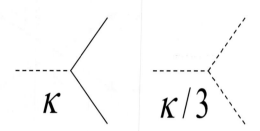

Fig. 4. Graphic representation of the two interaction vertexes of the φ^3 theory. Notice the relative factor of one third between them.

$G^R(t, t')^* \sim \theta(t - t')$ and thus their product has no support [1]. It is easy to see that, for exactly the same reasons, all higher order terms vanish and thus the normalization is unmodified (at least in a perturbative expansion).

As another example, consider the real boson field, Eq. (2.35), with the cubic nonlinearity:

$$S_{int} = \frac{\kappa}{6} \int dr \int_C dt \, \varphi^3 = \frac{\kappa}{6} \int dr \int_{-\infty}^{\infty} dt \, [\varphi_+^3 - \varphi_-^3] = \kappa \int dr \int_{-\infty}^{\infty} dt \, [\varphi_{cl}^2 \varphi_q + \frac{1}{3} \varphi_q^3]. \quad (3.3)$$

The causality condition, Eq. (2.32), is satisfied again. Diagrammatically the cubic nonlinearity generates two types of vertexes, Fig. 4: one with two classical fields (full lines) and one quantum field (dashed line), and the other with three quantum fields. The former vortex carries the factor κ, while the latter has a weight of $\kappa/3$. Notice that for a real field the direction of the lines is not specified by arrows.

Exercise: Show that there are no corrections of second order in κ to the partition function, $Z = 1$. Check that the same is true for the higher orders as well.

3.2. Saddle point equations

Before developing the perturbation theory further, one has to discuss the saddle points of the action. According to Eq. (2.32), there are no terms in the action

---

[1] Strictly speaking, $G^R(t', t)$ and $G^A(t', t)$ are both simultaneously non–zero on the diagonal: $t = t'$. The contribution of the diagonal to the integrals is, however, $\sim \delta_t^2 N \to 0$, when $N \to \infty$.

that have zero power of both $\bar{\phi}_q$ and ϕ_q. The same is obviously true regarding $\delta S / \delta \bar{\phi}_{cl}$ and therefore one of the saddle point equations:

$$\frac{\delta S}{\delta \bar{\phi}_{cl}} = 0 \tag{3.4}$$

may always be solved by

$$\phi_q = 0, \tag{3.5}$$

irrespectively of what the classical component, ϕ_{cl}, is. One may check that this is indeed the case for the action given by Eqs. (2.33) plus (3.2). Under condition (3.5) the second saddle point equation takes the form:

$$\frac{\delta S}{\delta \bar{\phi}_q} = \left(\left[G^R \right]^{-1} - \lambda \, |\phi_{cl}|^2 \right) \phi_{cl} = \left(i \partial_t + \frac{\nabla_r^2}{2m} - \lambda \, |\phi_{cl}|^2 \right) \phi_{cl} = 0. \tag{3.6}$$

This is the non–linear time–dependent (Gross–Pitaevskii) equation [18], that uniquely determines the classical field configuration, provided some initial and boundary conditions are specified.

The message is that among the possible solutions of the saddle–point equations for the Keldysh action, there is always one with a zero quantum component and with a classical component that obeys the classical (non–linear) equations of motion. We shall call such a saddle point – *"classical"*. Thanks to Eqs. (2.32) and (3.5), the action at the classical saddle–point field configurations is identically zero. As was argued above, the perturbative expansion in small fluctuations around the classical saddle point leads to a properly normalized partition function, $Z = 1$. This seemingly excludes the possibility of having any other saddle points. Yet, this conclusion is premature. The system may possess "non–classical" saddle points – such that $\phi_q \neq 0$. Such saddle points do not contribute to the partition function (and thus do not alter the fundamental normalization, $Z = 1$), however, they may contribute to the correlation functions. In general, the action at a *non–classical* saddle point is non–zero. Its contribution is thus associated with exponentially small (or oscillatory) terms. Examples include: tunneling, thermal activation (considered in the next chapter), Wigner-Dyson level statistics, etc...

Let us develop now a systematic perturbative expansion in deviations from the *classical* saddle point. As was discussed above, it does not bring any new information about the partition function. It does, however, provide information about the Green functions (and thus various observables). Most notably, it generates the kinetic equation for the distribution function. To simplify further consideration, let us assume that $\phi_{cl} = 0$ is the proper solution of the classical saddle–point equation (3.6) (i.e. there is no Bose condensate).

3.3. Dyson equation

The goal is to calculate the *dressed* Green function, defined as:

$$\mathbf{G}^{\alpha\beta}(t, t') = -i \int \mathcal{D}\bar{\phi}\phi \; e^{i(S_0 + S_{int})} \; \phi_\alpha(t)\bar{\phi}_\beta(t'), \tag{3.7}$$

where $\alpha, \beta = (cl, q)$ and the action is given by Eqs. (2.33) and (3.2) (or for real bosons: Eqs. (2.35) and (3.3), with $\phi \to \varphi$). To this end one may expand the exponent in deviations from the classical saddle point: $\phi_q \equiv 0$ and (in the simplest case) $\phi_{cl} = 0$. The functional integration with the remaining Gaussian action is then performed using the Wick theorem. This leads to the standard diagrammatic series. Combining all one–particle irreducible diagrams into the self–energy matrix $\hat{\Sigma}$, one obtains:

$$\hat{\mathbf{G}} = \hat{G} + \hat{G} \circ \hat{\Sigma} \circ \hat{G} + \hat{G} \circ \hat{\Sigma} \circ \hat{G} \circ \hat{\Sigma} \circ \hat{G} + \ldots = \hat{G} \circ \left(\hat{1} + \hat{\Sigma} \circ \hat{\mathbf{G}}\right), \tag{3.8}$$

where $\hat{G}$ is given by Eq. (2.20) and the circular multiplication sign implies integrations over intermediate times and coordinates as well as a 2×2 matrix multiplication. The only difference compared with the text–book [17] diagrammatic expansion is the presence of the 2×2 Keldysh matrix structure. The fact that the series is arranged as a sequence of matrix products is of no surprise. Indeed, the Keldysh index, $\alpha = (cl, q)$, is just one more index in addition to time, space, spin, etc. Therefore, as with any other index, there is a summation (integration) over all of its intermediate values, hence the matrix multiplication. The concrete form of the self–energy matrix, $\hat{\Sigma}$, is of course specific to the Keldysh technique and is discussed below in some details.

Multiplying both sides of Eq. (3.8) by $\hat{G}^{-1}$ from the left, one obtains the Dyson equation for the exact dressed Green function, $\hat{\mathbf{G}}$:

$$\left(\hat{G}^{-1} - \hat{\Sigma}\right) \circ \hat{\mathbf{G}} = \hat{1}, \tag{3.9}$$

where $\hat{1}$ is the unit matrix. The very non–trivial feature of the Keldysh technique is that the self energy matrix, $\hat{\Sigma}$, possesses the same causality structure as $\hat{G}^{-1}$, Eq. (2.29):

$$\hat{\Sigma} = \begin{pmatrix} 0 & \Sigma^A \\ \Sigma^R & \Sigma^K \end{pmatrix}, \tag{3.10}$$

where $\Sigma^{R(A)}$ are lower (upper) triangular matrices in the time domain, while Σ^K is an anti-Hermitian matrix. This fact will be demonstrated below. Since both $\hat{G}^{-1}$ and $\hat{\Sigma}$ have the same structure, one concludes that the dressed Green function, $\hat{\mathbf{G}}$, also possesses the causality structure, like Eq. (2.20). As a result, the Dyson equation acquires the form:

$$
\begin{pmatrix} 0 & [G^A]^{-1} - \Sigma^A \\ [G^R]^{-1} - \Sigma^R & -\Sigma^K \end{pmatrix} \circ \begin{pmatrix} \mathbf{G^K} & \mathbf{G^R} \\ \mathbf{G^A} & 0 \end{pmatrix} = \hat{1}, \tag{3.11}
$$

where one took into account that $[G^{-1}]^K$ is a pure regularization ($\sim i0F$) and thus may be omitted in the presence of a non–zero Σ^K. Employing the specific form of $[G^{R(A)}]^{-1}$, Eq. (2.34), one obtains for the retarded (advanced) components:

$$
\left(i\partial_t + \frac{1}{2m}\nabla_r^2 \right) \mathbf{G^{R(A)}} = \delta(t - t')\delta(r - r') + \Sigma^{R(A)} \circ \mathbf{G^{R(A)}}. \tag{3.12}
$$

Provided the self–energy component $\Sigma^{R(A)}$ is known (in some approximation), Eq. (3.12) constitutes a closed equation for the retarded (advanced) component of the dressed Green function. The latter carries the information about the spectrum of the interacting system.

To write down the equation for the Keldysh component, it is convenient to parameterize it as $\mathbf{G^K} = \mathbf{G^R} \circ \mathbf{F} - \mathbf{F} \circ \mathbf{G^A}$, where $\mathbf{F}$ is a Hermitian matrix in the time domain. The equation for the Keldysh component then takes the form: $([G^R]^{-1} - \Sigma^R) \circ (\mathbf{G^R} \circ \mathbf{F} - \mathbf{F} \circ \mathbf{G^A}) = \Sigma^K \circ \mathbf{G^A}$. Multiplying it from the right by $([G^A]^{-1} - \Sigma^A)$ and employing Eq. (3.12), one finally finds:

$$
\left[\mathbf{F}, \left(i\partial_t + \frac{1}{2m}\nabla_r^2 \right) \right]_- = \Sigma^K - \left(\Sigma^R \circ \mathbf{F} - \mathbf{F} \circ \Sigma^A \right), \tag{3.13}
$$

where the symbol $[\,,\,]_-$ stands for the commutator. This equation is the quantum kinetic equation for the distribution matrix $\mathbf{F}$. Its l.h.s. is called the *kinetic* term, while the r.h.s. is the *collision integral* (up to a factor). As is shown below, Σ^K has the meaning of an "incoming" term, while $\Sigma^R \circ \mathbf{F} - \mathbf{F} \circ \Sigma^A$ is an "outgoing" term. In equilibrium these two channels cancel each other (the kinetic term vanishes) and the self-energy has the same structure as the Green function: $\Sigma^K = \Sigma^R \circ \mathbf{F} - \mathbf{F} \circ \Sigma^A$. This is not the case, however, away from the equilibrium.

3.4. Self-energy

Let us demonstrate in the case of one specific example that the self-energy matrix $\hat{\Sigma}$ indeed possesses the causality structure, Eq. (3.10). To this end, we consider the real boson action, Eq. (2.35), with the $\kappa\varphi^3$ nonlinearity, Eq. (3.3), and per-

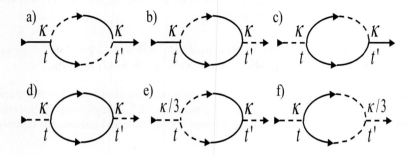

Fig. 5. Self-energy diagrams for the φ^3 theory.

form the calculations up to the second order in the parameter κ. Employing the two vertexes of Fig. 4 one finds that:

• *the cl − cl component* is given by the single diagram, depicted in Fig. 5a. The corresponding analytic expression is

$\Sigma^{cl-cl}(t, t') = 4i\kappa^2 D^R(t, t') D^A(t, t') = 0.$

Indeed, the product $D^R(t, t') D^A(t, t')$ has no support (see, however, the footnote in section 3.1).

• *the cl-q (advanced) component* is given by the single diagram, Fig. 5b. The corresponding expression is:

$$\Sigma^A(t, t') = 4i\kappa^2 D^A(t, t') D^K(t, t'). \tag{3.14}$$

Since $\Sigma^A(t, t') \sim D^A(t, t') \sim \theta(t' - t)$, it is, indeed, an advanced (upper triangular) matrix. There is a combinatoric factor of 4, associated with the diagram (4 ways of choosing external legs × 2 internal permutations × 1/(2!) for having two identical vertexes).

• *the q-cl (retarded) component* is given by the diagram of Fig. 5c:

$$\Sigma^R(t, t') = 4i\kappa^2 D^R(t, t') D^K(t, t'), \tag{3.15}$$

that could be obtained, of course, by the Hermitian conjugation of Eq. (3.14) with the help of Eq. (2.22): $\Sigma^R = [\Sigma^A]^\dagger$. Since $\Sigma^R(t, t') \sim D^R(t, t') \sim \theta(t - t')$, it is, indeed, a retarded (lower triangular) matrix.

• *the q-q (Keldysh) component* is given by the three diagrams, Fig. 5d–f. The corresponding expressions are:

$$\Sigma^K(t, t') = 2i\kappa^2 [D^K(t, t')]^2 + 6i\left(\frac{\kappa}{3}\right)\kappa [D^A(t, t')]^2 + 6i\kappa\left(\frac{\kappa}{3}\right)[D^R(t, t')]^2$$

$$= 2i\kappa^2\left([D^K(t, t')]^2 + [D^R(t, t') - D^A(t, t')]^2\right), \tag{3.16}$$

where the combinatoric factors are: 2 for diagram d and 6 for e and f. In the last equality, the fact that $G^R(t, t')G^A(t, t') = 0$, due to the absence of support in the time domain, has been used again. Employing Eq. (2.22), one finds $\Sigma^K = -[\Sigma^K]^\dagger$. This completes the proof of the statement that $\hat{\Sigma}$ possesses the same structure as $\hat{D}^{-1}$. One may check that the statement holds in higher orders as well. In Eqs. (3.14)–(3.16) one has omitted the spatial coordinates, that may be restored in an obvious way.

Exercise: Calculate the self–energy matrix for the $|\phi|^4$ theory to second order in λ. Show that it possesses the causality structure.

3.5. Kinetic term

To make further progress in the discussion of the kinetic equation it is convenient to perform the Wigner transformation (WT). The WT of a distribution function matrix, $\mathbf{F}(t, t'; r, r')$, is a function: $\mathbf{f}(\tau, \epsilon; \rho, k)$, where τ and ρ are the "center of mass" time and coordinate respectively. According to definition (2.28), the $\mathbf{F}$ matrix appears in a product with $G^R - G^A$ (or $D^R - D^A$). Since the latter is a sharply peaked function at $\epsilon = \omega_k$ (cf. Eq. (2.26) for free particles, while for interacting systems this is the condition for having well–defined quasi–particles), one frequently writes $\mathbf{f}(\tau, \rho, k)$, understanding that $\epsilon = \omega_k$.

To rewrite the kinetic term (the l.h.s. of Eq. (3.13)) in the Wigner representation, one notices that the WT of $i\partial_t$ is ϵ, while the WT of ∇_r^2 is $-k^2$. Then e.g. $[\mathbf{F}, \nabla_r^2]_- \rightarrow [k^2, \mathbf{f}]_- + i\nabla_k k^2 \nabla_\rho \mathbf{f} = 2ik\nabla_\rho \mathbf{f}$, where the commutator vanishes, since WT's commute. In a similar way: $[\mathbf{F}, i\partial_t]_- \rightarrow -i\partial_\tau \mathbf{f}$. If there is a scalar potential $V(r)a_r^\dagger a_r$ in the Hamiltonian, it translates into the term $-V(\bar{\phi}_{cl}\phi_q + \bar{\phi}_q\phi_{cl})$ in the action and thus $-V(r)$ is added to $[G^{R(A)}]^{-1}$. This, in turn, brings the term $-[\mathbf{F}, V]_-$ to the l.h.s. of the Dyson equation (3.13), or after the WT: $iE\nabla_k\mathbf{f}$, where $E \equiv -\nabla_\rho V$ is the electric field. As a result, the WT

Reminder: The Wigner transform of a matrix $A(r, r')$ is defined as
$a(\rho, k) \equiv \int dr_1 A\left(\rho + \frac{r_1}{2}, \rho - \frac{r_1}{2}\right) e^{ikr_1}$.
One may show that the Wigner transform of the matrix $C = A \circ B$ is equal to:
$c(\rho, k) = \iint dr_1 dr_2 \int \frac{dk_1 dk_2}{(2\pi)^{2d}} a\left(\rho + \frac{r_1}{2}, k + k_1\right) b\left(\rho + \frac{r_2}{2}, k + k_2\right) e^{i(k_1 r_2 - k_2 r_1)}$.
Expanding the functions under the integrals in k_i and r_i, one finds:
$c(\rho, k) = a(\rho, k) b(\rho, k) + (2i)^{-1}\left(\nabla_\rho a \nabla_k b - \nabla_k a \nabla_\rho b\right) + \dots$.

of the Dyson equation (3.9) takes the form:

$$\left(\partial_\tau - v_k \nabla_\rho - E \nabla_k\right) \mathbf{f}(\tau, \rho, k) = I_{col}[\mathbf{f}], \tag{3.17}$$

where $v_k \equiv k/m$ and $I_{col}[\mathbf{f}]$ is the WT of the r.h.s. of Eq. (3.13) (times i). This is the kinetic equation for the distribution function.

For real bosons with the dispersion relation $\epsilon = \omega_k$, the kinetic term is (cf. Eq. (2.36)): $[\epsilon^2 - \omega_k^2, \mathbf{F}]_- \rightarrow 2i\left(\epsilon \, \partial_\tau - \omega_k(\nabla_k \omega_k)\nabla_\rho\right)\mathbf{f} = 2i\epsilon\left(\partial_\tau - v_k \nabla_\rho\right)\mathbf{f}$, where $v_k \equiv \nabla_k \omega_k$ is the group velocity. As a result, the kinetic equation takes the form: $\left(\partial_\tau - v_k \nabla_\rho\right)\mathbf{f}(\tau, \rho, k) = I_{col}[\mathbf{f}]$, where the collision integral $I_{col}[\mathbf{f}]$ is the WT of the r.h.s. of Eq. (3.13), divided by $-2i\epsilon$.

3.6. Collision integral

Let us discuss the collision integral, using the φ^3 theory calculations of section 3.4 as an example. To shorten the algebra, let us consider a system that is spatially uniform and isotropic in momentum space. One, thus, focuses on the energy relaxation only. In this case the distribution function is $\mathbf{f}(\tau, \rho, k) = \mathbf{f}(\tau, \omega_k) = \mathbf{f}(\tau, \epsilon)$, where the dependence on the modulus of the momentum is substituted by the $\omega_k = \epsilon$ argument. Employing Eqs. (3.14)–(3.16), one finds for the WT of the r.h.s. of Eq. (3.13) [2]:

$$\Sigma^R \circ \mathbf{F} - \mathbf{F} \circ \Sigma^A \rightarrow -2i\, \mathbf{f}(\tau, \epsilon) \int d\omega\, M(\tau, \epsilon, \omega)\Big(\mathbf{f}(\tau, \epsilon - \omega) + \mathbf{f}(\tau, \omega)\Big);$$

$$\Sigma^K \rightarrow -2i \int d\omega\, M(\tau, \epsilon, \omega)\Big(\mathbf{f}(\tau, \epsilon - \omega)\mathbf{f}(\tau, \omega) + 1\Big), \tag{3.18}$$

where the square of the transition matrix element is given by:

$$M(\tau, \epsilon, \omega) = 2\pi\kappa^2 \sum_q \Delta_\mathbf{d}(\tau, \epsilon - \omega; k - q)\, \Delta_\mathbf{d}(\tau, \omega; q). \tag{3.19}$$

Here $\Delta_\mathbf{d} \equiv i(\mathbf{d}^R - \mathbf{d}^A)/(2\pi)$ and $\mathbf{d}^{R(A)}(\tau, \epsilon; k)$ is the WT of the retarded (advanced) Green function. One has substituted the dressed Green functions into Eqs. (3.14)–(3.16) instead of the bare ones to perform a partial resummation of the diagrammatic series. (This trick is sometimes called the *self–consistent Born approximation*. It still neglects the vertex corrections.) Assuming the existence of well defined quasi–particles at all times, one may regard $\Delta_\mathbf{d}(\tau, \epsilon, k)$ as a sharply peaked function at $\epsilon = \omega_k$. In this case Eq. (3.19) simply reflects the fact that

---

[2] Only products of WT's are retained, while all the gradient terms are neglected, in particular $\mathbf{D}^K \rightarrow \mathbf{f}(\mathbf{d}^R - \mathbf{d}^A)$. The energy–momentum representation is used, instead of the time–space representation as in Eqs. (3.14)–(3.16), and in the equation for $\Sigma^R \circ \mathbf{F} - \mathbf{F} \circ \Sigma^A$ one performs a symmetrization between the ω and $\epsilon - \omega$ arguments.

an initial particle with $\epsilon = \omega_k$ decays into two real (on mass-shell) particles with energies $\omega = \omega_q$ and $\epsilon - \omega = \omega_{k-q}$. As a result, one finally obtains for the kinetic equation:

$$\frac{\partial \mathbf{f}(\epsilon)}{\partial \tau} = \int d\omega \, \frac{M(\epsilon, \omega)}{\epsilon} \left[\mathbf{f}(\epsilon - \omega)\mathbf{f}(\omega) + 1 - \mathbf{f}(\epsilon)\big(\mathbf{f}(\epsilon - \omega) + \mathbf{f}(\omega)\big) \right], \quad (3.20)$$

where the time arguments are suppressed for brevity.

Due to the identity: $\coth(a-b)\coth(b)+1 = \coth(a)\big(\coth(a-b)+\coth(b)\big)$, the collision integral is identically nullified by

$$\mathbf{f}(\epsilon) = \coth \frac{\epsilon}{2T}. \qquad (3.21)$$

where T is temperature. This is the thermal equilibrium distribution function. According to the kinetic equation (3.20), it is stable for any temperature (the latter is determined either by an external reservoir, or, for a closed system, from the total energy conservation). Since the equilibrium distribution obviously nullifies the kinetic term, according to Eq. (3.13) the *exact* self–energy satisfies $\Sigma^K = \coth(\epsilon/(2T))(\Sigma^R - \Sigma^A)$. Since also the bare Green functions obey the same relation, Eq. (2.27), one concludes that in thermal equilibrium the *exact* dressed Green function satisfies:

$$\mathbf{D}^K = \coth \frac{\epsilon}{2T} \left(\mathbf{D}^R - \mathbf{D}^A \right). \qquad (3.22)$$

This is the statement of the *fluctuation–dissipation theorem* (FDT). Its consequence is that in equilibrium the Keldysh component does not contain any additional information with respect to the retarded one. Therefore, the Keldysh technique may be, in principle, substituted by a more compact construction – the Matsubara method. The latter does not work, of course, away from equilibrium.

Returning to the kinetic equation (3.20), one may identify "in" and "out" terms in the collision integral. Most clearly it is done by writing the collision integral in terms of the occupation numbers $\mathbf{n}_k$, defined as $\mathbf{f} = 1 + 2\mathbf{n}$. The expression in the square brackets on the r.h.s. of Eq. (3.20) takes the form: $4\left[\mathbf{n}_{\epsilon-\omega}\mathbf{n}_\omega - \mathbf{n}_\epsilon(\mathbf{n}_{\epsilon-\omega} + \mathbf{n}_\omega + 1)\right]$. The first term: $\mathbf{n}_{\epsilon-\omega}\mathbf{n}_\omega$, gives a probability that a particle with energy $\epsilon - \omega$ absorbs a particle with energy ω to populate a state with energy ϵ – this is the "in" term of the collision integral. It may be traced back to the Σ^K part of the self-energy. The second term: $-\mathbf{n}_\epsilon(\mathbf{n}_{\epsilon-\omega} + \mathbf{n}_\omega + 1)$, says that a state with energy ϵ may be depopulated either by stimulated emission of particles with energies $\epsilon - \omega$ and ω, or by spontaneous emission (unity). This is the "out" term, that may be traced back to the $\Sigma^{R(A)}$ contributions.

Finally, let us discuss the approximations involved in the Wigner transformations. Although Eq. (3.13) is formally exact, it is very difficult to extract any useful information from it. Therefore, passing to an approximate but much more tractable form like Eqs. (3.17) or (3.20) is highly desirable. In doing so, one has to employ the approximate form of the WT. Indeed, a formally infinite series in $\nabla_k \nabla_\rho$ operators is truncated, usually by the first non–vanishing term. This is a justified procedure as long as $\delta k \, \delta \rho \gg 1$, where δk is a characteristic microscopic scale of the momentum dependence of $\mathbf{f}$, while $\delta \rho$ is a characteristic scale of its spatial variations. One may ask if there is a similar requirement in the time domain: $\delta \epsilon \, \delta \tau \gg 1$, with $\delta \epsilon$ and $\delta \tau$ the characteristic energy and time scale of $\mathbf{f}$ respectively. Such a requirement is very demanding, since typically $\delta \epsilon \approx T$ and at low temperature it would allow to treat only very slow processes: with $\delta \tau \gg 1/T$. Fortunately, this is not the case. Because of the peaked structure of $\Delta_{\mathbf{d}}(\epsilon, k)$, the energy argument ϵ is locked to ω_k and does not have its own dynamics as long as the peak is sharp. The actual criterion is therefore that $\delta \epsilon$ is much larger than the width of the peak in $\Delta_{\mathbf{d}}(\epsilon, k)$. The latter is, by definition, the quasi–particle life–time, τ_{qp}, and therefore the condition is $\tau_{qp} \gg 1/T$. This condition is indeed satisfied by many systems with interactions that are not too strong.

4. Particle in contact with an environment

4.1. Quantum dissipative action

Consider a particle with coordinate $\Phi(t)$, living in a potential $U(\Phi)$ and attached to a harmonic string $\varphi(t; x)$. The particle may represent a collective degree of freedom, such as the phase of a Josephson junction or the charge on a quantum dot. On the other hand, the string serves to model a dissipative environment. The advantage of the one–dimensional string is that it is the simplest continuum system, having a constant density of states. Due to this property it mimics, for example, interactions with a Fermi sea. A continuous reservoir with a constant density of states at small energies is sometimes called an "Ohmic" environment (or bath). The environment is supposed to be in thermal equilibrium.

The Keldysh action of such a system is given by the three terms (cf. Eqs. (B.5) and (2.35)):

$$S_p[\hat{\Phi}] = \int\limits_{-\infty}^{\infty} dt \left[-2\,\Phi_q \frac{d^2 \Phi_{cl}}{dt^2} - U\left(\Phi_{cl} + \Phi_q\right) + U(\Phi_{cl} - \Phi_q) \right];$$

$$S_{str}[\hat{\varphi}] = \int\limits_{-\infty}^{\infty} dt \int dx \, \hat{\varphi}^T \, \hat{D}^{-1} \, \hat{\varphi}; \qquad\qquad (4.1)$$

$$S_{int}[\hat{\Phi}, \hat{\varphi}] = 2\sqrt{\gamma} \int\limits_{-\infty}^{\infty} dt \, \left. \hat{\Phi}^T(t)\, \hat{\sigma}_1 \, \nabla_x \hat{\varphi}(t, x) \right|_{x=0},$$

where we have introduced vectors of classical and quantum components, e.g. $\hat{\Phi}^T \equiv (\Phi_{cl}, \Phi_q)$ and the string correlator, $\hat{D}^{-1}$, is the same as in Eqs. (2.35), (2.36). The interaction term between the particle and the string is taken to be the local product of the particle coordinate and the string stress at $x = 0$ (so the force on the particle is proportional to the local stress of the string). In the time domain the interaction is instantaneous, $\Phi(t)\nabla_x\varphi(t, x)|_{x=0} \to \Phi_+\nabla\varphi_+ - \Phi_-\nabla\varphi_-$ on the Keldysh contour. Transforming to the classical–quantum notations leads to: $2(\Phi_{cl}\nabla\varphi_q + \Phi_q\nabla\varphi_{cl})$, that satisfies the causality condition, Eq. (2.32). In the matrix notations it takes the form of the last line of Eq. (4.1), where $\hat{\sigma}_1$ is the standard Pauli matrix. The interaction constant is $\sqrt{\gamma}$.

One may now integrate out the degrees of freedom of the Gaussian string to re-duce the problem to the particle coordinate only. According to the standard rules of Gaussian integration (see. Appendix A), this leads to the so–called dissipative action for the particle:

$$S_{diss} = -\gamma \iint\limits_{-\infty}^{\infty} dt dt' \, \hat{\Phi}^T(t) \, \underbrace{\hat{\sigma}_1^T \nabla_x \nabla_{x'} \hat{D}(t - t'; x - x')\Big|_{x=x'=0} \, \hat{\sigma}_1}_{-\hat{L}^{-1}(t-t')} \hat{\Phi}(t'). \quad (4.2)$$

The straightforward matrix multiplication shows that the dissipative correlator $\hat{L}^{-1}$ possesses the standard causality structure of the inverse Green function, e.g. Eq. (2.29). Fourier transforming its retarded (advanced) components, one finds:

$$\left[L^{R(A)}(\epsilon)\right]^{-1} = -\sum_k \frac{k^2}{(\epsilon \pm i0)^2 - k^2} = \pm \frac{i}{2}\epsilon + \text{const}, \quad (4.3)$$

where we put $v_s = 1$ for brevity. The ϵ–independent constant (same for R and A components) may be absorbed into the redefinition of the harmonic part of the potential $U(\Phi) = \text{const}\,\Phi^2 + \ldots$ and, thus, may be omitted. In equilibrium the Keldysh component of the correlator is set by the FDT:

$$\left[L^{-1}\right]^K(\epsilon) = \coth\frac{\epsilon}{2T}\left(\left[L^R\right]^{-1} - \left[L^A\right]^{-1}\right) = i\epsilon \coth\frac{\epsilon}{2T}. \quad (4.4)$$

It is an anti–Hermitian operator with a positive–definite imaginary part, rendering convergence of the functional integral over Φ.

In the time representation the retarded (advanced) component of the correlator takes a simple local form: $\left[L^{R(A)}\right]^{-1} = \mp\frac{1}{2}\delta(t - t')\,\partial_{t'}$. On the other hand, at low temperatures the Keldysh component is a non–local function, that may be found by the inverse Fourier transform of Eq. (4.4):

$$\left[L^{-1}\right]^K(t - t') = \frac{i\pi T^2}{\sinh^2(\pi T(t - t'))} \overset{T\to\infty}{\longrightarrow} i2T\delta(t - t'). \quad (4.5)$$

Finally, for the Keldysh action of the particle connected to a string, one obtains:

$$
S[\hat{\Phi}] = \int\limits_{-\infty}^{\infty} dt \left[-2\,\Phi_q \left(\frac{d^2\Phi_{cl}}{dt^2} + \frac{\gamma}{2} \frac{d\Phi_{cl}}{dt} \right) - U\left(\Phi_{cl} + \Phi_q \right) + U\left(\Phi_{cl} - \Phi_q \right) \right]
$$

$$
+ \; i\gamma \iint\limits_{-\infty}^{\infty} dt\,dt'\,\Phi_q(t) \frac{\pi T^2}{\sinh^2(\pi T (t - t'))} \Phi_q(t'). \tag{4.6}
$$

This action satisfies all the causality criteria listed in section 2.4. Notice that in the present case the Keldysh $(q - q)$ component is not just a regularization factor, but rather a quantum fluctuations damping term, originating from the coupling to the string. The other manifestation of the string is the presence of the friction term, $\sim \gamma \partial_t$ in the R and the A components. In equilibrium the friction coefficient and fluctuations amplitude are rigidly connected by the FDT. The quantum dissipative action, Eq. (4.6), is a convenient playground to demonstrate various approximations and connections to other approaches.

4.2. Saddle–point equation

The *classical* saddle point equation (the one that takes $\Phi_q(t) = 0$) has the form:

$$
-\frac{1}{2} \frac{\delta S[\hat{\Phi}]}{\delta \Phi_q} \bigg|_{\Phi_q=0} = \frac{d^2\Phi_{cl}}{dt^2} + \frac{\gamma}{2} \frac{d\Phi_{cl}}{dt} + \frac{\partial U(\Phi_{cl})}{\partial \Phi_{cl}} = 0. \tag{4.7}
$$

This is the deterministic classical equation of motion. In the present case it happens to be the Newton equation with the viscous force: $-(\gamma/2)\dot{\Phi}_{cl}$. This approximation neglects both *quantum* and *thermal* fluctuations.

4.3. Classical limit

One may keep the thermal fluctuations, while completely neglecting the quantum ones. To this end it is convenient to restore the Planck constant in the action (4.6) and then take the limit $\hbar \to 0$. For dimensional reasons, the factor $\hbar^{-1}$ should stand in front of the action. To keep the part of the action responsible for the classical equation of motion (4.7) free from the Planck constant it is convenient to rescale the variables as: $\Phi_q \to \hbar \Phi_q$. Finally, to have temperature in energy units, one needs to substitute $T \to T/\hbar$ in the last term of Eq. (4.6). The limit $\hbar \to 0$ is now straightforward: (i) one has to expand $U(\Phi_{cl} \pm \hbar \Phi_q)$ to the first order in $\hbar \Phi_q$ and neglect all higher order terms; (ii) in the last term of Eq. (4.6) the $\hbar \to 0$ limit is equivalent to the $T \to \infty$ limit, see Eq. (4.5). As a result, the classical limit of the dissipative action is:

$$S[\hat{\Phi}] = 2\int_{-\infty}^{\infty} dt \left[-\Phi_q \left(\frac{d^2\Phi_{cl}}{dt^2} + \frac{\gamma}{2}\frac{d\Phi_{cl}}{dt} + \frac{\partial U(\Phi_{cl})}{\partial \Phi_{cl}} \right) + i\gamma\, T\, \Phi_q^2 \right]. \quad (4.8)$$

Physically the limit $\hbar \to 0$ means that $\hbar\tilde{\Omega} \ll T$, where $\tilde{\Omega}$ is a characteristic classical frequency of the particle. This condition is necessary for the last term of Eq. (4.6) to take the time–local form. The condition for neglecting the higher order derivatives of U is $\hbar \ll \gamma\tilde{\Phi}_{cl}^2$, where $\tilde{\Phi}_{cl}$ is a characteristic classical amplitude of the particle motion.

4.4. Langevin equations

One way to proceed with the classical action (4.8) is to notice that the exponent of its last term (times i) may be identically rewritten in the following way:

$$e^{-2\gamma T\int dt\, \Phi_q^2(t)} = \int \mathcal{D}\xi(t)\, e^{-\int dt \left[\frac{1}{2\gamma T}\xi^2(t) - 2i\xi(t)\Phi_q(t) \right]}. \quad (4.9)$$

This identity is called the Hubbard–Stratonovich transformation, while $\xi(t)$ is an auxiliary Hubbard–Stratonovich field. The identity is proved by completing the square in the exponent on the r.h.s., performing the Gaussian integration at every instance of time and multiplying the results. There is a constant multiplicative factor hidden in the integration measure, $\mathcal{D}\xi$.

Exchanging the order of the functional integration over ξ and $\hat{\Phi}$, one finds for the partition function:

$$Z = \int \mathcal{D}\xi\, e^{-\frac{1}{2\gamma T}\int dt\, \xi^2} \int \mathcal{D}\Phi_{cl} \int \mathcal{D}\Phi_q\, e^{-2i\int dt\, \Phi_q \left(\frac{d^2\Phi_{cl}}{dt^2} + \frac{\gamma}{2}\frac{d\Phi_{cl}}{dt} + \frac{\partial U(\Phi_{cl})}{\partial \Phi_{cl}} - \xi \right)}. \quad (4.10)$$

Since the last (imaginary) exponent depends only linearly on $\Phi_q(t)$, the integration over $\mathcal{D}\Phi_q$ results in the δ–function of the expression in the round brackets. This functional δ–function forces its argument to be zero at every moment of time. Therefore, among all the possible trajectories $\Phi_{cl}(t)$, only those that satisfy the following equation contribute to the partition function:

$$\frac{d^2\Phi_{cl}}{dt^2} + \frac{\gamma}{2}\frac{d\Phi_{cl}}{dt} + \frac{\partial U(\Phi_{cl})}{\partial \Phi_{cl}} = \xi(t). \quad (4.11)$$

This is a Newton equation with a time dependent external force $\xi(t)$. Since the same arguments are applicable to any correlation function of the classical fields, e.g. $\langle \Phi_{cl}(t)\Phi_{cl}(t') \rangle$, a solution strategy is as follows: (i) choose some realization of $\xi(t)$; (ii) solve Eq. (4.11) (e.g. numerically); (iii) having its solution $\Phi_{cl}(t)$, calculate the correlation function; (iv) average the result over an ensemble of realizations of the force $\xi(t)$. The statistics of the latter are dictated by the weight factor in the $\mathcal{D}\xi$ functional integral. It states that $\xi(t)$ is a Gaussian short–range (white) noise with the correlators:

$$\langle \xi(t) \rangle = 0; \qquad\qquad \langle \xi(t)\xi(t') \rangle = \gamma T \delta(t - t'). \qquad (4.12)$$

Equation (4.11) with the white noise on the r.h.s. is called the Langevin equation. It describes classical Newtonian dynamics in the presence of stochastic thermal fluctuations. The fact that the noise amplitude is related to the friction coefficient γ and to the temperature is a manifestation of the FDT. The latter holds as long as the environment (string) is at thermal equilibrium.

4.5. Martin–Siggia–Rose

In section 4.4 one derived the Langevin equation for a classical coordinate, Φ_{cl}, from the action written in terms of Φ_{cl} and another field, Φ_q. An inverse procedure of deriving the effective action from the Langevin equation is known as the Martin–Siggia–Rose (MSR) [5] technique. It is sketched here in the form suggested by de-Dominicis [5].

Consider a Langevin equation:

$$\hat{\mathcal{O}}[\Phi] = \xi(t), \qquad (4.13)$$

where $\hat{\mathcal{O}}[\Phi]$ is a (non–linear) differential operator acting on the coordinate $\Phi(t)$ and $\xi(t)$ is a white noise force, specified by Eq. (4.12). Define the "partition function" as:

$$Z[\xi] = \int \mathcal{D}\Phi \, \mathcal{J}[\hat{\mathcal{O}}] \, \delta\big(\hat{\mathcal{O}}[\Phi] - \xi(t)\big) \equiv 1. \qquad (4.14)$$

It is identically equal to unity by virtue of the integration of the δ–function, provided $\mathcal{J}[\hat{\mathcal{O}}]$ is the Jacobian of the operator $\hat{\mathcal{O}}[\Phi]$. The way to interpret Eq. (4.14) is to discretize the time axis, introducing N–dimensional vectors $\Phi_j = \Phi(t_j)$ and $\xi_j = \xi(t_j)$. The operator takes the form: $\mathcal{O}_i = O_{ij}\Phi_j + \Gamma_{ijk}\Phi_j\Phi_k + \ldots$, where a summation is taken over repeated indices. The Jacobian, $\mathcal{J}$, is given by the absolute value of the determinant of the following $N \times N$ matrix: $J_{ij} \equiv \partial \mathcal{O}_i/\partial \Phi_j = O_{ij} + 2\Gamma_{ijk}\Phi_k + \ldots$. It is possible to choose a proper (retarded) regularization where the J_{ij} matrix is a lower triangular matrix with a unity main diagonal (coming entirely from the $O_{ii} = 1$ term). Clearly, in this case, $\mathcal{J} = 1$. Indeed, consider, for example, $\hat{\mathcal{O}}[\Phi] = \partial_t \Phi - W(\Phi)$. The retarded regularized version of the Langevin equation is: $\Phi_i = \Phi_{i-1} + \delta_t(W(\Phi_{i-1}) + \xi_{i-1})$. Clearly

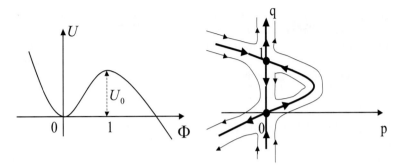

Fig. 6. a) A potential with a meta-stable minimum. b) The phase portrait of the Hamiltonian system, Eq. (4.19). Thick lines correspond to zero energy, arrows indicate evolution direction.

in this case $J_{ii} = 1$ and $J_{i,i-1} = -1 - W'(\Phi_{i-1})\delta_t$, while all other components are zero; as a result $\mathcal{J} = 1$.

Although the partition function (4.14) is trivial, it is clear that all the meaningful observables and the correlation functions may be obtained by inserting a set of factors: $\Phi(t)\Phi(t') \dots$ in the functional integral, Eq. (4.14). Having this in mind, let us proceed with the partition function. Employing the integral representation of the δ–function with the help of an auxiliary field $\Psi(t)$, one obtains:

$$Z[\xi] = \int \mathcal{D}\Phi \int \mathcal{D}\Psi \, e^{-2i\int dt \, \Psi(t)\left(\hat{O}^R[\Phi(t)]-\xi(t)\right)}, \tag{4.15}$$

where $\hat{O}^R$ stands for the retarded regularization of the $\hat{O}$ operator and thus one takes $\mathcal{J} = 1$. One may average now over the white noise, Eq. (4.12), by performing the Gaussian integration over ξ:

$$Z = \int \mathcal{D}\xi \, e^{-\frac{1}{2\gamma T}\int dt \, \xi^2} Z[\xi] = \int \mathcal{D}\Phi\Psi \, e^{-\int dt \left[2i \, \Psi(t)\hat{O}^R[\Phi(t)]+2\gamma T\Psi^2(t)\right]}, \tag{4.16}$$

The exponent is exactly the classical limit of the Keldysh action, cf. Eq. (4.8) (including the retarded regularization of the differential operator), where $\Phi = \Phi_{cl}$ and $\Psi = \Phi_q$. The message is that the MSR action is nothing, but the classical (high temperature) limit of the Keldysh action. The MSR technique provides a simple way to transform from a classical stochastic problem to its proper functional representation. The latter is useful for an analytical analysis. One example is given below.

4.6. Thermal activation

Consider a particle in a meta-stable potential well, plotted in Fig. 6a. The potential has a meta-stable minimum at $\Phi = 0$ and a maximum at $\Phi = 1$ with the amplitude U_0. Let us also assume that the particle's motion is over-damped, i.e. $\gamma \gg \sqrt{U''}$. In this case one may disregard the inertia term, leaving only viscous relaxation dynamics. The classical dissipative action (4.8) takes the form:

$$S[\hat{\Phi}] = 2 \int_{-\infty}^{\infty} dt \left[-\Phi_q \left(\frac{\gamma}{2} \frac{d\Phi_{cl}}{dt} + \frac{\partial U(\Phi_{cl})}{\partial \Phi_{cl}} \right) + i\gamma T \, \Phi_q^2 \right]. \tag{4.17}$$

The corresponding saddle point equations are:

$$\frac{\gamma}{2} \dot{\Phi}_{cl} = -\frac{\partial U(\Phi_{cl})}{\partial \Phi_{cl}} + 2i\gamma T \, \Phi_q; \tag{4.18}$$

$$\frac{\gamma}{2} \dot{\Phi}_q = \Phi_q \frac{\partial^2 U(\Phi_{cl})}{\partial \Phi_{cl}^2}.$$

These equations possess the *classical* solution: $\Phi_q(t) \equiv 0$ and $\Phi_{cl}(t)$ satisfies the classical equation of motion: $\frac{\gamma}{2} \dot{\Phi}_{cl} = -\partial U(\Phi_{cl})/\partial \Phi_{cl}$. For the initial condition $\Phi_{cl}(0) < 1$ the latter equation predicts the viscous relaxation towards the minimum at $\Phi_{cl} = 0$. According to this equation, there is no possibility to escape from this minimum. Therefore the classical solution of Eqs. (4.18) does *not* describe thermal activation. Thus one has to look for another possible solution of Eqs. (4.18), the one with $\Phi_q \neq 0$.

To this end let us make a simple linear change of variables: $\Phi_{cl}(t) = q(t)$ and $\Phi_q(t) = p(t)/(i\gamma)$. Then the dissipative action (4.17) acquires the form of a Hamiltonian action:

$$iS = -\int dt \left(p\dot{q} - H(p,q) \right); \qquad H(p,q) \equiv \frac{2}{\gamma} \left[-p \frac{\partial U(q)}{\partial q} + Tp^2 \right], \tag{4.19}$$

where the fictitious Hamiltonian, H, is introduced [3]. It is straightforward to see that in terms of the new variables the equations of motion (4.18) take the form of the Hamilton equations: $\dot{q} = \partial H/\partial p$ and $\dot{p} = -\partial H/\partial q$. One needs, thus, to investigate the Hamiltonian system with the Hamiltonian Eq. (4.19). To visualize it, one may plot its phase portrait, consisting of lines of constant energy $E = H(p(t), q(t))$ on the (p, q) plane, Fig. 6b. The topology is determined by the two lines of zero energy: $p = 0$ and $Tp = \partial U(q)/\partial q$, that intersect at the two stationary points of the potential: $q = 0$ and $q = 1$. The $p = 0$ line corresponds to the classical (without Langevin noise) dynamics (notice, that

---

[3] Amazingly, this trick of rewriting viscous (or diffusive) dynamics as a Hamiltonian one works in a wide class of problems. The price one has to pay is the doubling of the number of degrees of freedom: q and p in the Hamiltonian language, or "classical" and "quantum" components in the Keldysh language.

the action is identically zero for motion along this line) and thus $q = 0$ is the stable point, while $q = 1$ is the unstable one. Due to Liouville theorem, every fixed point must have one stable and one unstable direction. Therefore, along the "non-classical" line: $p = T^{-1}\partial U(q)/\partial q$, the situation is reversed: $q = 0$ is unstable, while $q = 1$ is stable. It is now clear that to escape from the bottom of the potential well, $q = 0$, the system must move along the non-classical line of zero energy until it reaches the top of the barrier, $q = 1$, and then continue to drop according to the classical equation of motion (moving along the classical line $p = 0$). There is a non-zero action associated with the motion along the non-classical line:

$$
iS = -\int dt\, p\dot{q} = -\int_0^1 p(q)dq = -\frac{1}{T}\int_0^1 \frac{\partial U(q)}{\partial q}\,dq = -\frac{U_0}{T}, \tag{4.20}
$$

where one has used that $H = 0$ along the integration trajectory. As a result, the thermal escape probability is proportional to $e^{iS} = e^{-U_0/T}$, which is nothing but the thermal activation exponent.

4.7. Fokker-Planck equation

Another way to approach the action (4.17) is to notice that it is quadratic in Φ_q and therefore the $\mathcal{D}\Phi_q$ integration may be explicitly performed. To shorten notations and emphasize the relation to the classical coordinate, we shall follow the previous section and denote $\Phi_{cl}(t) \equiv q(t)$. Performing the Gaussian integration over Φ_q of $e^{iS[\hat{\Phi}]}$, with $S[\Phi_{cl}, \Phi_q]$ given by Eq. (4.17), one finds the action, depending on $\Phi_{cl} \equiv q$ only:

$$
iS[q] = -\frac{1}{2\gamma T}\int_{-\infty}^{\infty} dt\,\left(\frac{\gamma}{2}\dot{q} + U_q'\right)^2. \tag{4.21}
$$

One may now employ the same trick that allows to pass from the Feynman path integral to the Schrödinger equation. Namely, let us introduce the "wave function", $\mathcal{P}(q, t)$, that is a result of the functional integration of $e^{iS[q]}$ over all trajectories that at time t pass through the point $q_N \equiv q$. Adding one more time step, δ_t, to the trajectory, one may write $\mathcal{P}(q_N, t + \delta_t)$ as an integral of $\mathcal{P}(q_{N-1}, t) = \mathcal{P}(q + \delta_q, t)$ over $\delta_q \equiv q_{N-1} - q$:

$$
\mathcal{P}(q, t + \delta_t) = C\int d\delta_q\, e^{-\frac{\delta_t}{2\gamma T}\left(\frac{\gamma}{2}\frac{-\delta_q}{\delta_t} + U_q'(q+\delta_q)\right)^2}\mathcal{P}(q + \delta_q, t) \tag{4.22}
$$

$$
= C\int d\delta_q\, e^{-\frac{\gamma}{8T}\frac{\delta_q^2}{\delta_t}}\left[e^{\frac{\delta_q}{2T}U_q'(q+\delta_q) - \frac{\delta_t}{2\gamma T}(U_q')^2}\mathcal{P}(q + \delta_q, t)\right],
$$

where the factor C from the integration measure is determined by the condition: $C\int d\delta_q \exp\{-\gamma\delta_q^2/(8T\delta_t)\} = 1$. Expanding the expression in the square brackets on the r.h.s. of the last equation to second order in δ_q and to first order in δ_t, one finds:

$$\mathcal{P}(t+\delta_t) = \left(1 + \frac{\langle\delta_q^2\rangle}{2T}U_{qq}'' + \frac{1}{2}\frac{\langle\delta_q^2\rangle}{4T^2}\left(U_q'\right)^2 - \frac{\delta_t}{2\gamma T}\left(U_q'\right)^2\right)\mathcal{P} + \frac{\langle\delta_q^2\rangle}{2T}U_q'\mathcal{P}_q'$$

$$+ \frac{\langle\delta_q^2\rangle}{2}\mathcal{P}_{qq}'' = \mathcal{P}(t) + \delta_t\left(\frac{2}{\gamma}U_{qq}''\mathcal{P} + \frac{2}{\gamma}U_q'\mathcal{P}_q' + \frac{2T}{\gamma}\mathcal{P}_{qq}''\right), \quad (4.23)$$

where $\langle\delta_q^2\rangle \equiv C\int d\delta_q \exp\{-\gamma\delta_q^2/(8T\delta_t)\}\delta_q^2 = 4T\delta_t/\gamma$. Finally, rewriting the last expression in the differential form, one obtains:

$$\frac{\partial\mathcal{P}}{\partial t} = \frac{2}{\gamma}\left[\frac{\partial}{\partial q}\frac{\partial U}{\partial q} + T\frac{\partial^2}{\partial q^2}\right]\mathcal{P} = \frac{2}{\gamma}\frac{\partial}{\partial q}\left[\frac{\partial U}{\partial q}\mathcal{P} + T\frac{\partial\mathcal{P}}{\partial q}\right]. \quad (4.24)$$

This is the Fokker–Planck (FP) equation for the evolution of the probability distribution function, $\mathcal{P}(q, t)$. The latter describes the probability to find the particle at the point $q(= \Phi)$ at time t. If one starts from an initially sharp (deterministic) distribution: $\mathcal{P}(q, 0) = \delta(q - q(0))$, then the first term on the r.h.s. of the FP equation describes the viscous drift of the particle in the potential $U(q)$. Indeed, in the absence of the second term ($T = 0$), the equation is solved by $\mathcal{P}(q, t) = \delta(q - q(t))$, where $q(t)$ satisfies the deterministic equation of motion $(\gamma/2)\dot{q}(t) = -\partial U(q(t))/\partial q$ [4]. The second term describes the diffusion spreading of the probability distribution due to the thermal stochastic noise $\xi(t)$. For a confining potential $U(q)$ (such that $U(\pm\infty) \to \infty$) the stationary solution of the FP equation is the equilibrium Boltzmann distribution: $\mathcal{P}(q) \sim \exp\{-U(q)/T\}$.

The FP equation may be considered as the (imaginary time) Schrödinger equation: $\dot{\mathcal{P}} = \hat{H}\mathcal{P}$, where the "Hamiltonian", $\hat{H}$, is nothing but the "quantized" version of the classical Hamiltonian, introduced in the previous section, Eq. (4.19). The "quantization" rule is $p \to \hat{p} \equiv -\partial/\partial q$, so the canonical commutation relation: $[q, \hat{p}]_- = 1$, holds. Notice that before applying this quantization rule, the corresponding classical Hamiltonian must be *normally ordered*. Namely, the momentum $\hat{p}$ should be to the left of the coordinate q, cf. Eq. (4.19). Us-

---

[4]To check this statement one may substitute $\mathcal{P}(q, t) = \delta(q - q(t))$ into the $T = 0$ FP equation: $\delta_q'(q - q(t))(-\dot{q}(t)) = (2/\gamma)\left[U_{qq}''\delta(q - q(t)) + U_q'\delta_q'(q - q(t))\right]$. Then multiplying both parts of this equation by q and integrating over dq (by performing integration by parts), one finds: $\dot{q}(t) = -(2/\gamma)U_q'(q(t))$.

ing the commutation relation, one may rewrite the quantized Hamiltonian as:
$\hat{H} = T\hat{p}^2 - \hat{p}U'_q = T\left(\hat{p} - U'_q/(2T)\right)\left(\hat{p} - U'_q/(2T)\right) - (U'_q)^2/(4T) + U''_{qq}/2$
(we took $\gamma/2 = 1$) and perform the canonical transformation: $Q = q$ and
$\hat{P} = \hat{p} - U'_q/(2T)$. In terms of these new variables the Hamiltonian takes the
familiar form: $\hat{H} = T\hat{P}^2 + V(Q)$, where $V(Q) = -(U'_Q)^2/(4T) + U''_{QQ}/2$,
while the "wave function" transforms as $\tilde{\mathcal{P}}(Q,t) = e^{U(Q)/(2T)}\mathcal{P}$.

4.8. From Matsubara to Keldysh

In some applications it may be convenient to derive an action in the equilibrium
Matsubara technique and change to the Keldysh representation at a later stage to
tackle out–of–equilibrium problems. This section intends to illustrate how such
transformation may be carried out. To this end consider the following bosonic
Matsubara action:

$$S[\Phi_m] = \gamma T \sum_{m=-\infty}^{\infty} \frac{1}{2} |\epsilon_m||\Phi_m|^2, \tag{4.25}$$

where $\Phi_m = \bar{\Phi}_{-m}$ are the Matsubara components of a real bosonic field, $\Phi(\tau)$.
Notice that due to the absolute value sign: $|\epsilon_m| \neq i\partial_\tau$. In fact, in the imaginary
time representation the action (4.25) has the non–local form:

$$S[\Phi] = -\frac{\gamma}{2} \iint_0^\beta d\tau \, d\tau' \, \Phi(\tau) \frac{\pi T^2}{\sin^2(\pi T(\tau - \tau'))} \Phi(\tau'). \tag{4.26}$$

This action is frequently named after Caldeira and Leggett [19], who used it to
investigate the influence of dissipation on quantum tunneling.

To transform to the Keldysh representation one needs to double the number
of degrees of freedom: $\Phi \to \hat{\Phi} = (\Phi_{cl}, \Phi_q)^T$. Then according to the causality structure, section 2.4, the general form of the time translationally invariant

---

Reminder: The Matsubara technique deals with the imaginary time τ confined to the interval
$\tau \in [0, \beta[$, where $\beta = 1/T$. All bosonic fields must be periodic in this interval: $\phi(\tau + \beta) = \phi(\tau)$,
whereas the fermionic fields are antiperiodic: $\psi(\tau + \beta) = -\psi(\tau)$.
It is convenient to introduce the discrete Fourier (Matsubara) transform, e.g.

$$\phi_m = \int_0^\beta d\tau \phi(\tau) \, e^{i\epsilon_m \tau},$$

where for bosons $\epsilon_m \equiv 2\pi m T$, while for fermions $\epsilon_m \equiv \pi(2m+1)T$ and $m = 0, \pm 1, \ldots$.

Keldysh action is:

$$S = \gamma \int \frac{d\epsilon}{2\pi} (\Phi_{cl}, \Phi_q)_\epsilon \begin{pmatrix} 0 & [L^A(\epsilon)]^{-1} \\ [L^R(\epsilon)]^{-1} & [L^{-1}]^K(\epsilon) \end{pmatrix} \begin{pmatrix} \Phi_{cl} \\ \Phi_q \end{pmatrix}_\epsilon , \qquad (4.27)$$

where $[L^{R(A)}(\epsilon)]^{-1}$ is the analytical continuation of the Matsubara correlator $|\epsilon_m|/2$ from the *upper (lower)* half–plane of the imaginary variable ϵ_m to the real axis: $-i\epsilon_m \to \epsilon$. As a result, $[L^{R(A)}(\epsilon)]^{-1} = \pm i\epsilon/2$. The Keldysh component follows from the FDT: $[L^{-1}]^K(\epsilon) = i\epsilon \coth \epsilon/(2\,T)$, cf. Eqs. (4.3) and (4.4). Therefore the Keldysh counterpart of the Matsubara action, Eqs. (4.25) or (4.26) is the already familiar dissipative action, Eq. (4.6), (without the potential terms, of course). One may now include external fields and allow the system to deviate from the equilibrium.

4.9. *Dissipative chains and membranes*

Instead of dealing with a single particle connected to a bath, let us now consider a chain or lattice of coupled particles, with *each one* connected to a bath. To this end, one (i) supplies a spatial index, r, to the field: $\Phi(t) \to \Phi(t; r)$, and (ii) adds the harmonic interaction potential between nearest neighbors particles: $\sim (\Phi(t, r) - \Phi(t, r + 1))^2 \to (\nabla_r \Phi)^2$ in the continuous limit. By changing to the classical–quantum components and performing the spatial integration by parts (cf. Eq. (B.9)), the gradient term translates to: $\Phi_q \nabla_r^2 \Phi_{cl} + \Phi_{cl} \nabla_r^2 \Phi_q$. Thus it modifies the retarded and advanced components of the correlator, but it does *not* affect the $(q - q)$ Keldysh component:

$$[L^{R(A)}]^{-1} = \frac{1}{2} \delta(t - t')\,\delta(r - r')(\mp \partial_{t'} + D\nabla_{r'}^2), \qquad (4.28)$$

where D is the rigidity of the chain or the membrane. In the Fourier representation: $[L^{R(A)}(\epsilon; k)]^{-1} = \frac{1}{2}(\pm i\epsilon - Dk^2)$. In equilibrium the Keldysh component is not affected by the gradient terms, and is given by Eq. (4.4) (in the real space representation it acquires the factor $\delta(r - r')$). In particular, its classical limit is (cf. Eq. (4.5)) $[L^{-1}]^K = i2T\delta(t - t')\delta(r - r')$. As a result, the action of a classical elastic chain in contact with a bath is:

$$S[\hat{\Phi}] = 2\int dr \int_{-\infty}^{\infty} dt \left[-\Phi_q \left(\dot{\Phi}_{cl} - D\nabla_r^2 \Phi_{cl} + \frac{\partial U(\Phi_{cl})}{\partial \Phi_{cl}} \right) + i2\,T\,\Phi_q^2 \right], (4.29)$$

where the inertia terms have been neglected and we put $\gamma/2 = 1$ for brevity.

One may introduce now an auxiliary Hubbard–Stratonovich field $\xi(t; r)$ and write the Langevin equation according to section 4.4:

$$\dot{\Phi}_{cl} - D\nabla_r^2\Phi_{cl} + \frac{\partial U(\Phi_{cl})}{\partial \Phi_{cl}} = \xi(t; r), \tag{4.30}$$

where ξ is a Gaussian noise: $\langle \xi(t; r)\xi(t'; r') \rangle = 2T\delta(t - t')\delta(r - r')$ with short–range correlations.

Let us consider an elastic chain sitting in the bottom of the (r–independent) meta-stable potential well, depicted in Fig. 6a. If a sufficiently large piece of the chain thermally escapes from the well, it may find it favorable to slide down the potential, pulling the entire chain out of the well. To find the shape of such an optimally large critical domain and its action, let us change to the Hamiltonian variables of section 4.6: $q(t; r) \equiv \Phi_{cl}(t; r)$ and $p(t; r) \equiv 2i\Phi_q(t; r)$. The action (4.29) takes the Hamiltonian form:

$$iS = -\iint dr dt \left(p\dot{q} - H(p, q) \right); \qquad H \equiv -p\frac{\partial U(q)}{\partial q} + pD\nabla_r^2 q + Tp^2, \tag{4.31}$$

and the corresponding equations of motion are:

$$\dot{q} = \frac{\delta H}{\delta p} = D\nabla_r^2 q - U_q'(q) + 2Tp; \tag{4.32}$$

$$\dot{p} = -\frac{\delta H}{\delta q} = -D\nabla_r^2 p + pU_{qq}''(q).$$

These are complicated partial differential equations, that cannot be solved in general. Fortunately, the shape of the optimal critical domain can be found. As was discussed in section 4.6, the minimal action trajectory corresponds to a motion with zero energy, $H = 0$. According to Eq. (4.31), this is the case if either $p = 0$ (classical zero–action trajectory), or $Tp = U_q'(q) - D\nabla_r^2 q$ (finite–action escape trajectory). In the latter case the equation of motion for $q(t; r)$ takes the form of the classical equation in the reversed time: $\dot{q} = -D\nabla_r^2 q + U_q'(q) = Tp$. Thanks to the last equality the equation of motion for $p(t; r)$ is automatically satisfied [5]. In the reversed time dynamics the $q(t; r) = 0$ configuration is unstable and therefore the chain develops a "tongue" that grows until it reaches the stationary shape:

$$-D\nabla_r^2 q + U_q'(q) = 0. \tag{4.33}$$

---

[5] Indeed, $T\dot{p} = \partial_t\dot{q} = -D\nabla_r^2\dot{q} + \dot{q}U_{qq}'' = T(-D\nabla_r^2 p + pU_{qq}'')$. This non–trivial fact reflects the existence of an accidental conservation law: $H\left(p(t; r), q(t; r)\right) = 0 - locally!$ While from general principles only the total global energy has to be conserved.

The solution of this equation gives the shape of the critical domain. Once it is formed, it may grow further according to the classical equation $\dot{q} = D\nabla_r^2 q - U_q'(q)$ and $p = 0$ with zero action. The action along the non–classical escape trajectory, paid to form the "tongue" is $(H(p, q) = 0)$:

$$
\begin{aligned}
-iS &= \iint dr\,dt\, p\dot{q} = \frac{1}{T}\iint dr\,dt \left(U_q'(q) - D\nabla_r^2 q\right)\dot{q} \\
&= \frac{1}{T}\int dr\left(U(q) + \frac{D}{2}(\nabla_r q)^2\right),
\end{aligned}
\tag{4.34}
$$

where in the last equality an explicit integration over time is performed. The escape action is given therefore by the static activation expression that includes both the potential and the elastic energies. The optimal domain, Eq. (4.33), is found by the minimization of this static action (4.34). One arrives, thus, at a thermodynamic Landau-type description of the first–order phase transitions. Notice that the effective thermodynamic description appears due to the assumption that $H(p, q) = 0$ and therefore that all the processes take an infinitely long time.

5. Fermions

5.1. Free fermion Keldysh action

Consider a single quantum state, with the energy ϵ_0. This state is populated by spinless fermions (particles obeying the Pauli exclusion principle). In fact, one may have either zero or one particle in this state. The secondary quantized Hamiltonian of such a system has the form:

$$
\hat{H} = \epsilon_0\, c^\dagger c,
\tag{5.1}
$$

where $c^\dagger$ and c are fermion creation and annihilation operators of the state ϵ_0. They obey standard *anti*commutation relations: $\{c, c^\dagger\}_+ = 1$ and $\{c, c\}_+ = \{c^\dagger, c^\dagger\}_+ = 0$, where $\{, \}_+$ stands for the anti-commutator.

One can now consider the evolution operator along the Keldysh contour, $\mathcal{C}$ and the corresponding "partition function", $Z = 1$, defined in exactly the same manner as for bosonic systems: Eq. (2.2). The trace of the equilibrium density

---

Reminder: The fermionic coherent state $|\psi\rangle \equiv (1 - \psi c^\dagger)|0\rangle$, parameterized by a Grassmann number ψ (such that $\{\psi, \psi'\}_+ = \{\psi, c\}_+ = 0$), is an eigenstate of the annihilation operator: $c|\psi\rangle = \psi|\psi\rangle$.

Similarly: $\langle\psi|c^\dagger = \langle\psi|\bar{\psi}$, where $\bar{\psi}$ is another Grassmann number, *unrelated* to ψ.

The matrix elements of a *normally ordered* operator, such as e.g. the Hamiltonian, take the form $\langle\psi|\hat{H}(c^\dagger, c)|\psi'\rangle = H(\bar{\psi}, \psi')\langle\psi|\psi'\rangle$.

The overlap between any two coherent states is $\langle\psi|\psi'\rangle = 1 + \bar{\psi}\psi' = \exp\{\bar{\psi}\psi'\}$.

The trace of an operator, $\hat{A}$, is calculated as: $\mathrm{Tr}\{\hat{A}\} = \iint d\bar{\psi}\,d\psi\, e^{-\bar{\psi}\psi}\langle-\psi|\hat{A}|\psi\rangle$, where the Grassmann integrals are *defined* as: $\int d\psi\, 1 = 0$ and $\int d\psi\,\psi = 1$.

matrix is $\text{Tr}\{\rho_0\} = 1 + \rho(\epsilon_0)$, where the two terms stand for the empty and the singly occupied state. One divides the Keldysh contour into $(2N - 2)$ time intervals of length $\delta_t \sim 1/N \to 0$ and introduces resolutions of unity in $2N$ points along $\mathcal{C}$, Fig. (1). The only difference from the bosonic case in section 2.1 is that now one uses a resolution of unity in the *fermionic* coherent state basis [17]:

$$1 = \iint d\bar{\psi}_j \, d\psi_j \; e^{-\bar{\psi}_j \psi_j} \, |\psi_j\rangle\langle\psi_j|, \tag{5.2}$$

where $\bar{\psi}_j$ and ψ_j are *mutually independent* Grassmann variables. The rest of the algebra goes through exactly as in the bosonic case, section 2.1. As a result, one arrives at:

$$Z = \frac{1}{\text{Tr}\{\rho_0\}} \iint \prod_{j=1}^{2N} \left[d\bar{\psi}_j \, d\psi_j\right] e^{i \sum_{j,j'=1}^{2N} \bar{\psi}_j \mathcal{G}_{jj'}^{-1} \psi_{j'}}, \tag{5.3}$$

where the $2N \times 2N$ matrix $\mathcal{G}_{jj'}^{-1}$ stands for:

$$i\,\mathcal{G}_{jj'}^{-1} \equiv \begin{bmatrix} -1 & & & & & -\rho(\epsilon_0) \\ 1-h & -1 & & & & \\ & 1-h & -1 & & & \\ \hline & & & 1 & -1 & \\ & & & 1+h & -1 & \\ & & & & 1+h & -1 \end{bmatrix}, \tag{5.4}$$

and $h \equiv i\epsilon_0\delta_t$. The only difference from the bosonic case is the negative sign before the $\rho(\epsilon_0)$ matrix element, originating from the minus sign in the $\langle-\psi_{2N}|$ coherent state in the expression for the fermionic trace. To check the normalization, let us evaluate the determinant of such a matrix:

$$\det\!\left[i\mathcal{G}^{-1}\right] = 1 + \rho(\epsilon_0)(1 - h^2)^{N-1} \approx 1 + \rho(\epsilon_0)\,e^{(\epsilon_0\delta_t)^2(N-1)} \to 1 + \rho(\epsilon_0). \tag{5.5}$$

Employing the fact that the fermionic Gaussian integral is given by the determinant (unlike the *inverse* determinant for bosons) of the correlation matrix, Appendix A, one finds:

$$Z = \frac{\det\!\left[i\mathcal{G}^{-1}\right]}{\text{Tr}\{\rho_0\}} = 1, \tag{5.6}$$

as it should be. Once again, the upper–right element of the discrete matrix, Eq. (5.4), is crucial to maintain the correct normalization.

Taking the limit $N \to \infty$ and introducing the continuous notations, $\psi_j \to \psi(t)$, one obtains:

$$Z = \int \mathcal{D}\bar\psi\psi \; e^{iS[\bar\psi,\psi]} = \int \mathcal{D}\bar\psi\psi \; \exp\left\{ i \int_C \left[\bar\psi(t)\,\mathcal{G}^{-1}\psi(t) \right] dt \right\}, \qquad (5.7)$$

where according to Eqs. (5.3) and (5.4) the action is given by

$$S[\bar\psi,\psi] = \sum_{j=2}^{2N} \left[i\bar\psi_j \frac{\psi_j - \psi_{j-1}}{\delta t_j} - \epsilon_0\bar\psi_j\psi_{j-1} \right] \delta t_j + i\,\bar\psi_1\left(\psi_1 + \rho(\epsilon_0)\psi_{2N} \right), \quad (5.8)$$

where $\delta t_j \equiv t_j - t_{j-1} = \pm\delta_t$. Thus the continuous form of the operator $\mathcal{G}^{-1}$ is the same as for bosons, Eq. (2.13): $\mathcal{G}^{-1} = i\partial_t - \epsilon_0$. Again the upper–right element of the discrete matrix (the last term in Eq. (5.8)), that contains information about the distribution function, is seemingly absent in the continuous notations.

Splitting the Grassmann field $\psi(t)$ into the two components $\psi_+(t)$ and $\psi_-(t)$ that reside on the forward and the backward parts of the time contour correspondingly, one may rewrite the action as:

$$S = \int_{-\infty}^{\infty} dt \; \bar\psi_+(t)[i\partial_t - \epsilon_0]\psi_+(t) - \int_{-\infty}^{\infty} dt \; \bar\psi_-(t)[i\partial_t - \epsilon_0]\psi_-(t), \qquad (5.9)$$

where the dynamics of ψ_+ and ψ_- are *not* independent from each other, due to the presence of non–zero off–diagonal blocks in the discrete matrix, Eq. (5.4).

The four fermionic Greens functions: $\mathcal{G}^{T(\tilde{T})}$ and $\mathcal{G}^{<(>)}$ are defined in the same way as their bosonic counterparts, Eq. (2.16):

$$\langle \psi_+(t)\bar\psi_-(t') \rangle \equiv i\mathcal{G}^<(t,t') = -\frac{\text{Tr}\{c^\dagger(t')c(t)\hat\rho_0\}}{\text{Tr}\{\hat\rho_0\}} = -n_F \, e^{-i\epsilon_0(t-t')};$$

$$\langle \psi_-(t)\bar\psi_+(t') \rangle \equiv i\mathcal{G}^>(t,t') = \frac{\text{Tr}\{c(t)c^\dagger(t')\hat\rho_0\}}{\text{Tr}\{\hat\rho_0\}} = (1 - n_F) \, e^{-i\epsilon_0(t-t')}; \quad (5.10)$$

$$\langle \psi_+(t)\bar\psi_+(t') \rangle \equiv i\mathcal{G}^T(t,t') = \theta(t-t')i\mathcal{G}^>(t,t') + \theta(t'-t)i\mathcal{G}^<(t,t');$$

$$\langle \psi_-(t)\bar\psi_-(t') \rangle \equiv i\mathcal{G}^{\tilde{T}}(t,t') = \theta(t'-t)i\mathcal{G}^>(t,t') + \theta(t-t')i\mathcal{G}^<(t,t');$$

The difference is in the minus sign in the expression for $\mathcal{G}^<$, due to the anti–commutation relations, and the Planck occupation number is exchanged for the Fermi one: $n \to n_F \equiv \rho(\epsilon_0)/(1 + \rho(\epsilon_0))$. Equations (2.17) and (2.18) hold for the fermionic Green functions as well.

5.2. Keldysh rotation

It is customary to perform the Keldysh rotation in the fermionic case in a different manner from the bosonic one. Define the new fields as:

$$\psi_1(t) = \frac{1}{\sqrt{2}}\big(\psi_+(t) + \psi_-(t)\big); \qquad \psi_2(t) = \frac{1}{\sqrt{2}}\big(\psi_+(t) - \psi_-(t)\big). \tag{5.11}$$

This line is exactly parallel to the bosonic one, Eq. (2.19). However, following Larkin and Ovchinnikov [20], it is agreed that the "bar" fields transform in a different way:

$$\bar{\psi}_1(t) = \frac{1}{\sqrt{2}}\big(\bar{\psi}_+(t) - \bar{\psi}_-(t)\big); \qquad \bar{\psi}_2(t) = \frac{1}{\sqrt{2}}\big(\bar{\psi}_+(t) + \bar{\psi}_-(t)\big). \tag{5.12}$$

The point is that the Grassmann fields $\bar{\psi}$ are *not* conjugated to ψ, but rather are completely independent fields, that may be chosen to transform in an arbitrary manner (as long as the transformation matrix has a non-zero determinant). Notice that there is no issue regarding the convergence of the integrals, since the Grassmann integrals are *always* convergent. We also avoid the subscripts cl and q, because the Grassmann variables *never* have a classical meaning. Indeed, one can never write a saddle–point or any other equation in terms of $\bar{\psi}$, ψ rather they must always be integrated out in some stage of the calculations.

Employing Eqs. (5.11), (5.12) along with Eq. (5.10), one finds:

$$-i \langle \psi_a(t)\bar{\psi}_b(t')\rangle \equiv \hat{\mathcal{G}}^{ab} = \begin{pmatrix} \mathcal{G}^R(t,t') & \mathcal{G}^K(t,t') \\ 0 & \mathcal{G}^A(t,t') \end{pmatrix}, \tag{5.13}$$

where hereafter $a, b = (1, 2)$. The presence of zero in the $(2, 1)$ element of this matrix is a manifestation of identity (2.18). The *retarded, advanced* and *Keldysh* components of the Green function are expressed in terms of $\mathcal{G}^{T(\bar{T})}$ and $\mathcal{G}^{<(>)}$ in exactly the same way as their bosonic analogs, Eq. (2.21), and therefore posses the same symmetry properties: Eqs. (2.22)–(2.24). An important consequence of Eqs. (2.23), (2.24) is:

$$\mathrm{Tr}\left\{\hat{\mathcal{G}}_1 \circ \hat{\mathcal{G}}_2 \circ \ldots \circ \hat{\mathcal{G}}_k\right\}(t,t) = 0, \tag{5.14}$$

where the circular multiplication sign involves integration over the intermediate times along with the 2×2 matrix multiplication. The argument (t, t) states that the first time argument of $\hat{\mathcal{G}}_1$ and the last argument of $\hat{\mathcal{G}}_k$ are the same.

Notice that the fermionic Green function has a different structure from its bosonic counterpart, Eq. (2.20): the positions of the R, A and K components in the matrix are exchanged. The reason, of course, is the different convention for transformation of the "bar" fields. One could choose the fermionic convention to be the same as the bosonic (but *not* the other way around!), thus having the same structure, Eq. (2.20), for fermions as for bosons. The rationale for the Larkin–Ovchinnikov choice, Eq. (5.13), is that the inverse Green function, $\hat{G}^{-1}$ and fermionic self energy $\hat{\Sigma}_F$ have the same appearance as $\hat{G}$:

$$\hat{G}^{-1} = \begin{pmatrix} [G^R]^{-1} & [G^{-1}]^K \\ 0 & [G^A]^{-1} \end{pmatrix}; \qquad \hat{\Sigma}_F = \begin{pmatrix} \Sigma_F^R & \Sigma_F^K \\ 0 & \Sigma_F^A \end{pmatrix}, \qquad (5.15)$$

whereas in the case of bosons $\hat{G}^{-1}$, Eq. (2.29), and $\hat{\Sigma}$, Eq. (3.10), look differently from $\hat{G}$, Eq. (2.20). This fact gives the form Eq. (5.13), (5.15) a certain technical advantage.

For the single fermionic state (see. Eq. (5.10)):

$$\begin{aligned} G^R(t, t') &= -i\theta(t - t')e^{-i\epsilon_0(t-t')} \rightarrow (\epsilon - \epsilon_0 + i0)^{-1}; \\ G^A(t, t') &= i\theta(t' - t)e^{-i\epsilon_0(t-t')} \rightarrow (\epsilon - \epsilon_0 - i0)^{-1}; \\ G^K(t, t') &= -i(1 - 2n_F)e^{-i\epsilon_0(t-t')} \rightarrow (1 - 2n_F(\epsilon))(-2\pi i)\delta(\epsilon - \epsilon_0). \end{aligned} \qquad (5.16)$$

where the r.h.s. provides also the Fourier transforms. In thermal equilibrium, one obtains:

$$G^K(\epsilon) = \tanh \frac{\epsilon}{2T}\left(G^R(\epsilon) - G^A(\epsilon)\right). \qquad (5.17)$$

This is the FDT for fermions. As in the case of bosons, the FDT statement is a generic feature of an equilibrium system, not restricted to the toy model. In general, it is convenient to parameterize the anti-Hermitian Keldysh Green function by a Hermitian matrix $\mathcal{F} = \mathcal{F}^\dagger$ as:

$$G^K = G^R \circ \mathcal{F} - \mathcal{F} \circ G^A, \qquad (5.18)$$

The Wigner transform of $\mathcal{F}(t, t')$ plays the role of the fermionic distribution function.

One may continue now to a system with many degrees of freedom, counted by an index k. To this end, one simply changes: $\epsilon_0 \rightarrow \epsilon_k$ and perform summations over k. If k is a momentum and $\epsilon_k = k^2/(2m)$, it is instructive to transform to the real space representation: $\psi(t; k) \rightarrow \psi(t; r)$ and $\epsilon_k = k^2/(2m) = -(2m)^{-1}\nabla_r^2$. Finally, the Keldysh action for a non–interacting gas of fermions takes the form:

$$S_0[\bar{\psi}, \psi] = \iint dx\, dx' \sum_{a,b=1}^{2} \bar{\psi}_a(x)[\hat{G}^{-1}(x, x')]_{ab}\, \psi_b(x'), \qquad (5.19)$$

where $x = (t; r)$ and the matrix correlator $[\hat{G}^{-1}]_{ab}$ has the structure of Eq. (5.15) with

$$[G^{R(A)}(x, x')]^{-1} = \delta(x - x')\left(i\partial_{t'} + \frac{1}{2m}\nabla_{r'}^2\right). \qquad (5.20)$$

Although in continuous notations the R and the A components look seemingly the same, one has to remember that in the discrete time representation, they are matrices with the structure below and above the main diagonal correspondingly. The Keldysh component is a pure regularization, in the sense that it does not have a continuum limit (the self–energy Keldysh component does have a non–zero continuum representation). All this information is already properly taken into account, however, in the structure of the Green function, Eq. (5.13).

5.3. *External fields and sources*

Let us introduce an external time–dependent scalar potential $-V(t)$ defined along the contour. It interacts with the fermions as: $S_V = \int_C dt\, V(t)\bar{\psi}(t)\psi(t)$. Expressing it via the field components, one finds:

$$S_V = \int_{-\infty}^{\infty} dt \left[V_+\bar{\psi}_+\psi_+ - V_-\bar{\psi}_-\psi_-\right] = \int_{-\infty}^{\infty} dt \left[V_{cl}\left(\bar{\psi}_+\psi_+ - \bar{\psi}_-\psi_-\right) + V_q\left(\bar{\psi}_+\psi_+ + \bar{\psi}_-\psi_-\right)\right]$$

$$= \int_{-\infty}^{\infty} dt \left[V_{cl}\left(\bar{\psi}_1\psi_1 + \bar{\psi}_2\psi_2\right) + V_q\left(\bar{\psi}_1\psi_2 + \bar{\psi}_2\psi_1\right)\right], \tag{5.21}$$

where the V_{cl} and the V_q components are defined in the standard way for real bosonic fields: $V_{cl(q)} = (V_+ \pm V_-)/2$. Notice that the physical fermionic density (symmetrized over the two branches of the contour): $\varrho = \frac{1}{2}\left(\bar{\psi}_+\psi_+ + \bar{\psi}_-\psi_-\right)$ is coupled to the *quantum* component of the source field, V_q. On the other hand, the classical source component, V_{cl}, is nothing but an external physical scalar potential, the same at the two branches.

Notations may be substantially compactified by introducing vertex gamma–matrices:

$$\hat{\gamma}^{cl} \equiv \begin{pmatrix} 1 & 0 \\ 0 & 1 \end{pmatrix}; \qquad \hat{\gamma}^q \equiv \begin{pmatrix} 0 & 1 \\ 1 & 0 \end{pmatrix}. \tag{5.22}$$

With the help of these definitions, the source action (5.21) may be written as:

$$S_V = \int_{-\infty}^{\infty} dt \sum_{a,b=1}^{2} \left[V_{cl}\,\bar{\psi}_a\gamma^{cl}_{ab}\psi_b + V_q\,\bar{\psi}_a\gamma^q_{ab}\psi_b\right] = \int_{-\infty}^{\infty} dt \; \hat{\bar{\psi}} V_\alpha \hat{\gamma}^\alpha \hat{\psi}, \tag{5.23}$$

where the summation index is $\alpha = (cl, q)$.

Let us define now the "generating" function as:

$$Z[V_{cl}, V_q] \equiv \left\langle e^{iS_V} \right\rangle, \tag{5.24}$$

where the angular brackets denote the functional integration over the Grassmann fields $\bar{\psi}$ and ψ with the weight of e^{iS_0} specified by the fermionic action (5.19). In the absence of the quantum component, $V_q = 0$, the source field is the same at both branches of the contour. Therefore, the evolution along the contour still brings the system back to its exact initial state. Thus one expects that the classical component alone does not change the fundamental normalization, $Z = 1$. As a result:

$$Z[V_{cl}, V_q = 0] = 1. \tag{5.25}$$

One may verify this statement explicitly by expanding the action in powers of V_{cl} and employing the Wick theorem. For example, in the first order one finds: $Z[V_{cl}, 0] = 1 + \int dt \, V_{cl}(t) \, \text{Tr}\{\hat{\gamma}^{cl}\hat{\mathcal{G}}(t, t)\} = 1$, where one uses that $\hat{\gamma}^{cl} = \hat{1}$ along with Eq. (5.14). It is straightforward to see that for exactly the same reasons all higher order terms in V_{cl} vanish as well.

A lesson from Eq. (5.25) is that one necessarily has to introduce *quantum* sources (that change sign between the forward and the backward branches of the contour). The presence of such source fields explicitly violates causality, and thus changes the generating function. On the other hand, these fields usually do not have a physical meaning and play only an auxiliary role. In most cases one uses them only to generate observables by an appropriate differentiation. Indeed, as was mentioned above, the physical density is coupled to the quantum component of the source. In the end, one takes the quantum sources to be zero, restoring the causality of the action. Notice that the classical component, V_{cl}, does *not* have to be taken to zero.

Let us see how it works. Suppose we are interested in the average fermion density ϱ at time t in the presence of a certain physical scalar potential $V_{cl}(t)$. According to Eqs. (5.21) and (5.24) it is given by:

$$\varrho(t; V_{cl}) = -\frac{i}{2} \frac{\delta}{\delta V_q(t)} Z[V_{cl}, V_q]\Big|_{V_q=0}. \tag{5.26}$$

The problem is simplified if the external field, V_{cl}, is weak in some sense. Then one may restrict oneself to the linear response, by defining the susceptibility:

$$\Pi^R(t, t') \equiv \frac{\delta}{\delta V_{cl}(t')} \varrho(t; V_{cl})\Big|_{V_{cl}=0} = -\frac{i}{2} \frac{\delta^2 Z[V_{cl}, V_q]}{\delta V_{cl}(t')\delta V_q(t)}\Big|_{V_q=V_{cl}=0}. \tag{5.27}$$

We add the subscript R anticipating on physical grounds that the response function must be *retarded* (causality). We shall demonstrate it momentarily. First, let us introduce the *polarization* matrix as:

$$\hat{\Pi}^{\alpha\beta}(t, t') \equiv -\frac{i}{2} \left. \frac{\delta^2 \ln Z[\hat{V}]}{\delta V_\beta(t')\delta V_\alpha(t)} \right|_{\hat{V}=0} = \begin{pmatrix} 0 & \Pi^A(t, t') \\ \Pi^R(t, t') & \Pi^K(t, t') \end{pmatrix}. \quad (5.28)$$

Due to the fundamental normalization, Eq. (5.25), the logarithm is redundant for the R and the A components and therefore the two definitions (5.27) and (5.28) are not in contradiction. The fact that $\Pi^{cl,cl} = 0$ is obvious from Eq. (5.25). To evaluate the polarization matrix, $\hat{\Pi}$, consider the Gaussian action, Eq. (5.19). Adding the source term, Eq. (5.23), one finds: $S_0 + S_V = \int dt \; \bar{\hat{\psi}} [\hat{G}^{-1} + V_\alpha \hat{\gamma}^\alpha]\psi$. Integrating out the fermion fields $\bar{\hat{\psi}}$, $\hat{\psi}$ according to the rules of fermionic Gaussian integration, Appendix A, one obtains:

$$Z[\hat{V}] = \frac{1}{\mathrm{Tr}\hat{\rho}_0} \det\left\{i\hat{G}^{-1} + iV_\alpha\hat{\gamma}^\alpha\right\} = \det\left\{1 + \hat{G} V_\alpha\hat{\gamma}^\alpha\right\} = e^{\mathrm{Tr}\ln(1 + \hat{G} V_\alpha\hat{\gamma}^\alpha)}, \quad (5.29)$$

where one used normalization, Eq. (5.6). Notice, that the normalization is exactly right, since $Z[0] = 1$. One may now expand $\ln(1 + \hat{G} V_\alpha\hat{\gamma}^\alpha)$ to the second order in V and then differentiate twice. As a result, one finds for the polarization matrix:

$$\Pi^{\alpha\beta}(t, t') = -\frac{i}{2} \, \mathrm{Tr}\left\{\hat{\gamma}^\alpha\hat{G}(t, t')\hat{\gamma}^\beta\hat{G}(t', t)\right\}. \quad (5.30)$$

Substituting the explicit form of the gamma-matrices, Eq. (5.22), and the Green functions, Eq. (5.13), one obtains for the *response* and the *correlation* components:

$$\Pi^{R(A)}(t, t') = -\frac{i}{2}\left[\mathcal{G}^{R(A)}(t, t')\mathcal{G}^K(t', t) + \mathcal{G}^K(t, t')\mathcal{G}^{A(R)}(t', t)\right]; \quad (5.31)$$

$$\Pi^K(t, t') = -\frac{i}{2}\left[\mathcal{G}^K(t, t')\mathcal{G}^K(t', t) + \mathcal{G}^R(t, t')\mathcal{G}^A(t', t) + \mathcal{G}^A(t, t')\mathcal{G}^R(t', t)\right].$$

>From the first line it is obvious that $\Pi^{R(A)}(t, t')$ is indeed a lower (upper) triangular matrix in the time domain, justifying their superscripts. Moreover, from the symmetry properties of the fermionic Green functions (same as Eq. (2.22)) one finds: $\Pi^R = [\Pi^A]^\dagger$ and $\Pi^K = -[\Pi^K]^\dagger$. As a result, the polarization matrix, $\hat{\Pi}$, possesses all the symmetry properties of the bosonic self-energy $\hat{\Sigma}$, Eq. (3.10).

Exercise: In the stationary case: $\hat{G}(t, t') = \hat{G}(t - t')$. Fourier transform to the energy domain and write down expressions for $\hat{\Pi}(\omega)$. Assume thermal equilibrium and, using Eq. (5.17), rewrite your results in terms of $\mathcal{G}^{R(A)}$ and the equilibrium distribution function. Show that in equilibrium the response, $\Pi^{R(A)}(\omega)$, and the correlation, $\Pi^K(\omega)$, functions are related by the bosonic FDT:

$$\Pi^K(\omega) = \coth\frac{\omega}{2T}\left(\Pi^R(\omega) - \Pi^A(\omega)\right). \quad (5.32)$$

Equation (5.31) for Π^R constitutes the Kubo formula for the density–density response function. In equilibrium it may be derived using the Matsubara technique. The Matsubara routine involves, however, the analytical continuation from discrete imaginary frequency ω_m to real frequency ω. This procedure may prove to be cumbersome in specific applications. The purpose of the above discussion is to demonstrate how the linear response problems may be compactly formulated in the Keldysh language. The latter allows to circumvent the analytical continuation and yields results directly in the real frequency domain.

5.4. Tunneling current

As a simple application of the technique, let us derive the expression for the tunneling conductance. Our starting point is the tunneling Hamiltonian:

$$\hat{H} = \sum_k \left[\epsilon_k^{(c)} c_k^\dagger c_k + \epsilon_k^{(d)} d_k^\dagger d_k \right] + \sum_{kk'} \left[T_{kk'} c_k^\dagger d_{k'} + T_{kk'}^* d_{k'}^\dagger c_k \right], \qquad (5.33)$$

where the operators c_k and $d_{k'}$ describe fermions in the left and right leads, while $T_{kk'}$ are tunneling matrix elements between the two. The current operator is:
$$\hat{J} = \frac{d}{dt} \sum_k c_k^\dagger c_k = i[\hat{H}, \sum_k c_k^\dagger c_k]_- = -i \sum_{kk'} \left[T_{kk'} c_k^\dagger d_{k'} - T_{kk'}^* d_{k'}^\dagger c_k \right].$$

To describe the system in the Keldysh formalism, one introduces the four–component spinor: $\hat{\bar{\psi}}_k = (\bar{\psi}_{1k}^{(c)}, \bar{\psi}_{2k}^{(c)}, \bar{\psi}_{1k}^{(d)}, \bar{\psi}_{2k}^{(d)})$, a similarly one for the fields without the bar, and the 4×4 matrices:

$$\hat{\mathcal{G}}_k = \begin{pmatrix} \hat{\mathcal{G}}_k^{(c)} & 0 \\ 0 & \hat{\mathcal{G}}_k^{(d)} \end{pmatrix}; \quad \hat{T}_{k,k'} = \begin{pmatrix} 0 & T_{kk'} \hat{\gamma}^{cl} \\ T_{kk'}^* \hat{\gamma}^{cl} & 0 \end{pmatrix}; \quad \hat{J}_{kk'} = \begin{pmatrix} 0 & i T_{kk'} \hat{\gamma}^q \\ -i T_{kk'}^* \hat{\gamma}^q & 0 \end{pmatrix}. \quad (5.34)$$

In addition to the already familiar Keldysh structure the spinors and matrices above possess the structure of the left–right space. In terms of these objects the action and the current operator take the form:

$$S = \int_{-\infty}^{\infty} dt \sum_{kk'} \hat{\bar{\psi}}_k \left[\delta_{kk'} \hat{\mathcal{G}}_k^{-1} - \hat{T}_{k,k'} \right] \hat{\psi}_{k'}; \qquad \hat{J}(t) = -\sum_{kk'} \hat{\bar{\psi}}_k \hat{J}_{k,k'} \hat{\psi}_{k'}. \qquad (5.35)$$

The current is expressed through the γ^q vertex matrix in the Keldysh space because any observable is generated by differentiation over the *quantum* component of the source field (the classical component of the source does not change the normalization, Eq. (5.25)).

One is now in a position to calculate the average tunneling current up to the second order in the matrix elements $T_{k,k'}$. To this end one expands the action up to the first order in $T_{k,k'}$, and applies the Wick theorem:

$$J(t) = i \int\limits_{-\infty}^{\infty} dt' \sum_{kk'} \mathrm{Tr} \left\{ \hat{J}_{kk'} \hat{\mathcal{G}}_{k'}(t,t') \hat{T}_{k'k} \hat{\mathcal{G}}_k(t',t) \right\} \tag{5.36}$$

$$= \int\limits_{-\infty}^{\infty} dt' \sum_{kk'} |T_{kk'}|^2 \mathrm{Tr} \left\{ \hat{\gamma}^q \hat{\mathcal{G}}_k^{(c)}(t,t') \hat{\gamma}^{cl} \hat{\mathcal{G}}_{k'}^{(d)}(t',t) - \hat{\gamma}^q \hat{\mathcal{G}}_k^{(d)}(t,t') \hat{\gamma}^{cl} \hat{\mathcal{G}}_k^{(c)}(t',t) \right\} =$$

$$\int\limits_{-\infty}^{\infty} dt' \sum_{kk'} |T_{kk'}|^2 \Big[\mathcal{G}_{k(t,t')}^{(c)R} \mathcal{G}_{k'(t',t)}^{(d)K} + \mathcal{G}_{k(t,t')}^{(c)K} \mathcal{G}_{k'(t',t)}^{(d)A} - \mathcal{G}_{k'(t,t')}^{(d)R} \mathcal{G}_{k(t',t)}^{(c)K} - \mathcal{G}_{k'(t,t')}^{(d)K} \mathcal{G}_{k(t',t)}^{(c)A} \Big].$$

Now let us assume a stationary situation so that all the Green functions depend on the time difference only. We shall also assume that each lead is in local thermal equilibrium and thus its Green functions are related to each other via the FDT: $\mathcal{G}_k^{(c)K}(\epsilon) = (1 - 2n_F^{(c)}(\epsilon))[\mathcal{G}_k^{(c)R}(\epsilon) - \mathcal{G}_k^{(c)A}(\epsilon)]$. Similarly for the "$d$"–lead with a different occupation function $n_F^{(d)}(\epsilon)$. As a result, one finds for the tunneling current:

$$J = \int\limits_{-\infty}^{\infty} \frac{d\epsilon}{\pi} \left[n_F^{(d)}(\epsilon) - n_F^{(c)}(\epsilon) \right] \sum_{kk'} |T_{kk'}|^2 [\mathcal{G}_k^{(c)R} - \mathcal{G}_k^{(c)A}][\mathcal{G}_{k'}^{(d)R} - \mathcal{G}_{k'}^{(d)A}]. \tag{5.37}$$

If the current matrix elements may be considered as approximately momentum independent: $|T_{kk'}|^2 \approx |T|^2$, the last expression is reduced to:

$$J = 4\pi |T|^2 \int\limits_{-\infty}^{\infty} d\epsilon \left[n_F^{(c)}(\epsilon) - n_F^{(d)}(\epsilon) \right] \nu^{(c)}(\epsilon) \, \nu^{(d)}(\epsilon), \tag{5.38}$$

where the density of states (DOS) is defined as:

$$\nu(\epsilon) \equiv \frac{i}{2\pi} \sum_k \left[\mathcal{G}_k^R(\epsilon) - \mathcal{G}_k^A(\epsilon) \right]. \tag{5.39}$$

5.5. Interactions

Consider a liquid of fermions that interact through instantaneous density–density interactions: $\hat{H}_{int} = -\frac{1}{2} \iint dr \, dr' : \hat{\varrho}(r) U(r - r') \hat{\varrho}(r') :$, where $\hat{\varrho}(r) = c_r^\dagger c_r$ is the local density operator and $: \ldots :$ stands for normal ordering. The corresponding Keldysh contour action has the form: $S_{int} = \frac{1}{2} \int_C dt \iint dr \, dr' U(r -$

$r')\bar{\psi}_r\bar{\psi}_{r'}\psi_r\psi_{r'}$. One may now perform the Hubbard–Stratonovich transformation with the help of a real boson field $\varphi(t; r)$, defined along the contour:

$$e^{\frac{i}{2}\int dt \iint dr dr' U(r-r')\bar{\psi}_r\bar{\psi}_{r'}\psi_r\psi_{r'}} = \int \mathcal{D}\varphi \, e^{i\int_c dt\left[\frac{1}{2}\iint dr dr'\varphi_r U_{rr'}^{-1}\varphi_{r'} + \int dr\varphi_r\bar{\psi}_r\psi_r\right]}, \quad (5.40)$$

where U^{-1} is a kernel, that is inverse to the interaction potential: $U^{-1} \circ U = 1$. One notices that the auxiliary bosonic field, φ, enters the fermionic action in exactly the same manner as a scalar source field. Following Eq. (5.21), one introduces $\varphi_{cl(q)} \equiv (\varphi_+ \pm \varphi_-)/2$ and rewrites the fermion–boson interaction term as $\bar{\psi}_a\varphi_\alpha\gamma_{ab}^\alpha\psi_b$, where summations are assumed over $a, b = (1, 2)$ and $\alpha = (cl, q)$. The free bosonic term takes the form of: $\frac{1}{2}\varphi U^{-1}\varphi \rightarrow \varphi_\alpha U^{-1}\hat{\sigma}_1^{\alpha\beta}\varphi_\beta$.

At this stage the fermionic action is Gaussian and one may integrate out the Grassmann variables in the same way it was done in Eq. (5.29). As a result, one finds for the generating function, Eq. (5.24), of the interacting fermionic liquid:

$$Z[\hat{V}] = \int \mathcal{D}\varphi \, e^{i\int_{-\infty}^{\infty} dt \iint dr dr' \hat{\varphi} U^{-1}\hat{\sigma}_1\hat{\varphi} + \text{Tr}\ln\left[1 + \hat{\mathcal{G}}(V_\alpha + \varphi_\alpha)\hat{\gamma}^\alpha\right]}. \quad (5.41)$$

Quite generally, thus, one may reduce an interacting fermionic problem to a theory of an effective non–linear bosonic field (longitudinal photons). Let us demonstrate that this bosonic theory possesses the causality structure. To this end, one formally expands the logarithm on the r.h.s. of Eq. (5.41). Employing Eq. (5.14) and recalling that $\hat{\gamma}^{cl} = \hat{1}$, one notices that for $\varphi_q = V_q = 0$ the bosonic action is zero. As a result, Eq. (2.32) holds.

To proceed we shall restrict ourselves to the, so called, *random phase approximation* (RPA). It neglects all terms in the expansion of the logarithm beyond the second order. The second order term in the expansion is conveniently expressed through the (bare) polarization matrix $\Pi^{\alpha,\beta}$ (see Eq. (5.30)) of the *non–interacting* fermions. The resulting effective bosonic theory is Gaussian with the action:

$$S_{RPA}[\hat{\varphi}, \hat{V}] = \iint_{-\infty}^{\infty} dt dt' \iint dr dr'\left[\hat{\varphi}\left(U^{-1}\hat{\sigma}_1 - \hat{\Pi}\right)\hat{\varphi} - 2\hat{\varphi}\hat{\Pi}\hat{V} - \hat{V}\hat{\Pi}\hat{V}\right]. \quad (5.42)$$

One notices that the bare polarization matrix plays exactly the same role as of the self-energy, $\hat{\Sigma}$, cf. Eqs. (3.9), (3.10), in the effective bosonic theory. As a result, the full bosonic correlator $(U^{-1}\hat{\sigma}_1 - \hat{\Pi})$ possesses all the causality properties, listed in section 2.4.

Finally, let us evaluate the *dressed* polarization matrix of the interacting fermi–liquid in the RPA. To this end one may perform the bosonic Gaussian integration in the RPA action (5.42) to find the logarithm of the generating function: $i \ln Z_{RPA}[\hat{V}] = \hat{V} \left(\hat{\Pi} + \hat{\Pi}(U^{-1}\hat{\sigma}_1 - \hat{\Pi})^{-1}\hat{\Pi} \right) \hat{V}$. Finally, employing the definition of the polarization matrix, Eq. (5.28), and performing simple matrix algebra, one finds:

$$\hat{\Pi}_{RPA} = \hat{\Pi} \circ \left(1 - \hat{\sigma}_1 U \circ \hat{\Pi} \right)^{-1}. \qquad (5.43)$$

It is straightforward to demonstrate that the dressed polarization matrix possesses the same causality structure as the bare one, Eq. (5.28). For the response component of the dressed polarization, $\hat{\Pi}^R_{RPA}$, the second factor on the r.h.s. of Eq. (5.43) may be considered as a modification of the applied field, V_{cl}. Indeed, cf. Eq. (5.27), $\varrho = \hat{\Pi}^R_{RPA} \circ V_{cl} = \hat{\Pi}^R \circ V^{scr}_{cl}$, where the screened external potential V^{scr}_{cl} is given by: $V^{scr}_{cl} = \left(1 - \hat{\sigma}_1 U \circ \hat{\Pi}^R \right)^{-1} \circ V_{cl}$. This is the RPA result for the screening of an external scalar potential.

5.6. Kinetic equation

According to Eq. (5.30) to evaluate the bare (and thus RPA dressed, Eq. (5.43)) polarization matrix, one needs to know the fermionic Green function, $\hat{\mathcal{G}}$. While it is known in equilibrium, it has to be determined self–consistently in an out–of–equilibrium situation. To this end one employs the same idea that was used in the bosonic theory of chapter 3. Namely, one writes down the Dyson equation for the dressed fermionic Green function:

$$\left(\hat{\mathcal{G}}_0^{-1} - \hat{\Sigma}_F \right) \circ \hat{\mathcal{G}} = \hat{1}, \qquad (5.44)$$

where the subscript "0" indicates the bare Green function. The fermionic self–energy, Σ_F turns out to have the same structure as $\hat{\mathcal{G}}^{-1}$, Eq. (5.15). Thus the R and A components of the Dyson equation take a simple form:

$$\left(i\partial_t + \frac{1}{2m}\nabla^2_r \right) \mathcal{G}^{R(A)} = \delta(t - t')\delta(r - r') + \Sigma^{R(A)}_F \circ \mathcal{G}^{R(A)}. \qquad (5.45)$$

Employing the parameterization $\mathcal{G}^K = \mathcal{G}^R \circ \mathcal{F} - \mathcal{F} \circ \mathcal{G}^A$, where $\mathcal{F}$ is a Hermitian matrix, along with Eq. (5.45), one rewrites the Keldysh component of the Dyson equation as:

$$\left[\mathcal{F}, \left(i\partial_t + \frac{1}{2m}\nabla^2_r \right) \right]_- = \Sigma^K_F - \left(\Sigma^R_F \circ \mathcal{F} - \mathcal{F} \circ \Sigma^A_F \right) = -i \, I_{col}[\mathcal{F}]. \qquad (5.46)$$

This equation is the quantum kinetic equation for the distribution matrix $\mathcal{F}$. Its l.h.s. is the *kinetic* term, while the r.h.s. is the *collision integral* with Σ_F^K having the meaning of an "incoming" term and $\Sigma_F^R \circ \mathcal{F} - \mathcal{F} \circ \Sigma_F^A$ that of an "outgoing" term.

The simplest diagram for the fermionic self-energy matrix, $\hat{\Sigma}_F^{ab}$, is obtained by expanding the Hubbard–Stratonovich transformed action, Eq. (5.40), to the second order in the fermion–boson interaction vertex, $\bar{\hat{\psi}}_a \varphi_\alpha \hat{\gamma}_{ab}^\alpha \hat{\psi}_b$, and applying the Wick theorem for both fermion and boson fields. As a result, one finds:

$$\hat{\Sigma}_F^{ab}(t, t') = \left(\hat{\gamma}_{ac}^\alpha \hat{\mathcal{G}}^{cd}(t, t') \hat{\gamma}_{db}^\beta\right) \langle \varphi_\alpha(t) \varphi_\beta(t') \rangle = \left(\hat{\gamma}^{cl} \hat{\mathcal{G}}(t, t') \hat{\gamma}^{cl}\right)^{ab} i D^K(t, t');$$

$$+\left(\hat{\gamma}^{cl} \hat{\mathcal{G}}(t, t') \hat{\gamma}^q\right)^{ab} i D^R(t, t') + \left(\hat{\gamma}^q \hat{\mathcal{G}}(t, t') \hat{\gamma}^{cl}\right)^{ab} i D^A(t, t'), \qquad (5.47)$$

where summations over all repeated indices are understood and the spatial arguments have the same general structure as the time ones. The boson Green function is denoted as $\langle \varphi_\alpha(t) \varphi_\beta(t') \rangle = i D^{\alpha\beta}(t, t')$. Finally one finds for the R, A (i.e. $(1, 1)$ and $(2, 2)$) and K (i.e $(1, 2)$) components of the fermionic self-energy:

$$\Sigma_F^{R(A)}(t, t') = i \left(\mathcal{G}^{R(A)}(t, t') D^K(t, t') + \mathcal{G}^K(t, t') D^{R(A)}(t, t')\right); \qquad (5.48)$$

$$\Sigma_F^K(t, t') = i \left(\mathcal{G}^K(t, t') D^K(t, t') + \mathcal{G}^R(t, t') D^R(t, t') + \mathcal{G}^A(t, t') D^A(t, t')\right)$$

$$= i \left(\mathcal{G}^K(t, t') D^K(t', t) + \left(\mathcal{G}^R(t, t') - \mathcal{G}^A(t, t')\right)\left(D^R(t, t') - D^A(t, t')\right)\right),$$

where in the last equality one had used that $\mathcal{G}^{R(A)}(t, t') D^{A(R)}(t, t') = 0$, since these expressions have no support in the time domain (see, however, the footnote in section 3.1). For the same reason: $\Sigma_F^{21}(t, t') = i\left(\mathcal{G}^A(t, t') D^R(t, t') + \mathcal{G}^R(t, t') D^A(t, t')\right) = 0$. As expected, the retarded and advanced components are lower and upper triangular matrices correspondingly, with $\Sigma^R = [\Sigma^A]^\dagger$, while $\Sigma^K = -[\Sigma^K]^\dagger$. Notice the close resemblance of expressions (5.48) to their bosonic counterparts, Eqs. (3.14)–(3.16).

If one understands the bosonic Green function, $\hat{D}$, as the bare *instantaneous* interaction potential (i.e. $D^R = D^A = U(r - r')\delta(t - t')$ and $D^K = 0$), one finds: $\Sigma_F^R = \Sigma_F^A = i U \mathcal{G}^K(t, t)\delta(t - t')$ and $\Sigma_F^K = 0$. In this approximation the r.h.s. of the kinetic equation (5.46) vanishes (since $\mathcal{F}$ is a symmetric matrix) and so there is no collisional relaxation. Thus one has to employ an approximation for $\hat{D}$ that contains some retardation. The simplest and most convenient one is the RPA, where $\hat{D} = (U^{-1}\hat{\sigma}_1 - \hat{\Pi})^{-1}$, cf. Eq. (5.42), with a matrix $\hat{\Pi}$ that is non–local in time. This relation may be rewritten as the Dyson equation for $\hat{D}$, namely $(U^{-1}\hat{\sigma}_1 - \hat{\Pi}) \circ \hat{D} = \hat{1}$. One may easily solve it for the three components of $\hat{D}$ and write them in the following way:

$$D^{R(A)} = D^R \circ \left(U^{-1} - \Pi^{A(R)}\right) \circ D^A; \quad D^K = D^R \circ \Pi^K \circ D^A. \quad (5.49)$$

Performing the Wigner transform following sections 3.5, 3.6, the kinetic term (the l.h.s. of Eq. (5.46)) is exactly the same as for the complex boson case (one has to take into account the gradient terms to obtain a non–zero result for the WT of the commutator). The result is (cf. Eq. (3.17)):

$$\left(\partial_\tau - v_k \nabla_\rho - E \nabla_k\right) \mathbf{f_F}(\tau, \rho, k) = I_{col}[\mathbf{f_F}], \quad (5.50)$$

where $v_k = \partial_k \epsilon_k$, E is an external electric field and the collision integral, I_{col}, is i times the WT of the r.h.s. of Eq. (5.46). On the r.h.s. one may keep only the leading terms (without the gradients). One also employs a parameterization of the Keldysh component of the fermionic Green function through the corresponding distribution function: $\mathcal{G}^K \to \mathsf{g}^K = \mathbf{f_F}(\mathsf{g}^R - \mathsf{g}^A)$, where $\mathbf{f_F}(\tau, \rho, k)$ is the WT of $\mathcal{F}$. Assuming, for brevity, a spatially uniform and momentum isotropic case, one may restrict oneself to $\mathbf{f_F}(\tau, \epsilon_k) = \mathbf{f_F}(\tau, \epsilon)$. As a result, one finds for the collision integral:

$$I_{col}[\mathbf{f_F}(\epsilon)] = i \int d\omega \sum_q D^R(\omega, q) D^A(\omega, q) \Delta_\mathsf{g}(\epsilon - \omega, k - q) \quad (5.51)$$

$$\times \left[\left(\Pi^R - \Pi^A\right)\left(1 - \mathbf{f_F}(\epsilon - \omega)\mathbf{f_F}(\epsilon)\right) - \Pi^K\left(\mathbf{f_F}(\epsilon) - \mathbf{f_F}(\epsilon - \omega)\right)\right],$$

where $\Pi^{\alpha\beta} = \Pi^{\alpha\beta}(\omega, q)$, while the time index, τ, is suppressed for brevity and the notation

$$\Delta_\mathsf{g}(\epsilon, k) \equiv \frac{i}{2\pi}\left(\mathsf{g}^R(\epsilon, k) - \mathsf{g}^A(\epsilon, k)\right) \quad (5.52)$$

is introduced. For free fermions $\Delta_\mathsf{g}(\epsilon, k) = \delta(\epsilon - \epsilon_k)$. At this stage one may observe that if the bosonic system is at equilibrium: $\Pi^K = \coth(\omega/2T)\left[\Pi^R - \Pi^A\right]$, then the fermionic collision integral is nullified by:

$$\mathbf{f_F}(\epsilon) = \tanh \frac{\epsilon}{2T} . \quad (5.53)$$

Indeed, $1 - \tanh(b - a)\tanh(b) = \coth(a)\left(\tanh(b) - \tanh(b - a)\right)$. One should take into account, however, that the bosonic degrees of freedom are *not* independent from the fermionic ones. Namely, components of the polarization matrix $\hat{\Pi}$

are expressed through the fermionic Green functions according to Eq. (5.31). In the WT representation these relations take the form:

$$\Pi^R - \Pi^A = i\pi \int d\epsilon' \sum_{k'} \Delta g(\epsilon', k') \Delta g(\epsilon' - \omega, k' - q) \left[\mathbf{f_F}(\epsilon' - \omega) - \mathbf{f_F}(\epsilon') \right];$$

$$\Pi^K(\omega, q) = i\pi \int d\epsilon' \sum_{k'} \Delta g(\epsilon', k') \Delta g(\epsilon' - \omega, k' - q) \left[\mathbf{f_F}(\epsilon' - \omega) \mathbf{f_F}(\epsilon') - 1 \right].$$

(5.54)

Due to the same trigonometric identity the equilibrium argument can be made self-consistent: if the fermionic system is in equilibrium, Eq. (5.53), then components of $\hat{\Pi}$ satisfy the bosonic FDT, Eq. (3.22).

One may substitute now Eqs. (5.54) into Eq. (5.51) to write down the closed kinetic equation for the fermionic distribution function. Most conveniently it is done in terms of the occupation numbers, defined as $\mathbf{f_F} \equiv 1 - 2\mathbf{n}$ [6]:

$$\frac{\partial \mathbf{n}_\epsilon}{\partial \tau} = \iint d\omega d\epsilon' M \left[\mathbf{n}_{\epsilon'} \mathbf{n}_{\epsilon - \omega} (1 - \mathbf{n}_\epsilon)(1 - \mathbf{n}_{\epsilon' - \omega}) - \mathbf{n}_\epsilon \mathbf{n}_{\epsilon' - \omega} (1 - \mathbf{n}_{\epsilon'})(1 - \mathbf{n}_{\epsilon - \omega}) \right], (5.55)$$

where the transition probability is given by:

$$M(\epsilon, \omega) = 4\pi \sum_{q,k'} |D^R(\omega, q)|^2 \Delta g(\epsilon - \omega, k - q) \Delta g(\epsilon', k') \Delta g(\epsilon' - \omega, k' - q).$$

(5.56)

Equation (5.55) is a generic kinetic equation with a "four–fermion" collisional relaxation. The first term in the square brackets on its r.h.s. may be identified as "in", while the second one as "out". Each of these terms consists of the product of four occupation numbers, giving a probability of having two initial states occupied and two final states empty. For $\mathbf{n}(\epsilon)$ given by the Fermi function the "in" and the "out" terms cancel each other. Therefore in thermal equilibrium the components of the *dressed* fermionic Green function must satisfy the FDT:

$$\mathcal{G}^K = \tanh \frac{\epsilon}{2T} \left(\mathcal{G}^R - \mathcal{G}^A \right).$$

(5.57)

The structure of the transmission probability M is illustrated in Fig. 7. The three factors of Δg enforce that all three intermediate fermionic particles must satisfy energy–momentum conservation (stand on mass–shell), up to the quasi-particle life–time. The real factor $|D^R|^2$ is associated with the square of the matrix element of the screened interaction potential (in the RPA).

---

[6] To derive this expression one should add and subtract $\mathbf{n}_\epsilon \mathbf{n}_{\epsilon' - \omega} \mathbf{n}_{\epsilon'} \mathbf{n}_{\epsilon - \omega}$ in the square brackets.

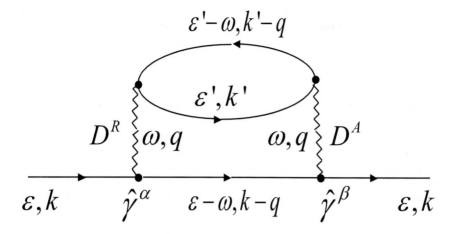

Fig. 7. Structure of the four–fermion collision integral. The full lines are fermionic Green functions; the wavy lines are the RPA screened interaction potential. The fermionic loop represents the polarization matrix, $\hat{\Pi}(\omega, q)$.

6. Disordered fermionic systems

6.1. Disorder averaging

We consider fermions in the field of a static (quenched) space–dependent scalar potential $U_{dis}(r)$. The potential is meant to model the effect of random static impurities, dislocations, etc. Since one does not know the exact form of the potential, the best one can hope for is to evaluate the statistical properties of various observables, assuming some statistics for $U_{dis}(r)$. It is usually a reasonable guess to prescribe a Gaussian distribution for the potential. Namely, one assumes that the relative probability for a realization of the potential to appear in nature is given by:

$$\mathcal{P}[U_{dis}] \sim e^{-\pi \nu \tau \int dr \, U_{dis}^2(r)}, \tag{6.1}$$

where ν is the bare fermionic DOS at the Fermi level and τ, called the *mean–free time*, measures the strength of the random potential.

In this chapter we concentrate on non–interacting fermions. We would like to evaluate, say, the response function, Π^R, in presence of the random potential and average it over the realizations of U_{dis} with the weight given by Eq. (6.1). The crucial observation is that the response function, Π^R, may be defined as variation of the generating function, Eq. (5.27), and *not the logarithm* of the generating function. More precisely, the two definitions with, Eq. (5.28), and without, Eq. (5.27), the logarithm coincide due to the fundamental normalization, Eq. (5.25). This is *not* the case in the equilibrium formalism, where the presence of the logarithm (leading to the factor Z^{-1} after differentiation) is unavoidable in order to have the correct normalization. Such a factor $Z^{-1} = Z^{-1}[U_{dis}]$ formidably complicates the averaging over U_{dis}. Two techniques were invented to perform the averaging: the replica trick [21] and the super-symmetry (SUSY) [22]. The first one utilizes the observation that $\ln Z = \lim_{n\to 0}(Z^n - 1)/n$, to perform calculations for an integer number, n, replicas of the same system and take $n \to 0$ at the end of the calculations. The second one is based on the fact that Z^{-1} of the *non–interacting* fermionic system equals to Z of a bosonic system in the same random potential. One thus introduces an additional bosonic replica of the fermionic system at hand. Both of these ideas have serious drawbacks: the replica technique requires analytical continuation, while the SUSY is not applicable to interacting systems.

The Keldysh formalism provides an alternative to these two methods by insuring that $Z = 1$ by construction. One may thus directly perform the averaging of the generating function, Eq. (5.24), over realizations of U_{dis}. Since the disorder potential possesses only the classical component (it is exactly the same on both branches of the contour), it is coupled only to $\hat{\gamma}^{cl} = \hat{1}$. The disorder–dependent term in the averaged generating function has the form:

$$\int \mathcal{D}U_{dis}\, e^{-\int dr\left[\pi \nu \tau U_{dis}^2(r) - iU_{dis}(r)\int_{-\infty}^{\infty} dt\, \hat{\bar{\psi}}_t \hat{\gamma}^{cl} \hat{\psi}_t\right]} \tag{6.2}$$

$$= e^{-\frac{1}{4\pi \nu \tau}\int dr \iint_{-\infty}^{\infty} dt dt'\, (\bar{\psi}_t^a \psi_t^a)(\bar{\psi}_{t'}^b \psi_{t'}^b)},$$

where $a, b = 1, 2$, and there is a summation over repeated indices. One can rearrange the expression in the exponent on the r.h.s. of the last equation as $(\bar{\psi}_t^a \psi_t^a)(\bar{\psi}_{t'}^b \psi_{t'}^b) = -(\bar{\psi}_t^a \psi_{t'}^b)(\bar{\psi}_{t'}^b \psi_t^a)$ [7] and then use the Hubbard–Stratonovich matrix field, $\hat{Q} = Q_{t,t'}^{ab}(r)$:

$$e^{\frac{1}{4\pi \nu \tau}\int dr \iint_{-\infty}^{\infty} dt dt'\, (\bar{\psi}_t^a \psi_{t'}^b)(\bar{\psi}_{t'}^b \psi_t^a)} = \int \mathcal{D}\hat{Q}\, e^{-\int dr\left[\frac{\pi \nu}{4\tau}\mathrm{Tr}\{\hat{Q}\circ\hat{Q}\} - \frac{i}{2\tau}\iint_{-\infty}^{\infty} dt dt'\, Q_{t,t'}^{ab} \bar{\psi}_{t'}^b \psi_t^a\right]}, \tag{6.3}$$

where the spatial coordinate, r, is suppressed in both $\hat{Q}$ and $\hat{\psi}$. At this stage the *average* action becomes quadratic in the Grassmann variables and they may be integrated out leading to the determinant of the corresponding quadratic form:

---

[7] The minus sign originates from commuting the Grassmann numbers.

$\hat{\mathcal{G}}_0^{-1} + V_\alpha \hat{\gamma}^\alpha + \frac{i}{2\tau}\hat{Q}$. All the matrices here should be understood as having a 2×2 Keldysh structure along with an $N \times N$ structure in discrete time. One thus finds for the *disorder averaged* generating function:

$$Z[\hat{V}] = \int \mathcal{D}\hat{Q} \; e^{iS[\hat{Q};\hat{V}]}, \tag{6.4}$$

where

$$iS[\hat{Q};\hat{V}] = -\frac{\pi \nu}{4\tau} \int dr \; \mathrm{Tr}\{\hat{Q}^2\} + \mathrm{Tr} \ln \left[\hat{\mathcal{G}}_0^{-1} + \frac{i}{2\tau}\hat{Q} + V_\alpha \hat{\gamma}^\alpha \right]. \tag{6.5}$$

As a result, one has traded the initial functional integral over the static field $U_{dis}(r)$ for the functional integral over the dynamic matrix field $\hat{Q}_{t,t'}(r)$. At first glance, it does not strike as a terribly bright idea. Nevertheless, there is a great simplification hidden in this procedure. The point is that the disorder potential, being δ–correlated, is a rapidly oscillating function. On the other hand, as one will see below, the $\hat{Q}$–matrix field is a slow (both in space and time) function. Thus it represents the true *macroscopic* (or hydrodynamic) degrees of freedom of the system, that happen to be the diffusively propagating modes.

6.2. Non–linear σ–model

To execute this program, one first looks for the stationary configurations of the action (6.5). Taking the variation over $\hat{Q}_{t,t'}(r)$, one obtains:

$$\underline{\hat{Q}}_{t,t'}(r) = \frac{i}{\pi \nu} \left[\hat{\mathcal{G}}_0^{-1} + \frac{i}{2\tau}\underline{\hat{Q}} \right]^{-1} \Bigg|_{t,t';r,r}, \tag{6.6}$$

where $\underline{\hat{Q}}$ denotes a stationary configuration of the fluctuating field $\hat{Q}$. For the purpose of finding the stationary configurations one has omitted the small source field, $\hat{V}$. It is important to notice that the spatially non–local operator $[\hat{\mathcal{G}}_0^{-1} + \frac{i}{2\tau}\underline{\hat{Q}}]^{-1}(t,t';r,r')$ on the r.h.s. is taken at coinciding spatial points $r' = r$.

The strategy is to find first a spatially uniform and time–translationally invariant solution of Eq. (6.6): $\underline{\hat{Q}}_{t-t'}$, and then consider space and time–dependent deviations from such a solution. This strategy is adopted from the theory of magnetic systems, where one first finds the uniform static magnetized configurations and then treats spin–waves as smooth perturbations on top of such static uniform solutions. From the structure of Eq. (6.6) one expects that the saddle–point configuration $\underline{\hat{Q}}$ possesses the same structure as the fermionic self–energy, Eq. (5.15) (more accurately, one expects that among the possible saddle points there is a "classical" one, that satisfies the causality structure, Eq. (5.15)). One looks, therefore, for a solution of Eq. (6.6) in the form of:

$$
\hat{\underline{Q}}_\epsilon \equiv \hat{\Lambda}_\epsilon = \begin{pmatrix} \Lambda^R_\epsilon & \Lambda^K_\epsilon \\ 0 & \Lambda^A_\epsilon \end{pmatrix}.
\tag{6.7}
$$

Substituting this expression in Eq. (6.6), one finds

$$
\Lambda^{R(A)}_\epsilon = \frac{i}{\pi\nu} \frac{1}{\left[\mathcal{G}^{R(A)}_0\right]^{-1} + \frac{i}{2\tau}\Lambda^{R(A)}_\epsilon}\Bigg|_{r,r} = \frac{i}{\pi\nu} \sum_k \frac{1}{\epsilon - \xi_k + \frac{i}{2\tau}\Lambda^{R(A)}_\epsilon} = \pm 1, \tag{6.8}
$$

where $\xi_k \equiv k^2/(2m) - \mu$ and one adopts $\sum_k \ldots = \nu \int d\xi_k \ldots$, where ν is the DOS at the Fermi surface. The summation over momentum appears because the matrix on the r.h.s. is taken at coinciding spatial points. The signs are chosen so as to respect causality: the retarded (advanced) Green function is analytic in the entire upper (lower) half-plane of complex energy ϵ. One has also assumed that $1/(2\tau) \ll \mu$. The Keldysh component, as always, may be parameterized through a Hermitian distribution function matrix: $\Lambda^K = \Lambda^R \circ \mathcal{F} - \mathcal{F} \circ \Lambda^A = 2\mathcal{F}_\epsilon$, where the distribution function $\mathcal{F}_\epsilon$ is not fixed by the saddle point equation (6.6) and must be determined through the boundary conditions. As a result one obtains:

$$
\hat{\Lambda}_\epsilon = \begin{pmatrix} 1 & 2\mathcal{F}_\epsilon \\ 0 & -1 \end{pmatrix}.
\tag{6.9}
$$

Transforming back to the time representation, one obtains $\Lambda^{R(A)}_{t-t'} = \pm\delta(t - t' \mp 0)$, where ∓ 0 indicates that the δ–function is shifted below (above) the main diagonal, $t = t'$. As a result, Tr $\hat{\Lambda}_\epsilon = 0$ and $S[\hat{\Lambda}] = 0$, as it should be, of course, for any purely *classical* field configuration, Eq. (6.7). There is, however, a wider class of configurations, that leave the action (6.5) invariant (zero). Indeed, any field configuration of the form:

$$
\hat{Q} = \hat{\mathcal{T}} \circ \hat{\Lambda} \circ \hat{\mathcal{T}}^{-1},
\tag{6.10}
$$

where $\hat{\mathcal{T}}_{t,t'}(r) = \hat{\mathcal{T}}_{t-t'}$, and thus commutes with $\hat{\mathcal{G}}_0$, obviously does not change the action (6.5). This is the zero–mode Goldstone manifold. The standard way to introduce the massless modes ("spin–waves") is to allow the deformation matrices $\hat{\mathcal{T}}$ to be slow functions of $t + t'$ and r. Thus the expression (6.10) parameterizes the soft modes manifold of the field $\hat{Q}$. One may thus restrict oneself *only* to the field configurations given by Eq. (6.10) and disregard all others (massive modes). An equivalent way to characterize this manifold is by the condition (cf. Eq. (6.9)):

$$
\hat{Q}^2 = \hat{1}.
\tag{6.11}
$$

Our goal now is to derive an action for the soft–mode field configurations given by Eqs. (6.10) or (6.11). To this end one substitutes $\hat{Q} = \hat{T} \circ \hat{\Lambda} \circ \hat{T}^{-1}$ into Eq. (6.5) and cyclically permutes the $\hat{T}$ matrices under the trace sign. This way one arrives at $\hat{T}^{-1} \circ \hat{\mathcal{G}}_0^{-1} \circ \hat{T} = \hat{\mathcal{G}}_0^{-1} + \hat{T}^{-1} \circ [\hat{\mathcal{G}}_0^{-1}, \hat{T}]_- = \hat{\mathcal{G}}_0^{-1} + i\hat{T}^{-1} \circ [\partial_t + v_F \nabla_r, \hat{T}]_-$, where one has linearized the dispersion relation near the Fermi surface $k^2/(2m) - \mu \approx v_F k \rightarrow iv_F \nabla_r$. As a result, the desired action has the form:

$$i S[\hat{Q}] = \mathrm{Tr} \ln \left[1 + i\hat{\mathcal{G}}\hat{T}^{-1}[\partial_t, \hat{T}]_- + i\hat{\mathcal{G}}\hat{T}^{-1}[v_F \nabla_r, \hat{T}]_- \right], \qquad (6.12)$$

where $\hat{\mathcal{G}}$ is the *impurity dressed* Green function, defined as: $(\hat{\mathcal{G}}_0^{-1} + \frac{i}{2\tau}\hat{\Lambda})\hat{\mathcal{G}} = \hat{1}$. For practical calculations it is convenient to write it as:

$$\hat{\mathcal{G}}_\epsilon(k) = \begin{pmatrix} \mathcal{G}_\epsilon^R(k) & \mathcal{G}_\epsilon^K(k) \\ 0 & \mathcal{G}_\epsilon^A(k) \end{pmatrix} = \frac{1}{2}\left(\mathcal{G}_\epsilon^R(k)\,[\hat{1} + \hat{\Lambda}_\epsilon] + \mathcal{G}_\epsilon^A(k)\,[\hat{1} - \hat{\Lambda}_\epsilon] \right), \quad (6.13)$$

with

$$\begin{aligned} \mathcal{G}_\epsilon^{R(A)}(k) &= (\epsilon - \xi_k \pm i/(2\tau))^{-1}; \\ \mathcal{G}_\epsilon^K(k) &= \mathcal{G}_\epsilon^R(k)\,\mathcal{F}_\epsilon - \mathcal{F}_\epsilon\,\mathcal{G}_\epsilon^A(k). \end{aligned} \qquad (6.14)$$

Notice that $\sum_k \hat{\mathcal{G}}_\epsilon(k) = -i\pi\nu\,\hat{\Lambda}_\epsilon$ and $\sum_k \mathcal{G}_\epsilon^R(k)\mathcal{G}_\epsilon^A(k) = 2\pi\nu\tau$, while the other combinations vanish: $\sum_k \mathcal{G}_\epsilon^R(k)\mathcal{G}_\epsilon^R(k) = \sum_k \mathcal{G}_\epsilon^A(k)\mathcal{G}_\epsilon^A(k) = 0$, due to the complex ξ_k–plane integration.

One can now expand the logarithm in Eq. (6.12) to the first order in the ∂_t term and to the second order in the ∇_r term (the first order term in ∇_r vanishes due to the angular integration) and evaluate traces using Eq. (6.14). For the ∂_t term one finds: $\pi\nu\,\mathrm{Tr}\{\hat{\Lambda}\hat{T}^{-1}[\partial_t, \hat{T}]_-\} = \pi\nu\,\mathrm{Tr}\{\partial_t \hat{Q}\}$, where one used that $\mathrm{Tr}\{\partial_t \hat{\Lambda}\} = 0$. For the ∇_r term, one finds: $-\frac{1}{4}\pi\nu D\mathrm{Tr}\{(\nabla_r \hat{Q})^2\}$, where $D \equiv v_F^2\tau/d$ is the diffusion constant and d is the spatial dimensionality [8]. Finally, one finds for the action of the soft–mode configurations [14]:

$$S[\hat{Q}] = i\pi\nu \int dr\, \mathrm{Tr}\left\{ \frac{1}{4} D\big(\nabla_r \hat{Q}(r)\big)^2 - \partial_t \hat{Q}(r) - iV_\alpha \hat{\gamma}^\alpha \hat{Q}(r) - \frac{i}{\pi}\hat{V}^T \hat{\sigma}_1 V \right\}, \quad (6.15)$$

where the trace is performed over the 2×2 Keldysh structure as well as over the $N \times N$ time structure. In the last expression we have restored the source term from Eq. (6.5). The last term, $\hat{V}^T \hat{\sigma}_1 V$ is the static compressibility of the electron gas. It originates from the second order expansion of Eq. (6.5) in $\hat{V}$,

---

[8] One uses that $v_F = k/m$ and $\sum_k \mathcal{G}_\epsilon^R(k)\frac{k}{m}\mathcal{G}_\epsilon^A(k)\frac{k}{m} = 2\pi\nu\tau v_F^2/d = 2\pi\nu D$, while the corresponding $R - R$ and $A - A$ terms vanish. Employing Eq. (6.13), one then arrives at $\frac{1}{4}\mathrm{Tr}\{[\hat{1} + \hat{\Lambda}_\epsilon](\hat{T}^{-1}\nabla_r\hat{T})[\hat{1} - \hat{\Lambda}_\epsilon](\hat{T}^{-1}\nabla_r\hat{T})\} = -\frac{1}{8}\mathrm{Tr}\{(\nabla_r(\hat{T}\hat{\Lambda}_\epsilon\hat{T}^{-1}))^2\} = -\frac{1}{8}\mathrm{Tr}\{(\nabla_r\hat{Q})^2\}$.

while keeping the high energy part of the $\mathcal{G}^R\mathcal{G}^R$ and $\mathcal{G}^A\mathcal{G}^A$ terms. Despite of the simple appearance, the action (6.15) is highly non–linear due to the condition $\hat{Q}^2 = 1$. The theory specified by Eqs. (6.11) and (6.15) is called the *matrix non–linear σ–model* (NLσM). The name came from the theory of magnetism, where the unit–length *vector*, $\sigma(r)$, represents a local (classical) spin, that may rotate over the sphere $\sigma^2 = 1$.

6.3. Usadel equation

Our goal is to investigate the physical consequences of the NLσM. As a first step, one wants to determine the most probable (stationary) configuration, $\underline{\hat{Q}}_{t,t'}(r)$, on the soft–modes manifold, Eq. (6.11). To this end one parameterizes deviations from $\underline{\hat{Q}}_{t,t'}(r)$ as $\hat{Q} = \hat{T} \circ \underline{\hat{Q}} \circ \hat{T}^{-1}$ and chooses $\hat{T} = e^{\hat{W}}$, where $\hat{W}_{t,t'}(r)$ is the generator of rotations. Expanding to the first order in $\hat{W}$, one finds: $\hat{Q} = \underline{\hat{Q}} + [\hat{W}\circ, \underline{\hat{Q}}]_-$. One may now substitute such a $\hat{Q}$–matrix into the action (6.15) and require that the term linear in $\hat{W}$ vanishes. This leads to the saddle–point equation for $\underline{\hat{Q}}$. For the first term in the curly brackets on the r.h.s. of Eq. (6.15) one obtains: $\frac{1}{2}\mathrm{Tr}\{\hat{W}\circ\nabla_r D(\nabla_r\underline{\hat{Q}}\circ\underline{\hat{Q}} - \underline{\hat{Q}}\circ\nabla_r\underline{\hat{Q}})\} = -\mathrm{Tr}\{\hat{W}\circ\nabla_r D(\underline{\hat{Q}}\circ\nabla_r\underline{\hat{Q}})\}$, where one employed that $\nabla_r\underline{\hat{Q}}\circ\underline{\hat{Q}} + \underline{\hat{Q}}\circ\nabla_r\underline{\hat{Q}} = 0$, since $\underline{\hat{Q}}^2 = \hat{1}$. For the second term one finds: $\mathrm{Tr}\{\hat{W}_{t,t'}(\partial_{t'}+\partial_t)\underline{\hat{Q}}_{t',t}\}$. It is written more compactly in the energy representation, where $\partial_t \to -i\epsilon$, and thus the second term is: $-i\mathrm{Tr}\{\hat{W}\circ[\epsilon, \underline{\hat{Q}}]_-\}$. Demanding that the linear term in $\hat{W}$ vanish, one finds:

$$\nabla_r(D\underline{\hat{Q}} \circ \nabla_r\underline{\hat{Q}}) + i[\epsilon, \underline{\hat{Q}}]_- = 0. \tag{6.16}$$

This is the Usadel equation for the stationary $\hat{Q}$–matrix, that must also satisfy $\underline{\hat{Q}}^2 = \hat{1}$. In the time representation $i[\epsilon, \underline{\hat{Q}}]_- \to -\{\partial_t, \underline{\hat{Q}}\}_+$.

If one looks for a solution of the Usadel equation (6.16) in the subspace of "*classical*" (having the causality structure) configurations, then the condition $\underline{\hat{Q}}^2 = \hat{1}$ restricts the possible solutions to $\hat{\Lambda}$, Eq. (6.9) (with a yet unspecified distribution matrix $\mathcal{F}_{t,t'}(r)$). Therefore, in the non–superconducting case the Usadel equation is reduced to a single equation for the distribution matrix $\mathcal{F}_{t,t'}(r)$. It contains much more information for the superconducting case (i.e. it also determines the local energy spectrum and superconducting phase). Substituting Eq. (6.9) into the Usadel equation (6.16), one finds:

$$\nabla_r(D\nabla_r\mathcal{F}) + i[\epsilon, \mathcal{F}]_- = 0. \tag{6.17}$$

Finally, performing the time Wigner transform, $\mathcal{F}_{t,t'}(r) \rightarrow \mathbf{f_F}(\tau, \epsilon; r)$, as explained in section 3.5, one obtains:

$$\nabla_r \left(D \nabla_r \mathbf{f_F} \right) - \partial_\tau \mathbf{f_F} = 0. \tag{6.18}$$

This is the kinetic equation for the fermionic distribution function $\mathbf{f_F}(\tau, \epsilon; r)$ of the disordered system. It happens to be the diffusion equation. Notice, that it is the same equation for any energy ϵ and different energies do not "talk" to each other (in the adiabatic case, where the WT works). This is a feature of non–interacting systems. In the presence of interactions, the equation acquires the collision integral on the r.h.s. that mixes different energies between themselves. It is worth mentioning that elastic scattering does not show up in the collision integral. It was already fully taken into account in the derivation of the Usadel equation and went into the diffusion term, $D\nabla_r^2$.

As an example, let us consider a disordered one–dimensional wire of length L [23], attached to two leads, that are kept at different voltages. There is a stationary current passing through the wire. We look for the space dependent distribution function, $\mathbf{f_F}(\epsilon; r)$, that satisfies $D\nabla_r^2 \mathbf{f_F} = 0$ in a stationary setup (for a space independent diffusion constant, D). As a result,

$$\mathbf{f_F}(\epsilon; r) = f_L(\epsilon) + (f_R(\epsilon) - f_L(\epsilon)) \frac{r}{L}, \tag{6.19}$$

where $f_{L(R)}(\epsilon)$ are the distribution functions of the left and right leads. The distribution function inside the wire interpolates the two distributions linearly. At low temperatures it looks like a two–step function, where the energy separation between the steps is the applied voltage, eV, while the height depends on position. Such a distribution was measured in a beautiful experiment [23]. Comparing equation (6.18) with the continuity equation, one notices that the current (at a given energy ϵ) is given by $J(\epsilon) = D\nabla_r \mathbf{f_F} = D\left(f_R(\epsilon) - f_L(\epsilon) \right)/L$. And thus the total current is $J = e \sum_k J(\epsilon) = e \frac{\nu D}{L} \int d\epsilon \left(f_R(\epsilon) - f_L(\epsilon) \right) = e^2 \frac{\nu D}{L} V$. This is the Drude conductivity: $\sigma_D = e^2 \nu D$.

6.4. Fluctuations

Our next goal is to consider fluctuations near the stationary solution, $\underline{\hat{Q}}_{t,t'}(r)$. We restrict ourselves to the soft–mode fluctuations that satisfy $\hat{Q}^2 = 1$ only, and neglect all massive modes that stay outside this manifold. As was already stated above these fluctuations of the $\hat{Q}$–matrix may be parameterized as

$$\hat{Q} = e^{-\mathcal{W}} \circ \underline{\hat{Q}} \circ e^{\mathcal{W}}. \tag{6.20}$$

The part of $\mathcal{W}$ that commutes with $\hat{\underline{Q}}$ does not generate any fluctuations, therefore one restricts $\mathcal{W}$ to satisfy: $\mathcal{W} \circ \hat{\underline{Q}} + \hat{\underline{Q}} \circ \mathcal{W} = 0$. Since $\hat{\underline{Q}}$ may be diagonalized according to:

$$\hat{\underline{Q}} = \begin{pmatrix} 1 & 2\mathcal{F} \\ 0 & -1 \end{pmatrix} = \begin{pmatrix} 1 & \mathcal{F} \\ 0 & -1 \end{pmatrix} \circ \begin{pmatrix} 1 & 0 \\ 0 & -1 \end{pmatrix} \circ \begin{pmatrix} 1 & \mathcal{F} \\ 0 & -1 \end{pmatrix}, \qquad (6.21)$$

any generator $\mathcal{W}$ that anticommutes with $\hat{\underline{Q}}$ may be parameterized as

$$\mathcal{W} = \begin{pmatrix} 1 & \mathcal{F} \\ 0 & -1 \end{pmatrix} \circ \begin{pmatrix} 0 & w \\ \overline{w} & 0 \end{pmatrix} \circ \begin{pmatrix} 1 & \mathcal{F} \\ 0 & -1 \end{pmatrix} = \begin{pmatrix} \mathcal{F} \circ \overline{w} & \mathcal{F} \circ \overline{w} \circ \mathcal{F} - w \\ -\overline{w} & -\overline{w} \circ \mathcal{F} \end{pmatrix}, \qquad (6.22)$$

where $\overline{w}_{t,t'}(r)$ and $w_{t,t'}(r)$ are arbitrary Hermitian matrices in time space. One, thus, understands the functional integration over $\hat{Q}$ as an integration over Hermitian $\overline{w}$ and w. The physical meaning of w is a deviation of the fermionic distribution function, $\mathcal{F}$, from its stationary value. At the same time, $\overline{w}$ has no classical interpretation. To a large extent it plays the role of the quantum counterpart of w, that appears only as the internal line in the diagrams.

One may now expand the action, Eqs. (6.15), in powers of $\overline{w}$ and w. Since $\hat{Q}$ was chosen to be a stationary point, the expansion starts from the second order. In a spatially uniform case one obtains:

$$i S^{(2)}[\mathcal{W}] = 2\pi \nu \int dr \iint \frac{d\epsilon_1 d\epsilon_2}{4\pi^2} \, \overline{w}_{\epsilon_1 \epsilon_2}(r) \left[-D\nabla_r^2 + i(\epsilon_1 - \epsilon_2) \right] w_{\epsilon_2 \epsilon_1}(r). \quad (6.23)$$

The quadratic form is diagonalized by transforming to the momentum representation. As a result, the propagator of small $\hat{Q}$–matrix fluctuations is given by:

$$\langle w_{\epsilon_2 \epsilon_1}(q) \overline{w}_{\epsilon_3 \epsilon_4}(-q) \rangle_{\mathcal{W}} = -\frac{1}{2\pi \nu} \frac{\delta_{\epsilon_1 \epsilon_3} \delta_{\epsilon_2 \epsilon_4}}{Dq^2 + i\omega} \equiv -\frac{\delta_{\epsilon_1 \epsilon_3} \delta_{\epsilon_2 \epsilon_4}}{2\pi \nu} D(\omega, q), \quad (6.24)$$

where $\omega \equiv \epsilon_1 - \epsilon_2$ and the object $D(\omega, q) = D(\epsilon_1 - \epsilon_2, q) = (Dq^2 + i(\epsilon_1 - \epsilon_2))^{-1}$ is called a *diffuson*. It is an advanced (retarded) function of its first (second) energy argument, $\epsilon_{1(2)}$, (or correspondingly $t_{1(2)}$). The higher order terms of the action's expansion describe non–linear interactions of the diffusons with vertices called *Hikami boxes*. These non–linear terms are responsible for the localization corrections. If the distribution function $\mathcal{F}$ is spatially non–uniform, there is an additional term in the quadratic action $i\tilde{S}^{(2)}[\mathcal{W}] = -2\pi \nu D \text{Tr}\{\overline{w}\nabla_r \mathcal{F} \overline{w} \nabla_r \mathcal{F}\}$. This term generates non–zero correlations of the type $\langle \overline{w}\overline{w} \rangle$ and is actually necessary for the convergence of the functional integral over $\overline{w}$ and w. In the spatially uniform case, such a convergence term is pure regularization (the situation that was already encountered before).

One can now derive the linear density response to the applied scalar potential. According to the general expression, Eq. (5.28), the retarded response is given by

$$\Pi^R(t, t'; r, r') = -\frac{i}{2} \left. \frac{\delta^2 Z[\hat{V}]}{\delta V_{cl}(t'; r')\delta V_q(t; r)} \right|_{\hat{V}=0}$$

$$= \nu\delta_{t,t'}\delta_{r,r'} + \frac{i}{2}(\pi\nu)^2 \left\langle \text{Tr}\{\hat{\gamma}^q Q_{t,t}(r)\}\text{Tr}\{\hat{\gamma}^{cl} Q_{t',t'}(r')\} \right\rangle, \tag{6.25}$$

where the angular brackets stand for the averaging over the action (6.15). In the Fourier representation the last expression takes the form:

$$\Pi^R(\omega; q) = \nu + \frac{i}{2}(\pi\nu)^2 \iint \frac{d\epsilon d\epsilon'}{(2\pi)^2} \left\langle \text{Tr}\{\hat{\gamma}^q Q_{\epsilon,\epsilon+\omega}(q)\}\text{Tr}\{\hat{\gamma}^{cl} Q_{\epsilon'+\omega,\epsilon'}(-q)\} \right\rangle. \tag{6.26}$$

Employing Eq. (6.22), one finds the linear in $\mathcal{W}$ terms:

$$\text{Tr}\{\hat{\gamma}^q Q_{\epsilon,\epsilon+\omega}(q)\} \sim 2 \left(\mathcal{F}_\epsilon \overline{w}_{\epsilon,\epsilon+\omega}(q) - \overline{w}_{\epsilon,\epsilon+\omega}(q)\mathcal{F}_{\epsilon+\omega}\right); \tag{6.27}$$

$$\text{Tr}\{\hat{\gamma}^{cl} Q_{\epsilon'+\omega,\epsilon'}(q)\} \sim 2\left(\mathcal{F}_{\epsilon'+\omega}\overline{w}_{\epsilon'+\omega,\epsilon'}(q)\mathcal{F}_{\epsilon'} - \overline{w}_{\epsilon'+\omega,\epsilon'}(q) + w_{\epsilon'+\omega,\epsilon'}(q)\right).$$

For a spatially uniform distribution $\langle\overline{w}w\rangle = 0$ and only the last term of the last expression contributes to the correlator. The result is:

$$\Pi^R(\omega; q) = \nu + \frac{i}{2}(\pi\nu)^2 4\int \frac{d\epsilon}{2\pi} (\mathcal{F}_\epsilon - \mathcal{F}_{\epsilon+\omega}) \left\langle w_{\epsilon'+\omega,\epsilon'}(-q)\overline{w}_{\epsilon,\epsilon+\omega}(q)\right\rangle$$

$$= \nu\left[1 + \frac{i\omega}{Dq^2 - i\omega}\right] = \nu\frac{Dq^2}{Dq^2 - i\omega}, \tag{6.28}$$

where we have used the fact that for any reasonable fermionic distribution $\mathcal{F}_{\pm\infty} = \pm 1$ and therefore $\int d\epsilon(\mathcal{F}_\epsilon - \mathcal{F}_{\epsilon+\omega}) = -2\omega$. The fact that $\Pi(\omega, 0) = 0$ is a consequence of the particle number conservation. One has obtained the diffusion form of the density–density response function. Also notice that this function is indeed retarded (analytic in the upper half–plane of complex ω), as it should be. The current–current response function, $K^R(\omega; q)$ may be obtained using the continuity equation $qj + \omega\varrho = 0$ and is $K^R(\omega; q) = \omega^2\Pi^R(\omega; q)/q^2$. As a result the conductivity is given by

$$\sigma(\omega; q) = \frac{e^2}{i\omega} K^R(\omega; q) = e^2\nu D \frac{-i\omega}{Dq^2 - i\omega}. \tag{6.29}$$

In the uniform limit $q \to 0$, one obtains the Drude result: $\sigma(\omega; 0) = e^2\nu D$.

6.5. Spectral statistics

Consider a piece of disordered metal of size L such that $L \gg l$, where $l \equiv v_F \tau$ is the elastic mean free path. The spectrum of the Schrödinger equation consists of a discrete set of levels, ϵ_n, that may be characterized by the *sample–specific* DOS, $\nu(\epsilon) \sim \sum_n \delta(\epsilon - \epsilon_n)$. This quantity fluctuates wildly and usually cannot (and need not) be calculated analytically. One may average it over the realizations of disorder to obtain a mean DOS: $\overline{\nu(\epsilon)}$. The latter is a smooth function of energy on the scale of the Fermi energy and thus at low temperature may be taken as a constant $\overline{\nu(\epsilon_F)} \equiv \nu$. This is exactly the DOS that was used in the previous sections.

One may wonder how to sense fluctuations of the sample–specific DOS $\nu(\epsilon)$ and, in particular, a given spectrum at one energy ϵ is correlated with itself at another energy ϵ'. To answer this question one may calculate the spectral correlation function:

$$R(\epsilon, \epsilon') \equiv \overline{\nu(\epsilon)\nu(\epsilon')} - \nu^2. \tag{6.30}$$

This function was calculated in the seminal paper of Altshuler and Shklovskii [24] in 1986. Here we derive it using the Keldysh NLσM.

The DOS is $\nu(\epsilon) = i \sum_k (\mathcal{G}_k^R(\epsilon) - \mathcal{G}_k^A(\epsilon))/(2\pi) = (\langle \psi_1 \bar{\psi}_1 \rangle - \langle \psi_2 \bar{\psi}_2 \rangle)/(2\pi) = -\bar{\vec{\psi}} \hat{\sigma}_3 \vec{\psi}/(2\pi)$, where the angular brackets denote quantum (as opposed to disorder) averaging and the indices are in Keldysh space. To generate the DOS at any given energy one adds a source term $- \int d\epsilon/(2\pi) J_\epsilon \int dr \, \bar{\vec{\psi}}_\epsilon(r) \hat{\sigma}_3 \vec{\psi}_\epsilon(r) = -\iint dt dt' \int dr \, \bar{\vec{\psi}}_t(r) J_{t-t'} \hat{\sigma}_3 \vec{\psi}_{t'}(r)$ to the fermionic action. Then the DOS is obtained by $\nu(\epsilon) = \delta Z[J]/\delta J_\epsilon$. After averaging over disorder and changing to the $\hat{Q}$–matrix representation in exactly the same manner as above, the source term is translated to $\pi \nu \int d\epsilon/(2\pi) J_\epsilon \int dr \text{Tr}\{\hat{Q}_{\epsilon,\epsilon}(r)\hat{\sigma}_3\}$. The derivation is the same as the derivation of Eq. (6.15). It is now clear that $\overline{\nu(\epsilon)} = \frac{1}{2}\nu\langle\text{Tr}\{\hat{Q}_{\epsilon,\epsilon}\hat{\sigma}_3\}\rangle_Q$. Substituting $\hat{Q}_{\epsilon,\epsilon} = \hat{\Lambda}_\epsilon$ one finds $\overline{\nu(\epsilon)} = \nu$, as it should be, of course. It is also easy to check that the fluctuations around $\hat{\Lambda}$ do not change the result (all the fluctuation diagrams cancel due to the causality constraints). We are now in the position to calculate the correlation function:

$$R(\epsilon, \epsilon') \equiv \frac{\delta^2 Z[J]}{\delta J_\epsilon \delta J_{\epsilon'}} - \nu^2 = \nu^2 \left[\frac{1}{4} \langle \text{Tr}\{\hat{Q}_{\epsilon,\epsilon}\hat{\sigma}_3\} \text{Tr}\{\hat{Q}_{\epsilon',\epsilon'}\hat{\sigma}_3\}\rangle_Q - 1 \right]. \tag{6.31}$$

Employing the parameterization of Eqs. (6.20)–(6.22), one finds up to the second order in $\mathcal{W}$:

$$\text{Tr}\{\hat{Q}\hat{\sigma}_3\} = 2\left[1 + \mathcal{F} \circ \overline{w} + \overline{w} \circ \mathcal{F} + w \circ \overline{w} + \overline{w} \circ w\right]. \tag{6.32}$$

Since $\langle \overline{w}w \rangle = 0$, the only non–vanishing terms contributing to Eq. (6.31) are those with no w and $\overline{w}$ at all (they cancel ν^2 term) and those of the type $\langle w\overline{w}w\overline{w} \rangle$. Collecting the latter terms one finds:

$$R(\epsilon, \epsilon') \tag{6.33}$$
$$= \nu^2 \int dr \iint \frac{d\epsilon_1 d\epsilon_2}{4\pi^2} \left\langle (w_{\epsilon,\epsilon_1} \overline{w}_{\epsilon_1,\epsilon} + \overline{w}_{\epsilon,\epsilon_1} w_{\epsilon_1,\epsilon})(w_{\epsilon',\epsilon_2} \overline{w}_{\epsilon_2,\epsilon'} + \overline{w}_{\epsilon',\epsilon_2} w_{\epsilon_2,\epsilon'}) \right\rangle_Q.$$

Finally, performing the Wick contractions according to Eq. (6.24) and taking into account that $\int d\epsilon_1 D^2(\epsilon - \epsilon_1; q) = 0$, due to the integration of a function that is analytic in the entire upper half–plane of ϵ_1, one finds:

$$R(\epsilon, \epsilon') = \frac{1}{(2\pi)^2} \sum_q \left[D^2(\epsilon - \epsilon'; q) + D^2(\epsilon' - \epsilon; q) \right], \tag{6.34}$$

where the q–summation stands for a summation over the discrete modes of the *diffusion* operator $D\nabla_r^2$ with the zero current (zero derivative) at the boundary of the metal. This is the result of Altshuler and Shklovskii for the unitary symmetry class. Notice that the correlation function depends on the energy difference $\omega = \epsilon - \epsilon'$ only.

For a small energy difference $\omega < E_{Thouless} \equiv D/L^2$ only the lowest homogenous mode, $q = 0$, of the diffusion operator (the so called zero–mode) may be retained and thus: $R(\omega) = -1/(2\pi^2\omega^2)$. This is the universal random matrix result. The negative correlations mean energy levels' repulsion. Notice that the correlations decay very slowly – as the inverse square of the energy distance. One may notice that the true random matrix result $R_{RMT}(\omega) = -(1 - \cos(2\pi\omega/\delta))/(2\pi^2\omega^2)$, where δ is the mean level spacing, contains also an oscillatory function of the energy difference. These oscillations reflect discreteness of the underlying energy spectrum. They *cannot* be found by the perturbation theory in small fluctuations near the $\hat\Lambda$ "point". However, they may be recovered once additional stationary points (not possessing the causality structure) are taken into account [25]. The saddle–point method and perturbation theory work as long as $\omega > \delta$. Currently it is not known how to work with the Keldysh NLσM at $\omega < \delta$.

In the opposite limit, $\omega > E_{Thouless}$, the summation over modes may be replaced by an integration and thus $R(\omega) = -c_d/\omega^{2-d/2}$, where c_d is a positive dimensionality dependent constant. This algebraic decay of the correlations is reflected by many experimentally observable phenomena generally known as *mesoscopic fluctuations*.

The purpose of these notes is to give the reader a general perspective of the Keldysh formalism, its structure, guiding principles, its strength and its limitations. Due to space limitations, I could not include many topics of contemporary

research interests into this introductory course. I hope to fulfill some of the gaps
on future occasions.

I am indebted to V. Lebedev, Y. Gefen, A. Andreev, A. I. Larkin, M. Feigelman, L. Glazman, I. Ussishkin, M. Rokni and many others for numerous discussions that shaped these notes. I am sincerely grateful to the school organizers for
their invitation. The work was supported by the A.P. Sloan foundation and the
NSF grant DMR–0405212.

Appendix A. Gaussian integration

For *any* complex $N \times N$ matrix A_{ij}, where $i, j = 1, \ldots N$, such that all its
eigenvalues, λ_i, have a positive real part: $\Re \lambda_i > 0$, the following statement
holds:

$$Z[J] = \iint_{-\infty}^{\infty} \prod_{j=1}^{N} \frac{d\Re z_j \, d\Im z_j}{\pi} \, e^{-\sum_{ij}^{N} \bar{z}_i A_{ij} z_j + \sum_{j}^{N} [\bar{z}_j J_j + \bar{J}_j z_j]} = \frac{e^{\sum_{ij}^{N} \bar{J}_i (A^{-1})_{ij} J_j}}{\det A}, \quad (A.1)$$

where J_j is an arbitrary complex vector. To prove it, one may start from a Hermitian matrix, that is diagonalized by a unitary transformation: $A = U^\dagger \Lambda U$,
where $\Lambda = \mathrm{diag}\{\lambda_j\}$. The identity is then easily proven by a change of variables (with unit Jacobian) to $w_i = U_{ij} z_j$. Finally, one notices that the r.h.s. of
Eq. (A.1) is an analytic function of both $\Re A_{ij}$ and $\Im A_{ij}$. Therefore, one may
continue them analytically to the complex plane to reach an arbitrary complex
matrix A_{ij}. The identity (A.1) is thus valid as long as the integral on its l.h.s. is
well defined, that is all the eigenvalues of A_{ij} have a positive real part.

The Wick theorem deals with the average value of a string $z_{a_1} \cdots z_{a_k} \bar{z}_{b_1} \cdots \bar{z}_{b_k}$
weighted with the factor $\exp\left\{ -\sum_{ij}^{N} \bar{z}_i A_{ij} z_j \right\}$. The theorem states that this average is given by the sum of all possible products of pair-wise averages. For
example,

$$\langle z_a \bar{z}_b \rangle \equiv \frac{1}{Z[0]} \frac{\delta^2 Z[J]}{\delta \bar{J}_a \delta J_b}\bigg|_{J=0} = \left(A^{-1} \right)_{ab}; \quad (A.2)$$

$$\langle z_{a_1} z_{a_2} \bar{z}_{b_1} \bar{z}_{b_2} \rangle \equiv \frac{1}{Z[0]} \frac{\delta^4 Z[J]}{\delta \bar{J}_{a_1} \bar{J}_{a_2} \delta J_{b_1} J_{b_2}}\bigg|_{J=0} = A^{-1}_{a_1 b_1} A^{-1}_{a_2 b_2} + A^{-1}_{a_1 b_2} A^{-1}_{a_2 b_1}, \quad (A.3)$$

etc.

The Gaussian identity for integration over real variables has the form:

$$Z[J] = \int_{-\infty}^{\infty} \prod_{j=1}^{N} \frac{dx_j}{\sqrt{\pi}} \; e^{-\sum_{ij}^{N} x_i A_{ij} x_j + 2 \sum_{j}^{N} x_j J_j} = \frac{e^{\sum_{ij}^{N} J_i (A^{-1})_{ij} J_j}}{\sqrt{\det A}}, \tag{A.4}$$

where A is a *symmetric* complex matrix with all its eigenvalues having a positive real part. The proof is similar to the proof in the case of complex variables: one starts from a real symmetric matrix, that may be diagonalized by an orthogonal transformation. The identity (A.4) is then easily proved by the change of variables. Finally, one may analytically continue the r.h.s. (as long as the integral is well defined) from a real symmetric matrix A_{ij}, to a *complex symmetric* one.

For an integration over two sets of *independent* Grassmann variables, $\bar{\xi}_j$ and ξ_j, where $j = 1, 2, \ldots, N$, the Gaussian identity is valid for *any invertible* complex matrix A:

$$Z[\bar{\chi}, \chi] \tag{A.5}$$

$$= \iint \prod_{j=1}^{N} d\bar{\xi}_j d\xi_j \; e^{-\sum_{ij}^{N} \bar{\xi}_i A_{ij} \xi_j + \sum_{j}^{N} [\bar{\xi}_j \chi_j + \bar{\chi}_j \xi_j]} = \det A \; e^{\sum_{ij}^{N} \bar{\chi}_i (A^{-1})_{ij} \chi_j}.$$

Here $\bar{\chi}_j$ and χ_j are two additional mutually independent (and independent from $\bar{\xi}_j$ and ξ_j) sets of Grassmann numbers. The proof may be obtained by e.g. brute force expansion of the exponential factors, while noticing that only terms that are linear in *all* $2N$ variables $\bar{\xi}_j$ and ξ_j are non–zero. The Wick theorem is formulated in the same manner as for the bosonic case, with the exception that every combination is multiplied by the parity of the corresponding permutation. E.g. the first term on the r.h.s. of Eq. (A.3) comes with a minus sign.

Appendix B. Single particle quantum mechanics

The simplest many–body system of a single bosonic state (considered in Chapter 2) is of course equivalent to a single–particle harmonic oscillator. To make this connection explicit, consider the Keldysh contour action Eq. (2.11) with the correlator Eq. (2.13) written in terms of the complex field $\phi(t)$. The latter may be parameterized by its real and imaginary parts as:

$$\phi(t) = \frac{1}{\sqrt{2\omega_0}} \left(p(t) - i\,\omega_0\,q(t) \right);$$

$$\bar{\phi}(t) = \frac{1}{\sqrt{2\omega_0}} \left(p(t) + i\,\omega_0\,q(t) \right). \tag{B.1}$$

In terms of the real fields $p(t)$ and $q(t)$ the action, Eq. (2.11), takes the form:

$$S[p, q] = \int_C dt \left[p\dot{q} - \frac{1}{2} \left(p^2 + \omega_0^2 q^2 \right) \right],$$

(B.2)

where the full time derivatives of p^2, q^2 and pq were omitted, since they contribute only to the boundary terms, not written explicitly in the continuous notation (they have to be kept for proper regularization). Equation (B.2) is nothing but the action of the quantum harmonic oscillator in the Hamiltonian form. One may perform the Gaussian integration over the $p(t)$ field to obtain:

$$S[q] = \int_C dt \left[\frac{1}{2} \dot{q}^2 - \frac{\omega_0^2}{2} q^2 \right].$$

(B.3)

This is the Feynman Lagrangian action of the harmonic oscillator, written on the Keldysh contour. It may be generalized for an arbitrary single particle potential $U(q)$:

$$S[q(t)] = \int_C dt \left[\frac{1}{2} \left(\dot{q}(t) \right)^2 - U\left(q(t) \right) \right].$$

(B.4)

One may split the $q(t)$ field into two components: $q_+(t)$ and $q_-(t)$, residing on the forward and backward branches of the contour, and then perform the Keldysh rotation: $q_\pm = q_{cl} \pm q_q$. In terms of these fields the action takes the form:

$$S[q_{cl}, q_q] = \int_{-\infty}^{\infty} dt \left[-2 q_q \frac{d^2 q_{cl}}{dt^2} - U\left(q_{cl} + q_q \right) + U(q_{cl} - q_q) \right],$$

(B.5)

where integration by parts was performed in the term $\dot{q}_q \dot{q}_{cl}$. This is the Keldysh form of the Feynman path integral. The omitted boundary terms provide a convergence factor of the form $\sim i0q_q^2$.

If the fluctuations of the quantum component $q_q(t)$ are regarded as small, one may expand the potential to the first order and find for the action:

$$S[q_{cl}, q_q] = \int_{-\infty}^{\infty} dt \left[-2 q_q \left(\frac{d^2 q_{cl}}{dt^2} + \frac{\partial U(q_{cl})}{\partial q_{cl}} \right) + i0 q_q^2 + O(q_q^3) \right].$$

(B.6)

In this limit the integration over the quantum component, q_q, may be explicitly performed, leading to a functional δ–function of the expression in the round brackets. This δ–function enforces the classical Newtonian dynamics of q_{cl} :

$$\frac{d^2 q_{cl}}{dt^2} = -\frac{\partial U\,(q_{cl})}{\partial q_{cl}}. \tag{B.7}$$

For this reason the symmetric (over forward and backward branches) part of the Keldysh field is called the classical component. The quantum mechanical information is contained in the higher order terms in q_q, omitted in Eq. (B.6). Notice, that for the harmonic oscillator potential the terms denoted as $O(q_q^3)$ are absent identically. The quantum (semiclassical) information resides, thus, in the convergence term, $i0q_q^2$, as well as in the retarded regularization of the $d^2/(dt^2)$ operator in Eq. (B.6).

One may generalize the single particle quantum mechanics onto a chain (or lattice) of harmonically coupled particles by assigning an index r to particle coordinates: $q_r(t)$, and adding the spring potential energy: $\frac{v_s^2}{2}(q_{r+1}(t) - q_r(t))^2$. Changing to spatially continuous notations: $\varphi(t;r) \equiv q_r(t)$, one finds for the Keldysh action of the real (phonon) field:

$$S[\varphi] = \int dr \int_C dt \left[\frac{1}{2}\dot{\varphi}^2 - \frac{v_s^2}{2}(\nabla_r\varphi)^2 - U(\varphi)\right], \tag{B.8}$$

where the constant v_s has the meaning of the sound velocity. Finally, splitting the field into (φ_+, φ_-) components and performing the Keldysh transformation: $\varphi_\pm = \varphi_{cl} \pm \varphi_q$, and integrating by parts, one obtains:

$$S[\varphi_+, \varphi_-] = \int dr \int_{-\infty}^{\infty} dt \left[2\varphi_q\left(v_s^2\nabla_r^2 - \partial_t^2\right)\varphi_{cl} - U(\varphi_{cl}+\varphi_q) + U(\varphi_{cl}-\varphi_q)\right]. \tag{B.9}$$

According to the general structure of the Keldysh theory the differential operator on the r.h.s., $\left(-\partial_t^2 + v_s^2\nabla_r^2\right)$, should be understood as the retarded one. This means it is a lower triangular matrix in the time domain. Actually, one may symmetrize the action by performing the integration by parts, and write it as:

$$\varphi_q\left(-\partial_t^2 + v_s^2\nabla_r^2\right)^R\varphi_{cl} + \varphi_{cl}\left(-\partial_t^2 + v_s^2\nabla_r^2\right)^A\varphi_q.$$

References

[1] L. V. Keldysh, Zh. Eksp. Teor. Fiz. **47**, 1515 (1964); [Sov. Phys. JETP **20**, 1018 (1965)].

[2] J. Schwinger, J. Math. Phys. **2**, 407 (1961).

[3] R. P. Feynman and F. L Vernon Jr., Ann. Phys. **24**, 118 (1963).

[4] H. W. Wyld, Ann. Phys. **14**, 143 (1961).

[5] P. C. Martin, E. D. Siggia and H. A. Rose, Phys. Rev. A**8**, 423 (1973); de Dominics, J. Physique (Paris), **37**, C1 (1976).

[6] E. M. Lifshitz and L. P. Pitaevskii, *Statistical Physics, part II*, Pergamon Press (1980).

[7] G. D. Mahan, *Many–particle physics*, Plenum Press, NY, 1990.

[8] J. Rammer and H. Smith, Rev. Mod. Phys. **58**, 323, (1986).

[9] L.S. Levitov, *The Statistical Theory of Mesoscopic Noise*, in *Quantum Noise*, edited by Yu. V. Nazarov and Ya. M. Blanter, Kluwer 2003.

[10] Yu. V. Nazarov, Ann. Phys. (Leipzig), **8**, 507 (1999); M. Kindermann and Yu. V. Nazarov, *Full counting statistics in electric circuits* in *Quantum Noise*, edited by Yu. V. Nazarov and Ya. M. Blanter, Kluwer 2003.

[11] H. Sompolinsky, Phys. Rev. Lett **47**, 935 (1981); H. Sompolinsky, and A. Zippelius, Phys. Rev. B **25**, 6860 (1982).

[12] V. S. Dotsenko, M. V. Feigelman, and L. B. Ioffe, *Spin glasses and related problems*, v. 15, pt. 1 *Soviet scientific reviews* **15** (Harwood Academic, New–York 1990).

[13] L. F. Cugliandolo, Lecture notes in *Slow Relaxation and non equilibrium dynamics in condensed matter*, Les Houches Session 77 July 2002, eds. J.-L. Barrat, J. Dalibard, J. Kurchan, M. V. Feigel'man, cond-mat/0210312.

[14] A. Kamenev, and A. Andreev, Phys. Rev. B **60**, 2218, (1999).

[15] C. Chamon, A. W. W. Ludwig, and C. Nayak, Phys. Rev. B **60**, 2239, (1999).

[16] A. A. Abrikosov, L. P. Gorkov, I. E. Dzyaloshinski, *Methods of quantum field theory in statistical physics*, Dover, NY, 1963.

[17] J.W. Negele and H. Orland, *Quantum Many-Particle Systems*, Adelison-Wesley, 1988.

[18] A. J. Leggett, Rev. Mod. Phys. **73**, 307 (2001).

[19] A. O. Caldeira and A. J. Leggett, Ann. Phys. (NY) **149**, 374 (1983).

[20] A. I. Larkin and Yu. N. Ovchinnikov, *Vortex motion in Superconductors*, in *Nonequilibrium Superconductivity*, eds. D. N. Langenberg, A. I. Larkin, Elsevier 1986.

[21] S. F. Edwards and P. W. Anderson, J. Phys. F. **5**, 89 (1975).

[22] K. B. Efetov, Adv. Phys. **32**, 53 (1983); K. B. Efetov, *Supersymmetry in Disorder and Chaos*, Cambridge University Press, 1997.

[23] F. Pierre, H. Pothier, D. Esteve, and M.H. Devoret, J. Low Temp. Phys. **118**, 437 (2000).

[24] B. L. Altshuler, and B. I. Shklovskii, Zh. Eksp. Teor. Fiz. **91**, 220 (1986) [Sov. Phys. JETP **64**, 127 (1986)].

[25] A. Altland and A. Kamenev, Phys. Rev. Lett. **85**, 5615 (2000).

Course 4

NON-LINEAR QUANTUM COHERENCE EFFECTS IN DRIVEN MESOSCOPIC SYSTEMS

V.E. Kravtsov

The Abdus Salam International Centre for Theoretical Physics, Strada Costiera 11, 34100 Trieste, Italy

Landau Institute for Theoretical Physics, 2 Kosygina Street, 117940 Moscow, Russia

H. Bouchiat, Y. Gefen, S. Guéron, G. Montambaux and J. Dalibard, eds.
Les Houches, Session LXXXI, 2004
Nanophysics: Coherence and Transport

Contents

1. Introduction

Over the last two decades theory and experiment on quantum disordered and chaotic systems have been an extremely successful field of research. The milestones on this route were:
- Weak Anderson localization in disordered metals
- Universal conductance fluctuations
- Application of random matrix theory to quantum disordered and chaotic systems

However the mainstream of research was so far limited to *linear* response of quantum systems where the kinetic and thermodynamic properties are calculated essentially at equilibrium. The main goal of this course is to develop a theory of *nonlinear* response to a time-dependent perturbation in the same way as the theory of Anderson localization and mesoscopic phenomena. It will require an extension of existing methods to *non-equilibrium phenomena*.

The four lectures will include the following topics:
- Perturbative theory of weak Anderson localization
- Keldysh formulation of nonlinear response theory
- Mesoscopic rings under AC pumping and quantum rectification
- Theory of weak dynamic localization in quantum dots

2. Weak Anderson localization in disordered systems

In this section we consider the main steps in the calculus of weak localization theory [1, 2]. The main object to study is the frequency-dependent conductivity

$$\sigma_{\alpha\beta}(\omega) = \int_{-\infty}^{+\infty} \frac{d\varepsilon}{4\pi\omega} [f(\varepsilon) - f(\varepsilon - \omega)] \tag{2.1}$$

$$\times \int \frac{d\mathbf{r}d\mathbf{r}'}{Vol} \langle e\hat{v}_\alpha (G^R - G^A)_{\mathbf{r},\mathbf{r}';\varepsilon} \, e\hat{v}_\beta (G^R - G^A)_{\mathbf{r}',\mathbf{r};\varepsilon-\omega} \rangle$$

where $G^{R/A}_{\mathbf{r}',\mathbf{r};\varepsilon}$ are retarded (advanced) electron Green's functions, $f(\varepsilon)$ is the Fermi energy distribution function and $\hat{v}_\alpha$ is the velocity operator. One can convince oneself using the representation in terms of exact eigenfunctions $\Psi_n(\mathbf{r})$ and exact eigenvalues E_n

$$G^{R/A}_{\mathbf{r}',\mathbf{r};\varepsilon} = \sum_n \frac{\Psi_n(\mathbf{r})\Psi_n^*(\mathbf{r}')}{\varepsilon - E_n \pm i0} \tag{2.2}$$

that Eq. (2.1) reduces to a familiar Fermi Golden rule expression:

$$\sigma_{\alpha\beta}(\omega) = 2\pi \sum_{E_n > E_m} \langle n|J_\alpha|m\rangle\langle m|J_\beta|n\rangle \, \delta(E_n - E_m - \omega) \frac{f_{E_m} - f_{E_n}}{E_n - E_m} \tag{2.3}$$

Eq. (2.2) is convenient to prove exact identities but is useless for practical calculus. In disordered systems with the momentum relaxation time τ the following expression for the disorder averaged Green's function is very useful:

$$\langle G^{R/A}\rangle_\varepsilon = \frac{1}{\varepsilon - \xi_{\mathbf{p}} \pm \frac{i}{2\tau}}. \tag{2.4}$$

where $\xi_{\mathbf{p}} = \varepsilon(\mathbf{p}) - E_F$ is related with the electron dispersion law $\varepsilon(\mathbf{p})$ relative the Fermi energy E_F. In what follows we will assume $\xi = \xi_p$ depending only on $|\mathbf{p}|$. Then ξ and the unit vector of the direction of electron momentum $\mathbf{n}$ constitute convenient variables of integration over $\mathbf{p}$:

$$\int ...d\mathbf{p} \rightarrow \int \nu(\xi)d\xi \int d\mathbf{n}... \tag{2.5}$$

For a typical metal with E_F far from the band edges and the energy scale of interest $\varepsilon \ll E_F$ the density of states $\nu(\xi)$ can be approximated by a constant $\nu = \nu(0)$ at the Fermi level and the integration over ξ can be extended over the entire real axis $(-\infty, +\infty)$.

2.1. Drude approximation

The simplest diagrams for the frequency-dependent conductivity are shown in Fig. 1. The diagrams Fig. 1a,b are given by the integrals

$$\int d\mathbf{n}\, v_\alpha v_\beta \int_{-\infty}^{+\infty} \frac{d\xi}{(\varepsilon - \xi \pm \frac{i}{2\tau})(\varepsilon - \omega - \xi \pm \frac{i}{2\tau})} = 0 \tag{2.6}$$

where $v_\alpha \approx v_F n_\alpha$.

It is important that the integral over ξ is the pole integral with all poles in the same complex half-plane. This is why such integrals are equal to zero. In contrast to that, the diagrams of Fig. 1c,d that contain both retarded (G^R) and advanced (G^A) Green's functions correspond to a similar integral with poles lying

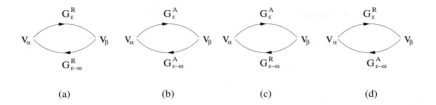

Fig. 1. Diagrams for the Drude conductivity.

in *different* half-planes of ξ. Such an integral over ξ can be done immediately using the residue theorem:

$$\int_{-\infty}^{+\infty} \frac{d\xi}{(\varepsilon - \xi \pm \frac{i}{2\tau})(\varepsilon - \omega - \xi \mp \frac{i}{2\tau})} = \frac{2\pi\tau}{(1 - i\omega\tau)}. \tag{2.7}$$

The angular integral is trivial:

$$\int d\mathbf{n}\, v_\alpha v_\beta = v_F^2 \frac{\delta_{\alpha\beta}}{d}, \tag{2.8}$$

where d is the dimensionality of space.

Now all we need to compute $\sigma_{\alpha\beta}$ using the simplest diagrams of Fig. 1 is the identity:

$$\int_{-\infty}^{+\infty} [f(\varepsilon) - f(\varepsilon - \omega)]\, d\varepsilon = -\omega. \tag{2.9}$$

This identity holds for all functions $f(\varepsilon)$ obeying the Fermi boundary conditions $f(\varepsilon) \to 0$ at $\varepsilon \to +\infty$ and $f(\varepsilon) \to 1$ at $\varepsilon \to -\infty$.

The result of the calculations is the Drude conductivity:

$$\sigma_{\alpha\beta}^{(D)} = \frac{e^2 v D_0}{1 + (\omega\tau)^2} \delta_{\alpha\beta}, \tag{2.10}$$

where $D_0 = v_F^2 \tau / d$ is the diffusion coefficient.

2.2. Beyond Drude approximation

The diagrams that were not included in the Drude approximation of Fig. 1 contain *cross-correlations* between the exact electron Green's functions G^R and G^A which arise because both of them see the *same* random impurity potential

$U(\mathbf{r})$. We make the simplest approximation of this potential as being the random Gaussian field with zero mean value and zero-range correlation function:

$$\langle U(\mathbf{r})U(\mathbf{r}')\rangle = \frac{\delta(\mathbf{r} - \mathbf{r}')}{2\pi\nu\tau}. \tag{2.11}$$

The simplest diagrams beyond the Drude approximation are the ladder series shown in Fig. 2. The dotted lines in these diagrams represent the momentum

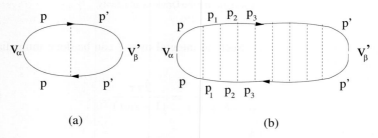

(a)   (b)

Fig. 2. Ladder diagrams.

Fourier transform of correlation function $\langle U(\mathbf{r})U(\mathbf{r}')\rangle$ which is a constant $1/2\pi\nu\tau$ independent of the momentum transfer $\mathbf{p} - \mathbf{p}'$. This makes the quantities v_α and v'_β completely independent of each other. Then the angular integration

$$\int d\mathbf{n}\, v_\alpha = 0 \tag{2.12}$$

results in vanishing of all the ladder diagrams of Fig. 2.

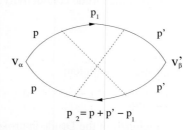

Fig. 3. The first of the "fan" diagrams.

The next diagram with cross correlations contains an intersection of dotted lines (see Fig. 3). In contrast to a diagram with two parallel dotted lines where all integrations over momenta are independent, here the momentum conservation imposes a constraint:

$$\mathbf{p} + \mathbf{p}' = \mathbf{p}_1 + \mathbf{p}_2. \tag{2.13}$$

At the same time the form Eq. (2.4) of the averaged Green's function suggests that the main contribution to the momentum integration is given by momenta confined inside a ring of radius p_F and width $1/\ell \ll p_F$ where ℓ is the elastic mean free path $\ell = v_F \tau$ [see Fig. 4a]. The constraint Eq. (2.13) implies that not

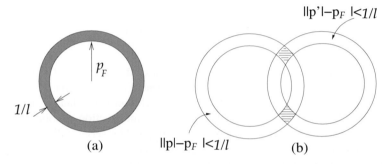

Fig. 4. Regions that make the main contribution to momentum integrals.

only $\mathbf{p}$ but also $-\mathbf{p} + \mathbf{p}_1 + \mathbf{p}_2 = \mathbf{p}'$ should be inside a narrow ring [see Fig. 4b]. Thus the effective region of $\mathbf{p}$ integration is an *intersection* of two narrow rings which volume is reduced by a large factor of $(p_F \ell)$ compared to an unconstrained case.

2.3. Weak localization correction

The lesson one learns from the above example is that any intersection of dotted lines in systems with dimensionality greater than one brings about a small parameter $1/(p_F \ell)$. However this does not mean that all such diagrams should be neglected. As a matter of fact they contain an important new physics of *quantum coherence* that changes completely the behavior of low dimensional systems with $d = 1, 2$ at small enough frequencies ω leading to the phenomenon of Anderson localization.

For this to happen there should be something that compensates for the small factor $1/(p_F \ell)$. We will see that this is a large return probability of a random walker or a particle randomly scattered off impurities in low-dimensional systems. On the formal level the corresponding diagrams are just the multiple scattering "fan" diagrams shown in Fig. 5. Using momentum conservation, one can

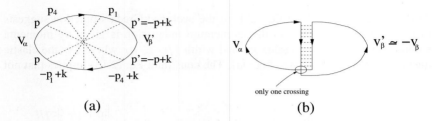

(a)   (b)

Fig. 5. Infinite series of "fan" diagrams.

see that such diagrams have only one constraint of the type Eq. (2.13) independently of the number of dotted lines. This statement becomes especially clear if one rewrites the fan series in a form Fig. 5b which contains a ladder series with all dotted lines parallel to each other and only one intersection of solid lines representing the electron Green's functions. It is this ladder series called Cooperon, which describes quantum interference, that leads to Anderson localization.

To make connection between random walks (or diffusion) in space and a quantum correction to conductivity and to see how a compensation of the small parameter $1/(p_F \ell)$ occurs we compute the Cooperon $C(\mathbf{k}, \omega)$ for a small value of $\mathbf{p_1} + \mathbf{p_2} = \mathbf{k}$ [see Fig. 6.]. This series is nothing but a geometric progression with

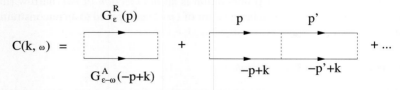

Fig. 6. Summation of the ladder series for a Cooperon.

the first term

$$\Pi_0 = \frac{1}{(2\pi\nu\tau)^2} \int d\mathbf{p}\, G_\varepsilon^R(\mathbf{p}) G_{\varepsilon-\omega}^A(-\mathbf{p} + \mathbf{k}) \tag{2.14}$$

and the denominator $q = 2\pi \nu \tau \Pi_0$. Thus we have

$$C(\mathbf{k}, \omega) = \frac{\Pi_0}{1 - q}. \tag{2.15}$$

To compute Π_0 we write

$$\Pi_0 = \frac{\nu}{(2\pi \nu \tau)^2} \int d\mathbf{n} \int_{-\infty}^{+\infty} \frac{d\xi}{(\varepsilon - \xi + \frac{i}{2\tau})(\varepsilon - \omega - \xi + v_F \mathbf{nk} - \frac{i}{2\tau})}. \tag{2.16}$$

In Eq. (2.16) we used an approximation $\xi_{-\mathbf{p}+\mathbf{k}} \approx \xi_{\mathbf{p}} - v_F \mathbf{nk}$ which is valid as long as $|\mathbf{k}| \ll p_F$. The pole integral in Eq. (2.16) can be done immediately with the result:

$$q = \int d\mathbf{n} \frac{1}{1 - i\omega\tau + i\ell \mathbf{nk}}. \tag{2.17}$$

One can see a remarkable property of Eq. (2.17): the denominator of the geometric series Eq. (2.15) is equal to 1 in the limit $\omega\tau, |\mathbf{k}|\ell \to 0$. At finite but small $\omega\tau, |\mathbf{k}|\ell \ll 1$ one can expand the denominator of Eq. (2.17) to obtain:

$$q \approx 1 + i\omega\tau - \ell^2 \mathbf{k}^2/d; \quad C(\mathbf{k}, \omega) \approx \frac{1}{2\pi \nu \tau^2} \frac{1}{D_0 \mathbf{k}^2 - i\omega}. \tag{2.18}$$

We immediately recognize an inverse diffusion operator in $C(\mathbf{k}, \omega)$ which is divergent at small ω and $\mathbf{k}$. This divergence is the cause of all the peculiar quantum-coherence phenomena in systems of dimensionality $d \leq 2$.

In particular the quantum correction to conductivity given by the diagram of Fig. 5 can be written as:

$$\delta\sigma_{\alpha\beta} = \frac{\sigma^{(D)}}{2\pi \nu D_0} \times \frac{1}{Vol} \sum_{\mathbf{k}} \Gamma_{\alpha\beta} C(\mathbf{k}, \omega). \tag{2.19}$$

where $\Gamma_{\alpha\beta}$ is the "Hikami box" shown in Fig. 7. Its analytic expression is given by:

$$\int d\mathbf{p} \, v_\alpha(\mathbf{p}) v_\beta(-\mathbf{p} + \mathbf{k}) \, G_\varepsilon^R(\mathbf{p}) G_{\varepsilon-\omega}^A(\mathbf{p}) G_\varepsilon^R(-\mathbf{p} + \mathbf{k}) G_{\varepsilon-\omega}^A(-\mathbf{p} + \mathbf{k}). \tag{2.20}$$

In the limit $\omega\tau \ll 1$ and $|\mathbf{k}|\ell \ll 1$ one can set $\omega = \mathbf{k} = 0$ in Eq. (2.20). Then $\mathbf{v}(-\mathbf{p}+\mathbf{k}) = -\mathbf{v}(\mathbf{p}) = -v_F\mathbf{n}$ and after the pole integration over ξ and the angular integration over $\mathbf{n}$ we obtain for $\Gamma_{\alpha\beta}$:

$$-\int d\mathbf{n} \, v_F^2 \mathbf{n}_\alpha \mathbf{n}_\beta \int_{-\infty}^{+\infty} \frac{\nu \, d\xi}{(\varepsilon - \xi + \frac{i}{2\tau})^2 (\varepsilon - \xi - \frac{i}{2\tau})^2} = -4\pi \nu \tau^2 D_0. \tag{2.21}$$

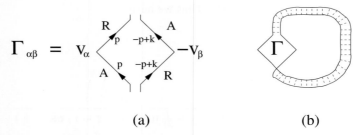

$$\Gamma_{\alpha\beta} = V_\alpha \quad \langle \quad \rangle \quad -V_\beta \qquad \langle \Gamma \rangle$$

(a)   (b)

Fig. 7. The diagram representation for the Hikami box (a) and for the quantum correction to conductivity (b).

Finally using Eq. (2.19) we get:

$$\sigma(\omega) = \sigma^{(D)} \left(1 - \frac{1}{\pi \nu} \frac{1}{Vol} \Re \sum_{\mathbf{k}} \frac{1}{D_0 \mathbf{k}^2 - i\omega} \right). \tag{2.22}$$

This is the celebrated formula for weak Anderson localization. One can see that the quantum correction is given by the sum over momenta of the diffusion propagator, that is, it is proportional to the *return probability* at a time $\sim 1/\omega$ for a random walker in the d-dimensional space. A remarkable property of random walks in low-dimensional space is that the return probability increases with time. That is why the quantum correction to conductivity increases with decreasing the frequency ω as $1/\sqrt{\omega}$ in a quasi-one dimensional wire and as $\log(1/\omega)$ in a two-dimensional disordered metal.

The structure of Eq. (2.22) suggests a qualitative picture of weak Anderson localization. It is an interference of two trajectories with a loop that contains trajectories with opposite directions [Fig. 8]. Although the phase that corresponds to each trajectory is large and random, the phase *difference* between them is zero because of the time-reversal invariance. As a result they can interfere contructively and amount to a creation of a *random standing wave* which is the paradigm of Anderson localization.

3. Non-linear response to a time-dependent perturbation

In this section we review the main steps of the Keldysh formalism [3, 4, 5] which are necessary to describe a quantum system of non-interacting electrons subject to external time-dependent perturbation $\hat{V}(t)$.

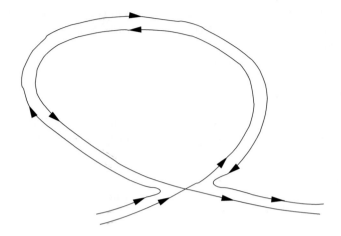

Fig. 8. Two interfering trajectories with loops.

3.1. General structure of nonlinear response function

The matrix Keldysh Green's function

$$\mathbf{G} = \begin{pmatrix} G^R & G^K \\ 0 & G^A \end{pmatrix} \tag{3.1}$$

contains – besides familiar retarded and advanced Green's functions $\mathbf{G}^{11} = G^R$ and $\mathbf{G}^{22} = G^A$ – also the third, Keldysh function $\mathbf{G}^{12} = G^K$. The latter is the only one needed to compute an expectation value of any operator $\hat{O}$ both in equilibrium and beyond:

$$O = -iTr(\hat{O}G^K). \tag{3.2}$$

The retarded and advanced Green's functions appear only at the intermediate stage to make it possible to write down Dyson equations in the matrix form. It is of principal importance that the component $\mathbf{G}^{21}$ is zero. One can show that this is a consequence of causality [6]. In equilibrium there is a relationship between G^K and $G^{R/A}$. It reads:

$$G^K_\varepsilon = (G^R - G^A)_\varepsilon \, \tanh\left(\frac{\varepsilon}{2T}\right). \tag{3.3}$$

As without time-dependent perturbation the system is in equilibrium, the corresponding components $G_{0,\varepsilon}^{R/A}$ and $G_{0,\varepsilon}^{K}$ of the matrix Keldysh Green's function obey the relationship Eq. (3.3).

In the presence of the perturbation one can write the Dyson equation:

$$\mathbf{G} = \mathbf{G}_0 + \mathbf{G}_0 \hat{\mathbf{V}} \mathbf{G}, \tag{3.4}$$

where the operator of *external* time-dependent perturbation $\hat{\mathbf{V}}(t)$ is proportional to a unit matrix in the Keldysh space. The structure of perturbation series for G^K

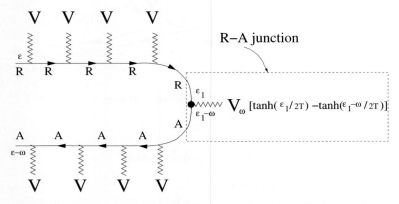

Fig. 9. The diagram representation of the "anomalous" term in the Keldysh function.

that corresponds to Eq. (3.4) is as follows:

$$G^K = G_0^{12} + G_0^{11}\hat{V}G_0^{12} + G_0^{12}\hat{V}G_0^{22} + G_0^{11}\hat{V}...G_0^{11}\hat{V}G_0^{12}\hat{V}G_0^{22}...G_0^{22} + ... \tag{3.5}$$

As $G_0^{21} = 0$ and $G_0^{12} = (G_0^R - G_0^A)\tanh(\varepsilon/2T)$ each term of the perturbation series Eq. (3.5) is a string of successive G_0^R functions followed by a string of G_0^A functions with only one switching point between them where the factor $\tanh(\varepsilon/2T)$ is attached to. For the same reason the components G^R and G^A are the series that contain only G_0^R or G_0^A, respectively:

$$G^{R/A} = G_0^{R/A} + G_0^{R/A}\hat{V}G_0^{R/A} + G_0^{R/A}\hat{V}G_0^{R/A}\hat{V}G_0^{R/A} + ... \tag{3.6}$$

Using Eq. (3.6) one can replace the string of $G_0^{R/A}$ functions in Eq. (3.5) by the exact retarded or advanced Green's function $G_{\varepsilon,\varepsilon'}^{R/A}$ that depends on two energy variables because of the breaking of translational invariance in the time domain by a time dependent perturbation. The result is:

$$G^K_{\varepsilon,\varepsilon'} = G^R_{\varepsilon,\varepsilon'} \tanh\left(\frac{\varepsilon'}{2T}\right) - \tanh\left(\frac{\varepsilon}{2T}\right) G^A_{\varepsilon,\varepsilon'} + \tag{3.7}$$

$$+ \int \frac{d\varepsilon_1}{2\pi} \int \frac{d\omega}{2\pi} G^R_{\varepsilon,\varepsilon_1} \hat{V}(\omega) G^A_{\varepsilon_1-\omega,\varepsilon'} \left[\tanh\left(\frac{\varepsilon_1}{2T}\right) - \tanh\left(\frac{\varepsilon_1-\omega}{2T}\right)\right].$$

The first two terms in Eq. (3.7) just reproduce the structure of the equilibrium Keldysh function Eq. (3.3). The most important for us will be the last, so called "anomalous" term G^{anom} in Eq. (3.7) as it contains both G^R_0 and G^A_0 functions which makes nonzero the pole integrals over ξ and allows to build a Cooperon. This term is graphically represented in Fig. 9.

One can see that of all the vertices with a perturbation operator $\hat{V}$ one is special: it is a switching point between the strings of retarded and advanced Green's functions, so called $R - A$ junction [6]. It is convenient to switch from the energy to the time domain. In this representation the anomalous part of G^K reads [5, 6]:

$$G^{\text{anom}}_{t,t'} = G^R_{t,t_1} G^A_{t_2,t'} (\hat{V}(t_2) - \hat{V}(t_1)) h_0(t_1 - t_2). \tag{3.8}$$

In Eq. (3.8) we denoted the Fourier transform of $\tanh(\varepsilon/2T)$ as:

$$h_0(t) = \int \frac{d\varepsilon}{2\pi} e^{i\varepsilon t} \tanh(\varepsilon/2T) = \frac{iT}{\sinh(\pi T t)}. \tag{3.9}$$

We also assume integration over repeated time variables.

The graphic representation of Eq. (3.8) is given in Fig. 10. Eqs. (3.7),(3.8) give a general structure of nonlinear response to a time-dependent perturbation. The main feature is the distinction between a regular $\hat{V}$ vertex that connects two Green's functions with the same analytical properties (both G^R or both G^A) and the $R - A$ junction. One can see that the time variables do not match at the R-A junction because of the attached energy distribution function, while they do match at any regular vertex $\hat{V}$.

3.2. Approximation of single photon absorption/emission

While an exact solution of the nonlinear response is a difficult task, there is an essentially nonlinear regime where a regular solution can be found. The corresponding approximation consists in expanding each *disorder averaged* retarded or advanced Green's function up to second order in $\hat{V}$:

$$\langle G^{R/A}\rangle \approx \langle G^{R/A}_0\rangle + \langle G^{R/A}_0\rangle \hat{V} \langle G^{R/A}_0\rangle + \langle G^{R/A}_0\rangle \hat{V} \langle G^{R/A}_0\rangle \hat{V} \langle G^{R/A}_0\rangle \tag{3.10}$$

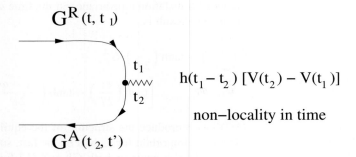

Fig. 10. Anomalous contribution to G^K in the time domain.

Yet, since the ladder series (a Cooperon) contains an *infinite* number of retarded (advanced) Green's functions, this approximation leads to an essentially non-linear result that does not reduce to a finite order response function. We will see that this approximation implies a *sequential* absorption/emission of photons rather than *multiple-photon* processes.

Another approximation which is similar to $\omega\tau \ll 1$ is based on the fact that the disorder averaged Green's function Eq. (2.4) decays exponentially in the time domain:

$$\langle G_0^{R/A}\rangle_t = \mp i\,\theta(\pm t)\,e^{-it\xi_p}\,e^{-t/\tau}. \tag{3.11}$$

Assuming that the momentum relaxation time is much shorter than the characteristic scale (e.g. the period) of a time-dependent perturbation one can approximate $\langle G_0^{R/A}\rangle_t$ by a δ-function and its derivative:

$$\langle G_0^{R/A}\rangle_t \approx \delta(t)\,\langle G_0^{R/A}\rangle_{\varepsilon=0} - i\,\partial_t\delta(t)\langle G_0^{R/A}\rangle_{\varepsilon=0}^2. \tag{3.12}$$

4. Quantum rectification by a mesoscopic ring

To proceed further, we need to make some assumptions about the time-dependent perturbation and specify an observable of interest.

In this section we consider the problem of rectification of an *ac signal* by a disordered metal. It was first considered in Ref. [10] in a single-connected geometry where the entire effect is due to mesoscopic fluctuations. Here we focus on the case of a quasi-one dimensional disordered metal ring pierced by a magnetic flux

$\phi(t)$ that contains both a constant part ϕ and an oscillating part $\phi_{ac}(t)$ [7, 8, 9]. In this case the topology of the ring and the presence of a constant magnetic flux make it meaningful to study the *disorder-averaged* rectification effect.

The time-dependent perturbation in this case is:

$$\hat{V} = -\hat{v}_x \, \varphi(t), \qquad \varphi(t) = \frac{2\pi}{L} \frac{\phi_{ac}(t)}{\phi_0}, \tag{4.1}$$

where L is the circumference of the ring, and $\phi_0 = hc/e$ is the flux quantum. We assume that the ring curvature is large compared to all microscopic lengths in the problem so that it can be replaced by quasi-one dimensional wire along the x-axis with the twisted boundary conditions $\Psi(L) = \Psi(0) \, \exp[2\pi i\phi/\phi_0]$.

To solve this problem we need to take into account the ac perturbation only in the denominator q of the geometric series Eq. (2.15) that determines the Cooperon. This is because $1 - q$ is governed by *small corrections* with energy scale much smaller than $1/\tau$. The corrections due to the ac flux perturbation are given by the diagrams in Fig. 11. Note that, in linear in $\hat{V}$ corrections, one should take into account the small momentum **k** whereas, in quadratic in $\hat{V}$ corrections, **k** can be set to zero. All the corrections of Fig. 11 can be computed by doing the pole integrals over ξ and angular integrals over **n** as in the previous section. Note also that $1 - q$ is essentially the *inverse* Cooperon operator. For an arbitrary time dependence of $\varphi(t)$ one should replace $-i\omega$ in Eq. (2.18) by the time derivative $\partial_{t_1} - \partial_{t_2}$ which cab be obtained from the second term in Eq. (3.12). Such a structure of the time derivatives implies that the sum of time arguments $t_1 + t_2 = t_1' + t_2'$ is conserved [see Fig. 12] which is the consequence of the constant density of states approximation. As a result, the equation for a time-dependent Cooperon takes the form [11, 12, 6]:

$$\left\{ \frac{\partial}{\partial \eta} + \frac{D_0}{2} \left[\varphi(t + \eta/2) + \varphi(t - \eta/2) - \mathbf{k}_x \right]^2 \right\} C_t(\eta, \eta'; \mathbf{k}_x) \tag{4.2}$$
$$= \frac{\delta(\eta - \eta')}{2\pi \nu \tau^2},$$

where the momentum $\mathbf{k}_x(\phi) = k_m(\phi) = (2\pi/L)(m - 2\phi/\phi_0)$ is quantized according to the twisted boundary conditions.

For completeness we give also an equation for the time-dependent *diffuson* [Fig. 13]:

$$\left\{ \frac{\partial}{\partial t} + D_0 \left[\varphi(t + \eta/2) - \varphi(t - \eta/2) - \mathbf{k}_x \right]^2 \right\} D_\eta(t, t'; \mathbf{k}_x) \tag{4.3}$$
$$= \frac{\delta(t - t')}{2\pi \nu \tau^2},$$

where $\mathbf{k}_x = (2\pi/L) m$ is independent of the DC flux ϕ.

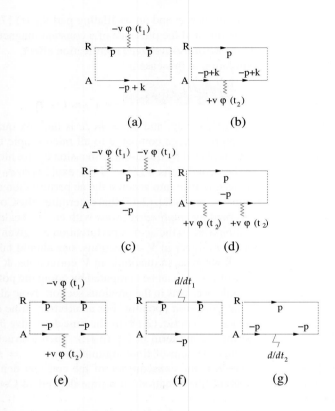

Fig. 11. The ac flux and time-derivative corrections to the inverse Cooperon operator.

In this case the structure of the time-derivative $\partial_{t_1} + \partial_{t_2}$ suggests that the *difference* of the time arguments is conserved $t_1 - t_2 = t_1' - t_2'$.

In general the current response $I(t)$ to the electric field $E(t - \tau)$ is given by:

$$I(t) = \int_0^\infty K(t, \tau) E(t - \tau) d\tau, \tag{4.4}$$

where in our particular case $E(t) = -\partial_t \varphi(t)$.

At small $1/p_F \ell$ the main contribution to the *nonlinear* response function is the quantum coherence correction shown in Figs. 5,7. All we have to do in order to compute this nonlinear response function is to substitute the time-dependent Cooperon into Eq. (2.19) and take care of all the time arguments [see Fig. 14]. Namely, (i) according to Eq. (3.12) the time arguments corresponding to the be-

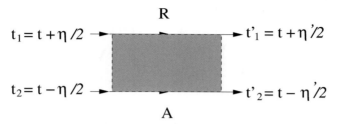

Fig. 12. The time arguments in the Cooperon.

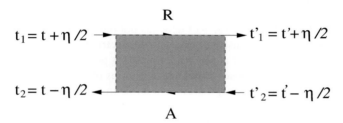

Fig. 13. The time-dependent Diffuson.

ginning and the end of any solid line representing the disorder averaged functions $\langle G_0^{R/A} \rangle$ should be the same and (ii) the sum (difference) of "incoming times" in the Cooperon (Diffuson) are equal to the sum (difference) of "outgoing times" as in Figs. 12,13. Note that application of rules (i),(ii) to the diagram of Fig. 14 leads to the coinciding time arguments $t_1 \to t_2 = t - \tau$ in the corresponding retarded-advanced junction (see Fig. 10). This ensures fulfillment of Eq. (4.4), as $h_0(t_1 - t_2)(V(t_2) - V(t_1)) \to v_x \, \partial_t \varphi(t - \tau) \propto E(t - \tau)$.

The disorder averaged current response function $K(t, \tau)$ is equal to:

$$K(t, \tau) = \sigma^D \tau^{-1} e^{-t/\tau} - \frac{4e^2 D_0}{h} \tilde{C}_{t-\tau/2}(\tau, -\tau), \tag{4.5}$$

where $\tilde{C}_t(\eta, \eta') = (2\pi v \tau^2 / Vol) \sum_{\mathbf{k}_x} C_t(\eta, \eta'; \mathbf{k}_x)$ and $C_t(\eta, \eta'; \mathbf{k}_x)$ is the solution to Eq. (4.2):

$$2\pi v \tau^2 C_t(\eta, \eta'; \mathbf{k}_x) = \theta(\eta - \eta') \times \tag{4.6}$$
$$\exp\left\{-\frac{D_0}{2} \int_{\eta'}^{\eta} [\varphi(t + \zeta/2) + \varphi(t - \zeta/2) - \mathbf{k_x}]^2 \, d\zeta\right\}.$$

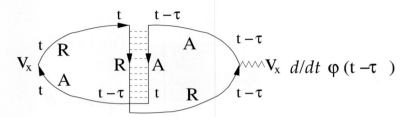

Fig. 14. Nonlinear current response function and the Cooperon.

The first term in Eq. (4.5) is just the linear response given by the diagrams Fig. 1c,d. The second term corresponds to the quantum-coherent contribution of Fig. 14. The fact that $\tilde{C}_{t-\tau/2}(\tau, -\tau)$ depends not only on τ but also on t makes it possible to have dc response caused by an ac electric field $E(t) = -\partial_t \varphi(t)$.

For the particular case of a ring with a constant plus an *ac* magnetic flux we note that the *dc* response arises only from *odd* terms of expansion of Eq. (4.6) in powers of the ac perturbation $\varphi(t)$. As odd terms in $\varphi(t)$ enter in a combination $\varphi(t) \mathbf{k}_x$, the *dc* response involves only *odd* terms in $\mathbf{k}_x = k_m = (2\pi/L)(m - 2\phi/\phi_0)$. If the constant magnetic flux $\phi = 0$ or if we neglect quantization of momentum and do an integral over $\mathbf{k}_x$ instead of a sum, the result for the odd in $\varphi(t)$ part of $\tilde{C}_t(\eta, \eta')$ is zero. Otherwise it is *periodic* in the flux ϕ with the period $\phi_0/2$, since $k_m(\phi+\phi_0/2) = k_{m-1}(\phi)$ and the summation is over all integer m. We see that the dc response to the ac perturbation, or the rectification of the ac flux by an ensemble of mesoscopic rings, is an essentially quantum, Bohm-Aharonov-like effect. Furthermore, it is *odd* in the flux Bohm-Aharonov effect that can be represented by the Fourier series:

$$\langle I_{dc}(\phi) \rangle = \sum_{n=1}^{\infty} I_n \sin\left(\frac{4\pi n\phi}{\phi_0}\right). \tag{4.7}$$

Expression Eq. (4.7) has the same symmetry and periodicity in magnetic flux ϕ as the disorder-averaged equilibrium persistent current [13, 14]. However, its magnitude may be much larger (in the grand-canonical ensemble considered here the disorder-averaged persistent current is strictly zero).

Applying the Poisson summation trick

$$\sum_m f(m - \phi) = \int dx\, f(x - \phi) \sum_m \delta(x - m) = \tag{4.8}$$

$$= \int dx\, f(x - \phi) \sum_n e^{2\pi i n x} = \sum_n e^{2\pi i n \phi} \int dx\, e^{2\pi i n x} f(x)$$

to Eq. (4.6) one obtains

$$I_n = \frac{4ieD_0}{\pi L} \int_0^\infty d\tau \, \overline{C_t^{(n)}(\tau) \partial_t \varphi(t - \tau/2)}, \tag{4.9}$$

where the overline means averaging over time t and

$$C_t^{(n)}(\tau) = \sqrt{\frac{\tau_D}{4\pi\tau}} \, e^{-\frac{n^2\tau_D}{4\tau}} \, e^{in S_1[\varphi]} \, e^{-\tau S_2[\varphi]}. \tag{4.10}$$

Here $\tau_D = L^2/D_0$ is the time needed for a diffusing particle to go around the ring, and $S_{1,2}$ are defined as follows:

$$S_1[\varphi] = 2L \left[\frac{1}{\tau} \int_{t-\tau/2}^{t+\tau/2} \varphi(t_1) \, dt_1 \right] \equiv 2L \, \langle \varphi_{t_1} \rangle_{t;\tau}, \tag{4.11}$$

$$S_2[\varphi] = 2D_0 \left[\langle \varphi_{t_1}^2 \rangle_{t;\tau} + \langle \varphi_{t_1} \varphi_{2t-t_1} \rangle_{t;\tau} - 2 \langle \varphi_{t_1} \rangle_{t;\tau}^2 \right]. \tag{4.12}$$

Eqs. (4.10)-(4.12) are valid for an arbitrary time-dependence of $\varphi(t)$. However, they take the simplest form for a noise-like ac flux with a small correlation time $\tau_0 \ll \tau_D$ (but we assume $\tau_0 \gg \tau$). In this case the second and the third terms in Eq. (4.12) can be neglected and the first term reduces to a *constant* that determines the *dephasing time* caused by ac noise:

$$\frac{1}{\tau_\varphi} = 2D_0 \overline{\varphi^2(t)}. \tag{4.13}$$

Then the time averaging in Eq. (4.9) reduces to

$$\overline{-i\partial_t \varphi(t - \tau/2) \exp\left\{ \frac{in}{\tau} 2L \int_{t-\tau/2}^{t+\tau/2} \varphi(t_1) \, dt_1 \right\}} = \frac{n}{\tau} 2L \, \overline{\varphi^2(t)}. \tag{4.14}$$

Since the time-average of the total time-derivative is zero, we can transfer the differentiation to the exponent. For the case of a white-noise ac flux only the lower limit of integration in the exponent should be differentiated, the quantity $\langle \varphi_{t_1} \rangle_{t;\tau}$ being set to zero afterwards. The remaining integral over τ is done exactly:

$$\int_0^\infty \frac{d\tau}{\tau^{3/2}} \exp\left[-\frac{n^2\tau_D}{4\tau} - \frac{\tau}{\tau_\varphi} \right] = \frac{2\sqrt{\pi}}{n\sqrt{\tau_D}} e^{-n\sqrt{\tau_D/\tau_\varphi}} \tag{4.15}$$

Finally we arrive at a remarkably simple result for the disorder-averaged dc current generated in mesoscopic rings by a white-noise ac perturbation [9]:

$$I_n = -\frac{4}{\pi} \left(\frac{e}{\tau_\varphi} \right) \exp\left[-n\frac{L}{L_\varphi} \right], \tag{4.16}$$

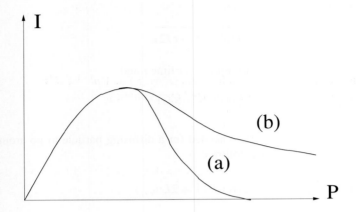

Fig. 15. The dependence of I_1 on the ac power $P = \overline{\varphi^2(t)}$ for a white-noise (a) and harmonic (b) perturbation.

where the *dephasing length* is equal to:

$$\frac{1}{L_\varphi^2} = \frac{1}{D_0 \tau_\varphi} = 2\overline{\varphi^2(t)}. \tag{4.17}$$

It follows from Eq. (4.16) that the weak ac white noise produces a net dc current in an ensemble of mesoscopic rings which is of the order of e/τ_φ where τ_φ is the dephasing time produced by the same ac noise. At an ac power large enough to produce a dephasing length smaller than the circumference of a ring times the winding number n, the destructive effect of ac perturbation prevails and the current I_n decreases exponentially [see Fig. 15].

Note that the tail of the dependence I_n on the ac power is very sensitive to the correlations in the ac perturbation at different times. For instance, in the case of harmonic perturbation where $\varphi(t)$ has an infinite range time-correlations, the current decreases very slowly, only as the inverse square-root of the ac power [7]. We will see later on that this is related with the phenomenon of *no-dephasing points*.

5. Diffusion in the energy space

In this section we apply the Keldysh formalism outlined above to the problem of *heating* by external time-dependent perturbation. The main object to study will be the *non-equilibrium* electron energy distribution function:

$$f(\varepsilon; t) = \frac{1}{2} - \frac{1}{2} \int d\eta \, h_t(\eta) \, e^{-i\varepsilon\eta} \tag{5.1}$$

which is related to the Keldysh function $G^K(t, t') \equiv Vol^{-1} \int d^d\mathbf{r} \, G^K(t, t'; \mathbf{r}, \mathbf{r})$ at coincident space variables (averaged over volume):

$$G^K(t + \eta/2, t - \eta/2) = -2\pi i\nu \, h_t(\eta). \tag{5.2}$$

Eqs. (5.1), (5.2) generalize Eq. (3.3) to the case of non-equilibrium, time-dependent energy distribution function. The total energy $\mathcal{E}(t)$ of the electron system can be expressed in terms of $f(\varepsilon, t)$:

$$\mathcal{E}(t) = \nu \, Vol \int d\varepsilon \, \varepsilon \, [f(\varepsilon, t) - \theta(-\varepsilon)] + const. \tag{5.3}$$

Then the time-dependent *absorption rate* $W(t) = \partial_t \mathcal{E}(t)$ is given by:

$$W(t) = -\frac{Vol}{2} \lim_{\eta \to 0} \partial_t \, \partial_\eta \, G^K \left(t + \frac{\eta}{2}, t - \frac{\eta}{2}\right) = \frac{i\pi}{\delta} \lim_{\eta \to 0} \partial_t \, \partial_\eta \, h_t(\eta), \tag{5.4}$$

where $\delta = (\nu \, Vol)^{-1}$ is the mean separation between electron energy levels (mean level spacing).

For tutorial reasons we consider in this section a specific model system described by the Hamiltonian:

$$\hat{H} = \varepsilon(\hat{\mathbf{p}}) + U(\mathbf{r}) + V(\mathbf{r}) \, \varphi(t), \tag{5.5}$$

where not only $U(\mathbf{r})$ given by Eq. (2.11) but also the perturbation potential $V(\mathbf{r})$ is a Gaussian random field which is statistically independent of $U(\mathbf{r})$ and is described by the correlation function:

$$\langle V(\mathbf{r})V(\mathbf{r}') \rangle = \frac{\Gamma}{\pi\nu} \, \delta(\mathbf{r} - \mathbf{r}'). \tag{5.6}$$

This model is the simplest example of a *potential ac source* in contrast to the *flux ac source* considered in the previous section. In the low-frequency $\omega\tau_D \ll 1$ and the modestly low ac intensity $\Gamma\tau \ll 1$ limits, this model is equivalent to the *random matrix theory* (RMT) with the time-dependent Hamiltonian [6, 12]:

$$\hat{H}_{RMT} = \hat{H}_0 + \hat{V} \, \varphi(t), \tag{5.7}$$

where both $\hat{H}_0$ and $\hat{V}$ are random real-symmetric $N \times N$ matrices from the independent Gaussian Orthogonal Ensembles described by the correlation functions $\langle (H_0)_{nm} \rangle = \langle V_{nm} \rangle = 0$:

$$\langle (H_0)_{nm} (H_0)_{n'm'} \rangle = N(\delta/\pi)^2 [\delta_{nn'}\delta_{mm'} + \delta_{nm'}\delta_{mn'}], \qquad (5.8)$$

$$\langle V_{nm} V_{n'm'} \rangle = (\Gamma\delta/\pi) [\delta_{nn'}\delta_{mm'} + \delta_{nm'}\delta_{mn'}],$$

with $\delta = 1/(\nu \, Vol)$ being the mean level spacing.

In this RMT limit *all* the disordered and chaotic systems described by a real, spin-rotational invariant Hamiltonian are believed to have a universal behavior in the limit $L, N \to \infty$ which is characterized by only two parameters δ and Γ and one dimensionless function $\varphi(t)$.

Let us first consider the disorder-averaged G^K in the non-crossing approximation. Since there are no vector vertices in the present model, the ladder diagrams (Loose Diffuson) similar to Fig. 2 make the main contribution. The corresponding diagrams are shown in Fig. 16 where the wavy line describes the $\langle VV \rangle$ correlator, Eq. (5.6). The condition

$$\Gamma\tau \ll 1 \qquad (5.9)$$

allows to keep only the linear in $\hat{V}$ term in the expansion Eq. (3.10) for the $\langle G^{R/A} \rangle$ function attached to the RA-junction. Note that the diagrams Fig. 16a,b have opposite signs because $\int d\xi \, (\xi \pm i/2\tau)^{-1} = \mp\pi i$. The result of calculation of both diagrams is:

$$\begin{aligned} \delta G^K &= (2\pi\nu\tau)^2 (i\pi\nu)(\Gamma/\pi\nu) \, D_\eta(t,t') \\ &\times [\varphi(t'+\eta/2) - \varphi(t'-\eta/2)]^2 \, h_0(\eta). \end{aligned} \qquad (5.10)$$

so that

$$h_t(\eta) = \left(1 - \tilde{D}_\eta(t,t') \, \Gamma \, [\varphi(t'+\eta/2) - \varphi(t'-\eta/2)]^2\right) h_0(\eta), \qquad (5.11)$$

where the integration over t' is assumed; $h_0(\eta)$ is determined by Eq. (3.9), and $\tilde{D}_\eta(t,t') = 2\pi\nu\tau^2 D_\eta(t,t'; \mathbf{k} = 0)$ is given by Eq. (4.3) with $D_0 \to \Gamma$. Using this equation, one can check that $h_t(\eta)$ obeys the equation:

$$\left\{ \partial_t + \Gamma \, [\varphi(t+\eta/2) - \varphi(t-\eta/2)]^2 \right\} h_t(\eta) = 0. \qquad (5.12)$$

This allows to give an explicit solution for $h_t(\eta)$:

$$h_t(\eta) = h_0(\eta) \exp\left\{-\Gamma \int_0^t [\varphi(\zeta+\eta/2) - \varphi(\zeta-\eta/2)]^2 \, d\zeta \right\} \qquad (5.13)$$

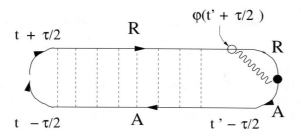

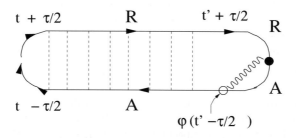

Fig. 16. The Loose Diffuson diagrams for the absorption rate: the dotted lines represent the $\langle UU \rangle$ correlation function while the wavy line corresponds to the $\langle VV \rangle$ correlator.

Note that a nontrivial dynamics of $h_t(\eta)$ and that of $f(\varepsilon; t)$ is hidden in the exponential factor which is a purely classical object. The Fermi statistics of electrons considered here is taken into account by an initial distribution $h_0(\eta)$. At $\eta\omega \ll 1$ where ω is the typical frequency of oscillations in $\varphi(t)$, or equivalently at the energy resolution $\delta\varepsilon \gg \omega$, one can approximate $\varphi(\zeta+\eta/2)-\varphi(\zeta-\eta/2) \approx (\partial_t\varphi)\,\eta$. Then the exponent in Eq. (5.13) reduces to $\exp\left[-t\,\Gamma\,\overline{(\partial_t\varphi)^2}\,\eta^2\right]$ which corresponds to Eq. (5.12) of the form:

$$\left\{\partial_t + D_E\,\eta^2\right\} h_t(\eta) = 0, \quad \left\{\partial_t - D_E\,\partial_\varepsilon^2\right\} f(\varepsilon, t) = 0. \tag{5.14}$$

In this approximation we come to the diffusion equation for the energy distribution function with the energy diffusion coefficient:

$$D_E = \Gamma\overline{(\partial_t\varphi)^2} \sim \Gamma\,\omega^2. \tag{5.15}$$

For a harmonic perturbation $\varphi(t) = \cos(\omega t)$ the form of energy diffusion co-efficient allows a simple interpretation. This is a random walk in the energy space due to a sequential absorption or emission of photons with the energy $\hbar\omega$. Then the size of an elementary step is $\pm\hbar\omega$ and the rate of making steps is $\Gamma/\hbar$. Now it is clear that the condition Eq. (5.9), that allows to use the approximation Eq. (3.10), has the physical meaning of a condition to absorb/emit *at most* one photon during the time of elastic mean free path. The opposite condition would mean that many photons can be absorbed/emitted during the time τ which implies inelastic processes being stronger than the elastic ones.

Using the diffusion equation we express $\partial_t h_t(\eta) = -D_E \eta^2 h_t(\eta)$. Then Eq. (5.4) gives for the net absorption rate:

$$W(t) = -i\pi \, (D_E/\delta) \lim_{\eta \to 0} \partial_\eta(\eta^2 h_t(\eta)). \tag{5.16}$$

It is important that

$$h_t \approx \frac{i}{\pi\eta}, \quad \eta \to 0. \tag{5.17}$$

for *all* $h_t(\eta)$ corresponding to the Fermi-like energy distribution function $f(\varepsilon; t) < 1$ with $f(\varepsilon \to +\infty; t) = 0$, $f(\varepsilon \to -\infty; t) = 1$. Thus the existence of the Fermi sea makes the net absorption rate non-zero despite the diffusion character of the energy distribution dynamics:

$$W(t) = W_0 = \frac{D_E}{\delta}. \tag{5.18}$$

6. Quantum correction to absorption rate

We see that the non-crossing, or zero-loop, approximation leads to the classical picture of a time-independent absorption rate, the so called *Ohmic absorption*. This is in line with the fact established below that the same approximation leads to the classical Drude conductivity. Let us go beyond and consider the diagrams for G^K that contain one crossing or one Cooperon loop. Different ways of presenting such a diagram are shown in Fig. 17. This is essentially the same fan diagram with the Loose Diffuson attached to it. However, because the perturbation does not contain a vector vertex, one should take special care about the Hikami box. It is shown in Fig. 18. Here a new element – besides the correlation between two $\hat{V}$ vertices that first appeared in Fig. 16– is the dotted line installed between two retarded or two advanced Green's functions. The rules of such an installation are (i) non-crossing of dotted-dotted and dotted-wavy lines and (ii)

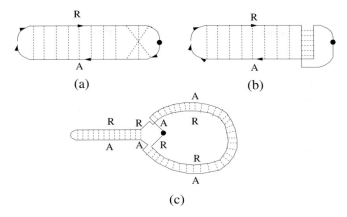

Fig. 17. Diagrams for weak dynamic localization.

the presence of both retarded and advanced functions in each "cell" separated by dotted or wavy lines (otherwise the ξ-integral is zero). One can see that the sum of three diagrams in Fig. 18a-c and in Fig. 18d-f are zero. The remaining diagrams Fig. 18g and Fig. 18h have opposite signs so that the R-A junction and the $\hat{V}(t)$ vertices (denoted by an open circle in Fig. 18) that appear as the first term of expansion of $\langle G^{R/A} \rangle$ compose a combination

$$\Gamma \quad (\varphi(t' + \eta/2) - \varphi(t' - \eta/2)) \, (\varphi(t' - t_1) - \varphi(t' - t_1 - \eta)) \, h_0(\eta) \quad (6.1)$$
$$\approx \quad \Gamma \, \partial_{t'}\varphi(t') \, \partial_{t'}\varphi(t' - t_1) \, \eta^2 h_0(\eta).$$

The contribution to the absorption rate that comes from the diagram of Fig. 17 is computed with the help of Eq. (5.4). The limit $\lim_{\eta \to 0} \partial_\eta \eta^2 h_0(\eta)$ results in a universal constant i/π, and the time derivative ∂_t of the Loose Diffuson gives a $\delta(t - t')$ function in the leading order in $\hat{V}$. So we obtain a quantum correction to the absorption rate [15]:

$$\frac{\delta W(t)}{W_0} = \frac{\Gamma \delta}{\pi D_E} \int_0^t \partial_t \varphi(t) \partial_t \varphi(t - t_1) \, \tilde{C}_{t - t_1/2}(t_1, -t_1) \, dt_1. \quad (6.2)$$

where the limits of integration are fixed by an assumption that a time-dependent perturbation has been switched on at $t = 0$; $C_t(\eta, \eta'; \mathbf{k})$ is the Cooperon, and

$$\tilde{C}_t(\eta, \eta') = (2\pi\nu\tau^2) \frac{1}{Vol} \sum_{\mathbf{k}} C_t(\eta, \eta'; \mathbf{k}). \quad (6.3)$$

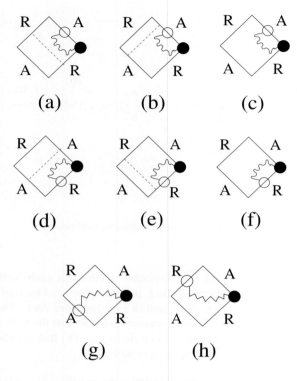

Fig. 18. Hikami box for a scalar vertex.

Here we have to make some remarks. Eq. (6.2) is general and gives the first quantum correction to the absorption rate in a disordered or chaotic system of any geometry. However, the quantity $\tilde{C}_t(\eta, \eta')$ depends on the type of perturbation and on the system geometry. For a closed system of finite volume Vol there always exists a regime where the geometry of the system does not play any role. This is the limit where the typical frequency ω of perturbation is much smaller than the so called Thouless energy E_{Th}. In disordered systems where the electronic transport is diffusive, the Thouless energy coincides with the gap between the *zero diffusion mode* that corresponds to $\mathbf{k} = 0$ and the first mode of dimensional quantization that corresponds to $k \sim 2\pi/L$, where L is the *largest* of the system sizes. In this case it is of the order of the inverse diffusion time $E_{Th} \sim 1/\tau_D$. For such small frequencies one can neglect the higher modes with nonzero values of $\mathbf{k}$ and consider the zero diffusion mode with $\mathbf{k} = 0$ that always exists in closed systems. In this *zero-mode*, or *ergodic* limit the actual shape of the system does not matter at all. One can show that this is exactly the limit where the results obtained using the Hamiltonian Eq. (5.5) are equivalent to the results obtained starting from the random matrix theory Eq. (5.7).

However, even in the ergodic limit the results for the Cooperon depend on the *topology* of the system. The equation for the Cooperon Eq. (4.2) that corresponds to the *non-simply connected* topology of a ring and a *global vector-potential* perturbation differs from that for the scalar potential perturbations with local correlations in space described by Eqs. (5.5),(5.6).

To see the difference we re-derive the Cooperon for the case of a scalar potential perturbation. The corresponding diagrams analogous to those shown in Fig. 11 for the case of the global vector-potential perturbation are given in Fig. 19. As a result of calculation of the corresponding ξ-integrals we obtain an equation

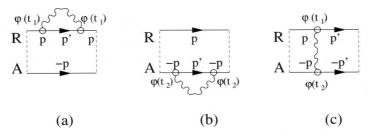

(a)   (b)   (c)

Fig. 19. The ac scalar potential corrections to the inverse Cooperon.

(in the ergodic limit $\mathbf{k}=0$):

$$\left\{ 2\frac{\partial}{\partial \eta} + \Gamma \left[\varphi(t+\eta/2) - \varphi(t-\eta/2) \right]^2 \right\} \tilde{C}_t(\eta, \eta') = 2\delta(\eta - \eta'). \qquad (6.4)$$

This equation contains the *difference* $\varphi(t+\eta/2) - \varphi(t-\eta/2)$ instead of the sum in Eq. (4.2). One can show [6] that the time-dependent random matrix theory that reproduces the Cooperon with the sum $\varphi(t+\eta/2) + \varphi(t-\eta/2)$ corresponds to the random perturbation matrix $\hat{V}$ which is pure imaginary *anti*-symmetric, rather than real symmetric as in Eq. (5.7).

The corresponding equation for a time-dependent Diffuson appears to be the same as Eq. (4.3):

$$\left\{ \frac{\partial}{\partial t} + \Gamma \left[\varphi(t+\eta/2) - \varphi(t-\eta/2) \right]^2 \right\} \tilde{D}_\eta(t, t') = \delta(t - t'). \qquad (6.5)$$

To complete the solution Eq. (6.2) we give a solution to Eq. (6.4) which is valid for an *arbitrary* time dependence of $\varphi(t)$:

$$\tilde{C}_t(\eta, \eta') = \theta(\eta - \eta') \exp\left\{-\frac{\Gamma}{2} \int_{\eta'}^{\eta} [\varphi(t + \zeta/2) - \varphi(t - \zeta/2)]^2 \, d\zeta\right\}. \quad (6.6)$$

7. Weak dynamic localization and no-dephasing points

In this section we concentrate on the *oscillating* time dependence of perturbation with $\overline{\varphi(t)} = 0$ and $\overline{\varphi^2(t)} = 1$. The simplest example is that of a *white-noise* with $\overline{\varphi(t)\varphi(t')} = 0$ for $t \neq t'$. Then Eq. (6.6) takes the form:

$$\tilde{C}_t(\eta, \eta') = \theta(\eta - \eta') \exp\left[-\Gamma(\eta - \eta')\right]. \quad (7.1)$$

The negative exponential factor describes *dephasing* by a white noise perturbation. Because of the factor $\partial_t \varphi(t) \partial_t \varphi(t - t_1)$ the effective range of integration in Eq. (6.2) is of the order of the correlation time $\tau_0 \ll 1/\Gamma$. This makes the correction to the absorption rate vanish $\delta W(t)/W_0 \to 0$ in the white-noise limit $\tau_0 \to 0$.

Let us now consider the simplest example of an *infinite − range* time correlations. This is the case of a harmonic perturbation $\varphi(t) = \cos(\omega t)$. Then for $\omega|\eta - \eta'| \gg 1$ the Cooperon takes again the form Eq. (7.1) but with the *time-dependent* dephasing rate:

$$\Gamma_t = \Gamma \sin^2 \omega t \quad (7.2)$$

One can see that there are certain times $\omega t_n = \pi n$ (n = integer) where the dephasing rate is zero. We will refer to these points t_n as *no-dephasing* points [16].

This remarkable phenomenon is generic to harmonic perturbation and is also present for the case of an ac vector-potential considered in the previous section. Its physical meaning is quite simple. Consider again a pair of electron trajectories with a closed loop shown in Fig. 8. In the presence of a time-dependent vector-potential $\mathbf{A}(t)$ the phase difference between these two trajectories is no longer zero but a random quantity:

$$\delta\Phi = \int_0^{\mathcal{T}} \mathbf{A}(t') [\mathbf{v}_1(t') - \mathbf{v}_2(t')] \, dt', \quad (7.3)$$

where $\mathbf{v}_{1,2}(t)$ are electron velocities at a time t and $\mathcal{T}$ is the time it takes for an electron to make a loop (the traversing time). For loops with the opposite directions of traversing $\mathbf{v}_1(t') = -\mathbf{v}_2(\mathcal{T} - t')$ [see Fig. 20]. Then we obtain

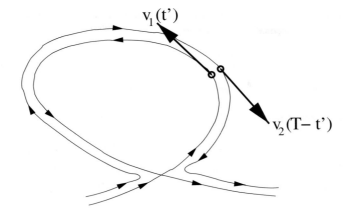

Fig. 20. Trajectories with loops traversed in opposite directions and synchronization.

$$\delta\Phi = \int_{-T/2}^{T/2} [\mathbf{A}(T/2 + t') + \mathbf{A}(T/2 - t')]\, \mathbf{v}_1(t' + T/2)\, dt' \tag{7.4}$$

Now assuming the period of oscillations of $\mathbf{A}(t)$ to be large compared to the velocity correlation time τ we obtain after averaging over $\mathbf{v}_1$:

$$\langle(\delta\Phi)^2\rangle = D_0 \int_{-T/2}^{T/2} [\mathbf{A}(T/2 + t') + \mathbf{A}(T/2 - t')]^2\, dt'. \tag{7.5}$$

We see that in the presence of a harmonic vector-potential $\mathbf{A}(t) \propto \sin\omega t$ the phase difference $\delta\Phi = 0$ vanishes for *all* the loop trajectories with the traversing time equal to the integer period of the time-dependent perturbation:

$$T = \frac{2\pi n}{\omega}, \quad n = 1, 2, 3\dots \tag{7.6}$$

Such loop trajectories with the traversing time *synchronized* with the period of perturbation play a crucial role in all the quantum coherence (interference) phenomena in disordered and chaotic systems (their effect on the universal conductance fluctuations has been studied in Ref. [16]). Any periodic in time perturbation leads to a *selection of trajectories*: for strong enough perturbation all trajectories that do not obey Eq. (7.6) do not interfere because of the random phase difference $\delta\Phi \gg 1$.

In the particular case of the quantum corrections to the energy absorption rate Eq. (6.2) the time-dependent Cooperon

$$\tilde{C}_{t-t_1/2}(t_1, -t_1) = \exp\left[-2t_1\Gamma_{t-t_1/2}\right] \tag{7.7}$$

is small everywhere except in the vicinity of no-dephasing points

$$\omega t_1^{(n)} = 2\omega t - 2\pi n, \quad n = 0, \pm 1, \pm 2... \tag{7.8}$$

For $\Gamma t_1 \sim \Gamma t \gg 1$ one can expand $\Gamma_{t-t_1/2} \approx (\omega/2)^2 \Gamma (t_1 - t_1^{(n)})^2$ in the vicinity of no-dephasing points, do the Gaussian integration over $\zeta = (t_1 - t_1^{(n)})$ and put $t_1 = t_1^{(n)}$ elsewhere. In particular $\partial_t\varphi(t - t_1^{(n)}) = -\partial_t\varphi(t)$. Then we arrive at:

$$\frac{\delta W(t)}{W_0} = -\frac{\delta}{\pi} (2\sin^2 \omega t) \sum_n \int_{-\infty}^{+\infty} d\zeta \, e^{-D_E\zeta^2 t_1^{(n)}}. \tag{7.9}$$

Replacing $\sum_n$ by the integral $(\omega/2\pi) \int_0^t dt_1$ and averaging over time intervals much larger than $2\pi/\omega$ we finally get:

$$\frac{W(t)}{W_0} = 1 - \frac{\delta\omega}{\pi} \int_0^t dt_1 \int_{-\infty}^{+\infty} \frac{d\zeta}{2\pi} e^{-D_E\zeta^2 t_1} = 1 - \sqrt{\frac{t}{t_*}}, \tag{7.10}$$

where

$$t_* = \frac{\pi^3\Gamma}{2\delta^2}. \tag{7.11}$$

We have obtained a remarkable result that the absolute value of the quantum correction to the absorption rate grows with time. This is the consequence of the existence of no-dephasing points, as otherwise the exponentially decaying Cooperon would lead to a saturation of the integral over t_1 in Eq. (6.2) at large times t. The negative sign of the correction implies that the absorption rate, or energy diffusion coefficient decreases with time. This phenomenon can be called "weak dynamic localization" in full analogy with the weak Anderson localization when the diffusion coefficient in space decreases with the system size. It has been first discovered [17, 18, 19] for a simple quantum system – quantum rotor subject to the periodic δ-function perturbation. At $t \sim t_*$ the quantum correction is of the order of the classical ohmic absorption, and we can expect strong *dynamic localization* to occur.

Eq. (7.10) suggests a more precise relationship between the dynamic localization for a quantum system in the ergodic (*zero-dimensional*) limit subject to a harmonic perturbation and the Anderson localization in a *quasi-one dimensional* disordered wire. We observe that

$$\partial_t W(t) \quad \propto \quad \int_{-\infty}^{+\infty} \frac{d\zeta}{2\pi} e^{D_E \zeta^2 t} \tag{7.12}$$

$$= \int_{-\infty}^{\infty} \frac{d\omega'}{2\pi} e^{-i\omega' t} \int_{-\infty}^{+\infty} \frac{d\zeta}{2\pi} \frac{1}{D_E \zeta^2 - i\omega'}.$$

At the same time, according to Eq. (2.22), the correction to the (complex) frequency dependent conductivity of an infinite quasi-one dimensional wire is proportional to:

$$\delta\sigma(\omega) = \int_{-\infty}^{+\infty} \frac{dk}{2\pi} \frac{1}{D_0 k^2 - i\omega}. \tag{7.13}$$

We see that the deviation from the no-dephasing points $\zeta = t_1 - t_1^{(n)}$ is an analogue of the momentum k. Then with a suitable choice of a time and frequency scale one obtains:

$$\frac{W(t)}{W_0} = \int_{-\infty}^{+\infty} \frac{d\omega'}{2\pi} \frac{e^{-i\omega' t}}{(-i\omega' + 0)} \left(\frac{\sigma(\omega')}{\sigma_0} \right). \tag{7.14}$$

Note that Eq. (7.14) is non-trivial as it establishes a relationship between an essentially non-equilibrium property of a zero-dimensional system and an equilibrium property of a quasi-one dimensional system.

This relationship can be proven for any diagram with an arbitrary number of Diffuson-Cooperon loops (for the case of two loops see Ref. [20]) and we conjecture that it is *exact* in the ergodic (or random matrix) regime for $t\Gamma, t\omega \gg 1$.

Eq. (7.14) helps to establish a character of decay of the absorption rate $W(t)$ for $t \gg t_*$. To this end we recall the Mott-Berezinskii formula for the frequency-dependent conductivity in the localized regime [21]:

$$\sigma(\omega) \propto \omega^2 \ln^2 \omega. \tag{7.15}$$

Then using Eq. (7.14) one obtains the corresponding absorption rate $W(t)$ at $t \gg t_*$:

$$W(t) \propto \frac{\ln t}{t^2}. \tag{7.16}$$

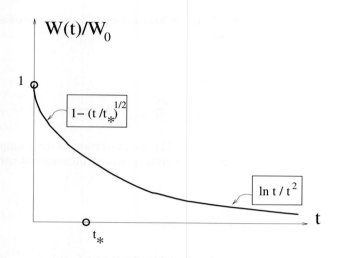

Fig. 21. Time dependence of the absorption rate.

8. Conclusion and open questions

The goal of this course was to give a unified picture and a unified theoretical tool to consider different quantum coherence effects in disordered metals. The unified picture is that of interference of electron trajectories with loops traversed in opposite directions. The corresponding theoretical machinery is the Diffuson-Cooperon diagrammatic technique. We have demonstrated that even the first quantum correction diagram with the Cooperon loop can describe such different and nontrivial phenomena as weak Anderson localization, quantum rectification and dynamic localization in energy space. We did not try to give a review of the development in the field but rather to concentrate on a few important examples and to demonstrate how the machinery works in different cases. That is why many related issues have not been discussed and the corresponding works have not been cited properly. We apologize for that.

There are few a open problems that are related with the main subjects of this school and in our opinion warrant a study. This is first of all a unified theory of energy absorption where both the Zener transitions picture [22] and the sequential photon absorption picture are incorporated. The suitable theoretical tool for that is believed to be a nonlinear sigma-model in the Keldysh representation derived in Refs. [15]. The perturbative treatment of this field theory reproduces the diagrammatic technique discussed in this course, and the non-perturbative consideration in the region Γ, $\omega < \delta$ should give the results obtained in the framework of the Zener transitions picture.

Another possible direction is the role of interactions in the quantum rectification and dynamic localization. Some of the interaction effects in dynamic localization have been recently considered in Refs. [23, 24, 25] using the Fermi Golden Rule approximation. However, an interesting regime of localization in the Fock space [26], where the Fermi Golden Rule does not apply, is awaiting an investigation.

References

[1] A. A. Abrikosov, L. P. Gorkov and I. E. Dzyaloshinskii, *Methods of Quantum Field Theory in Statistical Physics*, Pergamon Press, New York (1965).

[2] L. P. Gor'kov, A. I. Larkin and D. E. Khmel'nitskii, Pis'ma Zh. Eksp. Teor. Fiz. **30**, 248 (1979) [JETP Lett. **30**, 228 (1979)].

[3] L.V. Keldysh, Zh.Exp.Teor.Fiz., **47**, 515 (1964) [Sov.Phys.-JETP, **20**, 1018 (1965)].

[4] J.Rammer and H.Smith, Rev. Mod. Phys., bf 58, 323 (1986).

[5] A. Kamenev, *Many-body theory of non-equilibrium systems*, this Volume.

[6] V. I. Yudson, E. Kanzieper, and V. E. Kravtsov, Phys. Rev. B **64**, 045310 (2001).

[7] V.E. Kravtsov and V.I. Yudson, Phys. Rev. Lett., **70**, 210 (1993).

[8] A.G. Aronov and V.E. Kravtsov, Phys. Rev. B **47**, 13409 (1993).

[9] V.E. Kravtsov and B.L. Altshuler, Phys. Rev. Lett. **84**, 3394 (2000).

[10] V.I. Falko and D.E. Khmelnitskii, Zh. Exp. Teor. Fiz. **95**, 328 (1989) [Sov.Phys.-JETP, **68**,186 (1989)].

[11] B. L. Altshuler, A. G. Aronov and D. E. Khmelnitskii J. Phys. C **15**, 7367 (1982).

[12] M. G. Vavilov and I. L. Aleiner, Phys. Rev. B **60**, R16311 (1999); **64**, 085115 (2001); M. G. Vavilov, I. L. Aleiner and V. Ambegaokar, Phys. Rev. B **63**, 195313 (2001).

[13] M. Büttiker, Y. Imry and R. Landauer, Phys. Lett. A **96**, 365 (1983).

[14] L.P. Levy, G.Dolan, J.Dunmuir and H.Bouchiat, Phys. Rev. Lett. **64**, 2074 (1990).

[15] D.M. Basko, M.A. Skvortsov and V.E. Kravtsov, Phys. Rev. Lett. **90**, 096801 (2003).

[16] X.-B. Wang and V. E. Kravtsov, Phys. Rev. B **64**, 033313 (2001); V. E. Kravtsov, Pramana-Journal of Physics **58**, 183 (2002).

[17] G. Casati, B.V. Chirikov, J. Ford and F. M. Izrailev, in *Stochastic Behaviour in Classical and Quantum Hamiltonian Systems*, ed. by G. Casati and J. Ford, Lecture Notes in Physics, vol. 93 (Springer, Berlin, 1979).

[18] B. V. Chirikov, F. M. Izrailev and D. L. Shepelyansky, Physica (Amsterdam) **33** D, 77 (1988).

[19] A. Altland, Phys. Rev. Lett. **71**,69 (1993).

[20] M. A. Skvortsov, D. M. Basko and V. E. Kravtsov, JETP Lett. **80**, 60 (2004).

[21] N. F. Mott, Phil. Mag. **17**, 1259 (1968); V. L. Berezinskii, Zh. Eksp. Teor. Fiz.**65**, 1251 (1973) [Sov. Phys. JETP **38**, 620 (1974)].

[22] Y. Gefen and D. J. Thouless, Phys. Rev. Lett. **59**, 1752 (1987).

[23] D.M. Basko, Phys. Rev. Lett. **91**, 206801 (2003).

[24] D.M. Basko and V.E. Kravtsov, Phys. Rev. Lett. **93**, 056804 (2004).

[25] V.E. Kravtsov, cond-mat/0312316.

[26] B. L. Altshuler, Y. Gefen, A. Kamenev and L. S. Levitov, Phys. Rev. Lett. **78**, 2803 (1997).

Course 5

NOISE IN MESOSCOPIC PHYSICS

T. Martin

Centre de Physique Théorique et Université de la Méditerranée
case 907, 13288 Marseille, France

H. Bouchiat, Y. Gefen, S. Guéron, G. Montambaux and J. Dalibard, eds.
Les Houches, Session LXXXI, 2004
Nanophysics: Coherence and Transport
© *2005 Elsevier B.V. All rights reserved*

Contents

1. Introduction

When one applies a constant bias voltage to a conductor, a stationary current is typically established. However, if one carefully analyzes this current, one discovers that it has a time dependence: there are some fluctuations around the average value (Fig. 1). One of the ways to characterize these fluctuations is to compute the current-current correlation function and to calculate its Fourier transform: the noise. The information thus obtained characterizes the amplitude of the deviations with respect to the average value, as well as the frequency of occurrence of such fluctuations.

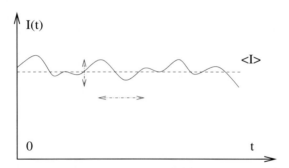

Fig. 1. The current as a function of time undergoes fluctuation around an average value, the height of the "waves" and their frequency are characterized by the noise.

Within the last decade and a half, there has been a resurgence of interest in the study of noise in mesoscopic devices, both experimentally and theoretically. In a general sense, noise occurs because the transport of electrons has a stochastic nature. Noise can arise in classical and quantum transport as well, but here we will deal mostly with quantum noise. Noise in solid state devices can have different origins: there is $1/f$ noise [1] which is believed to arise from fluctuations in the resistance of the sample due to the motion of impurities [2] and background charges. This course does not deal at all with $1/f$ noise. The noise I am about to describe considers the device/conductor as having no inelastic effects: a probability of transmission, or a hopping matrix element characterizes it for instance.

Typically in experiments, one cannot dissociate $1/f$ noise from this "other" noise contribution. $1/f$ noise can obviously not be avoided in low frequency measurements, so experimentalists typically perform measurements in the kHz to MHz range, sometimes higher.

If the sample considered is small enough that dephasing and inelastic effects can be neglected, equilibrium (thermal) noise and excess noise can be completely described in terms of the elastic scattering properties of the sample. This is the regime of mesoscopic physics, which is now described in several textbooks [3]. As mentioned above, noise arises as a consequence of random processes governing the transport of electrons. Here, there are two sources randomness: first, electrons incident on the sample occupy a given energy state with a probability given by the Fermi–Dirac distribution function. Secondly, electrons can be transmitted across the sample or reflected in the same reservoir they came from, with a probability given by the quantum mechanical transmission/ reflection coefficients.

Equilibrium noise refers to the case where no bias voltage is applied between the leads connected to the sample, where thermal agitation alone allows the electrons close to the Fermi level to tunnel through the sample [4]. Equilibrium noise is related to the conductance of the sample via the Johnson–Nyquist formula [5]. In the presence of a bias, in the classical regime, we expect to recover the full shot noise, a noise proportional to the average current. Shot noise was predicted by Schottky [6] and was observed in vacuum diodes [7].

If the sample is to be described quantum mechanically, a calculation of the noise should include statistical effects such as the Pauli principle: an electron which is successfully transmitted cannot occupy the same state as another electron incident from the opposite side, which is reflected by the potential barrier. The importance of the statistics of the charge carriers is a novelty in mesoscopic physics. After all, many experiments in mesoscopic physics – with electrons as carriers – have a direct optical analog if we interchange the carriers with photons. The conductance steps experiment [8] which shows the transverse quantization of the electron wave function is an example: this experiment has been successfully completed with photons [9]. Universal conductance fluctuations in mesoscopic wires and rings [10, 11] also has an analog when one shines a laser on white paint or cold atoms as one studies the retrodiffusion peak or the speckle pattern which is generated [12]. In contrast to these examples, a noise measurement makes a distinction between fermions and bosons.

Many approaches have been proposed to calculate noise. Some are quasi–classical. Others use a formulation of non-equilibrium thermodynamics based on the concept of reservoirs, introduced for the conductance formula [13]. Here, we shall begin with an intuitive picture [14, 15, 16], where the current passing through the device is a superposition of pulses, or electrons wave packets, which can be transmitted or reflected. We shall then proceed with scattering theory, and conclude with calculations of noise with the Keldysh formalism for a strongly correlated system.

The scattering approach based on operator averages [17, 18, 19] will be described in detail, because it allows to describe systematically more complex situations where the sample is connected to several leads. It also allows a generalization to finite frequency noise and to conductors which have an interface with a superconductor. Superconductors will be studied not only for their own sake, but also because they provide a natural source of entangled electrons. Noise correlation measurements can in principle be used to test quantum mechanical nonlocality. Starting from a microscopic Hamiltonian, one can use non-equilibrium formulations of field theory and thermodynamics [20] to compute the current, as well as the noise [21, 22, 23]. There are few systems in mesoscopic physics where the role of electronic interactions on the noise can be probed easily. Coulomb interaction give most of us a considerable headache, or considerable excitement. One dimensional systems are special, in the sense that they can be handled somewhat exactly. In the tunneling regime only, I will discuss the noise properties of Luttinger liquids, using the illustration of edge channels in the fractional quantum Hall effect. Even the lowest order tunneling calculation brings out an interesting result: the identification of anomalous, fractional charges via the Schottky formula.

Noise also enters in dephasing and decoherence processes, for instance, when a quantum dot with a sharp level is coupled electrostatically to the electrons which transit in a nearby wire [24].

Throughout the course, we will consider conductors which are connected to several terminals because one can build fermionic analogs of optical devices for photons [25].

Finally, I mention that there are some excellent publications providing reviews on noise. One book on fluctuations in solids is available [26], and describes many aspects of the noise – such as the Langevin approach – which I will not discuss here. Another is the very complete review article of Y. Blanter and M. Büttiker [27], which was used here in some sections. To some extent, this course will be complementary to these materials because it will present the Keldysh approach to noise in Luttinger liquids, and because it will discuss to which extent noise correlations can be probed to discuss the issue of entanglement.

2. Poissonian noise

Walter Schottky pioneered the field in 1918 by calculating the noise of a source of particles emitted in an independent manner [6]. Let τ be the mean time separating two tunneling events. In quantum and classical transport, we are dealing with situations where many particle are transmitted from one lead to another. Consider now the probability $P_N(t)$ for having N tunneling events during time t. This follows the probability law:

$$P_N(t) = \frac{t^N}{\tau^N N!} e^{-t/\tau}. \tag{2.1}$$

This can be proved as follows. The probability of N tunneling events can be expressed in terms of the probability to have $N - 1$ tunneling events:

$$P_N(t + dt) = P_{N-1}(t)\frac{dt}{\tau} + P_N(t)\left(1 - \frac{dt}{\tau}\right) \tag{2.2}$$

which leads (after multiplying by $\exp(t/\tau)$) to

$$\frac{d}{d(t/\tau)}\Pi_N = \Pi_{N-1}, \tag{2.3}$$

where $\Pi_N \equiv P_N \exp(t/\tau)$. The solution by induction is clearly

$$\Pi_N = (t/\tau)^N \Pi_0/N! \tag{2.4}$$

Since the solution for P_0 is given by:

$$\frac{dP_0}{dt} = -\frac{1}{\tau}P_0, \qquad P_0 = \exp(-t/\tau), \tag{2.5}$$

one finally obtains the above result for $P_N(t)$, which can be expressed in turn as a function of the mean number of particles transmitted during t.

$$P_N(t) = \frac{\langle N \rangle^N}{N!} e^{-\langle N \rangle}. \tag{2.6}$$

To show this, one considers the characteristic function of the distribution:

$$\phi = \sum_N \frac{e^{i\lambda N}(t/\tau)^N}{N!} e^{-t/\tau} = e^{(t/\tau)(e^{i\lambda}-1)}. \tag{2.7}$$

The characteristic function gives the average number of particle transmitted when differentiating it respect to λ once (and setting $\lambda = 0$), and the variance when differentiating with λ twice. We thus get:

$$\langle N \rangle = t/\tau, \qquad \langle N^2 \rangle - \langle N \rangle^2 = \langle N \rangle. \tag{2.8}$$

This has a fundamental consequence on the noise characteristic of a tunnel junction. The current is given by

$$\langle I \rangle = e\langle N \rangle / t = e/\tau. \tag{2.9}$$

The noise is proportional to the variance of the number of transmitted particles. The final result for the spectral density of noise is:

$$S = 2e^2(\langle N^2 \rangle - \langle N \rangle^2)/t = 2e\langle I \rangle. \tag{2.10}$$

This formula has the benefit that it applies to any tunneling situation. It applies to electrons tunneling between two metallic electrodes, but also to "strange" quasiparticles of the quantum Hall effect tunneling between two edge states. Below we shall illustrate this fact in discussing the detection of the quasiparticle charge. This formula will serve also as a point of comparison with the quantum noise derivations which apply to mesoscopic conductors.

3. The wave packet approach

We consider first a one dimensional sample connected to a source and a drain. The results presented here were first described in [28, 17] with other methods. The quantity we wish to calculate is the time correlation in the current:

$$C(t) = \frac{1}{T} \int_0^T dt' \, \langle I(t')I(t+t') \rangle. \tag{3.1}$$

The spectral density of noise $S(\omega)$ is related to the above quantity by a simple Fourier transform. The measurement frequencies which we consider here are low enough compared to the inverse of the time associated with the transfer of an electron from source to drain [29] and allow to neglect the self inductance of the sample. Using the Fourier representation for the current, this yields:

$$S(\omega) = \lim_{T \to \infty} \frac{2}{T} \langle |I(\omega)|^2 \rangle, \tag{3.2}$$

where the angular brackets denote some kind of average over electrons occupation factors. The wave packet approach views the current passing across the sample as a superposition of clocked pulses [16]: $I(t) = \sum_n j(t - n\tau)g_n$. In this expression, $j(t)$ is the current associated with a given pulse and g_n is an occupation factor which takes a value 1 if an electron has been transferred from the left side to the right side of the sample, -1 if the electron was transferred from right to left, and 0 when no electron is transferred at all. The quantum mechanics necessary to calculate the noise is hidden in g_n. The wave packets representing

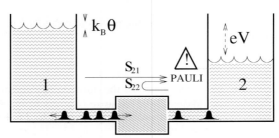

Fig. 2. The Landauer philosophy of quantum transport. Electrons travel as wave packets emitted from reservoirs (right and left). Reservoirs do not have the same chemical potential because of an applied bias. Thermal fluctuations (waves) exist in the reservoirs. The superposition of two electrons in the same scattering state cannot occur because of the Pauli principle.

the electrons are separated in time, but can overlap in space. An example of wave packet construction can be obtained if we consider states limited to a small energy interval ΔE [16]: choosing $\tau = h/\Delta E$ insures that successive pulses are orthogonal to each other. With the above definitions, the calculation of the noise spectral density in the energy interval $[E - \Delta E/2, E + \Delta E/2]$ reduces to the calculation of the fluctuation in the occupation factors:

$$S(\omega = 0) = \frac{2\Delta E e^2}{\pi \hbar}(\langle g^2 \rangle - \langle g \rangle^2), \tag{3.3}$$

where we have dropped the index n in g_n because all pulses contribute to the noise in the same fashion. Also, note that we have subtracted the average current in order to isolate the fluctuations. The calculation of the spectral density of noise is thus directly related to the statistics of the current pulses.

To obtain the correlator $\langle g^2 \rangle - \langle g \rangle^2$, we consider all possible pulse histories: first consider the case where two electrons are incident on the sample from opposite sides. In this situation, $g = 0$ because there will be no current if the two electrons are both reflected or both transmitted, and the situation where one electron is reflected and the other is transmitted is forbidden by the Pauli principle; two electrons (with the same spin) cannot occupy the same wave packet state. Secondly, there is the straightforward situation where both incident states are empty, with $g = 0$. The other possibilities where $g = 0$ follow if an electron is reflected from one side, when no electron was incident from the other side. In fact the only possibilities to have a current through the sample are when an electron incident from the right (left) is transmitted while no electron was present on the other side, giving the result $g = 1$ ($g = -1$) with respective weight $f_1(1 - f_2)T$ or $f_2(1 - f_1)T$. f_1 (f_2) is the Fermi–Dirac distribution associated with the left (right) reservoir, and T is the transmission probability. We therefore obtain:

$$\langle g^2 \rangle - \langle g \rangle^2 = T(f_1 + f_2 - 2f_1 f_2) - (f_1 - f_2)^2 T^2. \tag{3.4}$$

A note of caution here. By removing "by hand" the two processes in which the Pauli principle is violated, it looks as if the total probability for having either both electrons reflected or both electrons transmitted does not add up to one: from our argument it seems that this probability is equal to $T^2 + (1 - T)^2$. In reality, there is no such problem if one considers a wave function for incoming and outgoing states which is anti-symmetrized (as it should be). The processes where electrons are both reflected and both transmitted are not distinguishable, and their amplitudes should be added before taking the square modulus to get the total probability. Using the unitarity property of the S–matrix, one then obtain that this probability is indeed one. Summing now over all energy intervals, we thus obtain the total excess noise:

$$
\begin{aligned}
S(\omega = 0) \;=\; & \frac{4e^2}{h} \int dE \, T(E)[f_1(1 - f_1) + f_2(1 - f_2)] \\
& + \frac{4e^2}{h} \int dE \, T(E)[1 - T(E)](f_1 - f_2)^2.
\end{aligned}
\tag{3.5}
$$

In the absence of bias or at high temperatures Θ ($|\mu_1 - \mu_2| \ll k_B\Theta$), the two first terms on the right hand side dominate. The dependence on the distribution functions $f_i(1 - f_i)$ is typical of calculations of fluctuations in thermal equilibrium. Using the relation $f_i(1 - f_i) = -k_B\Theta \partial f_i/\partial E$, we recover the Johnson Nyquist [5] formula for thermal equilibrium noise [14]:

$$S(\omega = 0) = 4\frac{2e^2 T}{h} k_B\Theta = 4Gk_B\Theta, \tag{3.6}$$

where G is the Landauer conductance of the mesoscopic circuit. In the opposite limit, $|\mu_1 - \mu_2| \gg k_B\Theta$, we get a contribution which looks like shot noise, except that it is reduced by a factor $1 - T$:

$$S(\omega = 0) = 2e\langle I \rangle(1 - T), \tag{3.7}$$

which is called reduced shot noise or quantum shot noise. At this point, it is useful to define the Fano factor as the ratio between the zero frequency shot noise divided by the Poisson noise:

$$F \equiv \frac{S(\omega = 0)}{2e\langle I \rangle}, \tag{3.8}$$

which is equal to $1 - T$ in the present case. In the limit of poor transmission, $T \ll 1$, and we recover the full shot noise. For highly transmissive channels,

$T \sim 1$, and we can think of the reduction of shot noise as being the noise contribution associated with the poor transmission of holes across the sample. Because of the Pauli principle, a full stream of electrons which is transmitted with unit probability does not contribute to noise. Note that this is the effect seen qualitatively in point contact experiments [30, 31]. In the intermediate regime $|\mu_1 - \mu_2| \simeq k_B \Theta$, there is no clear separation between the thermal and the reduced shot noise contribution.

4. Generalization to the multi–channel case

We now turn to the more complex situation where each lead connected to the sample has several channels. Our concern in this case is the role of channel mixing: an outgoing channel on the right side collects electrons from all incoming channels transmitted from the left and all reflected channels on the right. We therefore expect that wave packets from these different incoming channels will interfere with each other. To avoid the issue of interference between channels and treat the system as a superposition of one dimensional contributions, we must find a wave packet representation where the mixing between channels is absent.

This representation is obtained by using a decomposition of the S– matrix describing the sample. Let us assume for simplicity that both leads have the same number of channels M. The S–matrix is then a block matrix containing four M by M sub-matrices describing the reflection from the right (left) side, $\mathbf{s}_{22}$ ($\mathbf{s}_{11}$), and the transmission from left to right (right to left), $\mathbf{s}_{12}$ ($\mathbf{s}_{21}$):

$$\mathbf{S} = \begin{pmatrix} \mathbf{s}_{11} & \mathbf{s}_{12} \\ \mathbf{s}_{21} & \mathbf{s}_{22} \end{pmatrix}. \tag{4.1}$$

From the unitarity of the S–matrix, which follows from current conservation, it is possible to write the sub-matrices in terms of two diagonal matrices and four unitary matrices:

$$\mathbf{s}_{11} = -i\mathbf{V}_1\mathbf{R}^{1/2}\mathbf{U}_1^{\dagger}, \quad \mathbf{s}_{12} = \mathbf{V}_1\mathbf{T}^{1/2}\mathbf{U}_2^{\dagger}, \tag{4.2}$$

$$\mathbf{s}_{21} = \mathbf{V}_2\mathbf{T}^{1/2}\mathbf{U}_1^{\dagger}, \quad \mathbf{s}_{22} = -i\mathbf{V}_2\mathbf{R}^{1/2}\mathbf{U}_2^{\dagger}, \tag{4.3}$$

where $\mathbf{R}^{1/2}$ and $\mathbf{T}^{1/2}$ are real diagonal matrices with diagonal elements $R_\alpha^{1/2}$ and $T_\alpha^{1/2}$ such that $T_\alpha + R_\alpha = 1$. R_α (T_α) are the eigenvalues of the matrices $\mathbf{s}_{11}^{\dagger}\mathbf{s}_{11}$ ($\mathbf{s}_{21}^{\dagger}\mathbf{s}_{21}$). $\mathbf{U}_1, \mathbf{U}_2, \mathbf{V}_1, \mathbf{V}_2$ are unitary transformations on the eigenchannels.

Using the unitary transformations, we can now choose a new basis of incoming and outgoing states on the left and the right side of the sample: $\mathbf{U}_1$ ($\mathbf{V}_1$) is the unitary transformation used to represent the incoming (outgoing) states on the left side, while $\mathbf{U}_2$ ($\mathbf{V}_2$) is the unitary transformation used to represent the incoming (outgoing) states on the right side of the sample. The effective S–matrix obtained in this new basis is thus a block matrix of four diagonal matrices. This corresponds to a situation where no mixing between channels occurs, effectively a superposition of one-dimensional (2 × 2) S–matrices which are totally decoupled (see Fig. 3). This is precisely the form which was assumed by Lesovik [17]

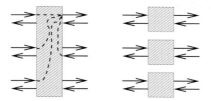

Fig. 3. A multichannel, two-terminal conductor mixes all channels in general (left). In the eigenchannel representation, it behaves like a set of one channel conductors which are decoupled from each other (right).

for the adiabatic point contact. The absence of correlations between the different incoming and outgoing wave packets allows us to write the noise as a superposition of one dimensional contributions:

$$S(\omega = 0) = \frac{4e^2}{h} \int dE \sum_\alpha T_\alpha(E)[f_1(1 - f_1) + f_2(1 - f_2)]$$

$$+ \frac{4e^2}{h} \int dE \sum_\alpha T_\alpha(E)[1 - T_\alpha(E)](f_1 - f_2)^2. \qquad (4.4)$$

This expression can be cast in terms of the block elements of the initial S–matrix using the properties of the trace, $\sum_\alpha T_\alpha^n = Tr[(s_{21}^\dagger s_{21})^n]$. In experiments using break junctions, the $T_\alpha's$ can actually be measured [32, 33]: they bear the name of "mesoscopic pin code".

5. Scattering approach based on operator averages

The method depicted in the previous section has an intuitive character because one is able to see directly the effects of the statistics when computing the noise. It can also be generalized to treat multi-terminal conductors. In contrast, in this section we adopt a more systematic procedure, which is due to G. Lesovik. This approach was generalized to multichannel, multi-terminal conductors in Ref. [19]. Here we chose to describe it because it allows to discuss noise correlations in the most straightforward way.

The philosophy of this method goes as follows. The reservoirs connected to the sample are macroscopic quantities. In a situation where electronic transport occurs, the chemical potential of each reservoir is only modified in a minor manner. One is thus tempted to apply equilibrium thermodynamics to quantities (operators) which involve only a given reservoir. This approach has the advantage that thermodynamical averages of operators can be computed in a systematic manner.

5.1. Average current

This is a classic result which can be found in Refs. [13, 34]. For simplicity, I will choose a situation where the conductor is connected to an arbitrary number of leads, but each lead connected to the sample has only one incoming/outgoing channel. The generalization for leads containing many channels gives additional algebra.

The current operator for terminal m reads in the usual way:

$$I_m(x_m) = \frac{\hbar e}{2mi} \sum_\sigma \left(\psi_{m\sigma}^\dagger(\mathbf{r}_m) \frac{\partial \psi_{m\sigma}(\mathbf{r}_m)}{\partial x_m} - \frac{\partial \psi_{m\sigma}^\dagger(\mathbf{r}_m)}{\partial x_m} \psi_{m\sigma}(\mathbf{r}_m) \right).$$

$$(5.1)$$

Here x_m is the coordinate in terminal m and $\psi_{m\sigma}^\dagger$ is the creation operator for a particle with spin σ in terminal m. In the following we shall consider conductors which transmit/reflect both spin species with the same amplitudes, and which do not scatter one spin into another along the way. In our calculations, this amounts to ignoring the spin index on the fermion operators and replacing the sum over spin by a factor 2. Spin indices will be restored in the discussion of mesoscopic superconductivity for obvious reasons. The fermion annihilation operator is expressed in terms of the scattering properties of the sample:

$$\psi_m(\mathbf{r}_m) = \sum_n \int \frac{dk_m}{\sqrt{2\pi}} \sqrt{\frac{k_m}{k_n}} (\delta_{mn} e^{ik_m x_m} + s_{mn} e^{-ik_m x_m}) c_n(k_n). \qquad (5.2)$$

s_{mn} is the scattering matrix amplitude for a state incident from n and transmitted into m. $c_m(k_m)$ is the annihilation operator for a scattering state incident from m. Momentum integrals are readily transformed into energy integrals with the substitution $\int dk_m \rightarrow m \int dE/\hbar^2 k_m$. Substituting the fermion operator in the expression for the current operator, we get:

$$I_m(x_m) = \int dE \int dE' \sum_{nn'} M_m(E, E', n, n') c_n^\dagger(k_n) c_{n'}(k_{n'}), \qquad (5.3)$$

where $M_m(E, E', n, n')$ is the current matrix element, which depends on position in general. In fact, it is the sum of two distinct contributions:

$$M_m(E, E', n, n') = M_m^{\Delta k}(E, E', n, n') + M_m^{\Sigma k}(E, E', n, n'), \tag{5.4}$$

which are defined as:

$$
\begin{aligned}
M_m^{\Delta k}(E, E', n, n') &= \frac{em}{2\pi\hbar^3} \left(\frac{k_m(E)k_m(E')}{k_n(E)k_{n'}(E')} \right)^{1/2} (k_m^{-1}(E) + k_m^{-1}(E')), \\
&\quad (e^{-i(k_m(E)-k_m(E'))x_m} \delta_{mn}\delta_{mn'} \\
&\quad -e^{i(k_m(E)-k_m(E'))x_m} s_{mn}^*(E)s_{mn'}(E'))
\end{aligned} \tag{5.5}
$$

$$
\begin{aligned}
M_m^{\Sigma k}(E, E', n, n') &= \frac{em}{2\pi\hbar^3} \left(\frac{k_m(E)k_m(E')}{k_n(E)k_{n'}(E')} \right)^{1/2} (k_m^{-1}(E) - k_m^{-1}(E')) \\
&\quad (-e^{-i(k_m(E)+k_m(E'))x_m} s_{mn'}(E')\delta_{mn} \\
&\quad +e^{i(k_m(E)+k_m(E'))x_m} s_{mn}^*(E)\delta_{mn'}).
\end{aligned} \tag{5.6}
$$

The current depends on the position where it is measured. This is quite unfortunate. Whether this is true or not depends on the working assumptions of our model. Here we anticipate a few results.

• When computing the average current, as our model does not take into account inelastic processes we get a delta function of energy and $M_m^{\Sigma k}(E, E', n, n') = 0$. As a result the average stationary current is constant.

• When computing the noise at zero frequency, the same thing will happen. The time integration will yield the desired delta function in energy. The zero frequency noise does not depend on where it is measured.

• When considering finite frequency noise, the term $M_m^{\Sigma k}(E, E', n, n')$ will once again be dropped out, "most of the time". Indeed, the typical frequencies of interest range from 0 to, say, a few $eV/\hbar$ in mesoscopic experiments. Yet in practical situations, the bias eV is assumed to be much smaller than the chemical potential μ_n of the reservoirs. Because most relevant momenta happen in the vicinity of the chemical potential within a few $eV/\hbar$, this implies that the momenta $k_m(E)$ and $k_m(E')$ are rather close. One then obtains a $M_m^{\Delta k}(E, E', n, n')$ with virtually no oscillations and a $M_m^{\Sigma k}(E, E', n, n')$ which oscillates rapidly with a wavelength π/k_F. The latter contribution has a smaller amplitude.

• About the wave vector dependences of $M_m(E, E', n, n')$. When calculating current or charge fluctuations at zero frequencies or at frequencies of the order of the bias voltage $eV/\hbar$, at the end of the day the factors containing these wave vectors cancel out in the expression of the current or noise. This is again due to the fact that $\mu_m \gg eV$.

• The question of whether such oscillations are real or not depends on how the current is measured. We can reasonably expect that a current measurement implicitly implies an average over many Fermi wavelengths along the wire. This is the case, for instance in a Gedanken experiment where the current is detected via the magnetic field it creates. In this case it is clear that only $M_m^{\Delta k}(E, E', n, n')$ survives.

• On the other hand, care must be taken when considering the density operator $\rho(x, t) = \Psi^\dagger(x, t)\Psi(x, t)$. This operator is related to the current operator via the continuity equation: $\partial_t \rho + \partial_x J_x = 0$. The density operator in frequency space can therefore be obtained either directly from $\Psi(x)$, or alternatively by taking derivatives of the current matrix elements. The oscillating contribution of the density operator has a matrix element of the order:

$$\rho^{\Delta k}(x_m, \omega) \sim (k_m^{-1}(E) + k_m^{-1}(E'))(k_m(E) - k_m(E')) \tag{5.7}$$

while the $2k_F$ contribution behaves like:

$$\rho^{\Sigma k}(x_m, \omega) \sim (k_m^{-1}(E) - k_m^{-1}(E'))(k_m(E) + k_m(E')). \tag{5.8}$$

As a result, they have the same order of magnitude. This is important when considering the issue of dephasing in a quantum dot close to a fluctuating current [24].

We now neglect the $2k_F$ oscillating terms and proceed. In order to compute the average current, it is necessary to consider the average:

$$\langle c_n^\dagger(k(E), t)c_{n'}(k(E'), t)\rangle \ = \ \frac{\hbar^2 k(E)}{m} f_n(E)\, \delta(E - E')\delta_{nn'}. \tag{5.9}$$

$f_n(E)$ is the Fermi-Dirac/Bose-Einstein distribution function associated with terminal n whose chemical potential is μ_n: $f_n(E) = 1/[\exp(\beta(E - \mu_n)) \pm 1]$. Consequently,

$$\langle I_m\rangle \ = \ \frac{2e}{h}\int dE\left(f_m(E) - \sum_n s_{mn}^*(E)s_{mn}(E)f_n(E)\right).$$

By virtue of the unitarity of the scattering matrix, the average current becomes

$$\langle I_m\rangle \ = \ \frac{2e}{h}\sum_n\int dE\,|s_{mn}(E)|^2\,(f_m(E) - f_n(E)), \tag{5.10}$$

where $|s_{mn}(E)|^2$ is the transmission probability from reservoir n to reservoir m. Eq. (5.10) is the Landauer formula, generalized to many channels and many terminals in Refs. [34].

5.2. Noise and noise correlations

The noise is defined in terms of current operators as:

$$S_{mn}(\omega) = \lim_{T \to +\infty} \frac{2}{T} \int_{-T/2}^{T/2} dt \int_{-\infty}^{+\infty} dt' e^{i\omega t'} \Big[\langle I_m(t) I_n(t+t') \rangle$$

$$- \langle I_m \rangle \langle I_n \rangle \Big]. \tag{5.11}$$

This definition will be justified in the finite frequency noise section. When $m = n$, S_{mm} corresponds to the (autocorrelation) noise in terminal m. When m and n differ, S_{mn} corresponds to the noise cross-correlations between m and n.

The calculation of noise involves products $I_m(t) I_n(t+t')$ of two current operators. It therefore involves grand canonical averages of four fermion operators, which can be computed with Wick's theorem:

$$\langle c_{p_1}^\dagger(k(E_1), t) c_{p_2}(k(E_2), t) c_{p_3}^\dagger(k(E_3), t+t') c_{p_4}(k(E_4), t+t') \rangle =$$

$$\frac{\hbar^4 k(E_1) k(E_3)}{m^2} f_{p_1}(E_1) f_{p_3}(E_3) \delta_{p_1 p_2} \delta_{p_3 p_4} \delta(E_1 - E_2) \delta(E_3 - E_4)$$

$$+ \frac{\hbar^4 k(E_1) k(E_2)}{m^2} f_{p_1}(E_1)(1 \mp f_{p_2}(E_3)) \delta_{p_1 p_4} \delta_{p_2 p_3}$$

$$\times \delta(E_1 - E_4) \delta(E_3 - E_2) e^{-i(E_1 - E_2)t'/\hbar}. \tag{5.12}$$

The first term represents the product of the average currents, while in the second term $f(1 \mp f)$, "$-$" corresponds to fermionic statistics, while the "$+$"corresponds to bosonic statistics. In the expression for the noise, only the irreducible current operator contributes, and the integral over time gives a delta function in energy (one of the energy integrals drops out). One gets the general expression for the finite frequency noise for fermions [27]:

$$S_{mn}(\omega) = \frac{4e^2}{h} \int dE \sum_{pp'} (\delta_{mp} \delta_{mp'} - s_{mp}^*(E) s_{mp'}(E - \hbar\omega))$$

$$\times (\delta_{np'} \delta_{np} - s_{np'}^*(E - \hbar\omega) s_{np}(E))$$

$$\times f_p(E)(1 - f_{p'}(E - \hbar\omega)). \tag{5.13}$$

Below, we discuss a few examples.

5.3. Zero frequency noise in a two terminal conductor

5.3.1. General case

We can recover the result previously derived from the wave packet approach. We emphasize that these results were first derived with the present approach [17].

Consider a conductor connected to two terminals (L et R). Considering only autocorrelation noise ($m = n$) and setting $\omega = 0$ in Eq. (5.13), we get:

$$
\begin{aligned}
S_{LL}(\omega = 0) &= \frac{4e^2}{h} \int dE \Big[T(E)^2 \left[f_L(1 \mp f_L) + f_R(1 \mp f_R) \right] \\
&\qquad + T(E)(1 - T(E)) \left[f_L(1 \mp f_R) + f_R(1 \mp f_L) \right] \Big], \\
&= \frac{4e^2}{h} \sum_{\alpha=1}^{N_c} \int dE \Big[T(E) \left[f_L(1 \mp f_L) + f_R(1 \mp f_R) \right] \\
&\qquad \pm T(E)(1 - T(E))(f_L - f_R)^2 \Big]. \quad (5.14)
\end{aligned}
$$

The result of Eq. (3.5) is thus generalized to describe fermions or bosons. These expressions, and the corresponding limits where the voltage is larger than the temperature (shot noise) or inversely when the temperature dominates (Johnson noise) have already been discussed.

5.3.2. Transition between the two noise regimes

In both limits for the noise, it was assumed for simplicity that the transmission probability did not depend significantly on energy. This is justified in practice if $T(dT/dE)^{-1} \gg eV$, which can be reached by choosing both a sufficiently small voltage bias, and choosing the chemical potentials such that there are no resonances in transmission within this energy interval. In this situation, the integrals over the two finite temperature Fermi functions can be performed.

Furthermore, although the single channel results are instructive to order the fundamental features of quantum noise, it is useful to provide now the general results for conductors containing several channels. Using the eigenvalues of the transmission matrix, one obtains [16]:

$$
\begin{aligned}
S_{LL}(0) &= \frac{4e^2}{h} \Bigg[2k_B \Theta \sum_\alpha T_\alpha^2 \\
&\qquad + eV \coth\left(\frac{eV}{2k_B \Theta} \right) \sum_\alpha T_\alpha(1 - T_\alpha) \Bigg]. \quad (5.15)
\end{aligned}
$$

Of interest for tunneling situations is the case where all eigenvalues are small compared to 1. Terms proportional to T_α^2 are then neglected. Eq. (5.15) becomes:

$$
S_{LL}(0) = 2e\langle I \rangle \coth\left(\frac{eV}{2k_B \Theta} \right). \quad (5.16)
$$

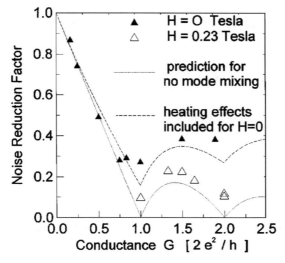

Fig. 4. Noise reduction factor versus conductance at zero magnetic field (filled triangles) and at $H = 0.23T$ (open triangles), from Ref. [36]. Predictions with (without) channel mixing correspond to the dashed (dotted) curve.

5.4. Noise reduction in various systems

Noise reduction in a point contact was observed in semiconductor point contacts [35, 36]. For a single channel sample, the noise has a peak for transmission $1/2$, a peak which was detected in [35], and subsequent oscillations are observed as the number of channels increases. A quantitative analysis of the $1 - T$ noise reduction was performed by the Saclay group [36]. Their data is displayed in Fig. 4. Noise reduction was subsequently observed in atomic point contacts [37, 38] using break junctions.

The $1 - T$ reduction of shot noise can be also observed in various mesoscopic systems, other than ballistic. It is most explicit in point contact experiment and break junctions experiment because one can tune the system in order to have the controlled opening of the first few conduction channels. But what happens in other systems such as cavities and multi-channel conductors?

5.4.1. Double barrier structures

Double barrier structures are interesting because there exists specific energies where the transmission probability approaches unity. Inside the well, one has quasi-bound states whenever the phase accumulated in a round trip equals 2π. The approximate energy dependence of the transmission coefficient for energies close to the nth resonance corresponds to a Breit-Wigner formula:

$$T(E) = T_n^{max} \frac{\Gamma_n^2/4}{(E - E_n)^2 + \Gamma_n^2/4}, \quad T_n^{max} = 4\frac{\Gamma_{Ln}\Gamma_{Rn}}{(\Gamma_{Ln} + \Gamma_{Rn})^2}, \quad (5.17)$$

where Γ_{Ln} and Γ_{Rn} are escape rates to the left and right. If the width of the individual levels are small compared to the resonance spacing, the total transmission coefficient is given by a sum of such transmission coefficients. Inserting this in the zero frequency result for the current and for the noise, one obtains current and noise contributions coming from the resonant level located in between the chemical potentials on the left and on the right. Of particular interest is the case of a single resonance:

$$\langle I \rangle = \frac{e}{\hbar} \frac{\Gamma_L \Gamma_R}{\Gamma_L + \Gamma_R}, \quad (5.18)$$

$$S = 2e\frac{(\Gamma_L^2 + \Gamma_R^2)}{(\Gamma_L + \Gamma_R)^2}\langle I \rangle = e\langle I \rangle, \quad (5.19)$$

where the last equality in Eq. (5.19) applies when the barriers are symmetric, giving a Fano factor $1/2$. Note that this approach assumes a quantum mechanical coherent treatment of transport. It is remarkable that the $1/2$ reduction can also be derived when transport is incoherent, using a master equation approach [39].

5.4.2. Noise in a diffusive conductor

The diffusive regime can be reached with the scattering approach. This point has been already emphasized in a previous les Houches summer school [40]. Until the late 1980's, results on diffusive metals where mainly obtained using diagrammatic techniques. Ensemble averages of non-equilibrium transport properties can also be reached using the scattering approach. Random matrix theory allows to study both the spectrum of quantum dots and the transport properties of mesoscopic conductors. A thorough review exist on this topic [41].

Consider a two-dimensional conductor whose transverse dimension W is larger than the Fermi wavelength: $W \gg \lambda_F$. The number of transverse channels is $N_\perp \sim Wk_F/\pi \gg 1$. When disorder is included in this wire, mixing occurs between these channels: a mean free path l separates each elastic scattering event. When the wire is in the diffusive regime, i.e. the Fermi wavelength $\lambda_F \ll l$, there are many collisions with impurities within the length of the sample $l \ll L$, while $L < L_\phi$, the phase coherence length of the sample. Due to disorder, electron states are localized within a length $L_\xi = N_\perp l$. In order to be in the metallic regime, $L \ll L_\xi$.

In quantum transport, the conductance of a diffusive metal is given by the Drude formula $\langle G \rangle = (n\,e^2\tau/m)(W/L)$. At the same time, this conductance can be identified with a Landauer conductance formula averaged over disorder:

$$\langle G \rangle = \frac{2e^2}{h}\langle \sum_n T_n \rangle = \frac{2e^2}{h}N_{\perp}\langle T \rangle. \tag{5.20}$$

Identification with the Drude formula yields $\langle T \rangle = l/L \ll 1$, where the mean free path has been multiplied by a numerical factor (this point is explained in Ref. [41]). One could think that this implies that all channels have a low transmission: in the diffusive regime, the mean free path is much smaller than the sample length. For the noise, this would then give a Poissonian regime. On the other hand, it is known that classical conductors exhibit no shot noise. Should the same be true for a mesoscopic conductor? The answer is that both statements are incorrect. The noise of a diffusive conductor exhibits neither full shot noise nor zero shot noise. The result is in between. The probability distribution of transmission eigenvalues is bimodal:

$$P(T) = (l/2L)[T\sqrt{1-T}]^{-1}. \tag{5.21}$$

There is a fraction l/L of open channels while all other channels have exponentially small transmission (closed channels). Consider now the noise, averaged over impurities:

$$\langle S(\omega = 0) \rangle = \frac{4e^3}{h}N_{\perp}\langle T(1-T) \rangle V. \tag{5.22}$$

The random matrix theory average yields [42] $\langle T(1-T) \rangle = l/3L$, so that the Fano factor is $F = 1/3$. This result has a universal character: it does not depend of the sample characteristics. It should also be noted that it can be recovered using alternative approaches, such as the Boltzmann-Langevin semi-classical approach [43, 44]. This shot noise reduction was probed experimentally in small metallic wires [45]. A careful tuning of the experimental parameters (changing the wire length and the geometry of the contacts) allowed to make a distinction between this disorder-induced shot noise reduction (1/3), and the $\sqrt{3}/4$ reduction associated with hot electrons.

5.4.3. Noise reduction in chaotic cavities

Chaotic cavities were studied by two groups [46, 47], under the assumption that the scattering matrix belongs to the so called Dyson circular ensemble. This is a class of random matrices which differs from the diffusive case, and the probability distribution for transmission eigenvalues is given by:

$$P(T) = 1/[\pi\sqrt{T(1-T)}]. \tag{5.23}$$

This universal distribution has the property that $\langle T \rangle = 1/2$ and $\langle T(1-T) \rangle = 1/8$, so that the Fano factor is $1/4$. This result applies to open cavities, connected symmetrically (same number of channels) to a source and drain. It is nevertheless possible [49] to treat cavities with arbitrary connections to the reservoirs, which allows to recover the $1/2$ suppression in the limit of cavities with tunneling barriers. The shot noise reduction in chaotic cavities was observed in gated two dimensional electron gases [48], where it was possible to analyze the effect of the asymmetry of the device (number of incoming/outgoing channels connected to the cavity).

6. Noise correlations at zero frequency

6.1. General considerations

It is interesting to look at how the current in one lead can be correlated to the current in another lead. Let us focus on zero frequency noise correlations for fermions, at zero temperature. In this case, the Fermi factors enforce in Eq. (5.13) that the contribution from $p = p'$ vanishes. The remaining contribution has matrix elements $M_m(E, p, p') \sim s_{mp}^* s_{mp'}$, because the delta function term drops out. The correlation then reads:

$$S_{mn}(\omega = 0) = \frac{2e^2}{h} \sum_{pp'} \int dE \, s_{mp}^* s_{mp'} s_{np'}^* s_{np} f_p(E)(1 - f_{p'}(E)). \tag{6.1}$$

The contribution $p = p'$ has been kept in the sums, although it vanishes, for later convenience. From this expression it is clear that the noise autocorrelation $m = n$ is always positive, simply because scattering amplitudes occur next to their hermitian conjugates.

On the other hand, if $m \neq n$, the correlations are negative. The unitarity of the S matrix implies in this case $\sum_p s_{mp} s_{np}^* = 0$, so that the terms linear in the Fermi functions drop out:

$$S_{mn}(\omega = 0) = -\frac{2e^2}{h} \int dE \left| \sum_p s_{np} s_{mp}^* f_p(E) \right|^2. \tag{6.2}$$

To measure the cross correlations between m and n, one could imagine fabricating a new lead "$m + n$" (assuming that both m and n are at the same chemical potential), and measuring the autocorrelation noise $S_{(m+n)(m+n)}$. The cross correlation is then obtained by subtracting the autocorrelation of each sub-lead, and dividing by two:

$$S_{(m+n)}(\omega = 0) = S_m(\omega = 0) + S_n(\omega = 0) + 2S_{mn}(\omega = 0). \tag{6.3}$$

This prescription was used in the wave packet approach to analyze the Hanbury-Brown and Twiss correlations for fermions [16], which are discussed below.

6.2. Noise correlations in a Y–shaped structure

In 1953, Hanbury–Brown and Twiss performed several experiments [25] where two detectors at different locations collected photons emitted from a light source. Their initial motivation was to measure the size of a distant star. This experiment was followed by another one where the source was replaced by a mercury arc lamp with filters. The filters insured that the light was essentially monochromatic, but the source was thermal. A semi-transparent mirror split the beam in two components, which were fed to two photo-multiplier tubes. The correlations between the two detectors were measured as a function of the distance separating the detectors, and were found to be always positive. The experiment was subsequently explained to be a consequence of the bunching effect of photons. Because the light source was thermal, several photons on average occupied the same transverse states of the beams, so the detection of one or several photons in one arm of the beam was typically accompanied by the detection of photons in the other arm.

This measurement, which can be considered one of the first in the field of quantum optics, can thus be viewed as a check that photons are indistinguishable particles which obey Bose–Einstein statistics. It turns out that this result is also fully consistent with a classical electromagnetism description: the photon bunching effect can be explained as a consequence of the superposition principle for light applied to noisy sources, because the superposition principle follows from Bose Einstein statistics.

Nowadays, low flux light sources – which are used for quantum communication purposes – can be produced such that photons are emitted one by one. The resulting correlation signal is then negative because the detection of a photon in one arm means that no photon is present in the other arm. This is in fact what happens for electrons.

It has been suggested by many authors that the analog experiment for fermions should be performed [50]. In this way the Fermi Dirac statistics of electrons could be diagnosed directly from measuring the noise correlations of fermions passing through a beam splitter. Because of the Pauli principle, two electrons cannot occupy the same transverse state. The correlations should then be negative. For technical reasons – the difficulty to achieve a dense beam of electrons in vacuum – the experiment proved quite difficult to achieve. However, it was suggested [16] that a similar experiment could be performed for fermions in nanostructures.

Consider a three terminal conductor, a "Y junction". Electrons are injected from terminal 3, which has a higher chemical potential than terminals 1 and 2,

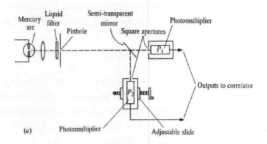

Fig. 5. The Hanbury-Brown and Twiss experiment.

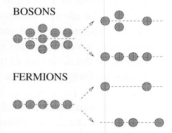

Fig. 6. Illustration of the bunching effect for bosons, and of the anti-bunching effect for electrons in a Hanbury-Brown and Twiss geometry.

where the correlations are measured. Here we focus on zero frequency noise (setting $\omega = 0$ in Eq. (5.13)). For simplicity we choose to work at zero temperature. The autocorrelation noise becomes:

$$S_{mm} = \frac{4e^2}{h} \sum_{p \neq p'} |s_{mp}|^2 |s_{mp'}|^2 f_p(E)(1 - f_{p'}(E)).$$

Assuming that $\mu_1 = \mu_2$, the voltage bias is then $eV = \mu_3 - \mu_{1,2}$. and the autocorrelation becomes $S_{11} = (4e^2/h)eVT_{13}(1 - T_{13})$, and similarly for terminal 2. On the other hand, the correlations between 1 and 2 yield:

$$S_{12} = = -\frac{4e^2}{h} eV |s_{13}|^2 |s_{23}|^2. \tag{6.4}$$

A natural way to summarize these results is to normalize the correlations by the square root of the product of the two autocorrelations:

$$S_{12}/\sqrt{S_{11} S_{22}} = -\sqrt{T_{13} T_{23}}/\sqrt{(1 - T_{13})(1 - T_{23})}, \tag{6.5}$$

where T_{13} and T_{23} are the transmission probabilities from 3 to 1 and 2. The correlations are therefore negative regardless of the transmission of the sample. Quantitative agreement with experiments has been found in the late nineties in two separate experiments. A first experiment [51] designed the electron analog of a beam splitter using a thin metallic gate on a two dimensional electron gas. A second experiment was carried out in the quantum Hall effect regime [52]: a point contact located in the middle of the Hall bar then plays the role of a controllable beam splitter, as the incoming edge state is split into two outgoing edge states at its location. Recently, negative noise correlations have been observed in electron field emission experiments [53].

7. Finite frequency noise

7.1. Which correlator is measured?

Finite frequency noise is the subject of a debate. What is actually measured in a finite frequency noise measurements? The current operator is indeed an hermitian operator, but the product of two current operators evaluated at different times is not hermitian. If one takes the wisdom from classical text books [54], one is told that when one is faced with the product of two hermitian operators, one should symmetrize the result in order to get a real, measurable quantity. This procedure has led to a formal definition of finite frequency noise [27]:

$$S_{sym}(\omega) = \int_{-\infty}^{+\infty} dt \, e^{i\omega t} \langle\langle I(t)I(0) + I(0)I(t)\rangle\rangle. \tag{7.1}$$

(the double bracket means the product of averaged current has been subtracted out). At the same time, one can define two unsymmetrized correlators:

$$S_+(\omega) = 2\int_{-\infty}^{+\infty} dt \, e^{i\omega t} \langle\langle I(0)I(t)\rangle\rangle. \tag{7.2}$$

$$S_-(\omega) = 2\int_{-\infty}^{+\infty} dt \, e^{i\omega t} \langle\langle I(t)I(0)\rangle\rangle. \tag{7.3}$$

Here, $P(i)$ is the probability distribution for initial states. The factor two has been added to be consistent with the Schottky relation. Note that in these expressions, the time dependence is specified by the Heisenberg picture. Assuming that one knows the initial (ground) states $|i\rangle$ and the final states $|f\rangle$, one concludes that:

$$S_+(\omega) = 4\pi \sum_{if} |\langle f|I(0)|i\rangle|^2 P(i)\delta(\omega + E_f - E_i). \tag{7.4}$$

$$S_-(\omega) = 4\pi \sum_{if} |\langle f|I(0)|i\rangle|^2 P(i)\delta(\omega + E_i - E_f). \tag{7.5}$$

Therefore, in $S_+(\omega)$ (in $S_-(\omega)$), positive (negative) frequencies correspond to an emission rate from the mesoscopic device, while negative (positive) frequencies correspond to an absorption rate. Since one does not expect to be able to extract energy from the ground state of this device, one concludes that the physically relevant frequencies for $S_+(\omega)$ (for $S_-(\omega)$) are $\omega > 0$ ($\omega < 0$).

7.2. Noise measurement scenarios

In ref. [55], the authors argue that one has to specify a measurement procedure in order to decide which noise correlator is measured. In their proposal (Fig. 7a), the noise is measured by coupling the mesoscopic circuit (the antenna) inductively to a LC circuit (the detector). By measuring the fluctuations of the charge on the capacitor, one obtains a measurement of the current-current fluctuations which is weighted by Bose-Einstein factors, evaluated at the resonant frequency $\omega = 1/\sqrt{LC}$ of the circuit:

$$\langle\langle Q^2(0)\rangle\rangle = K[S_+(\omega)(N_\omega + 1) - S_-(\omega)N_\omega], \qquad (7.6)$$

where K is a constant which depends on the way the two circuits (antenna and detector) are coupled and the double brackets imply that one measures the excess charge fluctuations. N_ω are bosonic occupation factors for the quantized LC circuit. It is therefore a mixture of the two unsymmetrized noise correlators which is measured in this case. This point has been reemphasized recently [56].

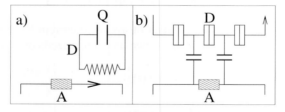

Fig. 7. Schematic description of noise measurement setups. A (antenna) is the mesoscopic circuit to be measured, while D is the detector circuit. a) inductive coupling with an LC circuit. b) capacitive coupling with a double dot.

Another proposal considers [57] a capacitive coupling between the mesoscopic circuit and the detector circuit (Fig. 7b). The detector circuit consists of a double dot system embedded in a circuit where current is measured. Each dot is coupled to each side of the mesoscopic device: current fluctuations in this antenna circuit are translated into voltage fluctuations – thus phase fluctuations – between the two dots. Indeed, it is understood since the early nineties that in nano-junctions, the phase is the canonical conjugate of the charge in the junction.

The Fermi golden rule calculation of the current across the junction needs to be revisited in order to take into account the effect of the environment. This is the so-called $P(E)$ theory of the dynamical Coulomb blockade [58]. $P(E)$ is the probability distribution for inelastic scattering: it is equal to a delta function for a low impedance environment. In the general case it is specified as follows:

$$P(\epsilon) = h^{-1} \int_{-\infty}^{+\infty} \exp\left[J(t) + i\frac{\epsilon}{\hbar}t\right] dt. \tag{7.7}$$

Here, $J(t)$ is the (unsymmetrized) correlation function of the phase. In the proposal of Ref. [57], the impedance environment coupled to the double dot does not only consist of the leads connected to these dots (these leads are assumed to be massive and well conducting): the environment is the mesoscopic circuit itself. Letting ϵ denote the level separation between the two dots, a DC inelastic current can circulate in the detection circuit only if the frequency $\omega = \epsilon/\hbar$ is provided by the mesoscopic circuit (antenna). The inelastic current is then given by

$$I_D(\epsilon) = \frac{e}{\hbar} T_c^2 P(\epsilon), \tag{7.8}$$

where T_c is the tunnel amplitude between the dots. In $P(\epsilon)$, the phase correlator $J(t)$ contains the trans-impedance $Z_{trans}(\omega)$ connecting the detector and the antenna circuit, as well as the unsymmetrized noise:

$$J(t) = \frac{e^2}{2\hbar^2} \int_{-\infty}^{+\infty} \frac{|Z_{trans}(\omega)|^2}{\omega^2} S_-(\omega)(e^{-i\omega t} - 1). \tag{7.9}$$

Under the assumption of a low impedance coupling, the inelastic current becomes:

$$I_D(\epsilon) \simeq 2\pi^2 \kappa^2 \frac{T_c^2}{e} \frac{S_-(\epsilon/\hbar)}{\epsilon}. \tag{7.10}$$

Here, κ depends on the parameters (capacitances, resistors,...) of both circuits. Given that the energy spacing between the dots can be controlled by a gate voltage, the measurement of I_D provides a direct measurement of the noise – non-symmetrized – of the mesoscopic device.

This general philosophy has been tested recently by the same group, who measured [59] the high frequency noise of a Josephson junction using another device – a superconductor-insulator-superconductor junction – as a detector: in this system, quasiparticle tunneling in the SIS junction can occur only if it is assisted by the frequency provided by the antenna.

7.3. Finite frequency noise in point contacts

Concerning the finite frequency noise, results are better illustrated by choosing the zero temperature case, and keeping once again the assumption that the transmission amplitudes are constant. In this case, only the last term in Eq. (5.13) contributes, and one obtains the result [60, 17]:

$$S_+(\omega) = \frac{4e^2}{h} T(1 - T)(eV - \hbar\omega) \Theta(eV - \hbar\omega). \qquad (7.11)$$

The noise decreases linearly with frequency until $eV = \hbar\omega$, and vanishes beyond that. The noise therefore contains a singularity: its derivative diverges at this point. Properly speaking, the above result is different from that of the first calculation of finite frequency noise [60], where the noise expression was symmetrized with respect to the two time arguments. It coincides only at the $\omega = 0$ point, and in the location of the singularity. This singularity was first detected in a photo-assisted transport measurement, using a metallic wire in the diffusive regime [61]. Photon-assisted shot noise is briefly discussed in the conclusion as an alternative to finite frequency noise measurements.

In what follows, when talking about finite frequency noise in superconducting–normal metal junctions, the unsymmetrized correlator $S_+(\omega)$ will always be considered.

8. Noise in normal metal-superconducting junctions

Before mesoscopic physics was born, superconductors already displayed a variety of phase coherent phenomena, such as the Josephson effect. Scientists began wondering what would happen if a piece of coherent, normal metal, was put in contact with a superconductor. The phenomenon of Andreev reflection [62, 63] – where an electron incident from a normal metal on a boundary with a superconductor is reflected as a hole (Fig. 8), now has its share of importance in mesoscopic physics. In electron language, Andreev reflection also corresponds to the absorption of two electrons and transfer as a Cooper pair in the superconductor. Here we analyze the noise of normal metal–superconductor junctions. Such junctions bear strong similarities with two terminal normal conductors, except that there are two types of carriers. Whereas charge is not conserved in an Andreev process, energy is conserved because the two normal electrons incident on the superconductor – one above the Fermi level and the other one symmetrically below – have the total energy for the creation of a Cooper pair, which is the superconductor chemical potential. An NS junction can be viewed as the electronic analog of a phase conjugation mirror in optics. Finally, spin is also conserved

at the boundary: the electron and its reflected hole have opposite spin, because
in the electron picture, the two electrons entering the superconductor do so as a
singlet pair because of the s wave symmetry of the superconductor. This fact will
be important later on when discussing entanglement.

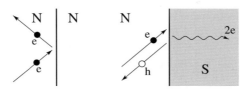

Fig. 8. Left, normal reflection: an electron is reflected as an electron, and parallel momentum is
conserved. Right, Andreev reflection: an electron is reflected as a hole whose momentum is exactly
opposite of that of the electron. A charge $2e$ is absorbed in the superconductor as a Cooper pair.

8.1. *Bogolubov transformation and Andreev current*

Consider a situation where a superconductor is connected to several normal metal
leads, in turn connected to reservoirs with chemical potentials μ_m. For simplicity,
we assume each lead to carry only one single transverse channel. All energies in
the leads are measured with respect to the superconductor chemical potential μ_S.
To describe transport from the point of view of scattering theory, the Bogolubov
de Gennes theory of inhomogeneous superconductors [63, 64, 65] is best suited.

The starting point is the mean field Hamiltonian of Bogolubov which describes a system of fermions subject to a scalar potential and an attractive interaction. This latter interaction contains in principle two creation operators as well
as two annihilation operators, making it a difficult problem to solve. Bogolubov
had the originality to propose an effective Hamiltonian which does not conserve
particles [65]. This Hamiltonian is diagonalized by the Bogolubov transformation:

$$\psi_{i\sigma}(x) = \sum_{j\beta} \int \frac{dk}{\sqrt{2\pi}} \left[u_{ij\beta}(x)\gamma_{j\beta\uparrow}(k) - \sigma v^*_{ij\beta}(x)\gamma^+_{j\beta\downarrow}(k) \right]. \tag{8.1}$$

The state $u_{ij\beta}$ ($v_{ij\beta}$) corresponds to the wave function of a electron-like (hole-
like) quasiparticle in terminal i injected from terminal j as a quasiparticle β
($\beta = e, h$). $\gamma(k)$ et $\gamma^\dagger(k)$ are fermionic quasiparticle operators. As before it
will be convenient to switch to energy integrals with the substitution $\int dk =
\int dE/\sqrt{\hbar v^j_{e,h}(E)}$, i.e., electrons and holes do not have the same velocities. The

corresponding Hamiltonian has a diagonal form:

$$H_{eff} = \sum_{j\beta\sigma} \int_0^{+\infty} dE \, E \, \gamma_{j\beta\sigma}^+(E)\gamma_{j\beta\sigma}(E), \tag{8.2}$$

provided that the electron and hole wave functions satisfy:

$$\begin{cases} Eu_{ij\beta} = \left(-\dfrac{\hbar^2}{2m}\dfrac{\partial^2}{\partial x^2} - \mu_S + V(x)\right) u_{ij\beta} + \Delta(x)v_{ij\beta}, \\[2ex] Ev_{ij\beta} = -\left(-\dfrac{\hbar^2}{2m}\dfrac{\partial^2}{\partial x^2} - \mu_S + V(x)\right) v_{ij\beta} + \Delta^*(x)u_{ij\beta}. \end{cases} \tag{8.3}$$

In principle, these equations need to be solved self-consistently since the gap parameter depends on $u_{ij\beta}$ and $v_{ij\beta}$. In most applications however, the gap is assumed to be a step-like function describing an abrupt transition from a super-conductor to a normal metal lead. On the normal metal side, the Bogolubov-de Gennes equations (8.3) can be solved easily assuming plane wave solutions for normal electrons and holes. For a given energy E, the corresponding wave numbers are $k_e^N = \sqrt{2m(\mu_S + E)}/\hbar$ and $k_h^N = \sqrt{2m(\mu_S - E)}/\hbar$. We are now dealing with an S-matrix which can either convert electrons from terminal j to terminal i, or electrons into holes in these terminals:

$$u_{ij\beta}(x) = \delta_{i,j}\delta_{e,\beta}e^{ik_e^N x} + s_{ije\beta}\sqrt{\frac{k_\beta^j}{k_e^N}}e^{-ik_e^N x}, \tag{8.4}$$

$$v_{ij\beta}(x) = \delta_{i,j}\delta_{h,\beta}e^{-ik_h^N x} + s_{ijh\beta}\sqrt{\frac{k_\beta^j}{k_h^N}}e^{ik_h^N x}. \tag{8.5}$$

A particular aspect of this formalism is that electrons and holes have opposite momenta. $s_{ij\alpha\beta}$ is the amplitude for getting a particle α in terminal i given that a particle β was incident from j. In comparison to the previous section, the spin index has to be restored in the definition of the current operator:

$$\begin{aligned} I_i(x) = {}& \frac{e\hbar}{2miv_F}\frac{1}{2\pi\hbar}\int_0^{+\infty} dE_1 \int_0^{+\infty} dE_2 \sum_{mn\sigma} \\ & \left[\left(u_{im}^*\partial_x u_{in} - (\partial_x u_{im}^*)u_{in}\right)\gamma_{m\sigma}^\dagger \gamma_{n\sigma}\right. \\ & \left. -\left(u_{im}^*\partial_x v_{in}^* - (\partial_x u_{im}^*)v_{in}^*\right)\sigma \, \gamma_{m\sigma}^\dagger \gamma_{n-\sigma}^\dagger \right. \end{aligned}$$

$$-\left(v_{im}\partial_x u_{in} - (\partial_x v_{im})u_{in}\right)\sigma \, \gamma_{m-\sigma}\gamma_{n\sigma}$$

$$+\left(v_{im}\partial_x v_{in}^* - (\partial_x v_{im})v_{in}^*\right)\gamma_{m-\sigma}\gamma_{n-\sigma}^\dagger\Big]. \qquad (8.6)$$

In order to avoid the proliferation of indices, we chose to replace sums over j (terminal number) and β (electron or hole) by a single index m. Expressions containing $m\,n$ also have an energy dependence E_1 (E_2). In the following calculations, the energy dependence of the wave numbers is neglected for simplicity. As before, this is justified by the fact that all chemical potentials (normal leads and superconductors) are large compared to the applied biases. Note that unlike the normal metal case, the current operator does not conserve quasiparticles. The evaluation of the average current implies the computation of the average $\langle \gamma_{m\sigma}^\dagger(E_1)\gamma_{n\sigma}(E_2)\rangle = f_{j\alpha}(E_1)\delta_{mn}\delta(E_1 - E_2)$. The distribution function f_m depends on which type of particle is considered: it is the Fermi Dirac distribution for electrons in a normal lead $f_{m(e)}(E) = 1/[1 + e^{\beta(E-\mu_j)}]$; for holes in the same lead it represents the probability for a state with energy $-E$ to be empty $f_{m(h)}(E) = 1 - f_{m(e)}(-E) = 1/[1 + e^{\beta(E+\mu_m)}]$; for electron or hole-like quasiparticles in the superconductor, it is simply $f_S(E) = 1/[1 + e^{\beta E}]$. The average current in lead i becomes:

$$\langle I_i(x)\rangle \;=\; \frac{e}{2\pi m i \, v_F} \int_0^{+\infty} dE \sum_m \Bigg[\left(u_{im}^*\partial_x u_{im} - (\partial_x u_{im}^*)u_{im}\right)f_m$$

$$+ \left(v_{im}\partial_x v_{im}^* - (\partial_x v_{im})v_{im}^*\right)(1 - f_m)\Bigg]. \qquad (8.7)$$

For a single channel normal conductor connected to a superconductor, the Andreev regime implies that the applied bias is much smaller than the superconducting gap, so that no quasiparticles can be excited in the transport process. It is also assumed that the scattering amplitudes have a weak energy dependence within the range of energies specified by the bias voltage. Using the unitarity of the S–matrix one obtains:

$$\langle I\rangle = \frac{4e^2}{h} R_A\, V, \qquad (8.8)$$

where $R_A = |s_{11he}|^2$ is the Andreev reflection probability. The conductance of a normal metal-superconductor junction in then doubled [66, 67] because of the transfer of two electron charges. Indeed, this result could have been guessed from a simple extension of the Landauer formula to NS situations.

8.2. Noise in normal metal–superconductor junctions

For the Andreev regime, in a single NS junction, noise can be calculated indeed using the wave packet approach [68], with the following substitutions from the normal metal case, Eq. (5.14):

• the transmission probability is replaced by the Andreev reflection probability: $T \to R_A$.

• The transfered charge is $2e$.

• Electrons have a Fermi distribution $f(E - eV)$, whereas holes have a Fermi distribution $f(E + eV)$.

• Although electrons with spin σ are converted into holes with spin $-\sigma$, the spin index is ignored, which is justified if the normal lead is non-magnetic. Here the spin only provides a factor two.

Using the general "wave packet" formula for the two terminal noise of normal conductors, one readily obtains:

$$
\begin{aligned}
S(0) \quad = \quad & \frac{8e^2}{h} \int dE \; \big[R_A(E)(1 - R_A(E))(f(E - eV) - f(E + eV))^2 \\
& + R_A(E)[f(E - eV)(1 - f(E - eV)) \\
& + f(E + eV)(1 - f(E + eV))]\big] ,
\end{aligned}
\tag{8.9}
$$

which (with the same standard assumptions) yields immediately the two known limits. For a voltage dominated junction $eV \gg k_B \Theta$,

$$
S(0) = \frac{16e^3}{h} R_A (1 - R_A) V \equiv 4e \langle I \rangle (1 - R_A),
\tag{8.10}
$$

while in a temperature dominated regime

$$
S(0) = \frac{16e^2}{h} R_A k_B \Theta \equiv 4 G_{NS} k_B \Theta,
\tag{8.11}
$$

with G_{NS} the conductance of the NS junction, and one recovers the fluctuation dissipation theorem.

One now needs a general description which can treat finite frequencies, and above gap processes, (when for instance an electron is transfered into the superconductor as an electron quasiparticle). The scattering formalism based on operator averages is thus used.

In order to guess the different contributions for the noise correlator, consider the expression for $I_i(x, t) I_j(x, t + t')$. It is a product of four quasiparticle creation/annihilation operators. It will have non-zero average only if two γ are paired with two $\gamma^{\dagger}$. The same is true for the noise in normal metal conductors.

Here, however, electron-like and hole-like contributions will occur, but the current operator of Eq. (8.6) also contains terms proportional to $\gamma\gamma$ and $\gamma^\dagger\gamma^\dagger$ which contribute to the noise. To compute the average $\langle I_i(x,t)I_j(x,t+t')\rangle$, it is useful to introduce the following matrix elements:

$$
\begin{aligned}
A_{imjn}(E,E',t) &= u_{jn}(E',t)\partial_x u^*_{im}(E,t) - u^*_{im}(E,t)\partial_x u_{jn}(E',t),\\
B_{imjn}(E,E',t) &= v^*_{jn}(E',t)\partial_x v_{im}(E,t) - v_{im}(E,t)\partial_x v^*_{jn}(E',t),\\
C_{imjn}(E,E',t) &= u_{jn}(E',t)\partial_x v_{im}(E,t) - v_{im}(E,t)\partial_x u_{jn}(E',t).
\end{aligned}
$$

The two first matrix elements involve products of either particle or hole wave

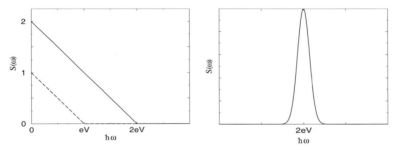

Fig. 9. Noise as a function of frequency. Left, full line: the noise (in units of $e\langle I\rangle(1-R_A)$)in an NS junction has a singularity at $\omega = 2eV/\hbar$. Left, dashed line: noise (in units of $e\langle I\rangle(1-T)$) for a junction between two normal metals with a singularity at $\hbar\omega = eV$. Right: noise in a Josephson junction, which presents a peak at the Josephson frequency. The line-width is due to radiation effects.

functions. Compared to the normal case, the last one, $C_{imjn}(E,E',t)$ is novel, because it involves a mixture of electrons and holes. It will be important in the derivation of the finite frequency spectrum of noise correlations. Computing the grand canonical averages, one obtains the difference $\langle I_i(t)I_j(t+t')\rangle - \langle I_i\rangle\langle I_j\rangle$, and the Fourier transform is performed in order to compute the noise. The integration over t' gives a delta function in energy as before. Note that because we are assuming positive frequencies, terms proportional to $(1-f_m)(1-f_n)$ vanish. The noise cross-correlations have the general form:

$$
S_{ij}(\omega) = \frac{e^2\hbar^2}{m^2 v_F^2}\frac{1}{2\pi\hbar}\int_0^{+\infty} dE \sum_{m,n}
$$

$$
\Big\{ \Theta(E+\hbar\omega)f_m(E+\hbar\omega)(1-f_n(E))
$$

$$
\times\big(A_{imin}(E+\hbar\omega,E)+B^*_{imin}(E+\hbar\omega,E)\big)
$$

$$
\times\big(A^*_{jmjn}(E+\hbar\omega,E)+B_{jmjn}(E+\hbar\omega,E)\big)
$$

$$+\Theta(\hbar\omega - E)f_m(\hbar\omega - E)f_n(E)C^*_{imjn}(\hbar\omega - E, E, t)$$

$$\times\left[C_{jnim}(E, \hbar\omega - E) + C_{imjn}(\hbar\omega - E, E)\right]\bigg\}. \tag{8.12}$$

Terms proportional to $f_n f_m$ or $(1 - f_n)(1 - f_m)$ disappear at zero frequency, because of the energy requirements. The zero frequency limit can thus be written in a concise form [69, 70, 27]:

$$
\begin{aligned}
S_{ij}(0) &= \frac{e^2\hbar^2}{m^2 v_F^2}\frac{1}{2\pi\hbar}\int_0^{+\infty} dE \sum_{m,n} f_m(E)(1 - f_n(E)) \\
&\quad \times\left(A_{imin}(E, E) + B^*_{imin}(E, E)\right) \\
&\quad \times\left(A^*_{jmjn}(E, E) + B_{jmjn}(E, E)\right)
\end{aligned} \tag{8.13}
$$

8.3. Noise in a single NS junction

From Eq. (8.12), choosing zero temperature and using the expressions of A_{mn}, B_{mn} and C_{mn}, one has to consider separately the three frequency intervals $\hbar\omega < eV$, $eV < \hbar\omega < 2eV$ and $\hbar\omega > 2eV$. The first two intervals give a noise contribution, whereas the last one yields $S(\omega) = 0$. Particularly puzzling is the fact that one needs to separate two regimes in frequency, whereas it is expected that the frequency $eV = \hbar\omega$ should not show any particular features in the noise.

8.3.1. Below gap regime

First consider the case where $eV \ll \Delta$. We also assume the scattering amplitudes to be independent on energy (this turns out to be a justified assumption in BTK model which we discuss shortly). One then obtains [71, 72]:

$$S(\omega) = \frac{8e^2}{h}(2eV - \hbar\omega)R_A(1 - R_A)\Theta(2eV - \hbar\omega). \tag{8.14}$$

Just as in the normal case, the noise decreases linearly with frequency, and vanishes beyond the Josephson frequency $\omega_J = 2eV/\hbar$ (left of Fig. 9). There is thus a singularity in the noise at that particular frequency. This result should be compared first to the normal case [60, 17], yet it should also be compared to the pioneering work on the Josephson junction [73, 74].

In the purely normal case (Fig. 10, center), wave functions have a time dependence $\psi_{1,2} \sim \exp[-i\mu_{1,2}t/\hbar]$, so whereas the resulting current is stationary, finite frequency noise contains the product $\psi_1\psi_2^*$ which gives rise to the singularity at $|\mu_2 - \mu_1|/\hbar = eV/\hbar$.

In the purely superconducting case (Fig. 10, right) a constant applied bias generates an oscillating current. $\psi_{1,2} \sim \exp[-i2\mu_{S_{1,2}}t/\hbar]$, so the order parameter oscillates as $2(\mu_{S_1} - \mu_{S_2})$, where μ_{S_1} and μ_{S_2} are the chemical potentials of the two superconductors. The noise characteristic has a peak at the Josephson frequency whose line-width was first computed in the sixties [74].

Turning back to the case of an hybrid NS junction, because here the bias is smaller than the gap, the Andreev process is the only process available to transfer charge, which allows Cooper pairs to be transmitted in or emitted from the superconductor. An electron incident from the normal side with energy $\mu_S + eV$ gets paired with another electron with energy $\mu_S - eV$, thus this second electron corresponds to a hole at $\mu_S - eV$. Both electrons have a total energy $2\mu_S$, which corresponds to the formation of a Cooper pair. The incoming electron wave function oscillates as $\psi_e \sim \exp[-i(\mu_S + eV)t/\hbar]$, whereas the hole wave function oscillates as $\psi_h \sim \exp[-i(\mu_S - eV)t/\hbar]$ (figure 10 left). The noise therefore involves the product $\psi_e \psi_h^*$ which oscillates at the Josephson frequency, thus giving rise to the singularity in the noise derivative. This can be considered as an analog of the Josephson effect observed in a single superconductor adjacent to a normal metal, but only in the noise.

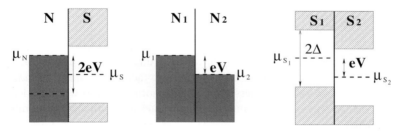

Fig. 10. Energy diagrams for three different types of junctions: left, normal metal (μ_N) connected to a superconductor (μ_S); center, 2 normal metals with chemical potentials μ_1 et μ_2; right, 2 superconductors (Josephson junction), with chemical potentials μ_{S_1} and μ_{S_2}.

8.3.2. Diffusive NS junctions

We have seen that the zero frequency shot noise of a tunnel junction is doubled [68, 28, 75, 76]. It is also interesting to consider a junction between a diffusive normal metal on one side, in perfect contact with a superconductor. The junction contains many channels, yet one can also find an eigenchannel representation in which the noise is expressed as a linear superposition of independent single-channel junctions. Consider here current and shot noise at zero temperature:

$$\langle I \rangle = \frac{4e^2}{h} \sum_n R_{A_n}, \tag{8.15}$$

$$S(\omega = 0) = \frac{16e^2}{h} \sum_n R_{A_n}(1 - R_{A_n}). \tag{8.16}$$

In order to compute the quantity $\langle R_{A_n}(1 - R_{A_n})\rangle$ using random matrix theory, a specific model for the NS junction has to be chosen. A natural choice [41] consists of a normal disordered region separated from a perfect Andreev interface. Because the ideal Andreev interface does not mix the eigenchannels, the Andreev reflection eigenvalues can be expressed in terms of the transmission eigenvalues of the normal metal scattering matrix which models the disordered region: $R_{A_n} = T_n^2/(2 - T_n)^2$. The noise can in turn be expressed in terms of these eigenvalues:

$$S(\omega = 0) = \frac{64e^2}{h} \sum_n \frac{T_n^2(1 - T_n)}{(2 - T_n)^4}. \tag{8.17}$$

Note that channels with either high or low transmission do not contribute to the shot noise. First, assume that all channels have the same transmission probability Γ. Γ represents the transparency per mode of the NS interface, but no mixing is assumed between the modes. The noise can be written in this case as:

$$S = \frac{8(1 - \Gamma)}{(2 - \Gamma)^2}(2e\langle I\rangle). \tag{8.18}$$

For $\Gamma \ll 1$, one obtains a Poissonian noise of uncorrelated charges $2e$. This means that the shot noise is doubled compared to the normal tunnel barrier shot noise result. Next, one considers a disordered normal region with an ideal interface. The averages over transmission eigenvalues are computed using random matrix theory:

$$S(\omega = 0) = \frac{2}{3}(2e\langle I\rangle), \tag{8.19}$$

thus giving a 2/3 reduction for the disordered NS interface.

The shot noise of normal metal–superconducting junctions has been studied in several challenging experiments. For diffusive samples, the 2/3 reduction was observed indirectly in S-N-S junctions [77]. Indeed, when the propagation in the normal region between the two superconductor interface is incoherent, one expects the noise signal to be associated to that of two NS junctions in series. If, on the other hand, one considers coherent propagation in the normal region, one has to take into account the effect of multiple Andreev reflections [78], a topic which is not treated here but rather mentioned in the conclusion.

Single NS junctions were first probed separately both in low frequency noise measurements [79] and also in photo-assisted transport measurements [80]. Although in both cases the metal on the normal side was diffusive, thus giving a 2/3 reduction rather than an enhancement of the noise by a factor 2, these experiments were the first to identify the charge $2e$ associated with the Andreev process. The "true" doubling of shot noise as described in the beginning of this section was measured experimentally only recently in normal-metal superconducting tunnel junctions containing an insulating barrier at the interface [81].

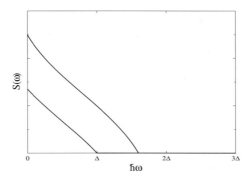

Fig. 11. Noise in an NS junction as a function of frequency, with a barrier with intermediate transparency ($Z = 1$), see text. The applied bias is lower than the gap (from bottom to top, $eV = 0.5\Delta$ and $eV = 0.8\Delta$). The behavior for low biases is essentially linear, see Eq. (8.14).

8.3.3. Near and above gap regime

To establish predictions on the noise characteristic of a single NS junction when the voltage bias lies in the vicinity of the gap, it is necessary to specify a concrete model for the junction. A generic model was introduced by Blonder, Tinkham and Klapwijk (BTK) [82]: it has the advantage that the energy dependent scattering amplitudes can be derived using connection formulas from the Bogolubov-de Gennes equations. In particular, it allows to monitor the crossover from below gap to above gap regime. At the location of the junction, there is a superposition between a scalar delta function potential and a gap potential. The scalar potential acts on electrons or holes, and represents either a potential barrier or mimics the presence of disorder. The gap parameter is assumed to be a step function:

$$V_B(x) = \quad V_B\delta(x), \quad \text{with} \quad V_B = Z\hbar^2 k_F/m, \quad (8.20)$$
$$\Delta(x) = \quad \Delta\Theta(x). \quad (8.21)$$

Z is the parameter which controls the transparency of the junction. $Z \gg 1$ corresponds to an opaque barrier.

In the preceding section, we assumed that the scattering amplitude had a weak energy dependence. These assumptions can be tested with the BTK model. Fig. 11 displays the finite frequency noise for a bias which is half the gap and for a bias which approaches the gap, assuming an intermediate value of the transparency $Z = 1$. In the first case, we are very close to the ideal case of energy independent scattering amplitudes, while in the second case the linear dependency gets slightly distorted.

More interesting for the present model is the analysis of the noise when the bias voltage exceeds the gap. The scattering amplitudes were computed [82]. Inserting these in the energy integrals, the finite frequency noise now displays additional features (singularities) at $\omega = (eV - \Delta)/\hbar$, $\omega = 2\Delta/\hbar$, $\omega = (eV + \Delta)/\hbar$, in addition to the existing singularity at ω_J. These frequency scales can be identified on the energy diagram of Fig. 13.

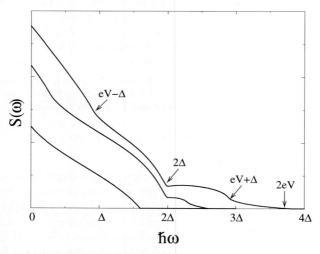

Fig. 12. Noise in an NS junction as a function of frequency, with a barrier with intermediate transparency ($Z = 1$), for biases below and above the gap $eV = 0.8\Delta$, $eV = 1.3\Delta$, $eV = 1.9\Delta$.

- Andreev reflection is still present. It leads to a singularity at $\omega_J = 2eV/\hbar$.
- Electrons can be transmitted as electron-like quasiparticles, involving wave functions $\psi_{N,e} \sim \exp[-i(\mu_S + eV)t/\hbar]$ and $\psi_{S,e} \sim \exp[-i(\mu_S + \Delta)t/\hbar]$, thus a singularity at $\omega = (eV - \Delta)/\hbar$ (likewise for holes transmitted as hole-like quasiparticles).
- Electrons from the normal side can be transmitted as hole-like quasiparticles (Andreev transmission) with associated wave functions $\psi_{N,e} \sim \exp[-i(\mu_S + eV)t/\hbar]$ and $\psi_{S,h} \sim \exp[-i(\mu_S - \Delta)t/\hbar]$, giving a singularity at $\omega = (eV + \Delta)/\hbar$ (likewise for holes incident from the normal side being transmitted as electron-like quasiparticles in the superconductor).
- Andreev reflection also occurs for electron (hole) quasiparticles incident from the superconductors, reflected as holes (electrons) quasiparticles. Wave function

are then $\psi_{S,e} \sim \exp[-i(\mu_S + \Delta)t/\hbar]$ and $\psi_{S,h} \sim \exp[-i(\mu_S - \Delta)t/\hbar]$, yielding a singularity at $2\omega = \Delta/\hbar$.

An extreme limit is the case where $\Delta \ll eV$. Transport is then fully dominated by the transfer of electrons and holes into quasiparticles, with a typical charge e because Andreev reflection is not as important as before. We indeed recover (not shown) a noise characteristic similar to Fig. 9 (dashed line) with an abrupt change of slope at $\omega_J/2 = eV/\hbar$ characteristic of a normal metal junction. Finally, we mention that finite temperatures smear out all the structures in the finite frequency noise.

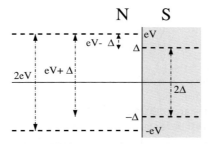

Fig. 13. Energy diagram of an NS junction when a bias is applied above the gap. Energy spacings in the frequency noise can be identified with the different processes: Andreev reflection (from either side), transmission of quasiparticles, and Andreev transmission.

8.4. Hanbury-Brown and Twiss experiment with a superconducting source of electrons

In this section, noise correlations are computed in a device which consists of two normal metal terminals (terminal 1 and 2, see Fig. 14) connected to an NS junction. The normal side of the NS junction is labelled 3, while the superconducting side is labelled 4. A junction playing the role of a beam splitter joins 1, 2 and 3. Let c_{ie}^+ (c_{ie}^-) label the state of an electron incident in (coming out from) terminal i. Likewise, incoming (outgoing) holes are labelled c_{ih}^- (c_{ih}^+) (see Fig. 14). A scattering matrix $\mathcal{S}$ (describing both the beam splitter and the NS boundary) connects incoming states to outgoing states:

$$
\begin{pmatrix} c_{1e}^- \\ c_{1h}^+ \\ c_{2e}^- \\ c_{2h}^+ \\ c_{4e}^- \\ c_{4h}^+ \end{pmatrix} = \mathcal{S} \begin{pmatrix} c_{1e}^+ \\ c_{1h}^- \\ c_{2e}^+ \\ c_{2h}^- \\ c_{4e}^+ \\ c_{4h}^- \end{pmatrix}. \tag{8.22}
$$

The noise correlations can be computed from the previous section, Eq. (8.13). In the limit of zero temperature, they can be shown to contain two contributions. The first one describes pure Andreev processes (involving one lead or both leads), while the second one involves above gap processes:

$$
S_{12}(0) = \frac{2e^2}{h} \int_0^{eV} dE \Big[\sum_{ij} \Big(s^*_{1iee} s_{1jeh} - s^*_{1ihe} s_{1jhh} \Big)
$$

$$
\times \Big(s^*_{2jeh} s_{2iee} - s^*_{2jhh} s_{2ihe} \Big)
$$

$$
+ \sum_{i\gamma} \Big(s^*_{1iee} s_{14e\gamma} - s^*_{1ihe} s_{14h\gamma} \Big)
$$

$$
\times \Big(s^*_{24e\gamma} s_{2iee} - s^*_{24h\gamma} s_{2ihe} \Big) \Big], \tag{8.23}
$$

where $i, j = 1, 2$ and $\gamma = e, h$. However, the sign of correlations cannot be determined uniquely from Eq. (8.23). In the regime where electron-like and hole-like quasiparticles are transmitted in the normal terminals, one expects that the noise correlation will be negative because this situation is quite similar to the fermionic Hanbury-Brown and Twiss experiments. But what about the sub-gap regime? Does it sustain positive or negative correlations? Here a minimal model is chosen to describe the combination of the beam splitter and the NS junction, using a Fabry-Pérot analogy (Fig. 14).

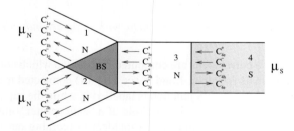

Fig. 14. Two normal terminals (1 and 2) are connected by a beam splitter (BS), itself connected to a superconductor (4) via a normal region (3).

8.4.1. S–matrix for the beam splitter
The electron part of the S matrix for the beam splitter only connects the states:

$$
\begin{pmatrix} c^-_{1e} \\ c^-_{2e} \\ c^-_{3e} \end{pmatrix} = S_{BS_e} \begin{pmatrix} c^+_{1e} \\ c^+_{2e} \\ c^+_{3e} \end{pmatrix}. \tag{8.24}
$$

A simple version for this matrix S_{BS_e}, which is symmetric between 1 and 2, and whose elements are real, has been used profusely in mesoscopic physics transport problems [83, 84]:

$$S_{BS_e} = \begin{pmatrix} a & b & \sqrt{\varepsilon} \\ b & a & \sqrt{\varepsilon} \\ \sqrt{\varepsilon} & \sqrt{\varepsilon} & -(a+b) \end{pmatrix}, \qquad (8.25)$$

with this choice, the beam splitter S matrix depends on only one parameter. Its unitarity imposes that $a = \left(\sqrt{1-2\varepsilon}-1\right)/2$, $b = \left(\sqrt{1-2\varepsilon}+1\right)/2$ where $\varepsilon \in [0, 1/2]$. $\varepsilon \ll 1/2$ means that the connection from 3 to 1 and 2 is opaque, whereas the opposite regime means a highly transparent connection to 1 and 2. Holes have similar scattering properties:

$$\begin{pmatrix} c_{1h}^+ \\ c_{2h}^+ \\ c_{3h}^+ \end{pmatrix} = S_{BS_h} \begin{pmatrix} c_{1h}^- \\ c_{2h}^- \\ c_{3h}^- \end{pmatrix}. \qquad (8.26)$$

This choice for the beam splitter does not couple electron and holes: the super-conductor boundary does that. The hole and the electron beam splitter S–matrix are related by $S_{BS_h}(E) = S^*_{BS_e}(-E)$ (in the absence of magnetic field). When one assumes that ε does not depend on energy, both matrices are the same.

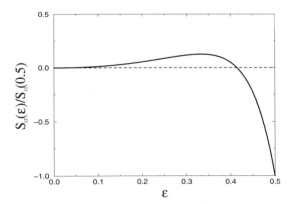

Fig. 15. Noise correlations between the two normal terminals (normalized to the autocorrelation noise in 1 or 2 for $\varepsilon = 1/2$) as a function of the beam splitter transparency ε. Correlations can either be positive or negative.

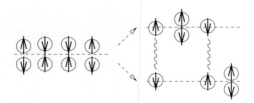

Fig. 16. (reverse) Andreev reflection emits pairs of electrons on the normal side. Correlations tend to be positive when the two electrons of the same pair go into opposite leads.

8.4.2. Sub-gap regime

When $eV \ll \Delta$, Andreev reflection is the only transport process at the NS boundary, and is described by the matrix [67]:

$$
\begin{pmatrix} c_{3e}^+ \\ c_{3h}^- \end{pmatrix} = \begin{pmatrix} 0 & \gamma \\ \gamma & 0 \end{pmatrix} \begin{pmatrix} c_{3e}^- \\ c_{3h}^+ \end{pmatrix},
\tag{8.27}
$$

where $\gamma = e^{-i \arccos(E/\Delta)}$. At the same time, $s_{14\alpha\beta} = s_{24\alpha\beta} = 0$ ($\alpha\beta = e, h$). Combining the S matrix of the beam splitter and that of the NS boundary, and defining $x = \sqrt{1 - 2\varepsilon}$, we obtain the elements of the S matrix of the combined system:

$$
s_{11ee} = s_{11hh} = s_{22ee} = s_{22hh} = \frac{(x-1)(1+\gamma^2 x)}{2(1-\gamma^2 x^2)},
\tag{8.28}
$$

$$
s_{21ee} = s_{21hh} = s_{12ee} = s_{12hh} = \frac{(x+1)(1-\gamma^2 x)}{2(1-\gamma^2 x^2)},
\tag{8.29}
$$

$$
\begin{aligned}
s_{11eh} = s_{21eh} &= s_{12eh} = s_{22eh} = s_{11he} = s_{21he} = s_{12he} = s_{22he} \\
&= \frac{\gamma(1-x)(1+x)}{2(1-\gamma^2 x^2)}.
\end{aligned}
\tag{8.30}
$$

Since all energies are much smaller than the gap, we further simplify $\gamma \to -i$. Because the amplitudes do not depend on energy, the energy integral in Eq. (8.23) is performed and one obtains [72, 85]:

$$
S_{12}(0) = \frac{2e^2}{h} eV \frac{\varepsilon^2}{2(1-\varepsilon)^4} \left(-\varepsilon^2 - 2\varepsilon + 1 \right).
\tag{8.31}
$$

Noise correlations vanish at $\varepsilon = 0$, which corresponds to a two terminal junction between 1 et 2. S_{12} also vanishes when $\varepsilon = \sqrt{2} - 1$. It is convenient to normalize S_{12} (for arbitrary ε) with the autocorrelation noise in 1 or 2 computed at $\varepsilon = 1/2$ (see Fig. 15). Correlations are *positive* if $0 < \varepsilon < \sqrt{2} - 1$ and negative for $\sqrt{2} - 1 < \epsilon < 1/2$. Minimal negative correlations (-1) are reached when the connection to 1 and 2 is optimal: this is the signature of a purely fermionic system. Eq. (8.31) predicts a maximum in the positive correlations at $\epsilon = 1/3$.

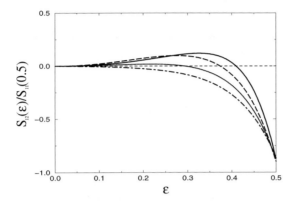

Fig. 17. Noise correlations between 1 and 2 (same normalization as before) as a function of ε. The NS junction is described within the BTK model, assuming a highly transparent barrier at the NS interface ($Z = 0.1$). Top to bottom, $eV/\Delta = 0.5, 0.95, 1.2, 1.8$.

Negative correlations correspond to Cooper pairs being distributed as a whole in the right or in the left arm.

Positive correlations have a simple interpretation. When a hole is reflected as an electron, this process can also be understood as a Cooper pair being emitted as two correlated electrons on the normal side [86]. It turns out that for an opaque beam splitter, the two electrons prefer to end up in opposite leads, giving a positive signal. This process is called the crossed Andreev process [87]. Other work, including full counting statistics approaches, describes in detail [88] why opaque barriers tend to favor positive correlations.

8.4.3. Near and above gap regime

The BTK model is chosen to describe the NS interface, in order to have the energy dependence of the scattering matrix elements. The integrals in Eq. (8.23) are computed numerically. As a first check, for a transparent interface ($Z = 0.1$) and a weak bias, one recovers the results of the previous section (see Fig. 17), except that the noise correlations do not quite reach the minimal value -1 for $\varepsilon = 1/2$, because of the presence of the barrier at the interface.

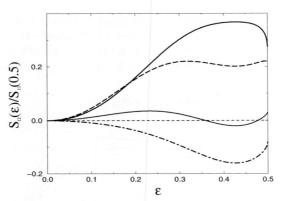

Fig. 18. Noise correlations between terminals 1 and 2 for a barrier with $Z = 1$ (same biases as in Fig. 17).

Note that noise correlations shift to negative values as one further increases the gap. For voltages beyond the gap, positive correlations disappear completely: the system behaves like a normal metal junction.

What happens next if the NS interface has an appreciable Schottky barrier? In Fig. 18, we consider a barrier with $Z = 1$, and the noise correlations change drastically: for small bias, the correlations are positive for all values of ε. This could be the regime where the positive correlations - in a fermionic system - are most likely to be observed. The presence of an oxide barrier reduces current flow at the junction, and thus a compromise has to be reached between signal detection and experimental conditions (opaque barrier) for observing the effect. For higher biases (par example $eV = 0.95\Delta$), one can monitor oscillations between positive and negative correlations. Yet, above the gap, the results are unchanged with respect to the high transparency case, as the correlations have a fermionic nature.

A number of approaches have shown the possibility of positive noise cross correlations in normal metal forks [89, 88].

9. Noise and entanglement

In quantum mechanics, a two-particle state is said to be entangled if a measurement on the state of one of the particles affects the outcome of the measurement of the state of the other particle. A celebrated example is the spin singlet:

$$\Psi_{12} = \frac{1}{\sqrt{2}} (|\uparrow_1, \downarrow_2\rangle| - |\downarrow_1, \uparrow_2\rangle) \tag{9.1}$$

Entanglement is a crucial ingredient in most information processing schemes for quantum computation or for quantum communication. Here, we enquire whether entangled states of electrons can be generated in the vicinity of an s–wave super-conductor, on the normal metal side.

9.1. Filtering spin/energy in superconducting forks

In the description of NS junction, we found that both positive and negative noise correlations were possible. Applying spin or energy filters to the normal arms 1 and 2 (Fig. 19), it is possible to generate positive correlations only [90]. For electrons emanating from a superconductor, it is possible to project either the spin or the energy with an appropriate filter, without perturbing the entanglement of the remaining degree of freedom (energy or spin). Energy filters, which are

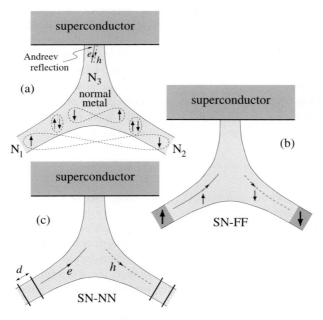

Fig. 19. Normal-metal–superconductor (NS) junction with normal-metal leads arranged in a fork geometry. (a) Without filters, entangled pairs of quasi-particles (Cooper pairs) injected in N_3 propagate into leads N_1 or N_2 either as a whole or one by one. The ferromagnetic filters in setup (b) separates the entangled spins, while the energy filters in (c) separate electron and hole quasi-particles.

more appropriate towards a comparison with photon experiments, will have reso-nant energies symmetric above and below the superconductor chemical potential which serve to select electrons (holes) in leads 1(2). The positive correlation signal then reads:

$$S_{12}(0) = \frac{e^2}{h} \sum_{\zeta} \int_0^{e|V|} d\varepsilon \mathcal{T}_\zeta^A(\varepsilon)[1 - \mathcal{T}_\zeta^A(\varepsilon)], \tag{9.2}$$

where the index $\zeta = h, \sigma, 2$ $(h, -\sigma, 2)$, $(\sigma = \uparrow, \downarrow)$ identifies the incoming hole state for energy filters (positive energy electrons with arbitrary spin are injected in lead 1 here). $\zeta = h, \uparrow, 1$ $(h, \downarrow, 2)$ applies for spin filters (spin up electrons – with positive energy – emerging from the superconductor are selected in lead 1). $\mathcal{T}_\zeta^A$ is then the corresponding (reverse) crossed-Andreev reflection probability for each type of setup: the energy (spin) degree of freedom is frozen, whereas the spin (energy) degree of freedom is unspecified. $eV < 0$ insures that the electrons of a Cooper pair from the superconductor are emitted into the leads without suffering from the Pauli exclusion principle. Moreover, because of such filters, the propagation of a Cooper pair as a whole in a given lead is prohibited. Note the similarity with the quantum noise suppression mentioned above. This is no accident: by adding constraints to our system, it has become a two terminal device, such that the noise correlations between the two arms are identical to the noise in one arm: $S_{11}(\omega = 0) = S_{12}(\omega = 0)$. The positive correlation and the perfect locking between the auto and cross correlations provide a serious symptom of entanglement. One can speculate that the wave function which describes the two-electron state in the case of spin filters reads:

$$|\Phi_{\varepsilon,\sigma}^{\text{spin}}\rangle = \frac{1}{\sqrt{2}} \left(|\varepsilon, \sigma; -\varepsilon, -\sigma\rangle - |-\varepsilon, \sigma; \varepsilon, -\sigma\rangle \right), \tag{9.3}$$

where the first (second) argument in $|\phi_1; \phi_2\rangle$ refers to the quasi-particle state in lead 1 (2) evaluated behind the filters, ε is the energy and σ is a spin index. Note that by projecting the spin degrees of freedom in each lead, the spin entanglement is destroyed, but energy degrees of freedom are still entangled, and can help to provide a measurement of quantum mechanical non-locality. A measurement of energy ε in lead 1 (with a quantum dot) projects the wave function so that the energy $-\varepsilon$ has to occur in lead 2. On the other hand, energy filters do preserve spin entanglement, and are appropriate to make a Bell test (see below). In this case the two-electron wave function takes the form:

$$|\Phi_{\varepsilon,\sigma}^{\text{energy}}\rangle = \frac{1}{\sqrt{2}} \left(|\varepsilon, \sigma; -\varepsilon, -\sigma\rangle - |\varepsilon, -\sigma; -\varepsilon, \sigma\rangle \right). \tag{9.4}$$

Electrons emanating from the energy filters (coherent quantum dots) could be analyzed provided that a measurement can be performed on the spin of the outgoing electrons with ferromagnetic leads.

9.2. Tunneling approach to entanglement

We recall a perturbative argument due to the Basel group [91] which supports the claim that two electrons originating from the same Cooper pair are entangled. Consider a system composed of two quantum dots (energies $E_{1,2}$) next to a superconductor. This system, called the entangler, was connected to two normal metal leads in Ref. [91], but for simplicity here we ignore these because we want to address what wave function is obtained when two electrons are transfered from the superconductor into the two dots.

An energy diagram is depicted in Fig. 20. The electron states in the superconductor are specified by the BCS wave function $|\Psi_{BCS}\rangle = \prod_k (u_k + v_k c^\dagger_{k\uparrow} c^\dagger_{-k\downarrow})|0\rangle$. Note that here one considers true electron creation operators, whereas previously we considered electron like and hole like quasiparticle operators. Tunneling to the dots is described by a single electron hopping Hamiltonian:

$$H_T = \sum_{kj\sigma} t_{jk} c^\dagger_{j\sigma} c_{k\sigma} + h.c., \tag{9.5}$$

where $c^\dagger_{k\sigma}$ creates an electron with spin σ, and $j = 1, 2$. Now let us assume that the transfer Hamiltonian acts on a single Cooper pair.

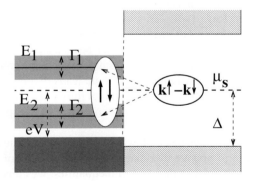

Fig. 20. Transfer of a Cooper pair on two quantum energy levels $E_{1,2}$ with a finite width $\Gamma_{1,2}$. The superconductor is located on the right hand side. The transfer of a Cooper pair gives an entangled state in the dots because it implies the creation and destruction of the same quasiparticle in the superconductor. The source drain voltage eV for measuring noise correlations is indicated.

Using the T-matrix to lowest (2nd) order, the wave function contribution of the two particle state with one electron in each dot and the superconductor back in its ground state reads:

$$
\begin{aligned}
|\delta\Psi_{12}\rangle &= H_T \frac{1}{i\eta - H_0} H_T |\Psi_{BCS}\rangle \\
&= [c_{1\uparrow}^\dagger c_{2\downarrow}^\dagger - c_{1\downarrow}^\dagger c_{2\uparrow}^\dagger] \sum_{jk} \frac{v_k u_k t_{1k} t_{2k}}{i\eta - E_k - E_j} |\Psi_{BCS}\rangle,
\end{aligned} \tag{9.6}
$$

where E_k is the energy of a Bogolubov quasiparticle. The state of Eq. (9.6) has entangled spin degrees of freedom. This is clearly a result of the spin preserving tunneling Hamiltonian. Given the nature of the correlated electron state in the superconductor in terms of Cooper pairs, H_T can only produce singlet states in the dots.

9.3. Bell inequalities with electrons

In photon experiments, entanglement is identified by a violation of Bell inequalities – which are obtained with a hidden variable theory. But in the case of photons, the Bell inequalities have been tested using photo-detectors measuring coincidence rates [92]. Counting quasi-particles one-by-one in coincidence measurements is difficult to achieve in solid-state systems where stationary currents and noise are the natural observables. Here, the Bell inequalities are re-formulated in terms of current-current cross-correlators (noise correlations) [93].

Note that the connection between noise correlations and entanglement has been pointed out in Ref. [94]. This works considers two beams of particles, which may be entangled or not, incident on a beam splitter. The noise (and noise correlations) at the output of the beam splitter bears a clear signature of the state of the two beams. An incoming singlet state leads to an enhancement of the noise (bunching effect), whhereas a triplet state leads to a noise reduction (antibunching). The measurement of the Fano factor thus allows to make a distinction between singlet/triplet entanglement and two beams of independent (classical) particles.

Because Bell inequalities tests allow to further quantify the degree of entanglement, we choose to operate this diagnosis in the context of electronic quantum transport [95]. In order to derive Bell inequalities, we consider that a source provides two streams of particles (labeled 1 and 2) as in Fig. 21a injecting quasiparticles into two arms labelled by indices 1 and 2. Filter $F_{1(2)}^d$ are transparent for electrons spin-polarized along the direction $\mathbf{a}(\mathbf{b})$.

Assuming separability and locality [96] the density matrix for joint events in the leads α, β is chosen to be:

$$
\rho = \int d\lambda f(\lambda) \rho_\alpha(\lambda) \otimes \rho_\beta(\lambda), \tag{9.7}
$$

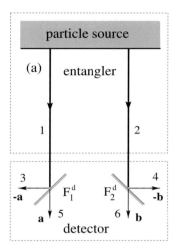
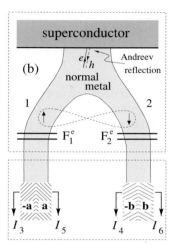

Fig. 21. a) Schematic setup for the measurement of Bell inequalities: a source emits particles into leads 1 and 2. The detector measures the correlation between beams labelled with odd and even numbers. Filters $F^d_{1(2)}$ select the spin: particles with polarization along the direction $\pm\mathbf{a}(\pm\mathbf{b})$ are transmitted through filter $F^d_{1(2)}$ into lead 5 and 3 (6 and 4). b) Solid state implementation, with superconducting source emitting Cooper pairs into the leads. Filters $F^e_{1,2}$ (e.g., Fabry-Perot double barrier structures or quantum dots) prevent Cooper pairs from entering a single lead. Ferromagnets with orientations $\pm\mathbf{a}$, $\pm\mathbf{b}$ play the role of the filters $F^d_{1(2)}$ in a); they are transparent for electrons with spin aligned along their magnetization axis.

where the lead index α is even and β is odd (or vice-versa); the distribution function $f(\lambda)$ is positive. $\rho_\alpha(\lambda)$ are standard density matrices for a given lead, which are Hermitian. The total density matrix ρ is the most general density matrix one can built for the source/detector system assuming no entanglement and only local correlations.

Consider an example of the solid-state analog of the Bell device where the particle source is a superconductor in Fig. 21b. The chemical potential of the superconductor is larger than that of the leads, which means that electrons are flowing out of the superconductor. Two normal leads 1 and 2 are attached to it in a fork geometry [90, 91] and the filters $F^e_{1,2}$ enforce the energy splitting of the injected pairs. $F^d_{1,2}$-filters play the role of spin-selective beam-splitters in the detector. Quasi-particles injected into lead 1 and spin-polarized along the magnetization $\mathbf{a}$ enter the ferromagnet 5 and contribute to the current I_5, while quasi-particles with the opposite polarization contribute to the current I_3.

Consider the current operator $I_\alpha(t)$ in lead $\alpha = 1, \ldots, 6$ (see Fig. 21) and the associated particle number operator $N_\alpha(t, \tau) = \int_t^{t+\tau} I_\alpha(t')dt'$. Particle-number correlators are defined as:

$$\langle N_\alpha(t, \tau) N_\beta(t, \tau) \rangle_\rho = \int d\lambda f(\lambda) \langle N_\alpha(t, \tau) \rangle_\lambda \langle N_\beta(t, \tau) \rangle_\lambda, \tag{9.8}$$

with indices α/β odd/even or even/odd. The average $\langle N_\alpha(t, \tau) \rangle_\lambda$ depends on the state of the system in the interval $[t, t + \tau]$. An average over large time periods is introduced in addition to averaging over λ, e.g.,

$$\langle N_\alpha(\tau) N_\beta(\tau) \rangle \equiv \frac{1}{2T} \int_{-T}^{T} dt \langle N_\alpha(t, \tau) N_\beta(t, \tau) \rangle_\rho, \tag{9.9}$$

where $T/\tau \rightarrow \infty$ (a similar definition applies to $\langle N_\alpha(\tau) \rangle$). Particle number fluctuations are written as $\delta N_\alpha(t, \tau) \equiv N_\alpha(t, \tau) - \langle N_\alpha(\tau) \rangle$. Let x, x', y, y', X, Y be real numbers such that:

$$|x/X|, |x'/X|, |y/Y|, |y'/Y| < 1. \tag{9.10}$$

Then $-2XY \leq xy - xy' + x'y + x'y' \leq 2XY$. Define accordingly:

$$x = \langle N_5(t, \tau) \rangle_\lambda - \langle N_3(t, \tau) \rangle_\lambda, \tag{9.11}$$

$$x' = \langle N_{5'}(t, \tau) \rangle_\lambda - \langle N_{3'}(t, \tau) \rangle_\lambda, \tag{9.12}$$

$$y = \langle N_6(t, \tau) \rangle_\lambda - \langle N_4(t, \tau) \rangle_\lambda, \tag{9.13}$$

$$y' = \langle N_{6'}(t, \tau) \rangle_\lambda - \langle N_{4'}(t, \tau) \rangle_\lambda, \tag{9.14}$$

where the subscripts with a 'prime' indicate a different direction of spin-selection in the detector's filter (e.g., let **a** denote the direction of the electron spins in lead 5 ($-$**a** in lead 3), then the subscript $5'$ means that the electron spins in lead 5 are polarized along **a**$'$ (along $-$**a**$'$ in the lead 3). The quantities X, Y are defined as

$$
\begin{aligned}
X &= \langle N_5(t, \tau) \rangle_\lambda + \langle N_3(t, \tau) \rangle_\lambda \\
&= \langle N_{5'}(t, \tau) \rangle_\lambda + \langle N_{3'}(t, \tau) \rangle_\lambda \\
&= \langle N_1(t, \tau) \rangle_\lambda,
\end{aligned} \tag{9.15}
$$

$$
\begin{aligned}
Y &= \langle N_6(t, \tau) \rangle_\lambda + \langle N_4(t, \tau) \rangle_\lambda \\
&= \langle N_{6'}(t, \tau) \rangle_\lambda + \langle N_{4'}(t, \tau) \rangle_\lambda \\
&= \langle N_2(t, \tau) \rangle_\lambda;
\end{aligned} \tag{9.16}
$$

The Bell inequality follows after appropriate averaging:

$$|F(\mathbf{a}, \mathbf{b}) - F(\mathbf{a}, \mathbf{b}') + F(\mathbf{a}', \mathbf{b}) + F(\mathbf{a}', \mathbf{b}')| \leq 2, \tag{9.17}$$

$$F(\mathbf{a}, \mathbf{b}) = \frac{\langle [N_1(\mathbf{a}, t) - N_1(-\mathbf{a}, t)][N_2(\mathbf{b}, t) - N_2(-\mathbf{b}, t)] \rangle}{\langle [N_1(\mathbf{a}, t) + N_1(-\mathbf{a}, t)][N_2(\mathbf{b}, t) + N_2(-\mathbf{b}, t)] \rangle}, \tag{9.18}$$

with $\mathbf{a}, \mathbf{b}$ the polarizations of the filters $F_{1(2)}$ (electrons spin-polarized along $\mathbf{a}$ ($\mathbf{b}$) can go through filter $F_{1(2)}$ from lead 1(2) into lead 5(6)). This is the quantity we want to test, using a quantum mechanical theory of electron transport. Here it will be written in terms of noise correlators, as particle number correlators at equal time can be expressed in general as a function of the finite frequency noise cross-correlations. The correlator $\langle N_\alpha(\tau) N_\beta(\tau) \rangle$ includes both reducible and irreducible parts. The irreducible correlator $\langle \delta N_\alpha(\tau) \delta N_\beta(\tau) \rangle$ can be expressed through the shot noise power $S_{\alpha\beta}(\omega) = 2 \int d\tau e^{i\omega\tau} \langle \delta I_\alpha(\tau) \delta I_\beta(0) \rangle$,

$$\langle \delta N_\alpha(\tau) \delta N_\beta(\tau) \rangle = \int_{-\infty}^{\infty} \frac{d\omega}{2\pi} S_{\alpha\beta}(\omega) \frac{4 \sin^2(\omega\tau/2)}{\omega^2}. \tag{9.19}$$

In the limit of large times, $\sin^2(\omega\tau/2)/(\omega/2)^2 \to 2\pi\tau\delta(\omega)$, and therefore:

$$\langle N_\alpha(\tau) N_\beta(\tau) \rangle \approx \langle I_\alpha \rangle \langle I_\beta \rangle \tau^2 + \tau S_{\alpha\beta}(\omega = 0) \tag{9.20}$$

where $\langle I_\alpha \rangle$ is the average current in the lead α, and $S_{\alpha\beta}$ denotes the shot noise. One then gets:

$$F(\mathbf{a}, \mathbf{b}) = \frac{S_{56} - S_{54} - S_{36} + S_{34} + \Lambda_-}{S_{56} + S_{54} + S_{36} + S_{34} + \Lambda_+}, \tag{9.21}$$

where $\Lambda_\pm = \tau(\langle I_5 \rangle \pm \langle I_3 \rangle)(\langle I_6 \rangle \pm \langle I_4 \rangle)$ comes from the reducible part of the number correlators (the average number product). For a symmetric device, $\Lambda_- = 0$.

So far we have only provided a dictionary from the number correlator language used in optical measurements to the stationary quantities encountered in nanophysics. We have provided absolutely no specific description of the physics which governs this beam splitter device. The test of the Bell inequality (9.17) requires information about the dependence of the noise on the mutual orientations of the magnetizations $\pm\mathbf{a}$ and $\pm\mathbf{b}$ of the ferromagnetic spin-filters. In the tunneling limit one finds the noise:

$$S_{\alpha\beta} = e \sin^2\left(\frac{\theta_{\alpha\beta}}{2}\right) \int_0^{|eV|} d\varepsilon \, T^A(\varepsilon), \tag{9.22}$$

which integral also represents the current in a given lead (we have dropped the subscript in $T^A(\varepsilon)$ assuming the two channels are symmetric). Here $\alpha = 3, 5,$

$\beta = 4, 6$ or vice versa; $\theta_{\alpha\beta}$ denotes the angle between the magnetization of leads α and β, e.g., $\cos(\theta_{56}) = \mathbf{a} \cdot \mathbf{b}$, and $\cos(\theta_{54}) = \mathbf{a} \cdot (-\mathbf{b})$. Below, we need configurations with different settings $\mathbf{a}$ and $\mathbf{b}$ and we define the angle $\theta_{\mathbf{ab}} \equiv \theta_{56}$. V is the bias of the superconductor.

The Λ-terms in Eq. (9.21) can be dropped if $\langle I_\alpha \rangle \tau \ll 1$, $\alpha = 3, \ldots, 6$, which corresponds to the assumption that only one Cooper pair is present on average. The resulting Bell inequalities Eqs. (9.17)-(9.21) then depend neither on τ nor on the average current but only on the shot-noise, and $F = -\cos(\theta_{\mathbf{ab}})$; the left hand side of Eq. (9.17) has a maximum when $\theta_{\mathbf{ab}} = \theta_{\mathbf{a'b}} = \theta_{\mathbf{a'b'}} = \pi/4$ and $\theta_{\mathbf{ab'}} = 3\theta_{\mathbf{ab}}$. With this choice of angles the Bell inequality Eq. (9.17) is *violated*, thus pointing to the nonlocal correlations between electrons in the leads 1,2 [see Fig. 21(b)].

If the filters have a width Γ the current is of order $eT^A\Gamma/h$ and the condition for neglecting the reducible correlators becomes $\tau \ll \hbar/\Gamma T^A$. On the other hand, in order to insure that no electron exchange between 1 and 2 one requires $\tau \ll \tau_{\text{tr}}/T^A$ (τ_{tr} is the time of flight from detector 1 to 2). The conditions for Bell inequality violation require very small currents, because of the specification that only one entangled pair at a time is in the system. Yet it is necessary to probe noise cross correlations of these same small currents. The noise experiments which we propose here are closely related to coincidence measurements in quantum optics [92].

If we allow the filters to have a finite line width, which could reach the energy splitting of the pair, the violation of Bell inequality can still occur, although violation is not maximal. Moreover, when the source of electron is a normal source, our treatment has to be revised. The low frequency noise approximation relating the number operators to the current operator breaks down at short times. Ref. [97] shows in fact that entanglement can exist in ballistic forks. It is also possible to violate Bell inequalities if the normal source itself, composed of quantum dots as suggested in Ref. [98, 99], could generate entangled electron states as the result of electron-electron interactions.

Spin entanglement from superconducting source of electrons relies on the controlled fabrication of an entangler [91] or a superconducting beam-splitter with filters. Since the work of Ref. [93], other proposals for entanglement have been proposed, which avoid using spin. Electron-hole entanglement for a Hall bar with a point contact [100] exploits the fact that an electron, which can occupy either one of two edge channels, can either be reflected or transmitted. Unitary transformations between the two outgoing channels play the role of current measurement in arbitrary spin directions. Orbital entanglement using electrons emitted from two superconductors has been suggested in Ref. [101]. The electron-hole proposal of Ref. [100] has also been revisited using two distinct electron sources in a Hanbury-Brown and Twiss geometry together with additional beam splitters for detection, resulting in a concrete experimental proposal [102].

10. Noise in Luttinger liquids

In the previous sections, the noise was computed essentially in non-interacting systems. Granted, a superconductor depends crucially on the attractive interactions between electrons. Yet, the description of the NS boundary is like that of a normal mesoscopic conductor with electron and hole channels which get mixed. Now we want the noise in a system with repulsive interactions. There are many possibilities for doing that. One could consider transport through quantum dots where double occupancy of a dot costs a charging energy [39]. Alternatively one could consider to treat interactions in a mesoscopic conductor using perturbative many body techniques.

Instead we chose a situation where the interactions provide a genuine departure from single electron physics. The standard credo about interactions in condensed matter systems is the Fermi liquid picture. In two and three dimensions, it has been known for a long time that the quasiparticle picture holds: the elementary excitations of a system of interacting fermions resemble the original electrons. The excitations are named quasiparticles because their dynamics can be described in a similar manner as electrons, except for the fact that their mass is renormalized and that these quasiparticles have a finite lifetime. Perturbation theory, when done carefully in such systems, works rather well.

It is therefore more of a challenge to turn to the case of one dimensional systems where the Fermi liquid picture breaks down. Indeed, whereas in 2 and 3 dimensions the distribution function retains a step at the Fermi level, interactions in a one dimensional system render the distribution function continuous, with only an infinite derivative at the Fermi energy [103]. But the most important feature of a one dimensional system is that the nature of the excitations changes drastically compared to its higher dimensional counterparts [104]. The excitations do not resemble electrons in any way: they consist of collective electron–hole excitations of the whole Fermi sea.

Luttinger liquid theory gives an account of the special properties of one-dimensional conductors [105]. For transport through an isolated impurity, the effect of interactions leads to a phase diagram [106]: in the presence of repulsive interactions, a weak impurity renders the wire insulating, whereas for positive interactions even a strong impurity is transparent.

The "easiest" type of Luttinger liquid arises on the boundaries of a sample which is put in the quantum Hall regime: a two dimensional electron gas (2DEG) under a high magnetic field. A classical description of such a system tells us that the electrons move along the edges, subject to the so called $\vec{E} \times \vec{B}$ drift. In mesoscopic physics, the electric field $\vec{E}$ comes from the confining electrostatic potential on the edges of the sample. Consider first a system of non-interacting electrons in the quantum regime (low temperature, high magnetic field). For an infinite sample, the quantum mechanical description gives the so called Landau levels, which are separated by the energy $\hbar\omega_c$, with $\omega_c = eB/mc$ the cyclotron frequency. Each Landau level has a degeneracy $N_D = N_\phi$, with $N_\phi = BS/\phi_0$ the number of flux tubes which can be fitted in a sample with area S. Given a magnetic field B one defines the filling factor ν as the ratio between the total number of electrons and the number of flux tubes, or equivalently the fraction of the Landau levels which are filled. In the integer quantum Hall effect [107], the Landau energy spectrum allows to explain the quantification of the Hall resistance and the simultaneous vanishing of the longitudinal resistance when the magnetic field is varied (or, equivalently when the density of electrons is varied).

What happens when one considers confinement? Landau levels bend upwards along the edges. So if one has adjusted the Fermi level of the system exactly between two landau levels, the highest populated states are precisely there. These are the quantum analogs of the classical skipping orbits. A very important feature is that they have a chiral character: they move only in one direction on one side of the sample, and in the opposite direction on the other side. This edge state description has allowed to explain in a rather intuitive manner the physics of the integer quantum Hall effect [34].

Interactions complicate things in a substantial manner, especially if one increases the magnetic field. When the lowest Landau level becomes partially filled, one reaches the fractional Hall effect regime [108, 109] (FQHE). The many body wave function for the electrons is such that it minimizes the effect of interaction. The Hall resistance exhibits plateaus, $R_H = h/\nu e^2$ when the inverse of the filling factor is an odd integer. Spectacular effects follow. The excitation spectrum of this fractional quantum Hall fluid has a gap. The quasiparticles have a fractional charge, and if one exchanges the position of two such objects, the phase is neither 0 (bosons) nor π (electrons) - quasiparticles have fractional statistics.

Here, we want to know what happens at the edge. There are several arguments which justify the action which we shall use below. One of them relies on field theory arguments: starting from the fact that one is dealing with a gaped system, and effective action can be derived for the fluctuating electromagnetic field. If one now considers a finite fractional quantum Hall fluid with boundaries, one finds out that a boundary term must be added in order to preserve gauge invariance. This term turns out to generate the dynamics of the edge excitations. Here however, I will use a more intuitive argument to describe the edges called the hydrodynamic approach.

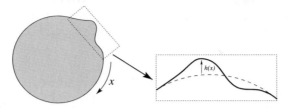

Fig. 22. Fractional quantum hole droplet. Excitations propagate along the edge.

10.1. Edge states in the fractional quantum Hall effect

The Hamiltonian which describes the edge modes is simply an electrostatic term:
[110]

$$H = \frac{1}{2} \int_0^L V(x)e\rho(x)dx, \tag{10.1}$$

with x a curvilinear coordinate along the edge and $V(x)$ the confining potential. This potential is related to the confining electric field as $E = -\nabla V \sim \partial_y V$, with y the coordinate perpendicular to the edge. E and B are related because the drift velocity is given by $|\vec{v}| = c|\vec{E} \times \vec{B}|/B^2$. The electrostatic potential can then be expressed in terms of the lateral displacement of the quantum Hall fluid $h(x)$, which is also expressed in terms of the linear charge density $\rho(x)$:

$$V(x) = Eh(x) = (vB/c)(\rho(x)/n_s). \tag{10.2}$$

Inserting this in the Hamiltonian, we find the remarkable property that the Hamiltonian is quadratic in the density. At this point it is useful to use the definition of the flux quantum $\phi_0 = hc/e$ in order to eliminate the 2D electron density n_s from the problem in favor of the filling factor v.

$$H = \frac{1}{2} \frac{h v}{v} \int_0^L \rho^2(x)\, dx. \tag{10.3}$$

So far we have used a purely classical argument. In order to obtain a quantum mechanical description, we need to impose quantification rules. First, it is convenient to transform this Hamiltonian into Fourier space using:

$$\rho(x) = \frac{1}{\sqrt{L}} \sum_k e^{-ikx} \rho_k, \tag{10.4}$$

$$H = \frac{1}{2} \frac{v h}{v} \sum_k \rho_k \rho_{-k}. \tag{10.5}$$

Quantification requires first to identify a set of canonical conjugate variables q_k and p_k which satisfy Hamilton's equations. Identifying $q_k = \rho_k$, one obtains

$$\dot{p}_k = -\frac{\partial H}{\partial \rho_k} = -\frac{1}{2}\frac{vh}{v}2\rho_{-k}. \tag{10.6}$$

The continuity equation for this chiral density reads $\dot{\rho}_{-k} = -vik\rho_{-k}$. Integrating over time, one thus get the canonical conjugate:

$$p_k = -i\frac{h}{v}\frac{\rho_{-k}}{k}. \tag{10.7}$$

Quantification is achieved by imposing the commutation relation

$$[q_k, p_{k'}] \quad = \quad i\hbar\, \delta_{kk'}. \tag{10.8}$$

Note that this is exactly the same procedure as one uses for phonons in conventional condensed matter physics. Replacing p_k by its expression in Eq. (10.7), one gets the Kac-Moody commutation relations:

$$[\rho_k, \rho_{k'}] = -\frac{vk}{2\pi}\delta_{k'\,-k}. \tag{10.9}$$

Computing the commutator of the Hamiltonian with the density yields

$$[H, \rho_{k'}] = v\hbar k' \rho_{k'}, \tag{10.10}$$

and one sees that the Heisenberg evolution equation $i\hbar\rho_k = [H, \rho_k]$ gives the continuity equation. We now turn to the definition of the electron operator. Because $\rho(x)$ is the charge density, we expect the electron creation operator to satisfy

$$[\rho(x'), \psi^{\dagger}(x)] \quad = \quad \delta(x - x')\psi^{\dagger}(x), \tag{10.11}$$

which is equivalent to saying that the measurement of the electronic density on a state on which $\psi^{\dagger}(x)$ is acting tells us that an electron has been added. This is the same commutation relation one uses in the derivation of the raising and lowering operators. For later purposes, the Luttinger bosonic field is introduced:

$$\phi(x) = \frac{\pi}{\sqrt{v}}\frac{1}{\sqrt{L}}\sum_k i\frac{e^{-a|k|/2}}{k}e^{-ikx}\rho_k. \tag{10.12}$$

Here the factor a takes the meaning of a spatial cutoff similar to that used in non-chiral Luttinger liquids. It insures the convergence of the integral. What is important about this definition is that the derivative of ϕ is proportional to the density:

$$\frac{\partial \phi}{\partial x} = \frac{\pi}{\sqrt{v}} \rho(x). \tag{10.13}$$

This allows to re-express the Hamiltonian in terms of ϕ [111]:

$$H = \frac{\hbar v}{\pi} \int_0^L (\frac{\partial \phi}{\partial x})^2 dx. \tag{10.14}$$

The form of the electron operators is found by an analogy with the properties of canonical conjugate variables $p(x)$ and $q(x)$:

$$[p(x), q(x')] = -i\delta(x - x') \rightarrow [p(x), e^{iq(x')}] = \delta(x - x')e^{iq(x)}. \tag{10.15}$$

Next, one can identify p as ρ and q as $\phi/\sqrt{v}$. Using the Kac-Moody commutation relations: $[\rho(x'), v^{-1/2}\phi(x)] = -i\delta(x-x')$ and comparing with the relation (10.15) and the definition (10.11), the annihilation operator takes the form:

$$\psi(x) = \frac{1}{\sqrt{2\pi a}} e^{-ikx} e^{i\frac{1}{\sqrt{v}}\phi(x)}, \tag{10.16}$$

with e^{ikx} giving the phase accumulated along the edge. This operator obviously depends on the filling factor. Fermion operators are known to anti-commute, so what are the constraints on this filling factor in order to insure anti-commutation relations $\{\psi(x), \psi(x')\} = 0$. The anti-commutator can be computed using the Baker-Campbell-Hausdorff formula: $e^A e^B = e^{A+B-\frac{[A,B]}{2}}$ which is only true of the commutator is a c-number. One thus needs the commutation relation of the bosonic field:

$$[\phi(x), \phi(x')] = -i\pi sgn(x - x'). \tag{10.17}$$

The two products of fermionic operators is then:

$$\psi(x)\psi(x') = \frac{1}{(2\pi a)^2} e^{ik(x+x')} e^{i\frac{1}{\sqrt{v}}\phi(x)+\phi(x')} e^{\frac{i\pi}{2v} sgn(x-x')}. \tag{10.18}$$

So one concludes that:

$$\psi(x)\psi(x') = e^{\pm i\frac{\pi}{v}} \psi(x')\psi(x). \tag{10.19}$$

In order to insure anti-commutation relations, one needs to set $v = 1/m$ with m an odd integer. This conclusion is consistent with the assumption that in the bulk, one is dealing with a fractional quantum Hall fluid.

In order to obtain information on the dynamics of electrons (or of fractional quasiparticles), one needs to specify the bosonic Green's function. It is thus convenient to derive the action for this bosonic field. The Lagrangian is obtained from a Legendre transformation on the Hamiltonian, taking as canonical conjugate variables $\phi(x)$ and $-i(\hbar/\pi)\partial_x\phi$ in accordance with the Kac-Moody relations. The Euclidean action then reads:

$$S_E = -\frac{\hbar}{\pi}\int d\tau \int dx\,[\partial_x\phi\,(v\partial_x + i\partial_\tau)\,\phi]. \tag{10.20}$$

The operator which is implicit in this quadratic action allows to define the Green's function $G(x, \tau) = \langle T_\tau\phi(x, \tau)\phi(0, 0)\rangle$ – the correlation function of the bosonic field. This Green's function is defined by the differential equation:

$$(i\partial_\tau + v\partial_x)\partial_x G(x, \tau) = 2\pi\delta(x)\delta(\tau). \tag{10.21}$$

The solution of this equation is obtained by setting $-\partial_x G = f$, and using the complex variables $z = x/v + i\tau$ et $\bar{z} = x/v - i\tau$. The equation for f becomes:

$$\partial_{\bar{z}} f = v\pi\delta(x)\delta(\tau). \tag{10.22}$$

From two dimensional electrostatics, it can be justified that $f(z) = 1/z$. Yet, one is dealing here with the thermal Green's function, which must be a periodic function of τ with period β, so a periodic extension of $f(z)$ is given:

$$f(z + i\beta) = f(z) = \frac{1}{z} + \sum_{n\neq 0}\frac{1}{z - in\beta} = \frac{\pi}{v\beta}\coth\left(\pi\frac{z}{\beta}\right). \tag{10.23}$$

The thermal Green's function is subsequently obtained by integrating over x:

$$G(x, \tau) = -\ln\left[\sinh\left(\pi\frac{x/v + i\tau}{\beta}\right)\right]. \tag{10.24}$$

10.2. Transport between two quantum Hall edges

The noise has been computed for a single point contact [112, 110, 113]. It is typically achieved by placing metallic gates on top of the 2DEG and applying a potential to deplete the electron gas underneath the gates (Fig. 23). By varying the gate potential, one can switch from a weak backscattering situation, where the Hall liquid remains in one piece, to a strong backscattering situation where the Hall fluid is split into two. In the former case, the entities which tunnel are edge quasiparticle excitations. In the latter case, the general convention is to say that in between the two fluids, only electrons can tunnel, because nothing can "dress" these electrons into strange quasiparticles like in the previous case.

Here we will focus mostly on the weak backscattering case, because this is the situation where the physics of FQHE quasiparticles is most obvious. Anyway, the description of the strong backscattering case can be readily obtained using a duality transformation.

a)   b)

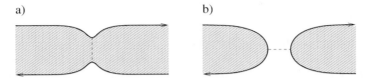

Fig. 23. Quantum transport in a quantum Hall bar: a) in the presence of a weak constriction, the quantum hall fluid stays as a whole (shaded area) and quasiparticles tunnel. b) the case of strong backscattering, where the quantum Hall fluid is broken in two and electrons tunnel between the two fluids.

The tunneling Hamiltonian describing the coupling between the two edges L and R is like a tight binding term, where for convenience we use a compact notation [58] to describe the two hermitian conjugate parts:

$$H_{int} = \sum_{\varepsilon=\pm} \left[\Gamma_0 \Psi_R^\dagger \Psi_L\right]^\varepsilon \text{ with } \begin{cases} \left[\Gamma_0 \Psi_R^\dagger \Psi_L\right]^+ = \Gamma_0 \Psi_R^\dagger \Psi_L \\ \left[\Gamma_0 \Psi_R^\dagger \Psi_L\right]^- = \Gamma_0^* \Psi_L^\dagger \Psi_R \end{cases} \quad (10.25)$$

Where the quasiparticle operators have the form:

$$\Psi_{R(L)}(t) = \frac{M_{R(L)}}{\sqrt{2\pi a}} e^{i\sqrt{\nu}\phi_{R(L)}(t)}. \quad (10.26)$$

The spatial cutoff is defined as $a = \nu\tau_0$, τ_0 is the temporal cutoff. $M_{R(L)}$ is a Klein factor, which insures the proper statistical properties: indeed the fact that the fermion operator is an exponential of a boson field does guarantee proper exchange statistics on a given edge, but not for electrons which belong to different edges. A derivation of the properties of Klein factors for the quasiparticle fields in the FQHE can be found in Ref. [114]. Nevertheless, for problems involving only two edges, it turns out to be irrelevant and it will be omitted. The quasiparticles are, in a sense, the $1/\nu$ root of the electron operators.

In scattering theory, the bias voltage was included by choosing appropriately the chemical potentials of the reservoirs, which enter in the energy representation of the Green's functions for the leads. Here it is more difficult, because the Green's function are defined in real time, and the density of states of FQHE quasiparticles diverges at the Fermi level. The trick is to proceed with a gauge transformation. Starting from a gauge where the electric field is solely described by the scalar potential, $A = 0$, we proceed to a gauge transformation such that the new scalar potential is zero:

$$\begin{cases} V' = V - \frac{1}{c}\partial_t \chi & = 0 \\ \vec{A}' = \vec{A} + \nabla\chi & \neq 0 \\ \Psi_i' = e^{i\frac{e^*\chi_i}{\hbar c}}\Psi_i & , e^* = ve \end{cases} \qquad \text{so } \nabla\chi = \vec{A}' \qquad (10.27)$$

For a constant potential along the edges, the gauge function χ depends only on time, and $\Delta\chi = \int_L^R \vec{A}'.\vec{dl} = cV_0 t$. Because we are dealing with quasiparticle transfer we *anticipate* that the quasiparticle charge is $e^* = ve$. Upon gauge transforming the quasiparticle operators, the tunneling amplitude becomes:

$$\Gamma_0 \to \Gamma_0 e^{i\omega_0 t}, \qquad (10.28)$$

where $\omega_0 = e^* V_0$. From this expression, the backscattering current operator is derived from the Heisenberg equation of motion for the density, or alternatively by calculating $I_B(t) = -c\partial H_B(t)/\partial\chi(t)$:

$$I_B(t) = ie^*\Gamma_0 \sum_\varepsilon \varepsilon e^{i\varepsilon\omega_0 t}[\Psi_R^\dagger(t)\Psi_L(t)]^{(\varepsilon)}. \qquad (10.29)$$

10.3. Keldysh digest for tunneling

In many body physics, it is convenient to work with a Wick's theorem (or one of its generalizations) in order to compute products of fermion and boson operators. It is encountered when one considers averages of Heisenberg operators ordered in time, and one is faced with the problem of translating this into interaction representation products. The problem with the Heisenberg representation is that the operators contains the "difficult" part (the interaction part) of the Hamiltonian. Consider the ground state average of a time-ordered product of Heisenberg operators:

$$\langle A_H(t_0)B_H(t_1)C_H(t_2)D_H(t_3)\ldots\rangle \text{ with } t_0 > t_1 > t_2 > t_3 > \ldots \qquad (10.30)$$

When translating to the interaction representation, the evolution operator reads:

$$S(t,t') = \hat{T}\exp\left\{-i\int_{t'}^t dt'' H_{int}(t'')\right\}, \qquad (10.31)$$

all operators such as H_{int} become $e^{iH_0 t}H_{int}e^{-iH_0 t}$ [115] in this language. The product of ordered operators then becomes:

$$\langle S(-\infty,+\infty)\hat{T}(A_I(t_0)B_I(t_1)C_I(t_2)D_I(t_3)\ldots S(+\infty,-\infty))\rangle, \qquad (10.32)$$

where $\hat{T}$ is the time ordering operator. When the system is at zero temperature or in equilibrium, the ground state (or thermal) expectation of this S–matrix is just a phase factor, because one assumes that the perturbation is turned on adiabatically. This means that $\langle S(+\infty, -\infty) \rangle = e^{i\gamma}$. One is therefore left with a T–product which is easily computed with the help of Wick's theorem.

However, if the system is out of equilibrium, one cannot a priori use the Wick theorem to compute the average: the S-matrix in front of the T–product spoils everything because particles are being transfered from one reservoir to the other, and the ground state at $t = +\infty$ does not look like anything like the ground state at $t = -\infty$ (both are no longer related by a phase factor). To remedy this problem, Keldysh proposed to invent a new contour, which goes from $t = -\infty$ to $t = +\infty$ and back to $t = -\infty$, and a corresponding new time ordering operator $\hat{T}_K$. Because times on the lower contour are "larger" than times on the upper contour, the product of operators can be written as:

$$\langle \hat{T}_K (A_I(t_0)B_I(t_1)C_I(t_2)D_I(t_3) \ldots S(-\infty, -\infty)) \rangle, \tag{10.33}$$

where the integral over the Keldysh contour K goes from $-\infty$ to $+\infty$ and then back to $-\infty$. Note that in general, the times appearing in the operator product $A_I(t_0)B_I(t_1)C_I(t_2)D_I(t_3)$ can be located either on the upper or on the lower contour. The Green's function associated with the two-branches Keldysh contour is therefore a 2×2 matrix:

$$\begin{aligned}
\tilde{G}(t - t') &= \begin{pmatrix} \tilde{G}^{++}(t - t') & \tilde{G}^{+-}(t - t') \\ \tilde{G}^{-+}(t - t') & \tilde{G}^{--}(t - t') \end{pmatrix} \\
&= \begin{pmatrix} \tilde{G}^{-+}(|t - t'|) & \tilde{G}^{-+}(t' - t) \\ \tilde{G}^{-+}(t - t') & \tilde{G}^{-+}(-|t - t'|) \end{pmatrix},
\end{aligned} \tag{10.34}$$

where $\tilde{G}^{-+}(t)$ can be computed from the thermal Green's function using a Wick rotation. Often one redefines the Green's function by subtracting from the initial Green's function its equal time arguments (see below).

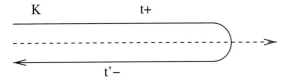

Fig. 24. The two-branches of the Keldysh contour.

10.4. Backscattering current

For the calculation of an operator which involves a single time argument, it does not matter on which branch of the Keldysh contour we assign the time. We therefore chose a symmetric combination:

$$\langle I_B(t)\rangle = \frac{1}{2}\sum_\eta \langle \hat{T}_K \{I_B(t^\eta)e^{-i\int_K dt_1 H_B(t_1)}\}\rangle. \tag{10.35}$$

To lowest order in the tunnel amplitude Γ_0, we have:

$$\langle I_B(t)\rangle = \frac{e^*\Gamma_0^2}{2}\sum_{\eta\eta_1\varepsilon\varepsilon_1}\varepsilon\eta_1 \int_{-\infty}^{+\infty} dt_1 e^{i\varepsilon\omega_0 t + i\varepsilon_1\omega_0 t_1}$$
$$\times \langle \hat{T}_K\{[\Psi_R^\dagger(t^\eta)\Psi_L(t^\eta)]^{(\varepsilon)}[\Psi_R^\dagger(t_1^{\eta_1})\Psi_L(t_1^{\eta_1})]^{(\varepsilon_1)}\}\rangle. \tag{10.36}$$

The correlator is different from zero only when $\varepsilon_1 = -\varepsilon$. This amounts to saying that quasiparticles are conserved in the tunneling process. The sum over ε gives, after inserting the chiral bosonic field $\phi_{R(L)}$:

$$\langle I_B(t)\rangle = \frac{e^*\Gamma_0^2 M_R^2 M_L^2}{8\pi^2 a^2}\sum_{\eta\eta_1}\eta_1 \int_{-\infty}^{+\infty} dt_1$$
$$\times \left(e^{i\omega_0(t-t_1)}\langle \hat{T}_K\{e^{-i\sqrt{\nu}\phi_R(t^\eta)}e^{i\sqrt{\nu}\phi_L(t^\eta)}e^{-i\sqrt{\nu}\phi_L(t_1^{\eta_1})}e^{i\sqrt{\nu}\phi_R(t_1^{\eta_1})}\}\rangle\right.$$
$$\left. - e^{-i\omega_0(t-t_1)}\langle \hat{T}_K\{e^{-i\sqrt{\nu}\phi_L(t^\eta)}e^{i\sqrt{\nu}\phi_R(t^\eta)}e^{-i\sqrt{\nu}\phi_R(t_1^{\eta_1})}e^{i\sqrt{\nu}\phi_L(t_1^{\eta_1})}\}\rangle\right). \tag{10.37}$$

We use $M_{R(L)}^2 = 1$, and introduce the chiral Green's function of the bosonic field $\tilde{G}^{\eta\eta'}(t-t') = \langle \hat{T}_K\{\phi_{R(L)}(t^\eta)\phi_{R(L)}(t'^{\eta'})\}\rangle - \frac{1}{2}\langle \hat{T}_K\{\phi_{R(L)}(t^\eta)^2\}\rangle - \frac{1}{2}\langle \hat{T}_K\{\phi_{R(L)}(t'^{\eta'})^2\}\rangle$ which does not depend on the chirality $R(L)$. We obtain the expression for the backscattering current:

$$\langle I_B(t)\rangle = \frac{ie^*\Gamma_0^2}{4\pi^2 a^2}\sum_{\eta\eta_1}\eta_1 \int_{-\infty}^{+\infty} dt_1 \sin(\omega_0(t-t_1))e^{2\nu\tilde{G}^{\eta\eta_1}(t-t_1)}. \tag{10.38}$$

Because the Green's function $\tilde{G}^{\eta\eta}$ is an even function (see Eq. (10.34)), the contributions $\eta = \eta_1$ vanish. We perform the change of variables: $\tau = t - t_1$ with $d\tau = -dt_1$, then:

$$\langle I_B(t)\rangle = -\frac{ie^*\Gamma_0^2}{4\pi^2 a^2}\sum_\eta \eta \int_{-\infty}^{+\infty} d\tau \sin(\omega_0\tau)e^{2\nu\tilde{G}^{\eta-\eta}(\tau)}. \tag{10.39}$$

At zero temperature, the off-diagonal Keldysh Green's function is $\tilde{G}^{\eta-\eta}(\tau) = -\ln(1 - i\eta v_F \tau/a)$. Thus, we have:

$$\langle I_B(t) \rangle = -\frac{ie^* \Gamma_0^2}{4\pi^2 a^2} \sum_\eta \eta \int_{-\infty}^{+\infty} d\tau \frac{\sin(\omega_0 \tau)}{(1 - i\eta\tau v_F/a)^{2\nu}}. \tag{10.40}$$

Performing the integration, we obtain the final result:

$$\langle I_B(t) \rangle = \frac{e^* \Gamma_0^2}{2\pi a^2 \Gamma(2\nu)} \left(\frac{a}{v_F} \right)^{2\nu} \text{sgn}(\omega_0) |\omega_0|^{2\nu-1}, \tag{10.41}$$

where Γ is the gamma function.

On the other hand, at finite temperatures, the Green's function is given by:

$$\tilde{G}^{\eta-\eta}(\tau) = -\ln \left(\sinh \left(\frac{\pi}{\beta} (\eta\tau + i\tau_0) \right) / \sinh \left(\frac{i\pi\tau_0}{\beta} \right) \right), \tag{10.42}$$

where $\tau_0 = a/v_F$. The average current is then given by the integral:

$$\langle I_B(t) \rangle = -\frac{ie^* \Gamma_0^2}{4\pi^2 a^2} \sum_\eta \eta \int_{-\infty}^{+\infty} d\tau \sin(\omega_0 \tau) \left(\frac{\sinh \left(\frac{i\pi\tau_0}{\beta} \right)}{\sinh \left(\frac{\pi}{\beta} (\eta\tau + i\tau_0) \right)} \right)^{2\nu}. \tag{10.43}$$

The change of variables $t = -\tau - i\eta\tau_0 + i\eta\beta/2$ with $dt = -d\tau$ is operated. The time integral now runs in the complex plane from $-\infty - i\eta\tau_0 + i\eta\beta/2$ to $+\infty - i\eta\tau_0 + i\eta\beta/2$. We can bring is back to $-\infty$ to $+\infty$ provided that there are no poles in the integrand, encountered when changing the contour. The poles are located at integer values of $i\pi$ and $i\pi/2$: for this reason the presence of the cutoff is crucial. Depending on the sign of η, one is always allowed to deform the contour to the real axis. The integral becomes:

$$\langle I_B(t) \rangle = \frac{e^* \Gamma_0^2}{2\pi^2 a^2} \left(\frac{\pi\tau_0}{\beta} \right)^{2\nu} \sinh \left(\frac{\omega_0 \beta}{2} \right) \int_{-\infty}^{+\infty} dt \frac{\cos(\omega_0 t)}{\cosh^{2\nu} \left(\frac{\pi t}{\beta} \right)}. \tag{10.44}$$

The integral can be computed analytically:

$$\langle I_B(t) \rangle = \frac{e^* \Gamma_0^2}{2\pi^2 a^2 \Gamma(2\nu)} \left(\frac{a}{v_F} \right)^{2\nu} \left(\frac{2\pi}{\beta} \right)^{2\nu-1}$$
$$\times \sinh \left(\frac{\omega_0 \beta}{2} \right) \left| \Gamma \left(\nu + i \frac{\omega_0 \beta}{2\pi} \right) \right|^2. \tag{10.45}$$

10.5. Poissonian noise in the quantum Hall effect

Using the symmetric combination of the noise correlators (we are interested in zero frequency noise):

$$
\begin{aligned}
S(t, t') &= \langle I_B(t) I_B(t') \rangle + \langle I_B(t') I_B(t) \rangle - 2 \langle I_B(t) \rangle \langle I_B(t') \rangle \\
&= \sum_\eta \langle \hat{T}_K \{ I_B(t^\eta) I_B(t'^{-\eta}) e^{-i \int_K dt_1 H_B(t_1)} \} \rangle - 2 \langle I_B \rangle^2,
\end{aligned}
$$

(10.46)

to lowest order in the tunnel amplitude Γ_0, it is not even necessary to expand the Keldysh evolution operator because the current itself contains Γ_0.

$$
\begin{aligned}
S(t, t') &= -e^{*2} \Gamma_0^2 \sum_{\eta \varepsilon \varepsilon'} \varepsilon \varepsilon' e^{i \varepsilon \omega_0 t} e^{i \varepsilon' \omega_0 t'} \\
&\quad \times \langle \hat{T}_K \{ [\Psi_R^\dagger(t^\eta) \Psi_L(t^\eta)]^{(\varepsilon)} [\Psi_R^\dagger(t'^{-\eta}) \Psi_L(t'^{-\eta})]^{(\varepsilon')} \} \rangle.
\end{aligned}
$$

(10.47)

The correlator is different from zero only when $\varepsilon' = -\varepsilon$. Such correlators have already been calculated for the current:

$$
S(t, t') = \frac{e^{*2} \Gamma_0^2}{2\pi^2 a^2} \sum_\eta \cos(\omega_0(t - t')) e^{2\nu G^{\eta - \eta}(t - t')} = S(t - t').
$$

(10.48)

From this expression, the Fourier transform at zero frequency is computed, first at zero temperature:

$$
\begin{aligned}
S(\omega = 0) &= \frac{e^{*2} \Gamma_0^2}{2\pi^2 a^2} \sum_\eta \int_{-\infty}^{+\infty} dt \, \cos(\omega_0 t) e^{2\nu G^{\eta - \eta}(t)} \\
&= \frac{e^{*2} \Gamma_0^2}{2\pi a^2 \Gamma(2\nu)} \left(\frac{a}{v_F} \right)^{2\nu} |\omega_0|^{2\nu - 1}.
\end{aligned}
$$

(10.49)

The Schottky relation applies, with a fractional charge $e^* = \nu e$:

$$
S(\omega = 0) = 2e^* |\langle I_B(t) \rangle|.
$$

(10.50)

At finite temperature, the noise is given by the integral:

$$
S(\omega = 0) = \frac{e^{*2} \Gamma_0^2}{2\pi^2 a^2} \sum_\eta \int_{-\infty}^{+\infty} d\tau \, \cos(\omega_0 \tau) \left(\frac{\sinh\left(\frac{i\pi \tau_0}{\beta} \right)}{\sinh\left(\frac{\pi}{\beta}(\eta \tau + i\tau_0) \right)} \right)^{2\nu}.
$$

(10.51)

Performing the same change of variables as for the current, this leads to:

$$S(\omega = 0) = \frac{e^{*2}\Gamma_0^2}{\pi^2 a^2} \left(\frac{\pi\tau_0}{\beta}\right)^{2\nu} \cosh\left(\frac{\omega_0\beta}{2}\right) \int_{-\infty}^{+\infty} dt\, \frac{\cos(\omega_0 t)}{\cosh^{2\nu}\left(\frac{\pi t}{\beta}\right)}. \quad (10.52)$$

Performing the integral:

$$S(\omega = 0) = \frac{e^{*2}\Gamma_0^2}{\pi^2 a^2 \Gamma(2\nu)} \left(\frac{a}{v_F}\right)^{2\nu} \left(\frac{2\pi}{\beta}\right)^{2\nu-1}$$

$$\times \cosh\left(\frac{\omega_0\beta}{2}\right) \left|\Gamma\left(\nu + i\frac{\omega_0\beta}{2\pi}\right)\right|^2. \quad (10.53)$$

The shot/thermal noise crossover is recovered in the tunneling limit:

$$S(\omega = 0) = 2e^* |\langle I_B \rangle| \coth(\omega_0\beta/2). \quad (10.54)$$

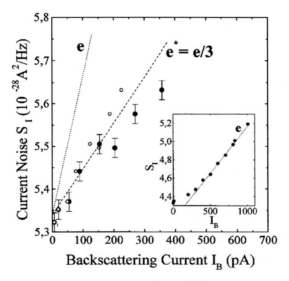

Fig. 25. Tunneling current noise at $\nu = 1/3$ versus backscattering current I_B (filled circles), from Ref. [116]. Data is taken at temperature $k_B\Theta = 25mK$, and the gate voltage is adjusted so as to make a weak constriction. Open circles take into account an empirical $(1 - T)$ reduction factor. Inset: same data taken in the integer quantum Hall regime, including the reduction factor.

The above theoretical predictions have been verified in remarkable point contact experiments at filling factor 1/3 in Saclay and at the Weizmann institute. Ref.

[116] was performed at low temperatures in the shot noise dominated regime, while Ref. [117] used a fit to the thermal-shot noise crossover curve to identify the fractional charge. The data of Ref. [116] is displayed in Fig. 25, and shows consistent agreement with the charge $1/3$ for a weak constriction. Subsequently, the Heiblum group also measured the fractional charge $e^* = e/5$ at filling factor $\nu = 2/5$ [118].

Experiments have also been performed in the strong backscattering regime, that is when the point contact splits the quantum Hall bar into two separate Luttinger liquids. The present calculation can be adapted to treat the tunneling of electrons instead of that of quasiparticles. This is achieved using the duality transformation $\nu \to 1/\nu$ and by replacing the anomalous charge $e^* = \nu e$ by the bare electron charge in the tunneling amplitude. On the experimental side, early reports suggested that the entities which tunnel are bare electrons, because they tunnel in a medium (vacuum) where Luttinger liquid collective excitations are absent.

There is now evidence that the noise deviates from the Poissonian noise of electrons. The noise at sufficiently low temperatures has been found to be super-poissonian [119], with an effective charge $2e$ or $4e$ suggesting that electrons tunnel in bunches. The data is displayed in Fig. 26. There is no clear theoretical explanation of this phenomenon so far.

On the theoretical side, an exact solution for both the current and the noise was found using the Bethe Ansatz solution of the boundary Sine-Gordon model [120]. It bridges the gap between the weak and the strong backscattering regimes. This work has also been extended to finite temperature [121], and careful comparison between theory and experiment has been motivated recently [122]. Noise at finite frequency has been computed in chiral Luttinger liquids using both perturbative techniques and using the exact solution at $\nu = 1/2$ [113]. The noise displays a singularity at the "Josephson" frequency $e^* V / \hbar$.

A Hanbury-Brown and Twiss proposal has been made to detect the statistics of the edge state quasiparticle in the quantum Hall effect [123]. Indeed, the quasiparticle fields obey fractional statistics, and a noise correlation measurement necessarily provides information on statistics. The geometry consists of three edge states (one injector and two detectors) which can exchange quasiparticles by tunneling through the fractional Hall fluid. To leading order in the tunnel amplitudes, one finds the zero-frequency noise correlations:

$$\tilde{S}_{12}(0) = (e^{*2}|\omega_0|/\pi) T_1^r T_2^r R(\nu), \tag{10.55}$$

where the renormalized transmission probabilities:

$$T_l^r = (\tau_0|\omega_0|)^{2\nu-2} [\tau_0 \Gamma_l / \hbar a]^2 / \Gamma(2\nu), \tag{10.56}$$

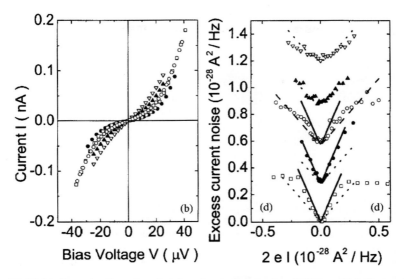

Fig. 26. Right: Current voltage characteristics of a pinched quantum Hall bar, which displays the zero bias anomaly associated with electron tunneling, from Ref. [119]. Left: shot noise versus current at high (top) and low temperature (bottom, $k_B\Theta < 32mK$). The low temperature data fits a super-oissonian noise formula $S = 2(2e)I$.

correspond to that of a non interacting system at $\nu = 1$. Except for some universal constants, the function $R(\nu)$ can be measured experimentally by dividing the noise correlations by the product of the two tunneling currents. By comparing the noise correlations with those of a non-interacting system ($R(1) = -1$), one finds that the noise correlations remain negative at $\nu = 1/3$ although their absolute value is substantially reduced: there is less anti-bunching is a correlated electron HBT experiment. For lower filling factors, the noise correlations are found to be positive, but this remains a puzzle because perturbation theory is less controlled for $\nu < 1/3$.

In the last few years, experiments studied the effect of depleting the edge state incoming on a point contact [124]. These experiments use a three edge state geometry as in Ref. [123]. Another point contact is in the path of this edge state in order to achieve dilution. If the former point contact is tuned as an opaque barrier, one normally expects poissonian noise from electrons. But because the edge state is dilute, quasiparticles seem to bunch together when tunneling. A noise diagnosis reveals that the effective charge predicted by Schottky's relation is in reality lower than the electron charge. These findings do not seem to find an appropriate theoretical explanation at this time [125].

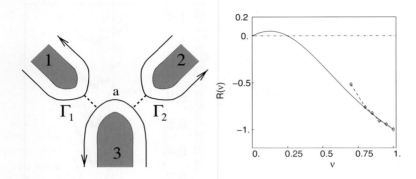

Fig. 27. Left: Hanbury-Brown and Twiss geometry with three edge states; quasiparticles are injected from 3 into 1 and 2 with tunnel hoppings Γ_1 and Γ_2. Right: normalized noise correlations as a function of filling factor; for comparison, the non-interacting value is $R(1) = -1$.

10.6. Effective charges in quantum wires

As mentioned above, the chiral Luttinger liquids of the FQHE are an excellent test-bed for probing the role of interactions in noise. In a one dimensional quantum wire [105], the Luttinger liquid is non-chiral: interaction between right and left going electrons are effective over the whole length of the wire: the notion of a backscattering current is ambiguous. Nevertheless, non–chiral Luttinger liquids also have underlying chiral fields [126]. Such chiral fields correspond to excitations with anomalous (non-integer) charge, which have eluded detection so far. Here we briefly mention how shot noise measurements can provide information on such anomalous charges.

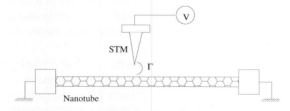

Fig. 28. Schematic configuration of the nanotube–STM device: electrons are injected from the tip at $x = 0$: current is measured at both nanotube ends, which are set to the ground.

Carbon nanotubes, given the appropriate helicity, can have a metallic behavior. Due to their one-dimensional character, they are good candidates to probe

Luttinger liquid behavior. Fig. 28 depicts a carbon nanotube with both ends grounded, but electrons are injected in the bulk of the tube with an STM tip. When electrons tunnel on the nanotube, they are not welcome because the excitations of this nanotube do not resemble electrons. This has been illustrated in tunneling density of states experiments [127]. The transport properties at the tunneling junction and in the nanotube can be computed using a Luttinger model for the nanotube, together with a perturbative treatment of the junction in the Keldysh formalism [128]. The tunneling electrons give rise to right and left moving quasiparticle excitations which carry charge $Q_+ = (1 + K_{c+})/2$ and $Q_- = (1 - K_{c+})/2$, where $K_{c+} < 1$ is the Luttinger liquid interaction parameter (in the absence of interactions or, equivalently, when the interactions in the nanotube are fully screened by a substrate, $K_{c+} = 1$). If the tunneling is purely local (say at $x = 0$), there is as much chance that Q_+ will propagate to the right while Q_- propagates to the left than the opposite. The state of the outgoing quasiparticle excitations is entangled between these two configurations.

How can one identify the anomalous charges of quasiparticle excitations? By performing a Hanbury-Brown and Twiss analysis of transport. Because bare electrons tunnel from the STM tip to the nanotube, the Schottky relation with charge e holds for the tunneling current and the tunneling noise:

$$S_T = 2e\langle I_T \rangle. \tag{10.57}$$

However, non-integer charges are found when calculating the autocorrelation noise $S_\rho(x, x, \omega = 0)$ on one side (say, $x > 0$) of the nanotube together with the cross correlations $S_\rho(x, -x, \omega = 0)$ between the two sides of the nanotube:

$$S(x, x, \omega = 0) = [1 + (K_{c+})^2]e \, |\langle I_\rho(x)\rangle| \sim (Q_+^2 + Q_-^2), \tag{10.58}$$

$$S(x, -x, \omega = 0) = -[1 - (K_{c+})^2]e \, |\langle I_\rho(x)\rangle| \sim Q_+ Q_-. \tag{10.59}$$

where $\langle I_\rho(x)\rangle$ is the average charge current at location x in the nanotube. Note that the cross correlations of Eq. (10.59) are negative because we have chosen a different convention from Sec. 6: there the noise is measured away from the junction between the three probes. We conclude that the noise correlations are positive, which makes sense because excitations are propagating in both directions away from the junction.

The above considerations apply for an infinite nanotube, without a description of the contacts connected to the nanotube. It is known that the presence of such contacts (modeled by an inhomogeneous Luttinger liquid with $K_{c+} = 1$ in the contacts) leads to an absence of the renormalization of the transport properties [129] at zero frequency. Because of multiple reflections of the quasiparticles at the interface, the zero frequency noise cross correlations vanish at $\omega = 0$ due to

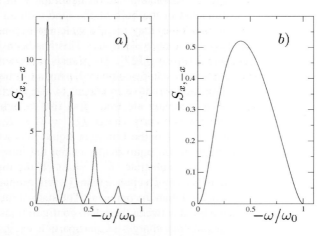

Fig. 29. Finite frequency noise correlations for a nanotube connected to leads. a) case where $\tau_L = 14\tau_V$; b) case where $\tau_L = \tau_V$.

the presence of leads. In order to retrieve information about anomalous charges, it is thus necessary to compute the finite frequency cross-correlations [130].

The result is depicted in Fig. 29. There are two competing time scales: $\tau_L = L/2v_{c+}$, the traveling time needed for Luttinger liquid excitations to reach the leads, and $\tau_V = \hbar/eV$, the time spread of the electron wave packet when the bias voltage between the tip and the nanotube is V. When $\tau_L \gg \tau_V$ (Fig. 29a), quasiparticle excitations undergo multiple bounces on the nanotube/contact interfaces, giving rise to multiple peaks in the autocorrelation noise. In the opposite case, the width of the tunneling electron wave packet is so large that there is only one maximum (Fig. 29b). To identify anomalous charges, one should specify τ_V, and the frequency should be tuned so that $\omega\tau_L$ is an odd multiple of $\pi/2$ (condition for a maximum). Then, one can measure experimentally the ratio $|S_{x,-x}/S_{x,x}| = (1 - K_{c+}^4)/(1 + K_{c+}^4)$ in order to extract the chiral charges $Q_\pm = (1 \pm K_{c+})/2$. Note that alternative finite frequency noise proposals to measure the anomalous charges in a Luttinger liquid with contacts have appeared recently in the literature [131].

11. Conclusions

The physics of noise in nanostructures is now one of the many exciting fields of mesoscopic physics. A motivation for this is the fact that compared to a current measurement, additional information can be extracted from the noise. Noise has been computed and measured experimentally in a variety of systems which are not all described in this review. For additional approaches for calculating the noise, readers are directed towards articles and reviews. The few examples of noise calculations presented in this course apply to situations where noise can either be characterized via scattering theory or, for the case of a Luttinger liquid, via perturbation theory in the tunneling Hamiltonian. Although this is a rather restrictive framework, these two situations can apply to a variety of nanodevices, possibly containing hybrid ferromagnetic or superconducting components.

An example is the Keldysh calculation of noise in junctions between two superconductors, a geometry which allows the multiple Andreev reflection (MAR) process [78]. It has been computed in various regimes, including point contacts [132, 133], and junctions with diffusive metals [134, 135]. In such situations, the noise acts as a diagnosis which allows to pinpoint the total charge which is transfered through the junction. This charge displays plateaus as a function of applied bias, as the latter fixes the number of reflections between the superconductors. Shot noise experiments were first performed using SIS junctions, where the enhancement of shot noise due to MAR was attributed to the presence of pinholes in the insulating junction [136]. More recently, noise experiments were performed using a more controlled geometry, using either superconducting point contacts for the ballistic regime [33] or in the diffusive regime. In the latter case, the regime of incoherent MAR was first investigated [137, 138]. In a recent work, it was shown that both the incoherent and the coherent MAR regime could be probed in disordered junctions with the same experiment by tuning the bias voltage above and below the Thouless energy [139].

There exists a connection between noise and the dynamical Coulomb blockade, which has been established using the Keldysh formalism [140]. It is well understood [58] that when an electron tunnels through a single junction, the electromagnetic environment surrounding the junction has to be reorganized. If the tunnel junction is in series with a large impedance, this causes a zero bias anomaly in the current voltage characteristics. For highly transparent barriers, this zero bias anomaly vanishes exactly in the same way as the $1 - T$ suppression of shot noise, and the connection between dynamical Coulomb blockade can be understood within a perturbation theory resummation scheme.

Another generalization includes the discussion of noise in non-stationary situations, for instance when an AC bias is superposed to the DC voltage bias: photo-assisted shot noise was computed for a two terminal scatterer [141] and measured experimentally in diffusive metals [61], in normal metal superconducting junctions [80, 71], as well as in point contacts [142]. Photo-assisted shot noise provides an alternative way for measuring the effective charge of carriers. It has been recently computed in the context of the FQHE [143]. To some extent, a measurement of photo-assisted shot noise performed at zero frequency can play the same role as a measurement of high frequency noise: the AC modulation imposed in the leads plays the role of a probing frequency.

Noise measurements are typically hard measurements because one is dealing with a very small current or a small voltage signal, which needs to be "squared", and it is difficult to isolate the wanted noise from the unwanted one. The ratio of theory to experimental noise publications is still a bit too large. Yet experimental detection is making fast progress. Conventional noise apparatus, which convert a quantum measurement to a classical signal using for instance cold amplifiers continues to be improved. On the other hand, new measurement techniques use a noise detector which is part of the same chip as the device to be measured [57]. In such situations, it will be necessary to analyze what is the effect of the back-action of the measuring device on the circuit to be measured.

On the theoretical side, while the interest in computing noise remains at a high level, there is an ongoing effort to study the higher moments of the current, and the generating function of all irreducible moments. This sub-field bears the name of full counting statistics and was pioneered in the context of scattering theory in Refs. [144], but it is now generalized to a variety of systems. We refer the reader to the final chapters of Ref. [145], see also Ref. reuletlect. Developments in this field include superconductor-hybrid systems as well as Luttinger liquids [147]. A nice seminar was presented during the les Houches school on this topic by A. Braggio, who applies full counting statistics to transport in quantum dots.

As discussed in the section on entanglement, more and more analogies can be found between nano-electronic and quantum optics, because both fields exploit the measurement of two-particle (or more) correlations. Bell inequality test allow to convince oneself that entanglement is at work in nano-devices. This has motivated several efforts to exploit this entanglement in a teleportation scenario. The entangler [91] can be used both as a generator of singlet pairs as well as a detector of such pairs, and Ref. [148] describes a cell which teleports the state of an electron spin in a quantum dot to another electron in another quantum dot. The electron-hole entanglement scenario of Ref. [100] also gave rise to a teleportation proposal [149]. Interestingly, in order to control the output of such quantum information proposals, it is necessary to analyze many-particle correlations – or generalized noise – at the input and at the output of such devices.

Finally, I would like to emphasize that this course is the result of an ongoing effort over the years, and I wish to thank all my collaborators on noise since the early 1990's. Foremost, I should mention the role played by Rolf Landauer, who introduced me to noise. I am very much indebted to Gordey Lesovik, for his input and collaborations. Next I would like to thank my close associates in Marseille: Julien Torres who started his thesis working on NS junctions; Ines Safi, for her passage here working on the FQHE; my present collaborators Adeline Crepieux who provided latex notes on Luttinger liquids and who kindly read the manuscript; Nikolai Chtchelkatchev, for his contribution on Bell inequalities; and Marjorie Creux for her nice Masters thesis on edge states physics.

References

[1] P. Dutta and P. M. Horn, Rev. Mod. Phys. **53**, 497 (1981); M. B. Weissman, Rev. Mod. Phys. **60**, 537 (1988).

[2] S. Feng, P. A. Lee and A. D. Stone, Phys. Rev. Lett. **56**, 1960 (1986).

[3] S. Datta, in *Electronic Transport in Mesoscopic Systems* (Cambridge University Press 1995), Y. Imry, *Introduction to Mesoscopic physics* (Oxford University Press, Oxford, 1997).

[4] R. Landauer, Phys. Rev. B **47**, 16427 (1993).

[5] J. B. Johnson, Phys. Rev. B **29**, 367 (1927); H. Nyquist, Phys. Rev. **32**, 110 (1928).

[6] W. Schottky, Ann. Phys. (Leipzig) **57**, 541 (1918).

[7] J. R. Pierce, Bell Syst. Tech. J. **27**, 15 (1948).

[8] B. J. van Wees *et al.*, Phys. Rev. Lett. **60**, 848 (1988).

[9] E.A. Montie *et al.*, Nature (London) **350**, 594 (1991); Physica B **175**, 149 (1991).

[10] J.C. Licini *et al.*, Phys. Rev. Lett. **55**, 2987 (1985); W.J. Skocpol *et al.*, Phys. Rev. Lett. **56**, 2865 (1986).

[11] S. Washburn and R. Webb, Adv. Phys. **35**, 375 (1986).

[12] M.P. van Albada, J.F. de Boer and A. Lagendijk, Phys. Rev. Lett. **64**, 2787 (1990); J.F. de Boer, M.P. van Albada and A. Lagendijk, Physica B **175**, 17 (1991).

[13] R. Landauer, Z. Phys. B **68**, 217 (1987); M. Büttiker, Y. Imry, R. Landauer and S. Pinhas, Phys. Rev. B **31**, 6207 (1985).

[14] R. Landauer, Physica **D38**, 226 (1987).

[15] R. Landauer and Th. Martin, Physica B **175**, 167 (1991); **182**, 288 (1992).

[16] Th. Martin and R. Landauer, Phys. Rev. B **45**, 1742 (1992).

[17] G. B. Lesovik, JETP Lett. **49**, 594 (1989).

[18] B. Yurke and G. P. Kochlanski, Phys. Rev. B **41**, 8141 (1990).

[19] M. Büttiker, Phys. Rev. Lett. **65**, 2901 (1990); Phys. Rev. B **46**, 12485 (1992).

[20] L. V. Keldysh, Sov. Phys. JETP **20**, 1018 (1965).

[21] C. Caroli, R. Combescot, P. Nozières, and D. Saint-James, J. Phys. C **4**, 916 (1971).

[22] J. Rammer and H. Smith, Rev. Mod. Phys. **58**, 323-359 (1986).

[23] G. D. Mahan, *Many Particle Physics* (Plenum Press, 1990).

[24] Y. Levinson, Europhys. Lett. **39**, 299 (1997).

[25] R. Hanbury Brown and R. Q. Twiss, Nature (London) **177**, 27 (1956); R. Hanbury Brown and R. Q. Twiss, Proc. Royal. Soc. London Ser. A **242**, 300 (1957); *ibid.* **243**, 291 (1957).

[26] Sh. M. Kogan, *Electronic Noise and Fluctuations in Solids* (Cambridge University Press, 1996).

[27] Ya. M. Blanter and M. Büttiker, Phys. Rep. **336**, 1 (2000).

[28] V. A. Khlus, Sov. Phys. JETP **66** 1243 (1987) (Zh. EKsp. Teor. Fiz. **93**, 2179 (1987)).

[29] M. Büttiker and R. Landauer, Phys. Rev. Lett. **49**, 1739 (1982).

[30] I.K. Yanson, A.I. Akimenko and A.B. Verki, Solid State Commun. **43**, 765 (1982); A.I. Akimenko, A.V. Verkin and I.K. Yanson, J. Low Temp. phys. **54** , 247 (1984).

[31] Y. P. Li, D. C. Tsui, J. J. Heremans, J. A. Simmons and G. W. Weimann, Appl. Phys. Lett. **57**, 774 (1990); Y. P. Li, A. Zaslavsky, D. C. Tsui, M. Santos and M. Shayegan, Phys. Rev. B **41**, 8388 (1990); *Resonant Tunneling in Semiconductors: Physics and Applications*, edited by L. L. Chang, E. E. Mendez and C. Tejedor (Plenum, New York 1991).

[32] E. Scheer, P. Joyez, D. Esteve, C. Urbina and M. H. Devoret, Phys. Rev. Lett. **78**, 3535 (1999).

[33] R. Cron, M. Goffman, D. Esteve and C. Urbina, Phys. Rev. Lett. **86**, 115 (2001).

[34] M. Büttiker, Phys. Rev. Lett. **57**, 1761 (1986); M. Büttiker, Phys. Rev. B **38**, 9375 (1988).

[35] M. Reznikov, M. Heiblum, H. Shtrikman and D. Mahalu, Phys. Rev. Lett. **75**, 3340 (1995).

[36] A. Kumar, L. Saminadayar, D. C. Glattli, Y. Jin and B. Etienne, Phys. Rev. Lett. **76**, 2778 (1996).

[37] H. E. van den Brom and J. M. van Ruitenbeek, Phys. Rev. Lett. **82**, 1526 (1999).

[38] H. Birk, M. J. M. de Jong and C. Schönenberger, Phys. Rev. Lett. **75**, 1610 (1995).

[39] A. N. Korotkov, Phys. Rev. B **49**, 10381-10392 (1994).

[40] *Mesoscopic quantum physics*, E. Akkermans, G. Montambaux, J.-L. Pichard and J. Zinn-Justin Editors, Les Houches session LXI, Nato Advanced Study Institute (North Holland 1994).

[41] C. W. J. Beenakker, Rev. Mod. Phys **69**, 731 (1997).

[42] C. W. J. Beenakker and M. Büttiker, Phys. Rev. B **46**, 1889 (1992).

[43] K. E. Nagaev, Phys. Lett. A **169**, 103 (1992).

[44] M. J. M. de Jong et C. W. J. Beenakker, Phys. Rev. B **51**, 16867 (1995).

[45] M. Henny, S. Oberholzer, C. Strunk and C. Schönenberger, Phys. Rev. B **59**, 2871 (1999).

[46] H. U. Baranger and P. Mello, Phys. Rev. Lett. **73**, 142 (1994).

[47] R. A. Jalabert, J.-L. Pichard and C.W.J. Beenakker, Europhys. Lett. **27**, 255 (1994).

[48] S. Oberholzer, E. V. Shukorukhov, C. Strunk and C. Schönenberger, Phys. Rev. Lett. **86**, 2114 (2001).

[49] P. W. Brouwer and C. W. J. Beenakker, J. Math. Phys. **37**, 4904 (1996).

[50] M.P. Silverman, Phys. Lett. A **120**, 442 (1987).

[51] W. D. Oliver, J. Kim, R. C. Liu and Y. Yamamoto, Science **284**, 299 (1999).

[52] M. Henny, S. Oberholzer, C. Strunk, T. Heinzel, K. Ensslin, M. Holland and C. Schönenberger, Science **284**, 296 (1999).

[53] H. Kiesel, A. Renz and F. Hasselbach, Nature (London) **218**, 393 (2002).

[54] L.D. Landau and E. Lifshitz, Statistical Physics (Pergamon).

[55] G. B. Lesovik and R. Loosen, Pis'ma Zh. Eksp. Theor. Fiz. **65**, 280 (1997) [JETP Lett. **65**, 295 (1997)].

[56] Y. Gavish, I. Imry and Y. Levinson, Phys. Rev. B **62**, 10637 (2000).

[57] R. Aguado and L. Kouvenhoven, Phys. Rev. Lett. **84**, 1986 (2000).

[58] in *Single charge tunneling*, M. Devoret and H. Grabert editors, Nato Advanced Study Institute (Kluwer 1991).

[59] R. Deblock, E. Onac, L. Gurevich and L. Kouvenhoven, Science **301**, 203 (2003).

[60] S.R. Eric Yang, Solid State Commun. **81**, 375 (1992).

[61] R. J. Schoelkopf, A. A. Kozhevnikov, D. E. Prober, and M. J. Rooks, Phys. Rev. Lett. **80**, 2437 (1998).

[62] A. F. Andreev, Sov. Phys. JETP **19**, 1228 (1964) (Zh. Eksp. Teor. Fiz. **46**, 1823 (1964)).

[63] F. Hekking, *Electron subgap transport in hybrid systems combining superconductors with normal or ferromagnetic metals*, this Volume.

[64] N. N. Bogolubov, V. V. Tolmachev and D. V. Shirkov, *A New Method in the Theory of Super-conductivity*, Consultant Bureau, New York, 1959.

[65] P.-G. De Gennes, *Superconductivity of Metals and Alloys* (Addison-Wesley, 1989).

[66] C. W. J. Beenakker, Phys. Rev. B **46**, 12841 (1992).

[67] C. W. J. Beenakker, in *Mesoscopic Quantum Physics*, E. Akkermans, G. Montambaux, J.-L. Pichard and J. Zinn-Justin editors., p. 279 (Les Houches LXI, North-Holland 1995).

[68] T. Martin, Phys. Lett. A **220**, 137 (1996).

[69] M. P. Anantram and S. Datta, Phys. Rev. B **53**, 16 390 (1996).

[70] S. Datta, P. Bagwell and M. P. Anantram, Phys. Low-Dim. Struct. **3**, 1 (1996).

[71] G. B. Lesovik, T. Martin and J. Torrès, Phys. Rev. B **60**, 11935 (1999).

[72] J. Torrès, T. Martin and G. B. Lesovik, Phys. Rev. B **63**, 134517 (2001).

[73] B. D. Josephson, Phys. Rev. Lett. **1**, 251 (1962).

[74] A. I. Larkin and Yu. N. Ovchinnikov, Sov. Phys. JETP **26**, 1219 (1968).

[75] M. J. M. de Jong and C. W. J. Beenakker, Phys. Rev. B **49**, 16070 (1994).

[76] B. A. Muzykantskii and D. E. Khmelnitskii, Phys. Rev. B **50**, 3982 (1994).

[77] X. Jehl, P. Payet-Burin, C. Baraduc, R. Calemczuk and M. Sanquer, Phys. Rev. Lett. **83**, 1660 (1999).

[78] T. M. Klapwijk, G.E. Blonder and M. Tinkham, Physica B **109-110**, 1657 (1982).

[79] X. Jehl, M. Sanquer, R. Calemczuk and D. Mailly, Nature (London) **405**, 50 (2000).

[80] A. A. Kozhevnikov, R. J. Schoelkopf and D. E. Prober, Phys. Rev. Lett. **84**, 3398 (2000).

[81] F. Lefloch, C. Hoffmann, M. Sanquer and D. Quirion Phys. Rev. Lett. **90**, 067002 (2003).

[82] G. E. Blonder, M. Tinkham, and T. M. Klapwijk, Phys. Rev. B **25**, 4515 (1982).

[83] Y. Gefen, Y. Imry and M. Ya. Azbel, Phys. Rev. Lett. **52**, 129 (1984).

[84] M. Büttiker, Y. Imry and M. Ya. Azbel, Phys. Rev. A, **30**, 1982 (1984).

[85] J. Torrès and T. Martin, Eur. Phys. J. B **12**, 319 (1999).

[86] B. Pannetier and H. Courtois, J. of Low Temp. Phys. **118**, 599 (2000).

[87] G. Falci, D. Feinberg and F. W. J. Hekking, Europhys. Lett. **54**, 255 (2001).

[88] J. Boerlin, W. Belzig and C. Bruder, Phys. Rev. Lett. **88**, 197001 (2002).

[89] P. Samuelson and M. Büttiker, Phys. Rev. Lett. **89**, 046601 (2002).

[90] G.B. Lesovik, T. Martin and G. Blatter, Eur. Phys. J. B **24**, 287 (2001).

[91] P. Recher, E.V. Sukhorukov and D. Loss, Phys. Rev. B **63**, 165314 (2001).

[92] A. Aspect, J. Dalibard and G. Roger, Phys. Rev. Lett. **49**, 1804 (1982); A. Aspect, Nature (London) **398**, 189 (1999).

[93] N. Chtchelkatchev, G. Blatter, G. Lesovik and T. Martin, Phys. Rev. B **66**, 161320 (2002).

[94] G. Burkard, D. Loss and E. V. Sukhorukov Phys. Rev. B **61**, 16303 (2000).

[95] S. Kawabata, J. Phys. Soc. Jpn. **70**, 1210 (2001).

[96] J.S. Bell, Physics (Long Island City, N.Y.) **1**, 195 (1965); J.S. Bell, Rev. Mod. Phys. **38**, 447 (1966).

[97] A. V. Lebedev, G. Blatter, C. W. J. Beenakker and G. B. Lesovik, Phys. Rev. B **69**, 235312 (2004); A. V. Lebedev, G. Blatter and G. B. Lesovik, cond-mat/0311423; G. B. Lesovik, A. V. Lebedev, and G. Blatter, cond-mat/0310020.

[98] W.D. Oliver, F. Yamaguchi and Y. Yamamoto, Phys. Rev. Lett. **88**, 037901 (2002).

[99] D.S. Saraga and D. Loss, Phys. Rev. Lett. **90**, 166803-1 (2003).

[100] C.W.J. Beenakker, C. Emary, M. Kindermann and J.L. van Velsen, Phys. Rev. Lett. **91**, 147901 (2003).

[101] P. Samuelsson, E.V. Sukhorukov and M. Buttiker Phys. Rev. Lett. 91, 157002 (2003).

[102] P. Samuelsson, E.V. Sukhorukov and M. Buttiker, Phys. Rev. Lett. 92, 026805 (2004).

[103] D. Maslov, *Fundamental aspects of electron correlations and quantum transport in one-dimensional systems*, this Volume.

[104] F. D. M. Haldane, J. Phys. C **14**, 2585 (1981).

[105] H. J. Schulz, in *Physique Quantique Mésoscopique*, édité par E. Akkermans, G. Montanbaux, J. L. Pichard and J. Zinn-Justin, Les Houches, 1994, Session LXI, (North Holland, Elsevier, 1995).

[106] C. L. Kane and M. P. A. Fisher Phys. Rev. Lett. **68**, 1220 (1992); C. L. Kane and M. P. A. Fisher Phys. Rev. B **46**, 15233 (1992).

[107] K. von Klitzing, G. Dorda and M. Pepper, Phys. Rev. Lett. **45**, 494 (1980).

[108] D. C. Tsui, H. L. Stormer and A. C. Gossard, Phys. Rev. Lett. **48**, 1559 (1982).

[109] R. B. Laughlin, Phys. Rev. Lett. **50**, 1395 (1983).

[110] C. de C. Chamon, D. E. Freed and X.-G. Wen, Phys. Rev. B **51**, 2363 (1995).

[111] X.G. Wen, Phys. Rev. B **43**, 11025 (1991); Phys. Rev. Lett. **64**, 2206 (1990); X.G. Wen, Int. J. Mod. Phys. B **6**, 1711 (1992); Adv. Phys. **44**, 405 (1995).

[112] C.L. Kane and M.P.A. Fisher, Phys. Rev. Lett. **72**, 724 (1994).

[113] C. de C. Chamon, D. E. Freed and X. G. Wen, Phys. Rev. B **53**, 4033 (1996).

[114] R. Guyon, P. Devillard, T. Martin, and I. Safi Phys. Rev. B **65**, 153304 (2002).

[115] A. A. Abrikosov, L. P. Gorkov et I. E. Dzyaloshinski, *Quantum Field Theoretical Methods in Statistical Physics*, (Permagon, 1965).

[116] L. Saminadayar, D. C. Glattli, Y. Jin, and B. Etienne, Phys. Rev. Lett. **79**, 2526 (1997).

[117] R. de-Picciotto, M. Reznikov, M. Heiblum, V. Umansky, G. Bunin, and D. Mahalu, Nature (London) **389**, 162 (1997).

[118] M. Reznikov, R. de-Picciotto, T. G. Griffiths, M. Heiblum, V. Umansky, Nature (London) **399**, 238 (1999).

[119] V. Rodriguez, P. Roche, D. C. Glattli, Y. Jin and B. Etienne, Physica E **12**, 88 (2002).

[120] P. Fendley, A. W. W. Ludwig, and H. Saleur, Phys. Rev. Lett. **75**, 2196 (1995).

[121] P. Fendley and H. Saleur Phys. Rev. B **54**, 10845 (1996).

[122] B. Trauzettel, P. Roche, D. C. Glattli, and H. Saleur Phys. Rev. B **70**, 233301 (2004).

[123] I. Safi, P. Devillard, and T. Martin Phys. Rev. Lett. **86**, 4628 (2001).

[124] E. Comforti, Y.C. Chung, M. Heiblum, V. Umansky, and D. Malahu, Nature **416**, 515 (2002); E. Comforti, Y.C. Chung, M. Heiblum, and V. Umansky, Phys. Rev. Lett. **89**, 066803 (2002) Y.C. Chung, M. Heiblum, and V. Umansky, Phys. Rev. Lett. **91**, 216804 (2003).

[125] C. L. Kane and M. P. A. Fisher Phys. Rev. B **67**, 045307 (2003).

[126] I. Safi, Ann. Phys. Fr. **22**, 463 (1997). K.-V. Pham, M. Gabay, and P. Lederer, Phys. Rev. B **61**, 16397 (2000).

[127] M. Brockrath, D.H. Cobden, J. Lu, A.G. Rinzler, R.E. Smalley, L. Balents, and P. McEuen, Nature (London) **397**, 598 (1999).

[128] A. Crépieux, R. Guyon, P. Devillard, and T. Martin, Phys. Rev. B **67**, 205408 (2003).

[129] I. Safi and H. Schulz, Phys. Rev. B **52**, 17040 (1995); D. Maslov and M. Stone, Phys. Rev. B **52**, 5539 (1995); V.V. Ponomarenko, Phys. Rev. B **52**, 8666 (1995).

[130] A. Lebedev, A. Crépieux, and T. Martin, cond-mat/0405325, Phys. Rev. B (2005).

[131] B. Trauzettel, I. Safi, F. Dolcini, and H. Grabert, Phys. Rev. Lett. **92**, 226405 (2004); F. Docini, B. Trauzettel, I. Safi and H. Grabert, cond-mat/0402479.

[132] D. Averin and H. T. Imam, Phys. Rev. Lett. **76**, 3814 (1996).

[133] J. C. Cuevas, A. Martín-Rodero, and A. L. Yeyati Phys. Rev. Lett. **82**, 4086 (1999).

[134] Y. Naveh and D. V. Averin, Phys. Rev. Lett. **82** 4090 (1990).

[135] E. V. Bezuglyi, E. N. Bratus, V. S. Shumeiko, and G. Wendin Phys. Rev. Lett. 83, 2050 (1999)

[136] P. Dieleman, H. G. Bukkems, T. M. Klapwijk, M. Schicke, and K. H. Gundlach Phys. Rev. Lett. **79**, 3486 (1997).

[137] T. Hoss, C. Strunk, T. Nussbaumer, R. Huber, U. Staufer, C. Schonenberger, Phys. Rev. B **62**, 4079 (2000).

[138] P. Roche, H. Perrin, D. C. Glattli, H. Takayanagi, T. Akazaki, Physica C **352**, 73 (2001).

[139] C. Hoffmann, F. Lefloch, M. Sanquer and B. Pannetier, Phys. Rev. B **70**, 180503 (2004).

[140] A. L. Yeyati, A. Martin-Rodero, D. Esteve, and C. Urbina Phys. Rev. Lett. **87**, 046802 (2001).

[141] G. Lesovik and L.S. Levitov, Phys. Rev. Lett. **72**, 538 (1994).

[142] L.H. Reydellet, P. Roche, D.C. Glattli, B. Etienne, and Y. Jin, Phys. Rev. Lett. **90**, 176803 (2003).

[143] A. Crépieux, P. Devillard and T. Martin, Phys. Rev. B **69**, 205302 (2004).

[144] L. S. Levitov, H. Lee, and G. B. Lesovik, J. Math. Phys. **37**, 4845 (1996).

[145] *Quantum Noise in Mesoscopic Physics*, Nato Science Series Vol. 97, Y. Nazarov, editor (Kluwer 2003).

[146] B. Reulet, *higher moments of noise*, this Volume.

[147] H. Saleur and U. Weiss, Phys. Rev. B 63, 201302 (2001).

[148] O. Sauret, D. Feinberg and T. Martin, Eur. Phys. J. B **32**, 545 (2003); ibid., Phys. Rev. B 69, 035332 (2004).

[149] C. W. J. Beenakker and M. Kindermann Phys. Rev. Lett. **92**, 056801 (2004).

Seminar 2

HIGHER MOMENTS OF NOISE

Bertrand Reulet

Departments of Applied Physics and Physics
Yale University, New Haven CT 06520-8284, USA
Laboratoire de Physique des Solides, UMR8502, bât 510
Université Paris-Sud, 91405 Orsay, France

H. Bouchiat, Y. Gefen, S. Guéron, G. Montambaux and J. Dalibard, eds.
Les Houches, Session LXXXI, 2004
Nanophysics: Coherence and Transport

Contents

1. Introduction

Transport studies provide a powerful tool for investigating electronic properties of a conductor. The $I(V)$ characteristic (or the differential resistance $R_{diff} = dV/dI$) contains partial information on the mechanisms responsible for conduction. A much more complete description of transport in the steady state, and further information on the conduction mechanisms, is given by the probability distribution of the current P, which describes both the dc current $I(V)$ and the fluctuations. Indeed, even with a fixed voltage V applied, $I(t)$ fluctuates, due to the discreteness of the charge carriers, the probabilistic character of scattering and the fluctuations of the population of energy levels at finite temperature T [1].

The current fluctuations are characterized by the moments of the probability distribution P of order two and higher. Experimentally, the average over P is obtained by time averaging. Thus, the average current is the dc current $I = \langle I(t) \rangle$, where $\langle . \rangle$ denotes time average. The second moment (the variance) of P, $\langle i^2 \rangle$, measures the amplitude of the current fluctuations, with $i(t) = I(t) - I$. The third moment $\langle i^3 \rangle$ (the skewness) measures the asymmetry of the fluctuations. Gaussian noise $P(i) \propto \exp(-\alpha i^2)$ is symmetric, so it has no third moment. The existence of the third moment is related to the breaking of time reversal symmetry by the dc current; at zero bias, $I = 0$ and positive and negative current fluctuations are equivalent, so $\langle i^3 \rangle = 0$.

In this article we present simple approaches to the calculation of $P(i)$ in a tunnel junction, and to the effect of the environment on noise measurements in terms of the modification of P. We do not provide rigorous calculations, but simple considerations that bear the essential ingredients of the phenomena. We also discuss the effect of a finite measurement bandwidth. We report experimental results of the first measurement of the third moment of voltage fluctuations in tunnel junctions, from room temperature down to 50mK. Then we discuss extensions of those measurements to finite frequencies and to the study of other systems. We show the first data of the third moment in the regime where the frequency is larger than the temperature. Finally we discuss a new quantity, the "noise thermal impedance", which links the second and third moment.

2. The probability distribution $P(i)$

2.1. A simple model for a tunnel junction

Let us consider a single channel tunnel junction of transmission probability t. For $t \ll 1$, the tunneling events are rare and well separated. Thus one can consider a time τ small enough such that there is at most one event during τ. The transport properties of the junction are given by the rates Γ_+ and Γ_- at which the electrons cross the barrier from left to right or vice-versa. One electron crossing the junction during a time τ corresponds to a current pulse during τ of average intensity $\bar{\imath} = e/\tau$. The probability for n electrons to cross the barrier during τ from left to right, giving rise to a current $n\bar{\imath}$ is:

$$\begin{cases} P(0) = 1 - (\Gamma_+ + \Gamma_-)\tau \\ P(\pm\bar{\imath}) = \Gamma_{\pm}\tau \\ P(n\bar{\imath}) = 0 \ \text{ for } \ n > 1 \end{cases} \tag{2.1}$$

Quantum mechanics enter in the calculation of the rates whereas the statistical mechanics of the junction is given by the probability $P(i)$. We deduce the pth moment of the distribution of the current:

$$\langle I^p \rangle = \sum_{n=\pm 1} P(n\bar{\imath})(n\bar{\imath})^p = \bar{\imath}^p(\Gamma_+ + (-1)^p\Gamma_-)\tau \tag{2.2}$$

Thus, all the odd moments are proportional to the dc current $I = \langle I \rangle = e(\Gamma_+ - \Gamma_-)$, and all the even moments are proportional to the second one $\langle I^2 \rangle = e^2(\Gamma_+ + \Gamma_-)\tau$. The values of $\Gamma_{\pm}$ are determined by Ohm's law $I = e(\Gamma_+ - \Gamma_-) = GV$ and the detailed balance $\Gamma_+/\Gamma_- = \exp(eV/k_BT)$. In particular at zero temperature and $V > 0$, $\Gamma_- = 0$, which gives $\langle I^2 \rangle = eI\tau^{-1}$.

One generally considers the moments of the *current fluctuations*, $M_p = \langle i^p \rangle$ with $i(t) = I(t) - \langle I \rangle$. One has for the first moments:

$$\begin{aligned} M_2 &= \langle i^2 \rangle = \langle I^2 \rangle - I^2 \\ M_3 &= \langle i^3 \rangle = \langle I^3 \rangle - 3I\langle I^2 \rangle + 2I^3 \end{aligned} \tag{2.3}$$

Since $\langle I^p \rangle \propto t$ for all p, one has for a tunnel junction with $t \ll 1$, $\langle i^p \rangle \simeq \langle I^p \rangle$ to leading order in t. In particular, $M_3 = e^2 I \tau^{-2}$. This result is valid even for the multichannel case and after integrating over energy [2]. It is remarkable that M_3 is totally temperature independent, in contrast with M_2, for which the fluctuation-dissipation theorem implies $M_2 \propto T$ at equilibrium. As expected, M_3 is an odd function of the dc current and is zero for $V = 0$.

One usually also defines the *cumulants* C_p of current fluctuations (often noted $\langle\langle I^p \rangle\rangle$). They are related to the moments through the Fourier transform of P

$\chi(q) = \sum_n P(n\bar{i}) \exp iqn$. The series expansion of χ gives the moments, whereas the series expansion of $\ln \chi$ give the cumulants. For a Gaussian distribution, $\ln \chi$ is a second degree polynomial, and thus all the cumulants of order ≥ 3 are zero. The cumulants measure the non-gaussian part of the noise. The C_p are linear combinations of products of the M_p. For example,

$$
\begin{aligned}
C_3 &= M_3 \\
C_4 &= M_4 - 3M_2^2
\end{aligned}
\tag{2.4}
$$

2.2. Noise in Fourier space

In the previous paragraph we used a simple model to calculate the probability of current fluctuations. Experimentally, the averaging procedure is usually done by integrating the desired quantity over time. However, the fluctuating current $i(t)$ contains Fourier components up to very high frequency which are usually not accessible experimentally. Thus, one rather measures the *spectral density* of the fluctuations around a certain frequencies. We introduce the spectral densities associated with the pth moment of the current fluctuations S_{I^p}, expressed in A^p/Hz^{p-1}. S_{I^p} depends on $p - 1$ frequencies $f_1 \ldots f_{p-1}$. However, it is convenient to express S_{I^p} as a function of p frequencies such that the sum of all the frequencies is zero. Introducing the Fourier components $i(f)$ of the current, one has, for a classical current:

$$
S_{I^p}(f_1, \ldots, f_{p-1}) = \langle i(f_1) \ldots i(f_p) \rangle \delta(f_1 + \ldots f_p)
\tag{2.5}
$$

In quantum mechanics, the current operators taken at different times do not commute; nor do they in Fourier space, and the question of how the operators have to be ordered is crucial [2, 3].

In the case of the second moment, one has $S_{I^2}(f) = \langle i(f)i(-f) \rangle = \langle |i(f)|^2 \rangle$. It measures the power emitted by the sample at the frequency f within a bandwidth of 1Hz. This is what a spectrum analyzer measures. Comparing this expression with the one we have calculated for $\langle \delta I^2 \rangle$, one sees that τ^{-1} roughly represents the full bandwidth of the current fluctuations.

Experimentally, the current emitted by the sample runs through a series of cables, filters and amplifiers before being detected (this can be avoided by an on-chip detection [14]). Thus, the measured quantity is a filtered current $j(f) = i(f)g(f)$ where $g(f)$ describes the filter function of the detection.

One is often interested in the total power emitted by the sample in a certain bandwidth. This is obtained by measuring the DC signal after squaring $j(t)$, i.e. $\int j^2(t)dt$. This quantity is related to S_{I^2} through:

$$
\begin{aligned}
\langle j^2 \rangle &= \int j^2(t)dt = \int \int_{-\infty}^{+\infty} g(f_1)g(f_2) \langle i(f_1)i(f_2) \rangle \delta(f_1 + f_2) \\
&= \int |g(f)|^2 S_{I^2}(f)df
\end{aligned}
\tag{2.6}
$$

It is remarkable that the frequency-dependent phase shift introduced by $g(f)$ has no influence, in agreement with the fact that S_{I2} has the meaning of a power. If the detection bandwidth extends from F_1 to F_2, i.e., $g(f) = 1$ for $F_1 < |f| < F_2$ and $g(f) = 0$ otherwise, and if S_{I2} is frequency independent between F_1 and F_2, then the total noise is given by $\langle j^2 \rangle = 2S_{I2}(F_2 - F_1)$.

Let us consider now S_{I3}, which depends on two frequencies f_1 and f_2. The third moment of the measured current $j(t)$ is given by:

$$\langle j^3 \rangle = \int j^3(t)dt = \iiint g(f_1)g(f_2)g(f_3)S_{I3}(f_1, f_2)\delta(f_1 + f_2 + f_3) \quad (2.7)$$

We see that now the phase of $g(f)$ matters. More precisely, S_{I3} measures how three Fourier components of the current can beat together to give a non-zero result, i.e., it measures the phase correlations between these three Fourier components. With the same hypothesis as before, for a detection between F_1 and F_2, one now has $\langle j^3 \rangle = 3S_{I3}(F_2 - 2F_1)^2$ if $F_2 > 2F_1$ and $\langle j^3 \rangle = 0$ otherwise. This shows how important it is to make a broadband measurement. This unusual dependence of the result on F_1 and F_2 comes from the fact that the lowest frequency being the sum of two others is $2F_1$ whereas the maximum frequency one can subtract to that in order to have a DC signal is F_2. As we show below, we have experimentally confirmed this unusual dependence of $\langle j^3 \rangle$ on F_1 and F_2, see fig. 1.

2.3. Consequences

Each moment of the distribution is affected in a different way by the finite bandwidth of the measurement. As a consequence, even if the case where the moments are supposed to be frequency-independent, the probability $P(i)$ measured within a finite bandwidth B depends on B. In particular, the higher moments, which are more sensitive to rare events (current spikes) are washed out by the finite bandwidth, since the spikes are broadened by the filtering. In the case where the moments are frequency-dependent (like the diffusive wire, [4]), the notion of counting statistics itself has to be taken with care.

3. Effect of the environment

Up till now we have considered the bias voltage V and the temperature T to be fixed, time-independent, external parameters. In practice, it is very hard to perfectly voltage-bias a sample at any frequency. The temperature of the sample is generally fixed by a connection to reservoirs, and thus the temperature is fixed

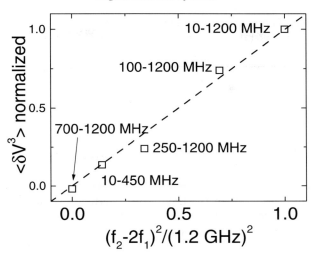

Fig. 1. Effect of the finite bandwidth on the measurement of $\langle \delta V^3 \rangle$. Each point corresponds to a different value of the frequencies F_1 and F_2, as indicated in the plot. The data shown here correspond to sample B at $T = 77$ K. (from Ref. [12]).

only at the ends of the sample. If V or T fluctuate, the probability $P(i)$ is modified. Let us call $P(i; V, T)$ the probability distribution of the current fluctuations around the dc current I when the sample is perfectly biased at voltage V and kept at fixed temperature T (as considered before), and $\tilde{P}(i)$ the probability distribution in the presence of an environment. R is the sample's resistance, taken to be independent of T and V.

3.1. Imperfect voltage bias

If the sample is biased by a voltage V through an impedance Z, the dc voltage across the sample is $V_s = VR/(R + Z)$. However, the current fluctuations in the sample flowing through the external impedance induce voltage fluctuations across the sample, given by:

$$\delta V_s(t) = -\int_{-\infty}^{+\infty} Z(f)i(f)e^{2i\pi ft}df \qquad (3.1)$$

Consequently, the probability distribution of the fluctuations is modified. This can be taken into account if the fluctuations are slow enough that the distribution $P(i)$ follows the voltage fluctuations. Under this assumption one has:

$$\tilde{P}(i) = P(i; V_s + \delta V_s, T) \qquad (3.2)$$

Supposing that the fluctuations are small ($\delta V_s \ll V_s, k_B T$), one can Taylor expand P_0 as:

$$\tilde{P}(i) \simeq P(i; V_s, T) + \delta V_s \frac{\partial P}{\partial V_s} + \dots \tag{3.3}$$

One deduces the moments of the distribution (to first order in δV_s) for a frequency-independent Z:

$$\tilde{M}_n(V, T) = M_n(V_s, T) - Z \frac{\partial M_{n+1}(V_s, T)}{\partial V_s} \tag{3.4}$$

This equation shows that the environmental correction to the moment of order n is related to the next moment of the sample perfectly voltage biased. It is a simplified version of the relation derived in refs. [5, 6]. Let us now apply the previous relation to the first moments.

3.1.1. dc current: dynamical Coulomb blockade
For $n = 1$ one gets a correction to the dc current given by:

$$\langle i \rangle = - \int_{-\infty}^{+\infty} Z(f) \frac{\partial M_2(f)}{\partial V_s} df \tag{3.5}$$

This is nothing but the environmental Coulomb blockade (within a factor 2) [7,8]. The bandwidth involved in M_2 is the intrinsic bandwidth of the sample, limited by the RC time, and not the detection bandwidth.

3.1.2. The second moment
Since the intrinsic third moment of a tunnel junction is linear in the applied voltage, to lowest order the imperfect voltage bias affects the second moment only by a constant term. There are however second order corrections [9].

3.1.3. The third moment
Similarly one obtains [10]:

$$\tilde{M}_3 = M_3 + 3 Z M_2 \frac{\partial M_2}{\partial V_s} \tag{3.6}$$

It is clear that the environmental correction to the third moment and the dynamical Coulomb blockade share the same physical origin, i.e. electron-electron interactions. However, since the third cumulant is a small quantity, the corrections can be as large as the intrinsic contribution, especially in a low impedance sample such as the one we have measured.

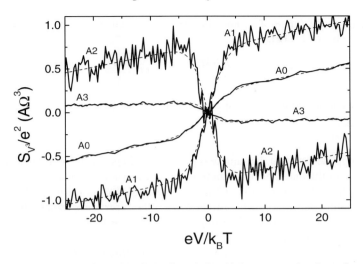

Fig. 2. Measurement of the spectral density of the third moment of voltage fluctuations, $S_{V3}(eV/k_BT)$ for sample A at T=4.2 K (solid lines). A0: no ac excitation (same as Fig. 4). A1: with an ac excitation at frequency $\Omega/2\pi$ such that $\cos 2\Omega\Delta t = +1$; A2: $\cos 2\Omega\Delta t = -1$; A3: no ac excitation but a 63 Ω resistor in parallel with the sample. The dashed lines corresponds to fits with Eq. (3.8). (from Ref. [12]).

3.1.4. Effect of an external fluctuating voltage

We consider in this paragraph the effect of an external source of noise $i_0(t)$ in parallel with the external impedance Z. This noise could be due to the Johnson noise of Z, to backaction of the current detector (e.g. the current noise of an amplifier), or an applied signal. The total current contains now i_0, and its moments involve correlations between i_0 and i. For the third moment, an additional term appears due to i_0: $\langle i_0 i^2 \rangle \simeq Z \langle i_0^2 \rangle (\partial M_2/\partial V_s)$. Our measurements verified the two effects of the environment (feedback and external noise) on the third moment, see fig. 2.

3.1.5. Voltage vs. current fluctuations

Instead of applying a voltage and measurig a current, one often prefers to apply a current and measure voltage fluctuations across the sample. The first two moments of V_S and i are related through:

$$V_S = R_D I$$
$$\langle \delta V_S^2 \rangle = R_D^2 \left(M_2 + \langle i_0^2 \rangle \right) \tag{3.7}$$

with $R_D = RZ/(R + Z)$. The situation is different for the third moment, for which we have [11, 12]:

$$\langle \delta V_S^3 \rangle = -R_D^3 M_3 + 3R_D^4 \langle i_0^2 \rangle \frac{\partial M_2}{\partial V_S} + 3R_D^4 M_2^0 \frac{\partial M_2}{\partial V_S} \tag{3.8}$$

We have confirmed this relation experimentally, see fig.2.

3.2. Imperfect thermalization

Let us suppose now that the sample is perfectly voltage biased, but that its temperature T_s can fluctuate, because there is a finite thermal impedance G_{th} between the sample and the reservoir. If the sample is biased at a finite voltage V, its average temperature might be different from the reservoir's temperature. Since the current flowing through the sample fluctuates, the Joule power P_J dissipated in the sample fluctuates as well, which induces temperature fluctuations. One has: $\delta T_s = G_{th}^{-1} \delta P_J = G_{th}^{-1} i V$. Since the probability distribution $P(i)$ depends on temperature, the temperature fluctuations modify in turn the current fluctuations. This *thermal* feedback is similar to the *electronic* feedback of the previous section. Thus, one has:

$$\tilde{P}(i) = P(i; V, T_s + \delta T_s) \simeq P(i; V, T_s) + \delta T_s \frac{\partial P}{\partial T_s} + \dots \tag{3.9}$$

This results in the following equation for the moments:

$$\tilde{M}_n(V, T) = M_n(V, T_s) + G_{th}^{-1} V \frac{\partial M_{n+1}}{\partial T_s} \tag{3.10}$$

Note that G_{th} is in fact complex, temperature- and frequency dependent.

Similarly to the case of imperfect voltage bias, the dc current and the third moment are affected by this feedback. One obtains a relative correction to the conductance $\delta G/G \propto k_B/C$ with C the heat capacitance of the sample. This correction is small, but diverges at low temperature. To our knowledge, this had never been predicted before.

For the third moment one obtains:

$$\tilde{M}_3 = M_3 + 6V G_{th}^{-1} M_2 \frac{\partial M_2}{\partial T} \tag{3.11}$$

In the case of a diffusive wire whose length is much longer than the electron-electron inelastic length, a local temperature can be defined. We designate T the (voltage-dependent) average temperature of the sample. In the absence of electron-phonon interactions, the electrical and thermal conductances are related

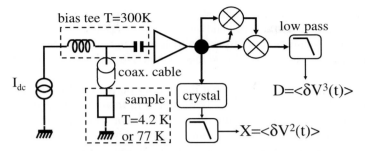

Fig. 3. Schematics of the experimental setup. (from Ref. [12])

through the Wiedemann-Franz law, $G_{th} \propto GT$ ($G = R^{-1}$), from which we deduce:

$$\tilde{M}_3 = M_3 + \alpha e^2 I B^2 \tag{3.12}$$

with α a numerical coefficient, and B the thermal bandwidth, $B \sim \tau_D^{-1}$ in a diffusive system. This result corresponds to the calculation of ref. [4] in the hot electron regime. We understand that at frequencies larger than the inverse of the diffusion time τ_D, the thermal conductance drops, and so does the thermal correction. The result of ref. [4], that the third moment vanishes at such frequencies, seems to imply that $M_3 = 0$. It might be an artifact of our oversimplified calculation.

Another interesting case is the SNS structure (a normal metal wire between two superconducting reservoirs), for which the cooling is due to electron-phonon interaction and not electron out-diffusion. This out-diffusion is suppressed exponentially at low temperature due to the superconducting gap of the reservoirs. The vanishing electron-phonon thermal conductance should lead to a divergence of the spectral density of the third moment, but the bandwidth also shrinks with lowering T.

4. Principle of the experiment

We will not discuss in full detail the experiment performed to measure the third moment of the voltage fluctuations in tunnel junctions. Those details can be found in refs [12, 13].

4.1. Possible methods

We present three methods that could be used to measure the non-Gaussian part of the noise. First, the simplest idea is to digitize in real time the voltage across the sample, and make histograms of the values. This method is very direct, but suffers from its limitations in bandwidth B. It is very difficult to acquire data with enough dynamics very fast and to treat them in real time. Since $M_3 \propto B^2$, the small bandwidth severely limits this method. No such measurement has been reported yet.

A second point of view is to put the priority on the bandwidth. That is the method we have chosen. Then it is very hard to treat the signal digitally, and we use analog mixers to compute the third power of $i(t)$, and average with the help of a low pass filter. We have implemented this method with a bandwidth of 1 GHz. This method is versatile, it works at any temperature and for many different kinds of samples; the main limitation comes from the necessity to have a sample's resistance close to 50Ω to ensure good coupling with the microwave circuits. The drawback of the method is the care needed to separate the real signal from any non-linearity due to the mixers and amplifiers.

Finally, another possibility, which offers huge bandwidth and great sensitivity is to couple the sample to an on-chip mesoscopic detector. This has been successfully realized to measure S_{V^2} [14]. One could even have access to the full statistics of the current [15]. The drawback of this method is the theoretical difficulty of extracting the intrinsic behavior of the sample when it is strongly coupled to another mesoscopic system [16].

4.2. Experimental setup

We have measured the third moment of the voltage fluctuations across a tunnel junction, by measuring $\delta V^3(t)$ in real time (see Fig. 3). For simplicity, we note V and δV the dc voltage and the voltage fluctuations *across the sample*. The sample is dc current biased through a bias tee. The noise emitted by the sample is coupled out to an rf amplifier through a capacitor so that only the ac part of the current is amplified. The resistance of the sample is close to 50 Ω, and thus is well matched to the coaxial cable and amplifier. After amplification at room temperature the signal is separated into four equal branches, each of which carries a signal proportional to $\delta V(t)$. A mixer multiplies two of the branches, giving $\delta V^2(t)$; a second mixer multiplies this result with another branch. The output of the second mixer, $\delta V^3(t)$, is then low pass filtered, to give a signal D proportional to S_{V^3}, where the constant of proportionality depends on mixer gains and frequency bandwidth. The last branch is connected to a square-law crystal detector, which produces a voltage X proportional to the the rf power it receives: the noise of the sample $\langle \delta V^2 \rangle$ plus the noise of the amplifiers. This detection scheme has the advantage of the large bandwidth it provides ($\sim$ 1 GHz), which is crucial for the measurement. Due to the imperfections of the mixers, D contains some contribution of X. Those contributions, even in I, are removed by calculating $(D(I) - D(-I)) \propto S_{V^3}$.

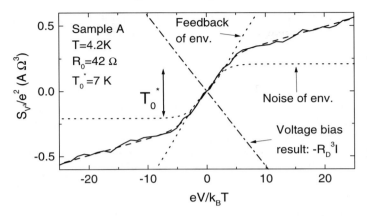

Fig. 4. Measurement of $S_{V^3}(eV/k_BT)$ for sample A (solid line). The dashed line corresponds to the best fit with Eq. (3.8). The dash dotted line corresponds to the perfect bias voltage contribution and the dotted lines to the effect of the environment. (from Ref. [13])

5. Experimental results

5.1. Third moment vs. voltage and temperature

$S_{V^3}(eV/k_BT)$ for sample A at $T = 4.2$K is shown in Fig. 4; these data were averaged for 12 days. $S_{V^3}(eV/k_BT)$ for sample B at $T = 4.2$K (top), $T = 77$K (middle) and $T = 290$K (bottom) is shown in Fig. 5. The averaging time for each trace was 16 hours. These results are clearly different from the voltage bias result (the dash-dotted line in Fig. 4). However, all our data are very well fitted by Eq. (3.8) which takes into account the effect of the environment (see the dashed lines of Fig. 4 and 5). The environment of the sample is made of the amplifier, the bias tee, the coaxial cable (~ 2 m long except at room temperature, where it is very short) and the sample holder. It is characterized by its impedance Z, that we suppose is real and frequency-independent (i.e., we model it by a resistor R_0 of the order of 50Ω), and a noise temperature T_0^* (the latter does not correspond to the real noise temperature of the environment, see below). Figs. 4 and 5 show the best fits to the theory, Eq. (3.8), for all our data. The four curves lead to $R_0 = 42$ Ω, a very reasonable value for microwave components, and in agreement with the fact that the electromagnetic environment was identical for the two samples.

We have measured directly the noise emitted by the room temperature amplifier; we find $T_0 \sim 100$ K. This is in disagreement with the parameters T_0^* deduced from the fit of the data, but is well explained by the finite propagation time along the coaxial cable between the sample and the amplifier.

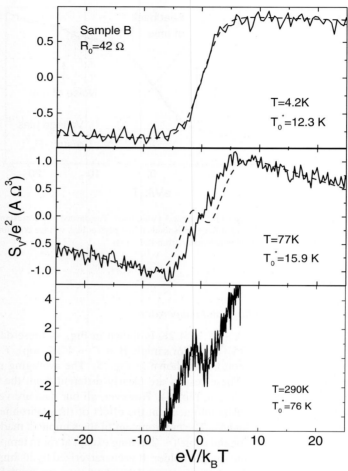

Fig. 5. Measurement of $S_{V^3}(eV/k_BT)$ for sample B (solid lines). The dashed lines corresponds to the best fit with Eq. (3.8). (from Ref. [13]).

5.2. *Effect of the detection bandwidth*

A powerful check that D really measures S_{V^3} is given by varying the bandwidth. The scaling of S_{V^3} with F_1 and F_2 ($S_{V^3} \propto (F_2 - 2F_1)^2$ if $F_2 > 2F_1$ and 0 otherwise) is characteristic of the measurement of a third order moment. We do not know of any experimental artifact that has such behavior. F_1 and F_2 are varied by inserting filters before the splitter. As can be seen in Fig. 1, our measurement follows the dependence on $(F_2 - 2F_1)^2$, which cannot be cast into a function of $(F_2 - F_1)$. Each point on the curve of Fig. 1 corresponds to a full $\langle \delta V^3(I) \rangle$ measurement (see figures 4 and 5).

6. Effect of the environment

In order to demonstrate more explicitly the influence of the environment on S_{V^3} we have modified the parameters T_0 and R_0 of the environment and measured the effect on S_{V^3}.

T_0 is a measure of the current fluctuations emitted by the environment towards the sample. Its influence on S_{V^3} is through the correlator $\langle i^2 i_0 \rangle$. This correlator does not require i_0 to be a randomly fluctuating quantity in order not to vanish. So we can modify it by adding a signal $A \sin \Omega t$ to i_0 (with Ω within the detection bandwidth). That way we have been able to modify T_0^* without changing R_0, as shown on Fig. 2. The current correlator involved in T_0^* contains interference between the current sent to the sample and what comes back after $2\Delta t$, with Δt the propagation time along the coaxial cable between the source and the sample. This correlator$\sim \langle i_0(t) i_0(t - 2\Delta t) \rangle$ oscillates Vs. Δt like $\cos 2\Omega \Delta t$, and thus one can enhance (curve A1 as compared to A0 in Fig. 2) or decrease T_0^*, and even make it negative (Fig. 2, A2). The curves A0–A2 are all parallel at high voltage, as expected, since the impedance of the environment remains unchanged; $R_0 = 42 \, \Omega$ is the same for the fit of the three curves.

Second, by adding a 63 Ω resistor in parallel with the sample (without the added ac excitation) we have been able to change the resistance of the environment R_0, and thus the high voltage slope of S_{V^3}. The fit of curve A3 gives $R_0 = 24.8 \, \Omega$, in excellent agreement with the expected value of 25.2 Ω (63 Ω in parallel with 42 Ω). The apparent negative T_0^* comes from the negative sign of the reflection coefficient of a wave on the sample in parallel with the extra resistor.

7. Perspectives

7.1. Quantum regime

We have chosen to discuss the statistics of the current fluctuations with a classical point of view. This seems enough to explain the properties we have shown until now. However, there are situations that need to be treated quantum mechanically, in particular, when the frequency is greater than voltage and/or temperature. It has been calculated that the third cumulant for a tunnel junction should be completely frequency independent [17]. This is in sharp contrast with the fact that

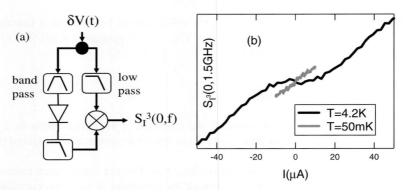

Fig. 6. (a) Setup for the measurement of $S_{V3}(0, \bar{f})$. The sample, bias tee and amplifiers have been omitted, see Fig. 1. (b) Measurement of $S_{V3}(0, \bar{f} \sim 1.5\text{GHz})$ on sample A at $T = 4.2\text{K}$ and $T = 50\text{mK}$.

the second moment S_{I2} is different for $\hbar\omega > eV$, due to the fact than no photon of energy greater than eV can be emitted. A picture of the third moment (and higher) in terms of photons is still missing. The effect of the environment, which involves S_{I2}, might also be different at high frequency. In particular, the distinction between emission and absorption might be relevant. For example, the zero point fluctuation of the voltage might modulate the noise, but not be detected, and thus some of the effect of the environment might vanish.

We illustrate this discussion by the first measurement of the third moment of voltage fluctuations $S_{V3}(0, f)$ across a tunnel junction (sample A) at finite frequency f with $hf > k_B T$. In order to perform such a measurement we have constructed the setup of Fig. 6a. The signal $\delta V(t)$ is split into two frequency bands, LF=]0, f_3] (dc excluded) and HF=[f_1, f_2]. The voltage is squared in the HF band (left branch of Fig. 3a) with a high speed tunnel diode, then low-pass filtered with a cutoff f_3. Thus one has products of the form $i(f)i(-f')$ at the end of the HF branch. This result is multiplied by a mixer to the LF branch, then low-pass filtered to get a dc signal. This signal corresponds to $S_{V3}(0, \bar{f})$ if $(f_2 - f_1)$ and f_3 are small enough, with $\bar{f} \sim (f_1 + f_2)/2$. f_1 and f_2 can be varied by changing the various filters.

The measurement of sample A at $T = 4.2$ K and $T = 50$ mK with the new setup operating in the LF=10-200MHz and HF=1-2.4 GHz bandwidths is presented in Fig. 6b. We find subtle but interpretable new results. The fit of these data with eq. (3.8) leads to $R_0 = 40\Omega$ and $T_0^*(T = 4.2K) = -0.4K$. The slope of S_{V3} at high voltage is found to be temperature independent, like it is between 300 K and 4.2 K (see Fig. 5). The negative T_0^* comes from the Johnson noise of the 12Ω contact resistance. The current fluctuations emitted by the contact

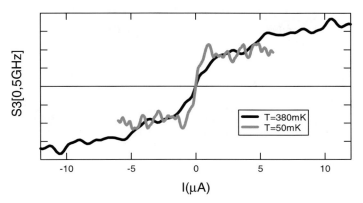

Fig. 7. Measurement of $S_{V^3}(0, \bar{f} \sim 5\text{GHz})$ on sample A at $T = 380\text{mK}$ (i.e., $h\bar{f}/k_B T \sim 0.6$ and $T = 50\text{mK}$ (i.e., $h\bar{f}/k_B T \sim 5$).

result in currents of opposite signs running through the sample and the amplifier. As a consequence, the contact contributes to T_0^* with a negative sign. Since we use in this new setup a cryogenic amplifier with low noise temperature T_0, and since the Johnson noise of the contact is not affected by the propagation time, its contribution dominates T_0^* at 4.2K. We indeed observe a sign reversal of T_0^* when cooling the sample below 1 K, since the noise of the amplifier dominates at low enough temperature (see Figs. 6 and 7). The non-linear behavior at low voltage at T=4.2 K is similar to the one observed at room temperature with the previous setup, see Fig. 5, i.e. when the noise emitted by the sample is larger than the noise emitted by the amplifier, revealing the contribution of the feedback of the environment.

In Fig. 7 we present the result for $h\bar{f} < k_B T$ and $h\bar{f} > k_B T$. Those are obtained with an HF bandwidth HF=4.5-5.5 GHz (5G Hz$\cong$ 250 mK). The slope at high voltage has changed whereas it was temperature independent up to 290 K. In the regime $h\bar{f} > k_B T$ the slope at high voltage is zero, as if only the term $\langle i_0 i^2 \rangle$ remained. Further experiments are required to see if this is a coincidence for the particular values of T or f. It is however clear that the result is different in the two regimes.

7.2. Noise thermal impedance

In the treatment of the effect of an external source of noise on the probability $P(i)$ we have supposed that P responds instantaneously to the external voltage fluctuations. In fact one may ask how fast can $P(i)$ react. We exemplify this discussion in the case of the second moment S_{I^2} of a diffusive wire.

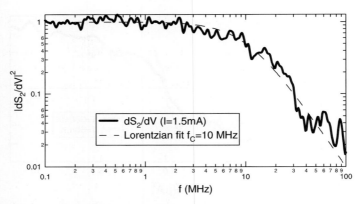

Fig. 8. $|dS_{V^2}/dV|^2$ renormalized to its value at low frequency, versus measurement frequency

For a macroscopic wire, $S_{I^2} = 4k_B TG$ is the equilibrium Johnson noise. A small voltage variation $\delta V \ll V$ induces a variation of the Joule power dissipated in the sample, $\delta P_J = 2GV\delta V$ which in turn induces a variation of the temperature $\delta T = G_{th}^{-1}\delta P_J$. This will take place as long as the time scale of the variations is much smaller than the thermalization time of the wire. For a phonon-cooled wire ($L \gg L_{e-ph}$ with L_{e-ph} the electron-phonon length) this time will be the electron-phonon interaction time. For a wire in the hot electron-diffusion cooled regime ($L_{e-ph} \gg L \gg L_{e-e}$ with L_{e-e} the electron-electron inelastic length), the thermalization is obtained by hot electrons leaving the sample, which occurs at a time scale given by the diffusion time. For a short wire ($L_{e-e} \gg L$), no thermalization occurs within the electron gas. The noise emitted by the sample depends on the distribution function of the electrons in the wire, whose dynamics is diffusive. Therefore, the relevant time scale in this regime of elastic transport is again set by the diffusion time. Note that in this regime the temperature is no longer defined, neither is the thermal impedance. However, one can define a "noise thermal impedance" that extends the definition of the usual thermal impedance to cover any case [18]. The noise thermal impedance clearly determines the electrical effect of the environment at finite frequency. According to our simple calculation of the thermal effect of the environment, it also determines S_{I^3} in the hot electron regime. The elastic case is less clear, but should be qualitatively the same [4].

In order to illustrate the noise thermal impedance, we show a preliminary measurement of dS_{V^2}/dV at high frequency in a $100\mu m$ long diffusive Au wire, see Fig.8. The sample is biased by $I(t) = I_{dc} + \delta I \cos \omega t$ with $\delta I \ll I_{dc}$. ω is varied between 100kHz and 100MHz. With a setup similar to the one of

Fig. 6a, we detect how much the rms amplitude of the noise, measured by the diode, oscillates at frequency ω, which is similar to the measurement of S_{V^3} in the presence of an external excitation. One clearly sees in the figure the cut-off at 10MHz, which probably corresponds to the electron-phonon time in the sample at $T = 4.2K$. The curve can be well fitted by a Lorentzian, as expected, from which we deduce $\tau_{e-ph} = 16$ns. Diffusion cooling would have led to a cut-off frequency of $\sim$ 800kHz. A study for different length at various temperatures is in progress.

7.3. Conclusion

In this article we have chosen to give simplified pictures of more complicated phenomena. We hope this will encourage the reader to go through more detailed literature. We have shown experimental evidence of the existence of non-gaussian shot noise. As we have sketched in the last section, this is not the end of the story, and we hope to motivate the emergence of new experiments in this domain.

Acknowledgements

This work has been accomplished at Yale University in the group of Prof. D. Prober, who I warmly thank for his confidence and enthusiasm, as well as for having given me the possibility to perform this research. This was possible thanks to the flexibility of the CNRS, which allowed me go to Yale during four years. I acknowledge J. Senzier for his help at the beginning of this adventure. L. Spietz, C. Wilson and M. Shen are fondly thanked for providing me with samples. I also acknowledge fruitful discussions with many colleagues, among them C. Beenakker, W. Belzig, M. Büttiker, A. Clerk, M. Devoret, M. Kindermann, L. Levitov, Y. Nazarov, S. Pilgram and P. Samuelsson. This work was supported by NSF DMR-0072022 and 0407082.

References

[1] Y.M. Blanter and M. Büttiker, Phys. Rep. **336**, 1 (2000).

[2] L.S. Levitov, H. Lee and G.B. Lesovik, J. Math. Phys. **37**, 4845 (1996).

[3] G.B. Lesovik and N.M. Chtchelkatchev, JETP Lett. **77**, 393 (2003).

[4] S. Pilgram, K. Nagaev and M. Büttiker, Phys. Rev. **B70** 045304 (2004).

[5] I. Safi and H. Saleur, Phys. Rev. Lett. **93**, 126602 (2004).

[6] A.V. Galaktionov, D.S. Golubev and A.D. Zaikin, Phys. Rev. **B68**, 085317 (2003).

[7] G.-L. Ingold and Yu. V. Nazarov in *Single Charge Tunneling*, edited by H. Grabert and M. H. Devoret (Plenum Press, New York, 1992).

[8] A. Levy Yeyati, A. Martin-Rodero, D. Esteve and C. Urbina, Phys. Rev. Lett. **87** 046802 (2001).

[9] D.B. Gutman and Y. Gefen, Phys. Rev. **B64**, 205317 (2001).

[10] M. Kindermann, Yu V. Nazarov and C.W.J. Beenakker, Phys. Rev. Lett. **90**, 246805 (2003).

[11] C.W.J. Beenakker, M. Kindermann and Yu V. Nazarov, Phys. Rev. Lett. **90**, 176802 (2003).

[12] B. Reulet, J. Senzier and D.E. Prober, Phys. Rev. Lett. **91**, 196601 (2003).

[13] B. Reulet *et al.*. Proceedings of SPIE, *Fluctuations and Noise in Materials*, 5469-33:244-56 (2004).

[14] R. Deblock, E. Onac, L. Gurevich and L.P. Kouwenhoven, Science **301**, 203 (2003).

[15] J. Tobiska and Yu V. Nazarov, Phys. Rev. Lett. **93**, 106801 (2004).

[16] P.K. Lindell *et al.*, Phys. Rev. Lett. **93**, 197002 (2004).

[17] A.V. Galaktionov, D.S. Golubev and A.D. Zaikin, Phys. Rev. **B68**, 235333 (2003).

[18] B. Reulet and D.E. Prober, unpublished (cond-mat/0501397).

Course 6

ELECTRON SUBGAP TRANSPORT IN HYBRID SYSTEMS COMBINING SUPERCONDUCTORS WITH NORMAL OR FERROMAGNETIC METALS

F.W.J. Hekking

Laboratoire de physique et modélisation des milieux condensés, Centre National de Recherche Scientifique and Université Joseph Fourier B.P. 166, 38042 Grenoble, France

H. Bouchiat, Y. Gefen, S. Guéron, G. Montambaux and J. Dalibard, eds.
Les Houches, Session LXXXI, 2004
Nanophysics: Coherence and Transport

383

Contents

1. Introduction

This chapter is devoted to some aspects of electronic transport in *hybrid structures*, containing a superconductor (S) connected to normal (N) metallic or to ferromagnetic (F) electrodes. We focus on properties associated with so-called *subgap conditions*, *i.e.*, temperature and applied voltages are smaller than the superconducting gap Δ.

Interest in hybrid systems containing superconductors goes back to the pioneering experimental work on tunnel junctions between a superconductor and a normal metal by Giaever in 1960 [1]. A milestone was achieved by Andreev [2], who discovered in 1964 the phenomenon now known as *Andreev reflection*. This discovery marked the start of an enormous activity dealing with subgap transport in NS hybrid structures. An important highlight was the development of the *Bogoliubov-de Gennes formalism* [3] that provides an adequate description of inhomogeneous systems in general and systems with NS interfaces in particular. Based on this formalism, Blonder, Tinkham, and Klapwijk (BTK) [4] formulated a transport theory that successfully describes subgap conductance across NS interfaces as long as these interfaces are macroscopically sized. Miniaturization of devices during the eighties lead to interest in mesoscopic NS interfaces, where low-energy properties are known to be sensitive to *disorder-induced interference phenomena*. Indeed, experiments done on diffusive NS systems in the early nineties reported significant deviations from the standard BTK-type of behavior. The associated *zero-bias anomalies* are relatively well-understood nowadays, with the development of powerful theoretical tools like scattering theory [5], quasi-classical non-equilibrium Green function techniques [6, 7, 8, 9, 10] and circuit theory [11, 12].

In these lecture notes, we will illustrate the peculiar behavior of subgap transport in mesoscopic diffusive NS hybrid systems focussing on *tunnel junctions*. Apart from being of experimental relevance, the main reason for studying tunnel junctions is that they are amenable to relatively simple calculations using perturbation theory in combination with standard Green function techniques for disordered metals. Specifically, we obtain the subgap current and conductance at small but finite voltage and temperature in a simple, closed form, that enables a direct physical interpretation [13]. This is in contrast to the aforementioned more elaborate techniques [5, 10, 11, 12] that are either restricted to zero energy only, or are too complicated to be discussed in simple terms.

We will start our considerations with junctions between N and S in the *clean limit*, which means that the effects of impurity scattering will be ignored. Then we will turn our attention to the *diffusive limit* and see that subgap transport is very sensitive to disorder-induced interference effects. We also consider *current noise* in the subgap regime, and show that it is affected by the same phenomenon. Finally, we discuss *multi-probe hybrid structures* involving ferromagnetic electrodes and show that additional correlated tunnelling processes must be taken into account to describe transport.

Despite the considerable amount of work on the Josephson effect in mesoscopic hybrid systems of the SNS or SFS type, we will not treat such systems in these notes. The reader is referred to a recent review for details [14].

2. NS junctions in the clean limit

2.1. Single particle tunnelling in a tunnel junction

2.1.1. Introduction
Tunnelling in hybrid systems has been known for many years already to be a sensitive tool enabling one to probe the energy-dependence of the *density of states* (DoS). Combining an ordinary normal metal, whose DoS is approximately constant close to the Fermi level, with an electrode whose DoS is energy-dependent (*e.g.*, due to the presence of interactions), yields non trivial tunnel current-voltage (I-V) characteristics. A careful analysis of these characteristics and in particular of the *differential conductance dI/dV* enables one to obtain information on the energy-dependent DoS. In 1960, Giaever [1] pioneered this technique, measuring the I-V characteristics of a tunnel junction between a normal metal and a superconductor (NIS junction). As temperature was decreased below the gap, the usual linear characteristics found at high temperatures became non-linear, developing a gap at low voltages. Giaever thus confirmed the energy-dependent density of states as predicted by the microscopic theory of superconductivity [15, 16]. It is instructive to analyze the NIS junction in some detail as a starting point of our study of hybrid systems.

2.1.2. Tunnel Hamiltonian
In order to describe the physics of NIS junctions, we will mainly make use of the *tunnel Hamiltonian formalism*, introduced in Ref. [17]. For a textbook treatment, see, *e.g.*, [18]. In this formalism, the total Hamiltonian of the junction is written

as the sum of three parts, $\hat{H} = \hat{H}_L + \hat{H}_R + \hat{H}_T$, where $\hat{H}_i$ with $i = L, R$ describes the left and right electrode, respectively, and $\hat{H}_T$ is the *tunnel Hamiltonian*

$$\hat{H}_T = \sum_{k,p,\sigma,\sigma'} t_{k,p;\sigma,\sigma'} \hat{a}_{k,\sigma}^{\dagger} \hat{b}_{p,\sigma'} + t_{k,p;\sigma,\sigma'}^* \hat{b}_{p,\sigma'}^{\dagger} \hat{a}_{k,\sigma}. \tag{2.1}$$

The operator $\hat{a}_{k,\sigma}^{\dagger}$ ($\hat{a}_{k,\sigma}$) creates (annihilates) an electron with quantum number k and spin σ in the left electrode, similarly for $\hat{b}^{\dagger}$ ($\hat{b}$) in the right electrode. Hence, the first term on the r.h.s. of Eq. (2.1) describes tunnelling from the right to the left electrode, the second term tunnelling from left to right. The creation and annihilation operators obey the usual fermionic anticommutation rules. The amplitude $t_{k,p;\sigma,\sigma'}$ is associated with the tunnel probability. Throughout these lectures we will assume spin to be conserved during tunnelling, hence the amplitude is independent of the spin indices. In this Section, for simplicity, we furthermore ignore the dependence of t on quantum numbers k and p and put $t_{k,p} = t_0$. This dependence will be restored below in Section 3, however, when we discuss the diffusive limit.

2.1.3. Perturbation theory: golden rule

In order to find the I-V characteristics of the junction, we use perturbation theory. According to Fermi's golden rule, the rate for tunnelling from a state k, σ in the left electrode to a state p, σ in the right electrode can be written as

$$\Gamma_{k,\sigma \rightarrow p,\sigma} = \frac{2\pi}{\hbar} |t_0|^2 f_L(\epsilon_k)[1 - f_R(\epsilon_p)]\delta(\epsilon_k - \epsilon_p). \tag{2.2}$$

Here, $f_i(\epsilon) = [e^{\beta(\epsilon - \mu_i)} + 1]^{-1}$ is the Fermi function for electrode $i = L, R$ kept at electrochemical potential μ_i and temperature T (we defined the inverse temperature $\beta = 1/k_B T$). The presence of the Fermi functions ensures that tunnelling only occurs if the initial state in L is occupied and the final state in R is empty. The δ-function imposes energy conservation during tunnelling. The energies ϵ_k constitute the continuous single particle spectrum of the electrodes.

We proceed by calculating the total rate $\Gamma = \Gamma_{L \rightarrow R} - \Gamma_{R \rightarrow L}$, where $\Gamma_{L \rightarrow R} = \sum_{k,p,\sigma} \Gamma_{k,\sigma \rightarrow p,\sigma}$, and similarly for $\Gamma_{R \rightarrow L}$. In the present, non-magnetic case the sum over spin yields just a factor of two. Finally, we change variables $\xi_k = \epsilon_k - \mu_L, \xi_p = \epsilon_p - \mu_R$, thus measuring energies with respect to the chemical potential of each electrode. Replacing the sums over k by an integral over the continuous spectrum, $\sum_{k,(p)} \ldots \rightarrow \int d\xi N_{L,(R)}(\xi) \ldots$, where $N_{L(R)}(\xi)$ is the density of states

(DoS) *per spin* per unit energy of the left (right) electrode, we obtain

$$
\begin{aligned}
\Gamma &= \frac{4\pi |t_0|^2}{\hbar} \int d\xi_1 d\xi_2 N_L(\xi_1) N_R(\xi_2) [f(\xi_1) - f(\xi_2)] \delta(\xi_1 - \xi_2 + eV) \\
&= \frac{4\pi |t_0|^2}{\hbar} \int d\xi N_L(\xi) N_R(\xi + eV)[f(\xi) - f(\xi + eV)],
\end{aligned}
\tag{2.3}
$$

where eV is the electrochemical potential difference between the left and right electrode.

The total electric current is then given by $I = e\Gamma$. Let us consider two cases of general interest:

NIN junction.   If both electrodes are ordinary normal metals, the DoS $N(\xi)$ can be considered as a constant, equal to the value N_0 at the Fermi energy. Integration over energy in (2.3) is straightforward. One finds that the I-V characteristic is linear and given by $I_{NIN} = G_T V$, where

$$
G_T = \frac{4\pi}{\hbar} e^2 |t_0|^2 N_0^2
\tag{2.4}
$$

is the normal state tunnel conductance of the junction.

NIS junction.   If one electrode is superconducting with a gap Δ, characterized by the standard BCS DoS

$$
N_{BCS}(\xi) = \begin{cases} N_0 \dfrac{|\xi|}{\sqrt{\xi^2 - \Delta^2}} & \text{for } |\xi| > \Delta; \\ 0 & \text{otherwise,} \end{cases}
\tag{2.5}
$$

we find

$$
I_{NIS} = \frac{G_T}{e} \int\limits_{|\xi| > \Delta} d\xi \frac{|\xi|}{\sqrt{\xi^2 - \Delta^2}} [f(\xi) - f(\xi + eV)].
\tag{2.6}
$$

We see that the I-V characteristics are strongly nonlinear. The differential conductance is given by

$$
G_{NIS} = \frac{dI_{NIS}}{dV} = \frac{G_T}{4k_B T} \int\limits_{|\xi| > \Delta} d\xi \frac{|\xi|}{\sqrt{\xi^2 - \Delta^2}} \frac{1}{\cosh^2[\beta(\xi + eV)/2]}.
\tag{2.7}
$$

At zero temperature, this reduces to

$$
G_{NIS} = \begin{cases} G_T \dfrac{|eV|}{\sqrt{(eV)^2 - \Delta^2}} & \text{for } |eV| > \Delta; \\ 0 & \text{otherwise,} \end{cases}
\tag{2.8}
$$

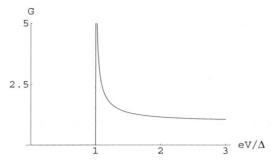

Fig. 1. Dimensionless zero-temperature tunnel conductance $G = G_{NIS}/G_T$, as a function of voltage eV/Δ.

hence this quantity is directly proportional to the superconducting DoS (2.5). In Fig. 1 we plot the zero-temperature differential conductance as a function of voltage. Note in particular the absence of a single particle contribution to the differential conductance for voltages $eV < \Delta$. This is directly related to the absence of quasi-particle states with energies below the gap in a superconductor.

2.1.4. Higher order processes

As we have seen above, a perturbation calculation up to order t_0^2 yields zero tunnel conductance at energies smaller than the superconducting gap Δ. Single particle transport is not possible below the gap since single quasiparticle states are not available in the superconducting electrode. Technically speaking this means that higher order tunnel processes should be considered. The next order pertains to transferring two particles at a time. This possibility was considered in 1969 by Wilkins [19]. Indeed, as we will detail in Section 3 below, an order t_0^4 calculation yields the transfer of electron pairs from the normal metal into the superconductor. This process is possible below the gap, as the two electrons from N can form a Cooper pair in S, hence this process will not suffer from the absence of final states.

In fact, the two-electron tunnelling process across an NIS junction corresponds to the tunnel limit of the well-known *Andreev reflection process* [2] that occurs in NS junctions with a clean interface. Below we will analyze Andreev reflection in more detail. In particular, we will treat Andreev reflection in the framework of so-called *Bogoliubov-de Gennes theory* [3], which was worked out for NS interfaces by Blonder, Tinkham and Klapwijk (BTK) [4]. The BTK approach enables one to treat arbitrary scattering at the NS interface. However, BTK theory does not take disorder in the electrodes into account. As we will see in Section 3 disorder plays a crucial role, and strongly influences subgap transport in NS hybrid systems. We will investigate these issues in particular in the tunnel limit.

2.2. Bogoliubov-de Gennes equations

2.2.1. BCS Hamiltonian and diagonalization
The starting point for the analysis developed by Bogoliubov and de Gennes [3] is the mean-field BCS Hamiltonian, which can be written as

$$\hat{H}_{BCS} = \sum_\sigma \int d^3r \, \hat{\Psi}_\sigma^\dagger(\vec{r}) \left[\frac{1}{2m} \left(\frac{\hbar}{i}\nabla - \frac{e}{c}\vec{A}(\vec{r}) \right)^2 + U(\vec{r}) - \mu \right] \hat{\Psi}_\sigma(\vec{r})$$
$$+ \int d^3r [\Delta(\vec{r})\hat{\Psi}_\uparrow^\dagger(\vec{r})\hat{\Psi}_\downarrow^\dagger(\vec{r}) + \Delta^*(\vec{r})\hat{\Psi}_\downarrow(\vec{r})\hat{\Psi}_\uparrow(\vec{r})]. \tag{2.9}$$

Here, m is the electron mass, $\vec{A}(\vec{r})$ is the vector potential, $U(\vec{r})$ an external potential and μ the chemical potential. The space-dependent gap is determined by the self-consistency equation

$$\Delta(\vec{r}) = -g(\vec{r})\langle\hat{\Psi}_\downarrow(\vec{r})\hat{\Psi}_\uparrow(\vec{r})\rangle, \tag{2.10}$$

where $g(\vec{r})$ is the interaction constant characterizing the attractive interaction between electrons. It depends on position $\vec{r}$, changing from its maximum value deep within superconducting regions to $g = 0$ in normal metallic parts. The fermionic field operators obey the usual anticommutation relations like, *e.g.*,

$$\{\hat{\Psi}_\sigma(\vec{r}), \hat{\Psi}_{\sigma'}^\dagger(\vec{r}')\}_+ = \delta_{\sigma,\sigma'}\delta(\vec{r} - \vec{r}'). \tag{2.11}$$

We now diagonalize $\hat{H}_{BCS}$ with the help of the *Bogoliubov transformation*

$$\hat{\Psi}_\uparrow(\vec{r}) = \sum_k u_k(\vec{r})\hat{\gamma}_{k,\uparrow} - v_k^*(\vec{r})\hat{\gamma}_{k,\downarrow}^\dagger; \tag{2.12}$$

$$\hat{\Psi}_\downarrow(\vec{r}) = \sum_k u_k(\vec{r})\hat{\gamma}_{k,\downarrow} + v_k^*(\vec{r})\hat{\gamma}_{k,\uparrow}^\dagger, \tag{2.13}$$

with new fermionic operators $\hat{\gamma}, \hat{\gamma}^\dagger$. According to this transformation, an electronic excitation can be annihilated either by annihilating a *Bogoliubov quasiparticle* of the same spin or by creating one with opposite spin. The Bogoliubov transformation leads to *diagonalization* of the BCS Hamiltonian,

$$\hat{H}_{BCS} \rightarrow \sum_{k,\sigma} E_k \hat{\gamma}_{k,\sigma}^\dagger \hat{\gamma}_{k,\sigma}, \tag{2.14}$$

under the condition that u, v satisfy the so-called *Bogoliubov-de Gennes equations*,

$$H_0(\vec{r})u(\vec{r}) + \Delta(\vec{r})v(\vec{r}) \ = \ Eu(\vec{r}); \tag{2.15}$$

$$-H_0^*(\vec{r})v(\vec{r}) + \Delta^*(\vec{r})u(\vec{r}) \ = \ Ev(\vec{r}), \tag{2.16}$$

with $H_0(\vec{r}) = \frac{1}{2m}[\frac{\hbar}{i}\nabla - \frac{e}{c}\vec{A}(\vec{r})]^2 + U(\vec{r}) - \mu$. These equations determine the eigenfunctions u, v and eigenenergies $E > 0$ corresponding to the excitations of the superconductor for given potentials $\vec{A}(\vec{r})$ and $U(\vec{r})$.

2.2.2. Simple examples

For simplicity, we work out the Bogoliubov-de Gennes (BdG) equations for a clean, one-dimensional geometry $\vec{r} \rightarrow x$, without applied fields, such that $U(x) = \vec{A}(x) = 0$.

Normal metal.   In a normal metal we have $g = 0$, hence $\Delta = 0$. The BdG equations therefore reduce to the simple set

$$-\left[\frac{\hbar^2}{2m}\frac{d^2}{dx^2} + \mu\right]u(\vec{r}) \ = \ Eu(\vec{r}); \tag{2.17}$$

$$\left[\frac{\hbar^2}{2m}\frac{d^2}{dx^2} + \mu\right]v(\vec{r}) \ = \ Ev(\vec{r}). \tag{2.18}$$

The excitations are plane wave solutions $u(x) = e^{\pm ik_+x}$ and $v(x) = e^{\pm ik_-x}$, where

$$k_\pm = \sqrt{2m(\mu \pm E)}/\hbar. \tag{2.19}$$

The spectrum is continuous. Defining the Fermi wave vector $k_F = \sqrt{2m\mu}/\hbar$, we see that the wave vector $k_+ > k_F$, i.e. the excitation u is outside the Fermi sea; we refer to it as a particle-like excitation. Similarly, $k_- < k_F$ corresponds to a hole-like excitation v within the Fermi sea. Thus in total there are four types of excitations: right- and left-moving particles $e^{\pm ik_+x}$ and holes $e^{\mp ik_-x}$, where the upper (lower) sign corresponds to right- (left-) movers. Note that holes are time-reversed particles and thus move *in opposite direction*. This can be seen explicitly by calculating the group velocity $v_{g,\pm}$ for particles and holes from Eq. (2.19), thereby showing that for holes $v_{g,-} \sim dE/dk_- \sim -k_-$.

Superconducting metal. In a uniform superconducting metal, we have $g = const \neq 0$, hence $\Delta(\vec{r}) = \Delta = |\Delta|e^{i\phi}$, where ϕ is the *superconducting phase*. The BdG equations therefore form a *coupled set*

$$-\left[\frac{\hbar^2}{2m}\frac{d^2}{dx^2} + \mu\right]u(\vec{r}) + \Delta v(\vec{r}) = Eu(\vec{r}); \tag{2.20}$$

$$\left[\frac{\hbar^2}{2m}\frac{d^2}{dx^2} + \mu\right]v(\vec{r}) + \Delta^* u(\vec{r}) = Ev(\vec{r}). \tag{2.21}$$

We find again four plane wave solutions for right- and left-moving particles and holes, $u(x) \sim v(x) \sim e^{\pm i\lambda_\pm x}$ where $\lambda_\pm = \sqrt{2m(\mu \pm \sqrt{E^2 - |\Delta|^2})}/\hbar$ (upper sign corresponds to particle, lower sign to hole). One immediately sees that there are two cases to be distinguished: $E > |\Delta|$ and $E < |\Delta|$. The first case corresponds to excitations with an energy larger than the gap. Their wave vector is purely real, thus these excitations correspond to extended states throughout the superconductor. The second case leads to wave vectors with a *nonzero imaginary part*, $\Im m [\lambda] \sim \sqrt{|\Delta|^2 - E^2}/\hbar v_F$. In other words, such excitations *decay* over a length scale $1/\Im m [\lambda]$. The smallest value of this length is obtained at the Fermi level ($E = 0$) and equals the *clean superconducting coherence length* $\xi_S = \hbar v_F/|\Delta|$. There are no extended state excitations with energies E less than $|\Delta|$.

We introduce amplitudes u_0 and v_0 such that

$$u_0^2 = 1 - v_0^2 = \frac{1}{2}\left(1 + \frac{\sqrt{E^2 - |\Delta|^2}}{E}\right). \tag{2.22}$$

Note that these amplitudes are complex for energies below the gap. The particle-like excitations can then be written as

$$\psi_p(x) = \begin{pmatrix} u_0 \\ v_0 \end{pmatrix} e^{\pm i\lambda_+ x}, \tag{2.23}$$

and similarly for hole-like ones

$$\psi_h(x) = \begin{pmatrix} v_0 \\ u_0 \end{pmatrix} e^{\mp i\lambda_- x}, \tag{2.24}$$

where the upper (lower) sign in the exponentials corresponds to right- (left-) moving excitations.

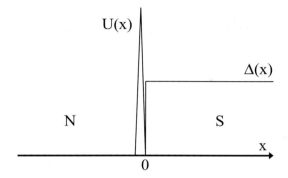

Fig. 2. NS interface as discussed in the text.

2.2.3. NS interface

We now turn to the discussion of the NS interface, as depicted in Fig. 2. We follow Ref. [4], and assume the gap $\Delta(x)$ to vary as a step function: $\Delta(x) = \Delta\Theta(x)$, with Δ real. This step function approximation implies in particular that we ignore spatial variations of the gap on the superconducting side induced by the presence of the interface with the normal metal (so-called *inverse proximity effect*). In addition, scattering may be present at the interface, represented by a non-vanishing potential energy $U(x)$ in the BdG equations. We thus can consider an incoming particle-like excitation,

$$\psi_{inc}(x) = \begin{pmatrix} 1 \\ 0 \end{pmatrix} e^{ik_+ x}, \tag{2.25}$$

that can be reflected either as a hole or as a particle,

$$\psi_{refl}(x) = a \begin{pmatrix} 0 \\ 1 \end{pmatrix} e^{ik_- x} + b \begin{pmatrix} 1 \\ 0 \end{pmatrix} e^{-ik_+ x}, \tag{2.26}$$

or transmitted as a particle or a hole,

$$\psi_{trans}(x) = c \begin{pmatrix} u_0 \\ v_0 \end{pmatrix} e^{i\lambda_+ x} + d \begin{pmatrix} v_0 \\ u_0 \end{pmatrix} e^{-i\lambda_- x}. \tag{2.27}$$

The coefficients a, b, c, d depend on the scattering at the interface and will be determined in some cases below.

Clean interface. A clean interface is one at which no backscattering occurs, thus we set $U(x) = 0$. Matching the solutions at the interface we find the following relations

$$1 + b = cu_0 + du_0, \qquad a = cv_0 + du_0,$$
$$(1 - b)k_+ = cu_0\lambda_+ - dv_0\lambda_-, \qquad k_-a = cv_0\lambda_+ - du_0\lambda_-. \qquad (2.28)$$

Making the *quasi-classical approximation* $k_+ \simeq k_- \simeq \lambda_+ \simeq \lambda_- \simeq k_F$, we immediately find $b = d = 0$, $a = v_0/u_0$, $c = 1/u_0$.

We thus see that, both for $E > \Delta$ and $E < \Delta$, the normal reflection probability $B = |b|^2 = 0$ for a clean interface. However, the particle can be *reflected as a hole*, with probability $A = |a^2| = |v_0/u_0|^2$. This probability is 1 for $E < \Delta$. Moreover, due to the complex nature of u_0 and v_0 at energies below the gap, the phase of the amplitude a is nonzero and given by $\phi_a = \arccos E/\Delta$. The maximum value of this phase $\phi_a = \pi/2$ is found for $E = 0$, and it decreases with increasing E. At the gap $E = \Delta$, the phase $\phi_a = 0$. For energies above the gap we find $A = (E - \sqrt{E^2 - \Delta^2})/(E + \sqrt{E^2 - \Delta^2})$, the phase ϕ_a remains 0. The probability to be reflected as a hole decreases with energy and tends to zero for $E \gg \Delta$.

The group velocity and the wave vector of a hole-like excitation have opposite sign, therefore the current carried by a reflected hole in the normal electrode is directed *towards* the interface. The total transmission associated to the particle flux is therefore given by $1 - B(E) + A(E)$. The *hole reflection channel below the gap thus increases current*: for $A = 1$ and $B = 0$ for each incoming particle there is a reflected hole and thus effectively two particles enter the superconductor, decaying eventually into a Cooper pair. This corresponds to the *Andreev reflection process*. Above the gap, $A = B = 0$, only usual particle transmission occurs with unity transmission.

Effect of scattering at the interface. Still following the paper by Blonder, Tinkham and Klapwijk [4] we model scattering at the interface with the help of a δ-function barrier located at $x = 0$, thus putting $U(x) = U_0\delta(x)$ in the BdG equations. The effect of this barrier can be conveniently expressed with the help of the dimensionless parameter $Z = mU_0/\hbar^2 k_F$ that measures the barrier strength with respect to the Fermi energy. Matching the wave functions at the interface yields the four coefficients a, b, c, d and the related probabilities A, B, C, D. For detailed calculations and results we refer the reader to Ref. [4]. In order to illustrate the general trend, we plot the behavior of $A(E)$ and $B(E)$ as a function of energy E for various values of the parameter Z in Fig. 3. If scattering at the interface is weak, we find the results of the previous paragraph: $B \approx 0$ at all energies and $A \approx 1$ below the gap, dropping to 0 above the gap. At energies below the gap,

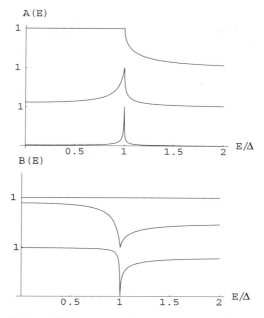

Fig. 3. Reflection coefficients $A(E)$ for Andreev reflection (upper panel) and $B(E)$ for normal reflection (lower panel) as a function E/Δ, for $Z = 0.01$, 1.0, and 2.0 (from top to bottom). Curves are off-set for clarity.

the only reflection process is Andreev reflection. Increasing the barrier strength increases normal backscattering thus reducing the subgap transmission, except for a narrow region close to the gap where Andreev reflection still dominates.

I-V characteristics. In order to discuss transport, we can use the standard *Landauer-Büttiker scattering formalism* [20, 21]. In this formalism, the usual, single-channel result for current through a conductor with transmission coefficient $T(E)$ is given by

$$I = \frac{e}{\pi \hbar} \int dE [f(E) - f(E + eV)] T(E). \tag{2.29}$$

The current I flows between two reservoirs kept at electrochemical potential difference eV. This result can be extended to the case of a one-dimensional, single-channel NS interface, replacing the transmission coefficient $T(E)$ by the quantity $1 - B(E) + A(E)$,

$$I_{NS} = \frac{e}{\pi \hbar} \int dE [f(E) - f(E + eV)][1 - B(E) + A(E)]. \tag{2.30}$$

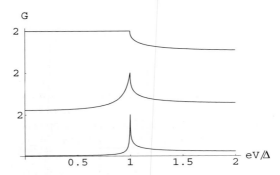

Fig. 4. Dimensionless zero-temperature differential conductance $G = dI_{NS}/G_Q dV$, where $G_Q = e^2/\pi\hbar$, as a function of voltage eV/Δ. Curves are off-set for clarity and correspond (from top to bottom) to $Z = 0.01, 1.0$, and 2.0.

Here, the distribution function $f(E)$ refers to the normal metallic electrode. As shown in [4], this expression can be easily generalized in the case of multichannel contacts, as long as the channels can be added independently. This is the case, *e.g.*, when considering a planar tunnel barrier. For our purpose, the single channel case is sufficient. In the zero-temperature limit, (2.30) becomes

$$I_{NS} = \frac{e}{\pi\hbar} \int_0^{eV} dE[1 - B(E) + A(E)]. \qquad (2.31)$$

Therefore, the differential conductance at $T = 0$ becomes simply $dI_{NS}/dV = (e^2/\pi\hbar)[1 + A(eV) - B(eV)]$. This result is plotted in Fig. 4, for the same parameters as in Fig. 3. We see that for weak backscattering the subgap conductance attains twice the normal state conductance: each incoming electron is accompanied by hole reflection. Thus two charges are transferred thereby yielding a Cooper pair in S. The doubling of the conductance with respect to the normal state situation leads to the phenomenon known as *excess current*. Upon increasing the backscattering strength, normal back reflection becomes possible, which decreases the conductance. For strong back-reflection the two-particle transfer becomes highly improbable and the conductance drops significantly below the gap.

Experimental results that are compatible with the analysis presented here were obtained by Blonder and Tinkham [22], measuring the I-V characteristics of point contact junctions between copper and niobium. However, as we will see below, experiments performed in the early nineties on mesoscopic diffusive NS systems show significant deviations from these results. These deviations can be understood, taking mesoscopic interference effects into account that so far have been ignored in our analysis.

3. Disordered NIS junctions

3.1. Introduction

In 1991, experimental work by Kastalsky *et al.* [23] demonstrated the existence of a large, narrow peak at zero bias in the subgap differential conductance of a semiconductor-based NS junction (Nb-InGaAs junction). The existence of this peak is clearly incompatible with the BTK theory developed above. This discovery therefore triggered a substantial activity, both experimentally and theoretically [24]. It was readily understood that the *phase-coherent motion of particles and holes* over distances $\xi_N = \sqrt{\hbar D/\max\{eV, k_B T\}}$ in the diffusive normal metal is responsible for the anomaly. Much experimental work was done on systems with relatively clean interfaces, *i.e.*, the parameter Z is not too large. A powerful theoretical formalism based on quasi-classical, non-equilibrium Green's functions [6, 7, 8] is generally used to describe such systems [9, 10]. Unfortunately, this formalism is not easily amenable to physical interpretation. However, the anomaly was also seen in tunnel systems [25, 26], which can be well described in the framework of the tunnel Hamiltonian. Indeed, as we will see below, the anomaly can be relatively simply obtained in this case, combining perturbation theory with well-known methods to describe disordered metals.

As a matter of fact, it has been shown in Ref. [13], that the subgap Andreev tunnel current is strongly affected by the coherent scattering of electrons by impurities near the junction region. Two electrons originating from the superconductor with a difference in energy of ϵ can propagate in the normal metal on a length scale of the order of $\xi_\epsilon = \sqrt{\hbar D/\epsilon}$ before dephasing (D being the diffusion coefficient). If the relevant energy scale of the problem, *i.e.* the voltage bias V and the temperature T are sufficiently small, the resulting coherence length $\xi_N = \sqrt{\hbar D/\max\{eV, k_B T\}}$ is much larger than the mean free path l. Thus at low temperatures and voltages the electron pairs are able to "see" the spatial layout on a length scale given by $\xi_N \gg l$. Due to *multiple scattering close to the barrier*, electron pairs attempt many times to tunnel into the superconductor before leaving the junction region. Interference thus substantially increases the current at low voltage bias and the resulting conductance depends strongly on the explicit layout.

3.2. Perturbation theory for NIS junction

3.2.1. Tunnel Hamiltonian and golden rule
We start our perturbative analysis again from the tunnel Hamiltonian

$$\hat{H}_T = \sum_{k,p,\sigma} t_{k,p}\hat{a}^\dagger_{k,\sigma}\hat{b}_{p,\sigma} + t^*_{k,p}\hat{b}^\dagger_{p,\sigma}\hat{a}_{k,\sigma}. \tag{3.1}$$

Note that as before we assume spin-independent tunnel matrix elements. However we keep the dependence on quantum numbers k and p as this enables us to properly treat the diffusive nature of the electronic motion in the electrodes. We will assume that $\hat{a}$-operators describe the normal electrode and $\hat{b}$-operators the superconducting electrode.

Indeed, the disorder in the electrodes gives rise to a nonzero, random impurity potential $U(\vec{r})$ which scatters electrons. In the absence of a magnetic field, we can introduce real one-electron wave functions that obey the associated Schrödinger equation

$$\left[-\frac{\hbar^2}{2m}\Delta + U(\vec{r}) - \mu \right] \psi_k(\vec{r}) = \xi_k \psi_k(\vec{r}). \tag{3.2}$$

We use these wave functions as a complete basis to decompose various space-dependent quantities. For example, in order to perform the Bogoliubov transformation given by Eqs. (2.12) and (2.13) in the presence of the disorder potential $U(\vec{r})$, we use the wave functions $\psi_k(\vec{r})$ and write $u_k(\vec{r}) = u_k \psi_k(\vec{r})$, $v_k(\vec{r}) = v_k \psi_k(\vec{r})$. Assuming the gap Δ to be uniform and imposing the normalization condition $|u_k|^2 + |v_k|^2 = 1$, the BdG equations (2.15) and (2.16) then imply $E_k = \sqrt{\xi_k^2 + |\Delta|^2}$ with

$$|u_k|^2 = 1 - |v_k|^2 = \frac{1}{2}\left(1 + \frac{\xi_k}{E_k}\right). \tag{3.3}$$

We also find the useful relation $u_k v_k^* = \Delta/2E_k$ which shows that the relative phase between u and v^* is just determined by the phase of the superconducting order parameter.

Comparing the transformation defined by Eqs. (2.12) and (2.13) with the usual decomposition

$$\hat{\Psi}_\sigma(\vec{r}) = \sum_k \hat{b}_{k,\sigma} \psi_k(\vec{r}) \tag{3.4}$$

in terms of ordinary electronic operators $\hat{b}$, we conclude that the Bogoliubov transformation can also be written as

$$\hat{b}_{p,\uparrow} = u_p \hat{\gamma}_{p,\uparrow} - v_p^* \hat{\gamma}_{p,\downarrow}^\dagger; \tag{3.5}$$

$$\hat{b}_{p,\downarrow} = u_p \hat{\gamma}_{p,\downarrow} + v_p^* \hat{\gamma}_{p,\uparrow}^\dagger. \tag{3.6}$$

The interest of this representation is that it enables us to express the tunnel Hamiltonian in terms of the Bogoliubov quasiparticle operators $\hat{\gamma}$ and $\hat{\gamma}^\dagger$.

Let us now consider the tunnel transfer of two electrons from the normal metal to the superconductor. This is a second order process in $\hat{H}_T$ that involves a virtual state with an excitation on both sides of the barrier. There are two ways to accomplish the second order process:

Possibility 1: As a first step, an electron with quantum number k and spin up is annihilated, thereby creating an excitation with spin up and quantum number p in the superconductor. This leads to a virtual state with energy $E_p - \xi_k$. During the second step this excitation is annihilated upon the transfer of the second electron (quantum number k' and spin down). The amplitude for this process is

$$A_1 = v_p t_{k',p}^* \frac{1}{E_p - \xi_k} u_p^* t_{k,p}^*. \tag{3.7}$$

Possibility 2: Alternatively, we can start with the transfer of the electron with quantum number k' and spin down, and then transfer the electron with quantum number k and spin up as a second step. Here, a virtual state with energy $E_p - \xi_k'$ is involved. The amplitude for this process is thus

$$A_2 = v_p t_{k,p}^* \frac{1}{E_p - \xi_{k'}} u_p^* t_{k',p}^*. \tag{3.8}$$

The total amplitude is then given by the sum $A_1 + A_2$, thereby summing over all possible intermediate states p,

$$A_{k\uparrow,k'\downarrow} = \sum_p t_{k,p}^* t_{k',p}^* u_p^* v_p \left[\frac{1}{E_p - \xi_k} + \frac{1}{E_p - \xi_{k'}} \right]. \tag{3.9}$$

The total rate for tunnelling from N to S is then given by the golden rule expression

$$\Gamma_{N \to S} = \frac{4\pi}{\hbar} \sum_{k,k'} |A_{k\uparrow,k'\downarrow}|^2 \delta(\xi_k + \xi_{k'} + 2eV) f(\xi_k) f(\xi_{k'}). \tag{3.10}$$

A similar expression can be written for the opposite rate $\Gamma_{S \to N}$, the total rate $\Gamma_{NIS} = \Gamma_{N \to S} - \Gamma_{S \to N}$.

Under subgap conditions, both eV and temperature $k_B T$ are small compared to the gap Δ. Therefore, the single particle energies ξ will be always small compared to the gap, implying that, when evaluating the rate Γ_{NIS}, we can approximate $\xi_k, \xi_{k'} \ll E_p$, leading to the following approximation for $A_{k\uparrow,k'\downarrow}$:

$$A_{k\uparrow,k'\downarrow} \simeq \sum_p t_{k,p}^* t_{k',p}^* \frac{2u_p^* v_p}{E_p}. \tag{3.11}$$

3.2.2. Real space representation

In order to make a connection with standard diagrammatic techniques for diffusive systems, as presented in the chapter by V. Kravtsov, we rewrite the above expressions in real space coordinates. We introduce the expansion of $t_{k,p}$ in terms of the complete set of wave functions for the disordered electrodes

$$t_{k,p} = \int d^3r d^3r' t(\vec{r}, \vec{r}') \psi_k(\vec{r}) \psi_p(\vec{r}'). \tag{3.12}$$

As a result, the amplitude $A_{k\uparrow,k'\downarrow}$ can be written as

$$A_{k\uparrow,k'\downarrow} = \int d^3r_1 d^3r_2 d^3r_1' d^3r_2' t^*(\vec{r}_1, \vec{r}_1') t^*(\vec{r}_2, \vec{r}_2') \times$$
$$\psi_k(\vec{r}_1) \psi_{k'}(\vec{r}_2) F^*(\vec{r}_1', \vec{r}_2'), \tag{3.13}$$

where we defined the quantity

$$F(\vec{r}_1, \vec{r}_2) = \sum_p \frac{2u_p v_p^*}{E_p} \psi_p(\vec{r}_1) \psi_p(\vec{r}_2), \tag{3.14}$$

which describes *anomalous propagation* between two points $\vec{r}_1$ and $\vec{r}_2$ in the superconductor. The amplitude $t(\vec{r}, \vec{r}')$ describes tunnelling between points $\vec{r}$ and $\vec{r}'$ on different sides of the barrier. We will assume that tunnelling predominantly occurs between neighboring points at the barrier, and thus $t(\vec{r}, \vec{r}') = t(\vec{r}) \delta(\vec{r} - \vec{r}') \delta(z)$. Here we assume that the barrier is planar and located at $z = 0$. Therefore,

$$A_{k\uparrow,k'\downarrow} = \int\limits_{barrier} d^2r_1 d^2r_2 t^*(\vec{r}_1) t^*(\vec{r}_2) F^*(\vec{r}_1, \vec{r}_2) \psi_k(\vec{r}_1) \psi_{k'}(\vec{r}_2), \tag{3.15}$$

leading to

$$|A_{k\uparrow,k'\downarrow}|^2 = \int\limits_{barrier} d^2r_1 d^2r_2 d^2r_3 d^2r_4 t^*(\vec{r}_1) t^*(\vec{r}_2) t(\vec{r}_3) t(\vec{r}_4) \times$$
$$F^*(\vec{r}_1, \vec{r}_2) F(\vec{r}_3, \vec{r}_4) \psi_k(\vec{r}_1) \psi_{k'}(\vec{r}_2) \psi_k(\vec{r}_3) \psi_{k'}(\vec{r}_4). \tag{3.16}$$

Further progress can be made introducing the retarded and advanced Green's function

$$G_\xi^{R,A}(\vec{r}_1, \vec{r}_2) = \sum_k \frac{\psi_k(\vec{r}_1) \psi_k(\vec{r}_2)}{\xi - \xi_k \pm i\eta}, \tag{3.17}$$

as well as the spectral function

$$K_\xi(\vec{r}_1, \vec{r}_2) = \frac{1}{2\pi i}[G_\xi^A - G_\xi^R]. \tag{3.18}$$

Using the definition (3.17), the latter can be rewritten

$$
\begin{aligned}
K_\xi(\vec{r}_1, \vec{r}_2) &= \frac{1}{2\pi i} \sum_k \psi_k(\vec{r}_1)\psi_k(\vec{r}_2) \left[\frac{1}{\xi - \xi_k - i\eta} - \frac{1}{\xi - \xi_k + i\eta} \right] \\
&= \sum_k \psi_k(\vec{r}_1)\psi_k(\vec{r}_2)\delta(\xi - \xi_k).
\end{aligned} \tag{3.19}
$$

For later use we note that in the limit $\vec{r}_1 = \vec{r}_2 = \vec{r}$, the spectral function is given by

$$
K_\xi(\vec{r}, \vec{r}) = \sum_k |\psi_k(\vec{r})|^2 \delta(\xi - \xi_k), \tag{3.20}
$$

which is by definition the *energy-dependent local DoS* $v(\vec{r}, \xi)$ per spin and per unit energy and unit volume.

We finally rewrite the anomalous propagator $F(\vec{r}_1, \vec{r}_2)$, Eq. (3.14), with the help of the spectral function $K_\xi(\vec{r}_1, \vec{r}_2)$. Using the relation $2u_p v_p^* = \Delta/E_p$, and inserting the identity $1 = \int d\zeta \, \delta(\zeta - \xi_p)$, we find that

$$
F(\vec{r}_1, \vec{r}_2) = \int d\zeta \, \frac{\Delta}{\zeta^2 + |\Delta|^2} K_\zeta(\vec{r}_1, \vec{r}_2). \tag{3.21}
$$

Substituting the results (3.16) and (3.21) into (3.10) and inserting the identity $1 = \int d\xi d\xi' \delta(\xi - \xi_k)\delta(\xi' - \xi_{k'})$ we can write, using representation (3.19),

$$
\begin{aligned}
\Gamma_{N \to S} =\ & \frac{4\pi}{\hbar} \int d\xi d\xi' d\zeta d\zeta' f_\Delta^*(\zeta) f_\Delta(\zeta')\delta(\xi + \xi' + 2eV)f(\xi)f(\xi') \times \\
& \int_{barrier} d^2r_1 d^2r_2 d^2r_3 d^2r_4 t^*(\vec{r}_1)t^*(\vec{r}_2)t(\vec{r}_3)t(\vec{r}_4) \times \\
& K_\zeta(\vec{r}_1, \vec{r}_2)K_{\zeta'}(\vec{r}_3, \vec{r}_4)K_\xi(\vec{r}_1, \vec{r}_3)K_{\xi'}(\vec{r}_2, \vec{r}_4),
\end{aligned} \tag{3.22}
$$

where we defined the function $f_\Delta(\zeta) = \Delta/(\zeta^2 + |\Delta|^2)$. This result is represented diagrammatically in Fig. 5. The crosses at $\vec{r}_i$, for $i = 1, \ldots, 4$ correspond to tunnelling with amplitude $t(\vec{r}_i)$. The lines describe propagation on the disordered N and S-side of the barrier and correspond to spectral functions K. The diagram can be evaluated by properly averaging over the disorder, as we will discuss below.

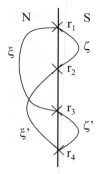

Fig. 5. Diagrammatic representation of two-electron tunnelling process.

3.2.3. Disorder averaging

We perform the disorder averaging in the standard way, using the following well-known facts (see also the chapter by V. Kravtsov):

– the impurity-averaged spectral function $\langle K_\xi(\vec{r}_1, \vec{r}_2)\rangle_{dis}$ is short-ranged in space (typical scale is the Fermi wavelength λ_F). It attains its maximum value for $\vec{r}_1 \simeq \vec{r}_2 = \vec{r}$,

$$\langle K_\xi(\vec{r}, \vec{r})\rangle_{dis} = \langle \nu(\vec{r}, \xi)\rangle = \nu_0 \tag{3.23}$$

where $\nu_0 \equiv N_0/\Omega$ is the DoS per spin and per unit energy and unit volume at the Fermi level. In other words, the density of states of a weakly disordered metal is practically disorder independent. This result has two important consequences.

First of all, the zero-temperature normal state tunnel conductance, which can be easily shown to be given by

$$G_T = \frac{4\pi e^2}{\hbar} \int_{barrier} d^2r_1 d^2r_2 t^*(\vec{r}_1) t(\vec{r}_2) \langle K_0(\vec{r}_1, \vec{r}_2)\rangle^2_{dis}, \tag{3.24}$$

can also be written as

$$G_T = \int_{barrier} d^2r\, g_T(\vec{r}), \tag{3.25}$$

where we defined the *local barrier tunnel conductance*

$$g_T(\vec{r}) = \frac{4\pi e^2}{\hbar} |t(\vec{r})|^2 \int_{barrier} d^2r' \langle K_0(\vec{r}, \vec{r}')\rangle^2_{dis} \sim \frac{4\pi e^2}{\hbar} |t(\vec{r})|^2 \nu_0^2 \lambda_F^2. \tag{3.26}$$

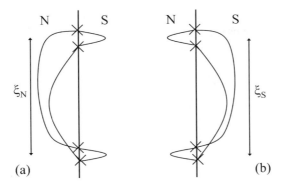

Fig. 6. Approximations discussed in the text.

Secondly, for coinciding arguments, the anomalous propagator F reads

$$\langle F(\vec{r}, \vec{r})\rangle_{dis} = \pi v_0 e^{i\phi(\vec{r})} \tag{3.27}$$

where we added a (weak) dependence of the gap on $\vec{r}$ through the superconducting phase $\phi(\vec{r})$. This enables one to describe, *e.g.*, the presence of a supercurrent or loop geometries with a magnetic flux.
– the average over impurities of the product of a retarded and advanced Green's functions yields

$$\langle G_\xi^R(\vec{r}_1, \vec{r}_2) G_{\xi'}^A(\vec{r}_2, \vec{r}_1)\rangle_{dis} \equiv 2\pi v_0 P_{\xi-\xi'}(\vec{r}_1 - \vec{r}_2), \tag{3.28}$$

where $P_{\xi-\xi'}(\vec{r}_1 - \vec{r}_2)$ is the Cooperon, which satisfies the equation

$$-\hbar D \Delta_{\vec{r}_1} P_\epsilon(\vec{r}_1 - \vec{r}_2) - i\epsilon P_\epsilon = \delta(\vec{r}_1 - \vec{r}_2), \tag{3.29}$$

with D the diffusion coefficient. This means that the impurity average

$$\langle K_\xi(\vec{r}_1, \vec{r}_2) K_{\xi'}(\vec{r}_2, \vec{r}_1)\rangle_{dis} = \frac{v_0}{2\pi}[P_{\xi-\xi'}(\vec{r}_1 - \vec{r}_2) + P_{\xi'-\xi}(\vec{r}_1 - \vec{r}_2)]. \tag{3.30}$$

This is a long-ranged object: as can be seen from Eq. (3.29), space-dependent diffusion occurs over a characteristic length scale $|\vec{r}_1 - \vec{r}_2|$ of the order of $\sqrt{\hbar D/\epsilon}$.

We are now in a position to calculate the total impurity averaged rate for transfer of electrons through the NIS junction. We use the fact that *single* impurity-averaged lines in diagram 5 are short-ranged but averaged *pairs* of lines are long-ranged. This leads us to the approximation depicted in Fig. 6. The long-ranged contribution stems from Cooperon propagation either in N or in S. Let us compare the spatial range of the Cooperon in N and in S. In N, the energy difference $\xi - \xi'$

is determined by the Fermi functions f and given by $\max(eV, k_BT)$. Therefore the range of the Cooperon is given by $\xi_N = \sqrt{\hbar D/\max(eV, k_BT)}$. As to S, the function $f_\Delta(\zeta)$ fixes $\zeta \sim \Delta$, therefore the range of the Cooperon is given by $\xi_S = \sqrt{\hbar D/\Delta}$. Under subgap conditions $k_BT, eV \ll \Delta$, and thus $\xi_N \gg \xi_S$. We conclude that the contribution from diagram 6a dominates the one stemming from 6b.

In accordance with the above statements, the total rate $\Gamma_{NIS} = \Gamma_{N\to S} - \Gamma_{S\to N}$ and hence the impurity-averaged current through the NIS junction, can be written as

$$I_{NIS} = 2e\Gamma_{NIS} = \int d\epsilon\, I(\epsilon)[f(\epsilon/2 - eV) - f(\epsilon/2 + eV)], \tag{3.31}$$

where we defined the *spectral current*

$$I(\epsilon) = \frac{\hbar}{8e^3v_0} \int_{barrier} d^2r_1 d^2r_2 \cos[\phi(\vec{r}_1) - \phi(\vec{r}_2)] g_T(\vec{r}_1) g_T(\vec{r}_2) \times$$
$$[P_\epsilon(\vec{r}_1 - \vec{r}_2) + P_{-\epsilon}(\vec{r}_1, \vec{r}_2)]. \tag{3.32}$$

We calculate the differential conductance directly from Eq. (3.31), finding

$$G_{NIS} = dI_{NIS}/dV = \frac{e}{4k_BT} \int d\epsilon \left[\frac{I(\epsilon + 2eV) + I(\epsilon - 2eV)}{\cosh^2(\beta\epsilon/4)}\right]. \tag{3.33}$$

These equations constitute the central results of this Section. They clearly show the interplay between phase coherence in a superconductor and a normal metal. The intrinsic coherence of the superconductor is reflected by the appearance of the phase difference $\phi(\vec{r}_1) - \phi(\vec{r}_2)$ between tunnel points $\vec{r}_1$ and $\vec{r}_2$. This phase difference plays a crucial role in the experiment [25], see also Section 3.4.3. below. In the normal metal the two incoming electrons undergo many elastic scattering events in the junction region, before they tunnel through the NIS interface, leading to electronic interference on a length scale given by ξ_N. These interference effects have been taken into account by averaging in the standard way over possible scattering events, leading to the appearance of the Cooperon P. Therefore, subgap transport in NIS junctions not only depends on properties of the barrier (the local tunnel conductance g_T), but also on its surroundings, over a distance ξ_N, due to interference occurring on this length scale.

3.3. Example: quasi-one-dimensional diffusive wire connected to a superconductor

3.3.1. Calculation of the spectral current

In order to illustrate the implications of the results just obtained we study an instructive example. A quasi-one-dimensional wire (length L and cross-section

S_{eff}) is connected to a massive superconducting electrode by a point-contact tunnel junction, placed at $\vec{r} = 0$. The spectral current is thus obtained by setting $\vec{r}_1 = \vec{r}_2$ in Eq. (3.32), hence its energy dependence is determined by the behavior of $P_\epsilon(0, 0)$. In order to obtain this quantity, we assume that $\xi_N \gg S_{eff}^{1/2}$ such that diffusion is one-dimensional and we can solve the one-dimensional diffusion equation

$$-\hbar D d^2 P/dx_1^2 P_\epsilon(x_1, x_2) - i\epsilon P_\epsilon(x_1, x_2) = \delta(x_1 - x_2), \qquad (3.34)$$

with boundary condition $dP/dx = 0$ at the barrier, indicating no current. Upon Fourier transforming this equation we obtain the Cooperon in momentum representation,

$$P(k, \epsilon) = \frac{1}{\hbar D k^2 - i\epsilon}, \qquad (3.35)$$

which can be transformed back to yield

$$
\begin{aligned}
P_\epsilon(0, 0) &= \frac{1}{S_{eff}} \frac{1}{L} \sum_k \frac{1}{\hbar D k^2 - i\epsilon} \\
&= \frac{1}{2\pi S_{eff}} \int dk \frac{1}{\hbar D k^2 - i\epsilon} \\
&= \frac{1}{2 S_{eff}} \sqrt{\frac{i}{\epsilon \hbar D}}.
\end{aligned}
\qquad (3.36)
$$

Therefore, the spectral current is given by

$$I(\epsilon) = \frac{\hbar G_T^2}{8 e^3 \nu_0 S_{eff}} \sqrt{\frac{1}{2|\epsilon| \hbar D}}. \qquad (3.37)$$

3.3.2. Zero-temperature limit

At $T = 0$ we can easily perform the integration over energy in Eqs. (3.31) and (3.33) with the result Eq. (3.37) to obtain, respectively

$$I_{NIS} = \frac{\hbar G_T^2}{2 e^2 \nu_0 S_{eff}} \frac{V}{\sqrt{|eV| \hbar D}} \equiv G_T^2 R_N(V) V, \qquad (3.38)$$

and

$$G_{NIS} = 4 e I(|2eV|) = \frac{\hbar G_T^2}{4 e^2 \nu_0 S_{eff}} \frac{1}{\sqrt{|eV| \hbar D}} \equiv G_T^2 R_N(V)/2. \qquad (3.39)$$

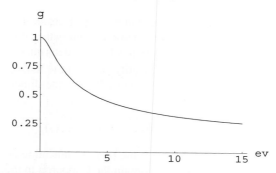

Fig. 7. Dimensionless zero-temperature differential subgap conductance $g = 2G_{NIS}/G_T^2 R_w$, as a function of dimensionless voltage $ev = eV/E_{th}$, in units of the Thouless energy E_{th}.

Here we defined the resistance $R_N(V) \equiv \xi_N/(\sigma_0 S_{eff})$ of the diffusive wire over a length $\xi_N = \sqrt{\hbar D/eV}$, using the Einstein relation $\sigma_0 = 2e^2 \nu_0 D$ to obtain the Drude conductivity σ_0 of the metal.

The length ξ_N, and hence the resistance $R_N(V)$, diverges for $V \to 0$. Physically, this is due to the fact that for $T, V = 0$ the two tunnelling electrons have the same energy, and thus their relative phase is conserved over infinite distances. However, the associated low-energy divergence of the subgap conductance is spurious and will be cut naturally by either of three mechanisms. First of all, the coherent propagation of two electrons in N is limited by decoherence effects. Indeed, electron-electron interactions are known to induce decoherence, and the range of Cooperon propagation is cut by the so-called *decoherence length* L_ϕ [27, 28]. Second, if the length L of the wire is finite, this also provides a cut-off for diffusion, as long as $L < L_\phi$. But even if both L and L_ϕ were infinite, the subgap conductance would not diverge at zero bias. In fact, since we performed a perturbation calculation, we must impose that the subgap conductance remains bound somehow. Indeed, perturbation theory can be shown to break down when the differential conductance becomes of the order of the normal state tunnel conductance, $G_{NIS} \sim G_T$, i.e., when $R_N(V) \sim 1/G_T$. A different charge transfer regime known as *reflectionless tunnelling* sets in, that can only be obtained nonperturbatively, see [5] for details.

In Fig. 7, we plot the differential conductance of the wire as a function of voltage. We assume that the finite length $L < L_\phi$ and thus cuts off the phase-coherent diffusion. Therefore, the maximum value of differential subgap conductance equals $G_{NIS} = G_T^2 R_w/2$, where $R_w = L/(\sigma_0 S_{eff})$ is the total resistance of the wire. The subgap conductance thus will saturate as soon as $R_N(V) \sim R_w$, *i.e.* for voltages V such that $eV < E_{th}$, where $E_{th} = \hbar D/L^2$ is the Thouless energy of the wire. The tunnel approach is correct as long as reflectionless tunnelling can be ignored. This means that $G_T R_w \ll 1$, *i.e.*, the resistance of the wire should be much less than the normal state tunnel resistance of the barrier.

3.4. Subgap noise of a superconductor-normal-metal tunnel interface

So far we have shown that the subgap conductance through a tunnel junction between a superconductor and a normal metal is strongly affected by interference of electron waves scattered by impurities. We conclude this Section by investigating how the same phenomenon affects the *low frequency current noise*, S, for voltages V and temperatures T much smaller than the superconducting gap Δ [29]. More results on subgap noise in mesoscopic NS junctions can be found in the chapter by T. Martin.

3.4.1. Current fluctuations in NS systems

The theoretical understanding of current fluctuations in mesoscopic normal metal-superconductor (N-S) hybrid systems has recently seen important advances [30, 31, 32]. An important reason for this is the development of simple techniques to calculate the *full counting statistics of quantum charge transfer* [33, 34, 35]. In particular, the current fluctuations in a diffusive wire in good contact with a superconductor have been calculated for any voltage and temperature below the gap [36], taking into account the space-dependent coherent propagation in the normal metal of electrons originating from the superconductor.

Interestingly, the case of a tunnel junction between a diffusive metal and a superconductor has been less investigated, partially because it has only very recently become possible to measure current noise in tunnel junctions [37]. Theoretically, current noise in a NIS junction has been considered by Khlus long ago [38], but neglecting the space-dependent coherent propagation of electrons in the normal metal. Later, de Jong and Beenakker included coherent propagation in the normal metal at vanishing temperature and voltage [39]. The effect of a finite voltage was studied using a numerical approach [40]. More complicated structures with several tunnel barriers have also been considered. In some limits these can be reduced to a single dominating NIS junction with a complex normal region [41, 42]. Based on the results obtained above, one may expect that the noise at finite voltage and temperature depends on the spatial layout when a tunnel barrier is present. Here we show indeed how the coherent diffusion of pairs of electrons determines the subgap current noise of a NIS tunnel junction.

In particular, at equilibrium we find that a *generalized Schottky relation* holds for voltages and temperatures smaller than the superconducting gap [29]:

$$S(V, T) = 4e \coth(eV/T) I(V, T). \tag{3.40}$$

As we have seen in Section 3.2 above, due to the interplay between the proximity effect and the presence of the barrier the current voltage characteristics is both non-universal and non-linear. However, according to Eq. (3.40), the *ratio* $F = S/(2eI) = 2\coth(eV/T)$, known as *Fano factor*, is *universal* as long as the electron distribution on the normal metal is at equilibrium. In particular, for $T \ll eV$ the Fano factor is 2, indicating that the elementary charge transfer is achieved by Cooper pairs. If the normal metal is driven out of equilibrium, within our formalism F can be calculated once the geometry is known. One finds then that the noise and the current remain both independent of each other *and* strongly dependent on the geometry. We discuss a realistic example inspired by [43], where we predict strong deviations from Eq. (3.40).

3.4.2. Current noise in tunnel systems

Charge transport in low transmission tunnel systems is characterized by the fact that subsequent events of charge transfer are *uncorrelated*. Therefore, tunnelling is a *stochastic process*, described by Poisson statistics. To be specific, let us assume that each tunnel event transfers a charge q. The average number of tunnel events per unit time from the left to right electrode is Γ_+; similarly, Γ_- for events from right to left. The probability that $N_{+(-)}$ tunnel events occur per unit time from left to right (right to left) is then given by the Poisson distribution

$$P(N_i) = e^{-\Gamma_i}\frac{\Gamma_i^{N_i}}{N_i!}, \tag{3.41}$$

where $i = \pm$.

The current $I_+(t)$ from left to right at time t is given by a sum over current pulses corresponding to tunnel events from left to right at times $t_1, \ldots t_{N_+}$,

$$I_+(t) = q\sum_{n=1}^{N_+}\delta(t - t_n); \tag{3.42}$$

a similar expression can be written for $I_-(t)$. The total current is given by $I(t) = I_+(t) - I_-(t)$. In order to find the average current we need to calculate

$$\langle I_i(t)\rangle = \sum_{N_i=0}^{\infty} P(N_i)\int dt_1 \ldots dt_{N_i}\, I_i(t) = q\Gamma_i. \tag{3.43}$$

We thus find that

$$\langle I(t)\rangle = q(\Gamma_+ - \Gamma_-), \tag{3.44}$$

independent of time t.

The noise $S(t, t')$ is defined as

$$S(t, t') = \langle I(t) I(t') \rangle - \langle I(t) \rangle^2. \tag{3.45}$$

Tunnelling from left to right is independent of tunnelling from right to left. Thus

$$S(t, t') = \sum_{i=\pm} \langle I_i(t) I_i(t') \rangle - \langle I_i(t) \rangle^2. \tag{3.46}$$

A straightforward calculation shows that

$$
\begin{aligned}
\langle I_i(t) I_i(t') \rangle &= \sum_{N_i=0}^{\infty} P(N_i) \int dt_1 \ldots dt_{N_i} \, I_i(t) I_i(t') \\
&= q^2 \Gamma_i^2 + q^2 \delta(t - t') \Gamma_i.
\end{aligned} \tag{3.47}
$$

Therefore, S is given by

$$S(t, t') = q^2 (\Gamma_+ + \Gamma_-) \delta(t - t'). \tag{3.48}$$

We thus conclude that knowledge of the tunnel rates Γ_i implies that not only the average current, but also the zero-frequency noise $S \equiv S(\omega = 0) = q^2(\Gamma_+ + \Gamma_-)$ can be immediately found.

3.4.3. Generalized Schottky relation

The results from Section 3.2 can be straightforwardly applied to our case of noise in an NS tunnel interface under subgap conditions. Identifying the charge $q = 2e$ per tunnel event and the rates $\Gamma_+ = \Gamma_{N \to S}$, $\Gamma_- = \Gamma_{S \to N}$, we obtain immediately, using Eq. (3.10),

$$I = \frac{4\pi e}{\hbar} \sum_{k,k'} |A_{k\uparrow,k'\downarrow}|^2 \delta(\xi_k + \xi_{k'} + 2eV) H_-(\xi_k, \xi_{k'}), \tag{3.49}$$

$$S = \frac{16\pi e^2}{\hbar} \sum_{kk'} |A_{k\uparrow,k'\downarrow}|^2 \delta(\xi_k + \xi_{k'} + 2eV) H_+(\xi_k, \xi_{k'}), \tag{3.50}$$

where

$$H_{\pm}(\xi, \xi') = f(\xi) f(\xi') \pm [1 - f(\xi)][1 - f(\xi')]. \tag{3.51}$$

The impurity averaging procedure can be carried through, both for I and S, as above in Section 3.2. We thus obtain:

$$\begin{pmatrix} I \\ S \end{pmatrix} = \int_{-\infty}^{\infty} d\epsilon \, I(\epsilon) \begin{pmatrix} H_-(\frac{\epsilon}{2} + eV, -\frac{\epsilon}{2} + eV) \\ 4e H_+(\frac{\epsilon}{2} + eV, -\frac{\epsilon}{2} + eV) \end{pmatrix}, \tag{3.52}$$

where the spectral current $I(\epsilon)$ is given by Eq. (3.32).

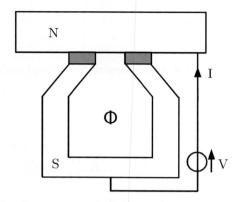

Fig. 8. Experimental set up similar to Ref. [25]. The magnetic flux dependence of the current and noise can be used to measure the doubling of the Fano factor with high accuracy. (From Ref. [29].)

If the normal metal is in thermal equilibrium, $f(\xi)$ is the Fermi distribution. In that case H_+ and H_- are related to each other by the simple relation:

$$\frac{H_+(\epsilon/2 + eV, -\epsilon/2 + eV)}{H_-(\epsilon/2 + eV, -\epsilon/2 + eV)} = \coth\left(\frac{eV}{T}\right), \tag{3.53}$$

and this way we have demonstrated the validity of Eq. (3.40).

In order to emphasize the generality of Eq. (3.40) let us consider a non-trivial example. In Ref. [25] the low-voltage anomalies due to the proximity effect have been detected in a superconducting fork connected to a normal metal through two tunnel junctions as shown in Fig. (8). The superconducting phase difference $\phi_1 - \phi_2$ between the two NIS junctions at the ends of the fork could be tuned by magnetic flux Φ, applied perpendicular to the fork. If the distance between these junctions is less than ξ_N, according to Eq. (3.32) a contribution to subgap current exists involving both junctions; moreover this contribution is proportional to $\cos(\phi_1 - \phi_2)$ and should therefore oscillate with flux Φ. Indeed, it was found experimentally that a part $I(\Phi)$ of the subgap current through the structure is periodically modulated by the external magnetic field. Using Eq. (3.40) one can predict the following noise dependence on the magnetic flux Φ through the ring: $S(\Phi) = 4e \coth(eV/T)I(\Phi)$. Since the Φ dependent part is given by the coherent propagation of *two electrons* between the two junctions any *single particle* contribution is eliminated from the outset. This experiment could thus be used to test accurately the validity of Eq. (3.40) and in particular the doubling of the charge at low temperature.

If $f(\xi)$ is not given by the equilibrium Fermi distribution, Eq. (3.53) does not hold and S and I become independent. Indeed, the Fano factor ceases to be

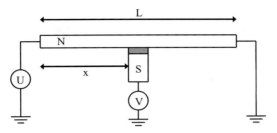

Fig. 9. Normal metallic wire (N) of length L, connected at position x to a superconducting electrode (S) via a tunnel junction. The potential difference across the wire is U, the superconductor is kept at potential V.

universal and gives new information on the system. The technique developed so far enables calculation of F also in this case, as we will now show.

3.4.4. An explicit example: a wire out of equilibrium

The results obtained so far in this Section do not crucially depend on the normal electrode being at thermal equilibrium. We therefore may well consider the effects on transport and noise induced by a *non-equilibrium distribution f* in the normal electrode. Let us consider a realistic example: In Ref. [43] the non-equilibrium electron distribution for a small metallic wire of length L has been measured by using a tunnel junction between the wire (at different positions x along the wire) and a superconductor, see also Fig. 9. In that case the quasiparticle current was measured, but the subgap current and noise in the same configuration could also be measured. When L is shorter than the inelastic mean free path the electron distribution $f(x, \xi)$ becomes space-dependent along the wire and is obtained by solving the diffusion equation [43]:

$$f(x, \xi) = f_F(\xi + eU)[1 - x/L] + f_F(\xi)x/L \tag{3.54}$$

where f_F denotes the equilibrium Fermi distribution imposed by normal reservoirs at the two ends of the wire, kept at potential difference U. The distribution $f(x, \xi)$ has a double discontinuity as a function of ξ at $T = 0$.

Since the transverse dimension $S_{eff}^{1/2}$ of the wire is much smaller than ξ_N, we can use the one-dimensional diffusion equation, as we have seen in Section 3.3.1. Following Ref. [28] we impose the boundary condition on the propagator P corresponding to a good contact with a reservoir. This means that $P_\epsilon(x_1, x_2)$ should vanish when $x_1 = 0$ or L. In the dimensionless variables $u = x/L$, Eq. (3.34) becomes

$$\frac{d^2 P}{du_1^2}(u_1, u_2) + i\frac{\epsilon}{E_{th}} P(u_1, u_2) = -\frac{L}{\hbar D S_{eff}}\delta(u_1 - u_2) \tag{3.55}$$

with the boundary conditions $P(0, u_2) = P(1, u_2) = 0$ and where $E_{th} = \hbar D/L^2$ is the Thouless energy of the wire. The solution can be written as a sum of a special solution of the complete equation plus a linear combination of the two linearly independent solutions of the homogeneous equation. The coefficients of this combination can be chosen in such a way as to fulfill boundary conditions. The solution reads:

$$
P(u_1, u_2) = \frac{L}{2\hbar D S_{eff} z} \left[e^{-z|u_1 - u_2|} \right.
$$
$$
\left. - \frac{e^{z(u_1 + u_2)} + e^{z(u_1 + u_2 - 2)} - e^{z(u_1 - u_2)} - e^{-z(u_1 - u_2)}}{e^{2z} - 1} \right],
$$

(3.56)

where, for $\epsilon > 0$, $z = \sqrt{\epsilon/E_{the}} e^{-i\pi/4}$. We can use Eq. (3.56) to calculate the spectral current Eq. (3.32), thereby assuming that the size of the junction is small,

$$
I(\epsilon) = \frac{G_T^2 L}{4\pi e^3 v_0 D S_{eff}} \phi(\epsilon),
$$

(3.57)

where

$$
\phi(\epsilon) = \mathrm{Re} \left[\frac{\sinh(z u_0) \sinh(z(1 - u_0))}{z \sinh(z)} \right],
$$

(3.58)

and u_0 indicates the position of the junction along the wire. Introducing again the resistance of the wire $R_w = L/2S_{eff} e^2 D v_0$ and using Eq. (3.52), the expression for the current simplifies to

$$
I = \frac{G_T^2 R_w}{2\pi e} \int d\epsilon \, \phi(\epsilon) H_-(\epsilon/2 - eV, -\epsilon/2 - eV).
$$

(3.59)

The noise obeys an identical expression with $H_- \rightarrow 4e H_+$. Note that in the limit of an infinite wire the function $\phi(\epsilon) \sim 1/\sqrt{\epsilon}$ diverges at low energy, as expected for one dimensional diffusion. Substituting Eq. (3.51) with Eq. (3.54) into Eq. (3.59) one obtains current and noise for any temperature, voltage, and position along the wire. Let us consider for simplicity one specific case: zero temperature and voltage $0 \leq V \leq U/2$. It is not difficult to show that in this case

$$
I = \frac{G_T^2 R_w}{\pi e} \phi_-, \quad \text{and} \quad S = \frac{4 G_T^2 R_w}{\pi} \phi_+
$$

(3.60)

with

$$
\phi_\pm = (1 - u_0) \int_{2eV}^{2e(U-V)} \phi(\epsilon) d\epsilon + [(1 - u_0)^2 \pm u_0^2] \int_0^{2eV} \phi(\epsilon) d\epsilon.
$$

(3.61)

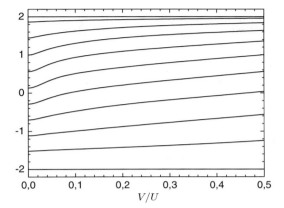

Fig. 10. Differential Fano factor $\mathcal{F}$ as a function of V/U. Different curves represent different values of u_0 going from 0 to 1 in step of 0.1 starting from the top. Departure from the value 2 or -2 is due to the energy dependence of the tunnelling matrix element. (From Ref. [29].)

Differentiating with respect to V Eqs. (3.60) gives the differential Fano factor

$$\mathcal{F}(eV) \equiv \frac{(dS/dV)}{(2e\,dI/dV)}. \tag{3.62}$$

In our case $\mathcal{F}$ is given by the following expression

$$\mathcal{F}(eV) = 2\frac{(1 - u_0)\phi(2eU - 2eV) - u_0(2u_0 - 1)\phi(2eV)}{(1 - u_0)\phi(2eU - 2eV) + u_0\phi(2eV)}. \tag{3.63}$$

Note that normally the Fano factor is positive since it corresponds to the (positive) current noise divided by the absolute value of the current. The *differential Fano factor* is then usually defined in such a way that a factor sign(I) is implicit in the definition. This makes it usually positive, since current and noise are supposed to increase with the voltage bias. But in our case the wire is off-equilibrium, thus the voltage V has not the usual meaning of bias voltage between two systems at local equilibrium. Along the wire it is not even possible to define a chemical potential, since fermions are not at equilibrium. Thus the concept of difference of potential between the superconductor and the wire at the position of the contact is ill-defined. The potential V is nevertheless well-defined and can be used to study the evolution of the conductance or of the differential noise. Clearly the sign of the current needs not be the same as that of V. For this reason we use the definition (3.62) as it is, without changing the sign according to the direction of the current. This allows a more simple representation in Fig. 10.

One should keep in mind that the change of sign has no special meaning in this case, since increasing the voltage bias V can well decrease the current and the noise through the SIN junction.

Let us discuss briefly some simple limiting cases of Eq. (3.63). For $u_0 = 0$ and 1, when the junction is at an extremity of the wire, Eq. (3.63) gives $\mathcal{F} = \pm 2$. This is expected from Eq. (3.40) since in this case the normal metal is actually at equilibrium. The sign change is simply due to fact that for $x = 0$ the potential of the wire U is greater than the potential of the superconducting point V, while for $x = L$ the situation is reversed. For $0 < u_0 < 1$, $\mathcal{F}$ is in general different from 2. An interesting point is the middle of the wire $u_0 = 1/2$. In this case we have:

$$\mathcal{F}(eV, u_0 = 1/2) = 2\frac{\phi(2eU - 2eV)}{\phi(2eV) + \phi(2eU - 2eV)}. \tag{3.64}$$

For $V = U/2$ we then obtain $\mathcal{F} = 1$ exactly for any form of $\phi(\epsilon)$, while for $V = 0$ $\mathcal{F}$ should become very small on general grounds, since we have seen that for $\epsilon \to 0$ the function $\phi(\epsilon)$ is expected to diverge in the infinite system.

For arbitrary values of u_0 one can see a crossover between these limiting behaviors. We report in Fig. 10 the results predicted by Eq. (3.63) for $U = 200E_{th}$ which is a typical value for experiments. We would like to emphasize that the whole energy dependence seen in this plot is due to interference of electronic waves, since it stems from the ϵ-dependence of $\phi(\epsilon)$.

4. Tunnelling in a three-terminal system containing ferromagnetic metals

In this Section we will consider tunnelling in a hybrid system consisting of a superconductor with two probe electrodes which can be either normal metals or polarized ferromagnetic metals [44]. In particular we study transport at subgap voltages and temperatures. Besides the Andreev pair tunnelling at each contact that we discussed in Section 3, in multi-probe structures subgap transport involves *additional channels*, which are due to coherent propagation of two particles (electrons or holes), each originating from a *different probe electrode*. The relevant processes here are *electron co-tunnelling* through the superconductor and *conversion into a Cooper pair* of two electrons stemming from different probes. These processes are *non-local* and decay when the distance between the pair of involved contacts is larger than the superconducting coherence length. The conductance matrix of a three terminal hybrid structure is calculated. The multi-probe processes *enhance* the conductance of each contact. If the contacts are magnetically polarized the contribution of the various conduction channels can be separately detected.

4.1. Introduction

Superconductor-normal metal (SN) contacts at the mesoscopic scale are of primary importance in view of the interplay between coherence effects in the metal and intrinsic coherence of the superconducting condensate which is probed by Andreev reflection [2, 4]. The situation becomes even more interesting with ferromagnetic metals (F). In such metals, due to short-range Coulomb repulsion, there is an imbalance between the number of spin up and spin down electrons. As a result, the DoS is spin-dependent, and a mismatch exists between Fermi wave vectors for given spin: $k_F^{\uparrow(\downarrow)} = k_F \pm q$, where $q = E_{ex}/\hbar v_F$. We defined the characteristic energy of the effect, the *exchange energy* E_{ex}, which is usually large compared to Δ, but still small compared to E_F. *Single SF interfaces* are not expected to give rise to the phenomena discussed in Section 3 above. The subgap Andreev conductance tends to be hindered by magnetic polarization [45, 46, 47] of the F electrode, since not every electron from the spin up band can find a spin down partner to be converted in a Cooper pair. Moreover, the characteristic scale $\xi_F = \sqrt{\hbar D/E_{ex}}$ is too small to give rise to significant diffusive anomalies in subgap transport. In this Section we therefore consider *multi-terminal FS structures*.

Generally, multi-terminal NS structures offer the possibility of manipulating phase coherent transport, similarly to multi-terminal NS structures [48]. An example is the *Andreev interferometer* consisting of a mesoscopic N sample, made of two arms connected to two S "mirrors" on one side and to a single reservoir on the other, where electron propagation in N is sensitive to the difference of the phase of the two superconductors. In general the two SN contacts are separated by a distance larger than the superconducting coherence length $\xi_S = \hbar v_F/\pi \Delta$. A dual configuration, consisting of a superconductor connected to two N probe electrodes separated by a distance *smaller* than ξ_S (see Fig. 11a), with two *independent* reservoirs, was considered in Ref. [49, 50]. It has been proposed that correlations between the N probes could be established across the superconductor, by a process where two electrons *each originating from a different N electrode*, are converted in a Cooper pair. Recently, it has been shown that such correlations could be used to build *entangled states of electrons* [51, 52]. If the probe electrodes are ferromagnetic, the conductance due to these *crossed Andreev* (CA) processes is sensitive to the relative magnetic polarization, being maximal for opposite polarization of the F probes [50].

Here, we investigate in detail the dependence of the conductance on the distance between the probes and on their spin polarization. We consider a three-terminal device A/S/B (see Fig. 11a) where S is an s-wave superconductor and the electrodes A and B can be either normal or ferromagnetic metals. We study the linear conductance matrix at subgap temperature and voltages $k_B T, e V_i \ll \Delta$.

Then, as we have seen, no single electron channel is left and all the relevant channels involve simultaneous (on a time $\sim \hbar/\Delta$) tunnelling of two electrons or holes (Fig. 11b): (a) single-contact Andreev reflection ($2A$ and $2B$), where two electrons, both originating from the same electrode, A or B, are converted in a Cooper pair giving rise to the currents $G_{2A}V_A$ and $G_{2B}V_B$; (b) CA processes, where two electrons, each originating from a *different* A/B electrode are converted into a Cooper pair, with associated current $G_{CA}(V_A + V_B)$ from each A/B probe to S; (c) cotunnelling (EC), which is easily visualized in the tunnelling limit [53] as processes in which, for instance, an electron from A tunnels to B via a virtual state in S; EC processes yield a current from A to B, $G_{EC}(V_B - V_A)$. Then the subgap currents can be presented in a matrix form

$$
\begin{pmatrix} I_A \\ I_B \end{pmatrix} =
$$
$$
\begin{pmatrix} G_{2A} + G_{CA} + G_{EC} & G_{CA} - G_{EC} \\ G_{CA} - G_{EC} & G_{2B} + G_{CA} + G_{EC} \end{pmatrix} \begin{pmatrix} V_A \\ V_B \end{pmatrix}. \tag{4.1}
$$

The multi-contact processes, CA and EC, increase the "diagonal" conductances $dI_A/dV_A, dI_B/dV_B$ and give rise to off-diagonal terms $dI_A/dV_B, dI_B/dV_A$, i.e., the current at probe A (B) depends also on the voltage at probe B (A). The generalization of Eq. (4.1) to more complicated structures (see Fig. 11c) is straightforward.

For illustrative purposes we calculate the conductance matrix for contacts being tunnel junctions. The single-junction conductances G_{2A} and G_{2B} were calculated above in Section 3. Here we find that EC and CA conductances depend on the relative position $\vec{R}$ of the contacts, vanishing exponentially for $|\vec{R}| \gg \xi$. For multichannel tunnel junctions they are found to depend on propagation in S in a way which is sensitive to the geometry of the sample. This is due to interference effects between different channels, which, as we have seen, also play a role in determining the single-junction NS conductances. In multichannel junctions with normal A/B electrodes we find $G_{CA} = G_{EC}$, thus the off-diagonal terms in the conductance matrix Eq. (4.1) vanish. This symmetry is broken if A/B are polarized ferromagnetic metals. Indeed G_{EC} is suppressed if A and B have opposite polarization, because EC processes preserve the spin of the involved electron (in absence of any magnetic scattering whatsoever) and take advantage from parallel A/B polarization. On the other hand CA tunnelling is suppressed for parallel A/B polarization since it is difficult to find two partner electrons with opposite spin to pair up. Then if magnetically polarized probes are used, off-diagonal conductances in Eq. (4.1) are non vanishing, due to the presence of unbalanced CA and EC processes, for inter-contact distance $|\vec{R}| \sim \xi$.

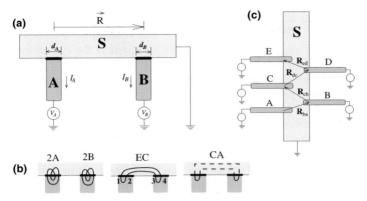

Fig. 11. (a) Schematics of the three-terminal A/S/B device; (b) diagrammatic representation of the processes leading to subgap conductance: single-contact two-particle tunnelling (2A and 2B), elastic cotunnelling (EC), which probes the normal Green's function (full lines) of S and "crossed Andreev" (CA), which probes the anomalous propagator (dashed line); (c) a simple experimental design for measuring the excess EC and CA currents: various electrodes (A,B,C,D,E) allow to probe different distances $\vec{R}_{ij}$. (From Ref. [44].)

4.2. Co-tunnelling and crossed Andreev tunnelling rates

4.2.1. Tunnel Hamiltonian
We describe contacts in the tunnelling regime by standard tunnel Hamiltonians

$$\hat{H}_{TA} = \sum_{kp\sigma} t_{kp}^A \, \hat{a}_{k\sigma}^\dagger \hat{b}_{p\sigma} + t_{kp}^{(A)*} \, \hat{b}_{p\sigma}^\dagger \hat{a}_{k\sigma}; \tag{4.2}$$

$$\hat{H}_{TB} = \sum_{pq\sigma} t_{pq}^B \, \hat{b}_{p\sigma}^\dagger \hat{a}_{q\sigma} + t_{pq}^{(B)*} \, \hat{a}_{q\sigma}^\dagger \hat{b}_{p\sigma}, \tag{4.3}$$

where t_{kp}^A and t_{qp}^B are matrix elements between single electron states k in A, p in S and q in B. Quasiparticle states in S are defined as usual by operators $\hat{\gamma}$ and $\hat{\gamma}^\dagger$ via the transformation (3.5) and (3.6).

We now classify processes which appear in perturbation theory in $\hat{H}_{Ti}$. Single electron processes are absent in the subgap regime. In lowest non-vanishing order we have to consider only "elastic" processes, where the quasiparticle created in the intermediate state is destroyed [53]. "Inelastic" processes leave an excitation of energy $\sim \Delta$ in the superconductor, so under subgap conditions they can be neglected. The relevant processes, represented in Fig. 11b, are single-contact Andreev reflection in the tunnelling limit (2A and 2B), see Section 3, CA tunnelling, and "elastic" cotunnelling [53], where *the same* electron tunnels at once from A to B (B to A). One notices that EC and CA are non-local probes of electron propagation in the superconductor[49]. This is apparent if one considers the gedanken case of single channel tunnel junctions, where the size of the contacts is $d_A, d_B \sim \lambda_F$, the Fermi wavelength. EC probes the "normal" propagator in the superconductor whereas CA processes probe the *anomalous* propagator (see Fig. 11b) and both processes are exponentially suppressed for $R \gg \xi$.

4.2.2. Calculation of spin-dependent tunnel rates

We now turn to the calculation of the spin dependent tunnel rates, using Fermi's golden rule. We consider electrodes A and B which can be magnetically polarized (parallel or antiparallel) along the same axis, resulting in different spectral properties. We neglect here the influence of ferromagnetic electrodes on the superconductor's spectrum. This a reasonable assumption for small tunnel contacts on a massive superconductor. In this case, the inverse proximity effect should not be too serious, *e.g.*, superconductivity is not destroyed in the contact vicinity. This is in contrast with Ref. [54] where a sandwich geometry is considered, and where conduction above the gap is studied.

By proceeding along the lines of Section 3 and writing the EC rate $A\sigma \to B\sigma$ as

$$\Gamma^{\sigma}_{A \to B} = \frac{2\pi}{\hbar} \int d\epsilon d\epsilon' d\zeta d\zeta' \, \delta(\epsilon - \epsilon') \, f(\epsilon - eV_A) \, [1 - f(\epsilon' - eV_B)]$$
$$F_{EC}(\zeta, \epsilon) \, F_{EC}(\zeta', \epsilon') \, \Xi^{\sigma}_{EC}(\epsilon - eV_A, \epsilon' - eV_B, \zeta, \zeta'), \quad (4.4)$$

where $f(\epsilon)$ is the Fermi function, $F_{EC}(\zeta, \epsilon) = (\zeta + \epsilon)/(\zeta^2 + \Delta^2 - \epsilon^2)$ contains information on energies of virtual quasiparticles in S. The rate for CA processes, $(A\sigma, B - \sigma) \to S$, is given by a similar expression

$$\Gamma^{\sigma}_{AB \to S} = \frac{2\pi}{\hbar} \int d\epsilon d\epsilon' d\zeta d\zeta' \, \delta(\epsilon + \epsilon') \, f(\epsilon - eV_A) \, f(\epsilon' - eV_B)$$
$$F_{CA}(\zeta, \epsilon) \, F_{CA}(\zeta', \epsilon') \, \Xi^{\sigma}_{CA}(\epsilon - eV_A, \epsilon' - eV_B, \zeta, \zeta'), \quad (4.5)$$

where $F_{CA}(\zeta, \epsilon) = \Delta/(\zeta^2 + \Delta^2 - \epsilon^2)$. Information about propagation in the specific geometry is contained in the functions $\Xi(\epsilon, \epsilon', \zeta, \zeta')$. Here we give the explicit expression for planar uniform tunnel junctions and local tunnelling, $t(\vec{r}, \vec{r}') = t \, \delta(z) \delta(\vec{r} - \vec{r}')$. In this Section, we consider the case of *ballistic propagation* of plane wave states. The generalization to diffusive conductors can be made following Section 3, this was done in Ref. [55]. The functions Ξ_{EC} and Ξ_{CA} can be expressed as

$$\Xi^{\sigma}_{EC}(\epsilon, \epsilon', \zeta, \zeta') = |t_A t_B|^2 \times$$
$$\int_A d\vec{r}_1 d\vec{r}_2 \int_B d\vec{r}_3 d\vec{r}_4 \, J^{\sigma}_A(12, \epsilon) \, J^{\sigma}_S(31, \zeta) \, J^{\sigma}_S(24, \zeta') \, J^{\sigma}_B(43, \epsilon'); \quad (4.6)$$

$$\Xi^{\sigma}_{CA}(\epsilon, \epsilon', \zeta, \zeta') = |t_A t_B|^2 \times$$

$$\int_A d\vec{r}_1 d\vec{r}_2 \int_B d\vec{r}_3 d\vec{r}_4 \ J^{\sigma}_A(12, \epsilon) \ K^{\sigma}_S(31, \zeta) \ K^{*\sigma}_S(24, \zeta') J^{-\sigma}_B(43, \epsilon'), \quad (4.7)$$

where the spectral functions are denoted J, K and defined as in Section 3, e.g., $J^{\sigma}_A(12, \omega) \equiv J^{\sigma}_A(\vec{r}_1, \vec{r}_2, \omega) = \sum_k \delta(\omega - \epsilon_{k\sigma}) \psi_{k\sigma}(\vec{r}_1) \psi^*_{k\sigma}(\vec{r}_2)$, $K^{\sigma}_S(31, \omega) \equiv K^{\sigma}_A(\vec{r}_3, \vec{r}_1, \omega) = \sum_k \delta(\omega - \epsilon_{k\sigma}) \psi_{-k-\sigma}(\vec{r}_3) \psi_{k\sigma}(\vec{r}_1)$. The space integrals in Eqs. (4.6) and (4.7) run on the contact surfaces. The diagrammatic representation is given in Fig. 11b.

At low temperature and voltages the main contribution to the rates (4.4) and (4.5) is due to electrons close to the Fermi level in the A/B probes, $\epsilon = \pm\epsilon' \approx 0$, and the leading dependence of the rates on the voltages comes from the Fermi functions. One can put $\Xi(\epsilon, \epsilon', \zeta, \zeta') \approx \Xi(0, 0, \zeta, \zeta')$ and perform the ϵ and ϵ' integrations in Eqs. (4.4) and (4.5). At $T = 0$ this gives $I^{\sigma}_{EC} = e\Gamma^{\sigma}_{A \to B} = G^{\sigma}_{EC}(V_B - V_A)$ and $I^{\sigma}_{CA} = 2e\Gamma^{\sigma}_{AB \to S} = 2G^{\sigma}_{CA}(V_A + V_B)$, which define the spin-dependent conductances.

4.2.3. Spin-dependent conductance matrix

Single channel junctions. It is instructive to consider first *single channel junctions* $(d_A, d_B \sim \lambda_F)$. The conductances are calculated by putting $\vec{r}_1 = \vec{r}_2$, $\vec{r}_3 = \vec{r}_4$, $\vec{r}_1 - \vec{r}_3 = \vec{R}$ in Eqs. (4.6) and (4.7). By performing the ζ and ζ' integrations the result is obtained

$$\begin{pmatrix} G^{\sigma}_{EC} \\ G^{\sigma}_{CA} \end{pmatrix} \approx \frac{2\pi^3 e^2}{\hbar} |t_A d^2_A|^2 \, |t_B d^2_B|^2 \, v^2_S v^{\sigma}_A \times$$

$$\frac{e^{-2R/\pi\xi}}{(k_S R)^2} \begin{pmatrix} v^{\sigma}_B \, \cos^2(k_S R) \\ v^{-\sigma}_B \, \sin^2(k_S R) \end{pmatrix}, \quad (4.8)$$

where k_S and v_S are the Fermi wave vector and the normal state DoS of the superconductor. Propagation in the superconductor is characterized by factors which depend on $|\vec{R}|$, are periodic with period π/k_S and are suppressed for $|\vec{R}| \gg \xi$. The magnetic polarization of A/B electrodes enters only via the spin-dependent DoS $v^{\sigma}_{A,B}$, assuming that $v^{\sigma}_{A,B} = v_{A,B}(\pm\mu_e h)$ where $v_{A,B}(\xi)$ is the normal state DoS at energy ξ in absence of magnetism, $h = E_{ex}/\mu_e$ is the exchange field (μ_e being the magnetic moment of the electron), and $\mp$ sign stands for spin σ (anti)parallel to the magnetization. If the DoS for minority spin can be neglected, then for parallel (antiparallel) polarized contacts only EC (CA) processes will yield a conductance.

Multichannel junctions. For the more realistic case of *multichannel tunnel junctions*, interference between different channels has to be considered. This has been studied in Section 3 for single contact processes (*A*1 and *A*2) in a NS tunnel junction, which probe propagation in the *normal* electrode (it is in general diffusive and depends on the geometry). As for EC and CA, we notice that in clean superconductors the single particle propagator is rapidly oscillating ($\sim k_S$), so it would average out on a length $\xi \ll 1/k_S$ in multichannel junctions. This is not the case for two-particle propagators involved in EC and CA processes, as we show explicitly below by considering a specific geometry.

At this stage one could assume that the overall conductance is given by independent single-channel contributions, start from Eq. (4.8) and argue that the factors $\cos^2(k_S R)$ and $\sin^2(k_S R)$ are averaged over distances $\sim d_A, d_B$. The resulting expressions for the EC and CA conductances would be nearly identical yielding, in the special case of normal A/B electrodes, $G_{EC} = G_{CA}$. This conclusion turns out to be correct, even if the actual expression for the conductances involves interference between different channels. To show that we start the calculation from Eqs. (4.6) and (4.7), accounting also for different spin-dependent Fermi wave vectors k_A^σ (k_B^σ) of the A (B) electrode and k_S. We perform the coordinate integrations, and at this stage it becomes apparent that the terms leading to the special dependence $\cos^2(k_S R)$ and $\sin^2(k_S R)$ drop out. Because of interference between different channels, the result depends on the geometry and on the mismatch between the spin-dependent Fermi wave vectors. The simplest case is a geometry with the two junctions belonging to the same plane (see Fig. 11a), where the results depend on the distance $R > d_A, d_B$,

$$\begin{pmatrix} G_{EC}^\sigma \\ G_{CA}^\sigma \end{pmatrix} \approx \frac{h}{8e^2}\, \mathcal{F}_A^\sigma\, G_A^\sigma \begin{pmatrix} \mathcal{F}_B^\sigma\, G_B^\sigma \\ \mathcal{F}_B^{-\sigma}\, G_B^{-\sigma} \end{pmatrix} \frac{e^{-2R/\pi\xi}}{(k_S R)^2}. \tag{4.9}$$

Here the factors $\mathcal{F}_{A,B}^\sigma$ contain information on the geometry and $G_{A,B}^\sigma$ are the spin-dependent one-electron conductances for each junction (S being in the normal state), for instance

$$G_A^\sigma \approx \frac{4\pi e^2}{\hbar}\, v_A^\sigma v_S\, \frac{|t_A|^2\, S}{k_A^\sigma k_S}\, \mathcal{F}(\kappa_A^2, \sqrt{k_S k_A^\sigma} d_A) \tag{4.10}$$

where $\kappa_{A,B} = (k_{A,B}^\sigma/k_S)^{1/2}$ and S is the area of the junction. Here the single junction geometry factor is given by

$$\mathcal{F}(\kappa^2, y) = 2\pi \int_0^y (dx/x)\, \sin(\kappa x)\, \sin(x/\kappa) \tag{4.11}$$

and the two-junction factors in Eq. (4.9) are, *e.g.*,

$$
\begin{aligned}
\mathcal{F}_A &= [\mathcal{F}(\kappa_A^2, \sqrt{k_S k_A} d_A)]^{-1} \times \\
&\quad \Re \int_0^{2\pi} d\theta \, [1 - e^{i k_S d_A (\kappa_A^2 + \cos\theta)}]/(\kappa_A^2 + \cos\theta)
\end{aligned}
\tag{4.12}
$$

for the geometry we consider. It is important to point out only some general property. For $\kappa = 1$ the factor $\mathcal{F}(1, y) \propto \ln y$ depends weakly [56] on the reduced size y; the factors $\mathcal{F}_{A,B}$ are even more weakly dependent on y and substantially of order one. A slight asymmetry $\kappa \neq 1$ makes all the factors $\mathcal{F}$ independent on the size of the junctions, if d_A, d_B are large enough, $y|\kappa - \kappa^{-1}| \gg 1$. Still $\mathcal{F}_{A,B}$ are of order one, so EC and CA processes determine an appreciable conductance.

4.3. Discussion

We can now discuss the full conductance matrix in equation (4.1) by defining the total EC and CA conductivities $G_{EC} = G_{EC}^\sigma + G_{EC}^{-\sigma}$ and $G_{CA} = G_{CA}^\sigma + G_{CA}^{-\sigma}$. The EC and CA conductances appear both in the diagonal and in the off-diagonal conductance matrix elements in Eq. (4.1).

4.3.1. Nonmagnetic probes

Let us first consider the case of non magnetic contacts, where we can drop the spin dependence. For multichannel contacts Eq. (4.9) shows that $G_{EC} = G_{CA}$, so the off diagonal conductances *vanish*. Coherent tunnelling processes involving two distant contacts enter only in the diagonal terms, and provide an extra contribution with respect to the standard Andreev conductances G_{2A} and G_{2B}. The extra current depends on the distance R between the two contacts. A simple setup where the R dependence of the extra current can be studied is shown in Fig. 11c (alternatively one may use a STM tip as a mobile contact). Another signature of CA and EC processes can be found if one considers contacts of very different transparency, say $|t_A| \gg |t_B|$. In this case $G_{2A} \propto |t_A|^4$ dominates $G_{EC} + G_{CA} \propto |t_A|^2 |t_B|^2$, which itself is much larger than $G_{2B} \approx |t_B|^4$. Thus the conductance at the less transparent probe (lower right diagonal element in Eq. (4.9)) is given by $G_{EC} + G_{CA}$, so it is essentially due to two-contact processes: if we bias contact B a current will flow because of correlations of superconductive nature with contact A. If $R < \xi$, the crossed conductances are still affected by the factor $(k_S R)^{-2}$ which can be very small [52]. This problem can be partially overcome for instance by choosing one contact A to be of size $d_A \gg \xi$ or even a semi-infinite interface.

4.3.2. Spin-polarized probes

Let us now consider the case of spin polarized probes. As it is apparent from
Eq. (4.9) $G_{EC} \neq G_{CA}$ so the off diagonal elements in the conductance matrix,
Eq. (4.1), are finite and the current in one contact can be manipulated by the
voltage bias of the other contact. The sign of this effect depends on the mutual
polarization of the electrodes. This generalizes the result of Ref. [50]. More
specifically, spin polarization enters in two ways in the result (4.8) and (4.9):
first, in the spin-dependent densities of states; second, in the shift of the Fermi
momenta $k_{A,B}^{\sigma}$, which modifies the factors $\mathcal{F}$ in Eq. (4.9). To fix the ideas, let us
for simplicity neglect the latter and concentrate on the effect of the DoS. Defining
the contact polarizations $P_{A,B} = \frac{v_{A,B}^{\sigma} - v_{A,B}^{-\sigma}}{v_{A,B}^{\sigma} + v_{A,B}^{-\sigma}}$, one simply finds that G_{EC} is proportional to $(1 + P_A P_B)$ and G_{CA} to $(1 - P_A P_B)$. Therefore the off-diagonal conduc-
tance is roughly proportional to $(-P_A P_B)$. This shows a striking consequence of
the competition between cotunnelling and crossed Andreev processes, via their
spin-dependence: non only the amplitude, but also the sign of the conductances
can be controlled by spin polarizations. In the extreme case of parallel complete
polarizations the only possible process is cotunnelling, with $I_A = -I_B$, while for
antiparallel polarization crossed Andreev tunnelling prevails, with $I_A = I_B$.

So far we have discussed the zero-temperature case. Direct generalization to
finite temperature leads formally to a divergence of the tunnelling rate. It is due to
the finite, though very small ($\propto e^{-\Delta/T}$), probability of exciting an electron from
$A(B)$ to a quasiparticle state in S, and to the divergence of the quasiparticle DoS
in S. The divergence in the rates disappears if the latter is rounded off at Δ. The
EC and CA conductances acquire an additional contribution $\propto e^{-\Delta/T} \ln(\Delta/\Gamma)$
where Γ is a scale related to the mechanism of broadening of the quasiparticle
levels in the superconductor.

In the present Section we have demonstrated the non-local character of cotunnelling and Andreev reflections on a superconductor and we have also studied
the role of magnetic polarization. We have discussed possible schemes to detect
these effects in devices with three or more terminals. A recent experiment [57]
obtained results that are consistent with the predictions made above. Further
theoretical analysis would require the self-consistent analysis of the mutual effects of superconductivity, diffusive propagation and ferromagnetism in the hybrid system. For both high and low transparency contacts, propagation in the
specific geometry has to be taken into account.

Acknowledgments

I am indebted to G. Bignon, G. Falci, D. Feinberg, Yu. V. Nazarov, and F. Pistolesi
for fruitful collaboration on the various topics presented in this chapter. I wish to

thank H. Bouchiat, G. Montambaux, and F. Pistolesi for a critical reading of the manuscript. I acknowledge financial support from CNRS/ATIP-JC 2002, from Institut Universitaire de France and from EU STREP *SFINx* NMP2-CT-2003-505587.

References

[1]   I. Giaever, Phys. Rev. Lett. **5**, 147 (1960).

[2]   A.F. Andreev, Zh. Eksp. Teor. Fiz. **46**, 1823 (1964) [Sov. Phys. JETP **19**, 1228 (1964)].

[3]   P.G. de Gennes, *Superconductivity of Metals and Alloys* (W.A. Benjamin, New York, 1966; reprinted by Addison-Wesley, Reading MA, 1989).

[4]   G.E. Blonder, M. Tinkham and T.M. Klapwijk, Phys. Rev. B **25**, 4515 (1982).

[5]   C.W.J. Beenakker in *Mesoscopic quantum physics*, Les Houches Session LXI, edited by E. Akkermans, G. Montambaux, J.-L. Pichard, and J. Zinn-Justin (Elsevier, Amsterdam, 1995).

[6]   G. Eilenberger, Z. Phys. **214**, 195 (1968).

[7]   A.I. Larkin and Yu.N. Ovchinnikov, Zh. Eksp. Teor. Fiz. **46**, 1823 (1968) [Sov. Phys. JETP **26**, 1200 (1968)].

[8]   K.D. Usadel, Phys. Rev. Lett. **25**, 507 (1970).

[9]   A.F. Volkov, A.V. Zaitsev, and T.M. Klapwijk, Physica C **59**, 21 (1993).

[10]   W. Belzig, F.K. Wilhelm, C. Bruder, G. Schön, and A.D. Zaikin, Superlattices and Microstructures **25**, 1251 (1999).

[11]   Yu. V. Nazarov, Phys. Rev. Lett. **73**, 134 (1994).

[12]   Yu. V. Nazarov, Superlattices and Microstructures **25**, 1221 (1999).

[13]   F.W.J. Hekking and Yu.V. Nazarov, Phys. Rev. Lett. **71**, 1625 (1993); F.W.J. Hekking and Yu.V. Nazarov, Phys. Rev. B **49** 6847 (1994).

[14]   A.A. Golubov, M.Yu. Kupryanov, and E. Il'ichev, Rev. Mod. Phys. **76**, 411 (2004).

[15]   J. Bardeen, L.N. Cooper, and J.R. Schrieffer, Phys. Rev. **108**, 1175 (1957).

[16]   M. Tinkham, *Introduction to Superconductivity* (McGraw-Hill, Singapore, 1996).

[17]   M.H. Cohen, L.M. Falicov, and J.C. Phillips Phys. Rev. Lett. **8**, 316 (1962).

[18]   G.D. Mahan, *Many-Particle Physics* (Kluwer Academic/Plenum Publishers, New York, 2000).

[19]   J.W. Wilkins in *Tunnelling Phenomena in Solids*, edited by E. Burstein and S. Lundqvist (Plenum, New York, 1969).

[20]   S. Datta, *Electronic transport in mesoscopic systems* (Cambridge University Press, Cambridge, 1997).

[21]   Y. Imry, *Introduction to mesoscopic physics* (Oxford University Press, Oxford, 2002).

[22]   G.E. Blonder and M. Tinkham, Phys. Rev. B**27**, 112 (1983).

[23]   A. Kastalsky, A.W. Kleinsasser, L.H. Greene, R. Bhat, F.P. Milliken, and J.P. Harbison, Phys. Rev. Lett. **67**, 3026 (1991).

[24]   See various articles in *Mesoscopic Superconductivity*, edited by F.W.J. Hekking, G. Schön, and D.V. Averin, Physica B **203**, Nos. 3 & 4 (1994).

[25]   H. Pothier, S. Guéron, D. Esteve, M.H. Devoret, Phys. Rev. Lett. **73**, 2488, (1993).

[26]   J.M. Hergenrother, M.T. Tuominen, and M. Tinkham, Phys. Rev. Lett. **72**, 1742 (1994).

[27]   B.L. Altshuler, A.G. Aronov, and D.E. Khmelnitskii, J. Phys. C **15**, 7367 (1982).

[28] B. L. Altshuler and A. G. Aronov, in *Electron-electron interactions in disordered systems*, Eds. A. L. Efros and M. Pollak, (North-Holland, Amsterdam) (1985).

[29] F. Pistolesi, G. Bignon, and F.W.J. Hekking, Pys. Rev. B **69** , 214518 (2004).

[30] For a recent review see Y.M. Blanter and M. Büttiker, Phys. Rep. **336**, 1 (2000).

[31] *Quantum noise in mesoscopic systems*, edited by Y. V. Nazarov, Kluwer, Dordrecht (2003).

[32] W. Belzig and Y.V. Nazarov, Phys. Rev. Lett. **87**, 197006 (2001); Y. V. Nazarov and D.A. Bagrets, ibid **88**, 196801 (2002).

[33] L.S. Levitov, H.W. Lee, and G.B Levitov, J. Math. Phys. **37**, 4845 (1996).

[34] Y. V. Nazarov, Ann. Phys. (Leipzig) **8**, SI-193 (1999).

[35] B.A. Muzykantskii and D.E. Khmelnitskii, Phy. Rev. B **50**, 3982 (1994).

[36] W. Belzig and Y.V. Nazarov, Phys. Rev. Lett. **87**, 067006 (2001).

[37] F. Lefloch, C. Hoffmann, M. Sanquer, and D. Quirion, Phys. Rev. Lett. **90**, 067002 (2003).

[38] V.A. Khlus, Sov. Phys. JETP **66**, 1243 (1987).

[39] M.J.M. de Jong and C.W.J. Beenakker, Phys. Rev. B **49**, 16070 (1994).

[40] M.P.V. Stenberg and T.T. Heikkilä, Phys. Rev. B, **66**, 144504 (2002).

[41] J. Börlin, W. Belzig, and C. Bruder, Phys. Rev. Lett. **88** 197001 (2002); P. Samuelsson and M. Büttiker, ibid., **89**, 046601 (2002).

[42] P. Samuelsson, Phys. Rev. B 67, 054508 (2003).

[43] H. Pothier, S. Guéron, N.O. Birge, D. Esteve, M.H. Devoret, Z. Phys. B **104**, 178, (1997).

[44] G. Falci, D. Feinberg, and F.W.J. Hekking, Europhys. Lett. **54**, 255 (2001).

[45] M. J. M. de Jong and C. W. J. Beenakker, Phys. Rev. Lett. **74**, 1657 (1995).

[46] R. J. Soulen Jr, J. M. Byers, M. S. Osofsky, B. Nadgorny, T. Ambrose, S. F. Cheng, P. R. Broussard, C. T. Tanaka, J. Nowak, J. S. Moodera, A. Barry and J. M. D. Coey, Science **282**, 85 (1998).

[47] S. K. Upadhyay, A. Palanisami, R. N. Louie and R. A. Buhrman, Phys. Rev. Lett. **81**, 3247 (1998).

[48] C. J. Lambert and R. Raimondi, J. Phys. Condens. Matter 10, 901 (1998).

[49] J. M. Byers and M. E. Flatté, Phys. Rev. Lett. **74**, 306 (1995).

[50] G. Deutscher and D. Feinberg, Appl. Phys. Lett. **76**, 487 (2000).

[51] G. B. Lesovik, Th. Martin and G. Blatter, Eur. Phys. J. B **24**, 287 (2001).

[52] M. S. Choi, C. Bruder and D. Loss, Phys. Rev. B**62**, 13569 (2000); P. Recher, E. V. Sukhorukov and D. Loss, Phys. Rev. B **63**, 165314 (2001).

[53] D.V. Averin and Yu.V. Nazarov, in *"Single Charge Tunneling"*, Chap. 6, edited by H. Grabert and M.H. Devoret (Plenum Press, New York, 1992).

[54] S. Takahashi, H. Imamura and S. Maekawa, Phys. Rev. Lett. **82**, 3911 (1999).

[55] D. Feinberg, Eur. Phys. J. B **36**, 419 (2003).

[56] F.W.J. Hekking in *Coulomb and interference effects in small electronic structures*, edited by D.C. Glattli, M. Sanquer, and J. Trân Thanh Vân (Editions Frontières, Gif-sur-Yvette, 1994).

[57] D. Beckmann, H.B. Weber, and H. v. Löhneysen, Phys. Rev. Lett. **93**, 197003 (2004).

Course 7

LOW-TEMPERATURE TRANSPORT THROUGH A QUANTUM DOT

Leonid I. Glazman[1] and Michael Pustilnik[2]

[1] *William I. Fine Theoretical Physics Institute, University of Minnesota, Minneapolis, MN 55455, USA*
[2] *School of Physics, Georgia Institute of Technology, Atlanta, GA 30332, USA*

H. Bouchiat, Y. Gefen, S. Guéron, G. Montambaux and J. Dalibard, eds.
Les Houches, Session LXXXI, 2004
Nanophysics: Coherence and Transport

Contents

Abstract

We review mechanisms of low-temperature electronic transport through a quantum dot weakly coupled to two conducting leads. Transport in this case is dominated by electron-electron interaction. At temperatures moderately lower than the charging energy of the dot, the linear conductance is suppressed by the Coulomb blockade. Upon further lowering of the temperature, however, the conductance may start to increase again due to the Kondo effect. We concentrate on lateral quantum dot systems and discuss the conductance in a broad temperature range, which includes the Kondo regime.

1. Introduction

In quantum dot devices [1] a small droplet of electron liquid, or just a few electrons are confined to a finite region of space. The dot can be attached by tunneling junctions to massive electrodes to allow electronic transport across the system. The conductance of such a device is determined by the number of electrons on the dot N, which in turn is controlled by varying the potential on the gate - an auxiliary electrode capacitively coupled to the dot [1]. At sufficiently low temperatures, N is an integer at almost any gate voltage V_g. Exceptions are narrow intervals of V_g in which an addition of a single electron to the dot does not change much the electrostatic energy of the system. Such a degeneracy between different charge states of the dot allows for an activationless electron transfer through it, whereas for all other values of V_g the activation energy for the conductance G across the dot is finite. The resulting oscillatory dependence $G(V_g)$ is the hallmark of the Coulomb blockade phenomenon [1]. The contrast between the low- and high-conductance regions (Coulomb blockade valleys and peaks, respectively) gets sharper at lower temperatures. The pattern of periodic oscillations in the G vs. V_g dependence is observed down to the lowest attainable temperatures in experiments on tunneling through small metallic islands [2].

Conductance through quantum dots formed in semiconductor heterostructures exhibits a richer behavior [1]. In larger dots (in the case of GaAs heterostructures, such dots may contain hundreds of electrons), the fluctuations of the heights of

431

the Coulomb blockade peaks become apparent already at moderately low temperatures. Characteristic of mesoscopic phenomena, the heights are sensitive to the shape of a dot and to the magnetic flux threading it. The separation in gate voltage between the Coulomb blockade peaks, and the conductance in the valleys also exhibit mesoscopic fluctuations. However, the pattern of sharp conductance peaks separating the low-conductance valleys of the $G(V_g)$ dependence persists. Smaller quantum dots (containing few tens of electrons in the case of GaAs) show yet another feature [4]: in some Coulomb blockade valleys the dependence $G(T)$ is not monotonic and has a minimum at a finite temperature. This minimum is similar in origin [5] to the well-known non-monotonic temperature dependence of the resistivity of a metal containing magnetic impurities [6] – the *Kondo effect*. Typically, the valleys with anomalous temperature dependence correspond to an odd number of electrons in the dot. In an ideal case, the low-temperature conductance in such a valley is of the order of conductance at peaks surrounding it. Thus, at low temperatures the two adjacent peaks merge to form a broad maximum.

The number of electrons on the dot is a well-defined quantity as long as the conductances of the junctions connecting the dot to the electrodes is small compared to the conductance quantum e^2/h. In quantum dot devices formed in semiconductor heterostructures the conductances of junctions can be tuned continuously. With the increase of the conductances, the periodic pattern in $G(V_g)$ dependence gradually gives way to mesoscopic conductance fluctuations. Yet, electron-electron interaction still affects the transport through the device. A strongly asymmetric quantum dot device with one junction weakly conducting, while another completely open, provides a good example of that [7, 8, 9]. The differential conductance across the device in this case exhibits zero-bias anomaly i.e. suppression at low bias. Clearly, Coulomb blockade is not an isolated phenomenon, but is closely related to interaction-induced anomalies of electronic transport and thermodynamics in higher dimensions [3].

The emphasis of these lectures is on the Kondo effect in quantum dots. We will concentrate on the so-called *lateral quantum dot systems* [1, 4], formed by gate depletion of a two-dimensional electron gas at the interface between two semiconductors. These devices offer the highest degree of tunability, yet allow for relatively simple theoretical treatment. At the same time, many of the results presented below are directly applicable to other systems as well, including vertical quantum dots [10, 11, 12], Coulomb-blockaded carbon nanotubes [12, 13], single-molecule transistors [14], and magnetic atoms on metallic surfaces [15].

The Kondo effect emerges at relatively low temperature, and we will follow the evolution of the conductance upon reduction of temperature. On the way to Kondo effect, we encounter also the phenomena of Coulomb blockade and of mesoscopic conductance fluctuations.

2. Model of a lateral quantum dot system

The Hamiltonian of interacting electrons confined to a quantum dot has the following general form,

$$H_{\text{dot}} = \sum_s \sum_{ij} h_{ij} d_{is}^\dagger d_{js} + \frac{1}{2} \sum_{ss'} \sum_{ijkl} h_{ijkl} d_{is}^\dagger d_{js'}^\dagger d_{ks'} d_{ls}. \tag{2.1}$$

Here an operator $d_{is}^\dagger$ creates an electron with spin s in the orbital state $\phi_i(r)$ (the wave functions are normalized according to $\int dr \phi_i^*(r)\phi_j(r) = \delta_{ij}$); $h_{ij} = h_{ji}^*$ is an Hermitian matrix describing the single-particle part of the Hamiltonian. The matrix elements h_{ijkl} depend on the potential $U(r - r')$ of electron-electron interaction,

$$h_{ijkl} = \int dr\, dr'\, \phi_i^*(r)\phi_j^*(r')U(r - r')\phi_k(r')\phi_l(r). \tag{2.2}$$

The Hamiltonian (2.1) can be simplified further provided that the quasiparticle spectrum is not degenerate near the Fermi level, that the Fermi-liquid theory is applicable to the description of the dot, and that the dot is in the metallic conduction regime. The first of these conditions is satisfied if the dot has no spatial symmetries, which implies also that the motion of quasiparticles within the dot is chaotic.

The second condition is met if the electron-electron interaction within the dot is not too strong, i.e. the gas parameter r_s is small,

$$r_s = (k_F a_0)^{-1} \lesssim 1, \quad a_0 = \kappa \hbar^2/e^2 m^* \tag{2.3}$$

Here k_F is the Fermi wave vector, a_0 is the effective Bohr radius, κ is the dielectric constant of the material, and m^* is the quasiparticle effective mass.

The third condition requires the ratio of the Thouless energy E_T to the mean single-particle level spacing δE to be large [16],

$$g = E_T/\delta E \gg 1. \tag{2.4}$$

For a ballistic two-dimensional dot of linear size L the Thouless energy E_T is of the order of $\hbar v_F/L$, whereas the level spacing can be estimated as

$$\delta E \sim \hbar v_F k_F/N \sim \hbar^2/m^* L^2. \tag{2.5}$$

Here v_F is the Fermi velocity and $N \sim (k_F L)^2$ is the number of electrons in the dot. Therefore,

$$g \sim k_F L \sim \sqrt{N}, \tag{2.6}$$

so that having a large number of electrons $N \gg 1$ in the dot guarantees that the condition Eq. (2.4) is satisfied.

Under the conditions (2.3), (2.4) the *Random Matrix Theory* (for a review see, e.g., [17, 18, 19, 20]) is a good starting point for the description of non-interacting quasiparticles within the energy strip of width E_T about the Fermi level [16]. The matrix elements h_{ij} in Eq. (2.1) belong to a Gaussian ensemble [19, 20]. Since the matrix elements do not depend on spin, each eigenvalue ϵ_n of the matrix h_{ij} represents a spin-degenerate energy level. The spacings $\epsilon_{n+1} - \epsilon_n$ between consecutive levels obey the Wigner-Dyson statistics [19]; the mean level spacing $\langle \epsilon_{n+1} - \epsilon_n \rangle = \delta E$.

We now discuss the second term in the Hamiltonian (2.1), which describes electron-electron interaction. It turns out [21, 22, 23] that the vast majority of matrix elements h_{ijkl} are small. Indeed, in the lowest order in $1/g \ll 1$, the wave functions $\phi_i(r)$ are Gaussian random variables with zero mean, statistically independent of each other and of the corresponding energy levels [24]:

$$\overline{\phi_i^*(r)\phi_j(r')} = \frac{\delta_{ij}}{\mathcal{A}} F(|r - r'|), \qquad \overline{\phi_i(r)\phi_j(r')} = \frac{\delta_{\beta,1}\delta_{ij}}{\mathcal{A}} F(|r - r'|). \quad (2.7)$$

Here $\mathcal{A} \sim L^2$ is the area of the dot, and the function F is given by

$$F(r) \sim \langle \exp(i\mathbf{k} \cdot \mathbf{r}) \rangle_{\mathrm{FS}}. \quad (2.8)$$

where $\langle \ldots \rangle_{\mathrm{FS}}$ stands for the average over the Fermi surface $|\mathbf{k}| = k_F$. In two dimensions, the function $F(r)$ decreases with r as $F \propto (k_F r)^{-1/2}$ at $k_F r \gg 1$, and saturates to $F \sim 1$ at $k_F r \ll 1$.

The parameter β in Eq. (2.7) distinguishes between the presence ($\beta = 1$) or absence ($\beta = 2$) of time-reversal symmetry. The symmetry breaking is driven by the orbital effect of the magnetic field and is characterised by the parameter

$$\chi = (\Phi/\Phi_0)\sqrt{g},$$

where Φ is the magnetic flux threading the dot and $\Phi_0 = hc/e$ is the flux quantum, so that the limits $\chi \ll 1$ and $\chi \gg 1$ correspond to, respectively, $\beta = 1$ and $\beta = 2$. Note that in the case of a magnetic field $H_\perp$ applied perpendicular to the plane of the dot, the crossover (at $\chi \sim 1$) between the two regimes occurs at such weak field that the corresponding Zeeman energy B is negligible[1].

After averaging with the help of Eqs. (2.7)-(2.8), the matrix elements (2.2) take the form

$$\overline{h_{ijkl}} = \left(2E_C + E_S/2\right)\delta_{il}\delta_{jk} + E_S\delta_{ik}\delta_{jl} + \Lambda\left(2/\beta - 1\right)\delta_{ij}\delta_{kl}.$$

---

[1] For example, in the experiments [25] the crossover takes place at $H_\perp \sim 10\,mT$. The Zeeman energy in such a field $B \sim 2.5\,mK$, which is by an order of magnitude lower than the base temperature in the measurements.

We substitute this expression into Hamiltonian (2.1), and rearrange the sum over the spin indices with the help of the identity

$$2\,\delta_{s_1 s_2}\delta_{s_1' s_2'} = \delta_{s_1 s_1'}\delta_{s_2 s_2'} + \boldsymbol{\sigma}_{s_1 s_1'}\cdot\boldsymbol{\sigma}_{s_2' s_2},\tag{2.9}$$

where $\boldsymbol{\sigma} = (\sigma^x, \sigma^y, \sigma^z)$ are the Pauli matrices. This results in a remarkably simple form [22, 23]

$$H_{\text{int}} = E_C\hat{N}^2 - E_S\hat{\mathbf{S}}^2 + \Lambda(2/\beta - 1)\hat{T}^\dagger\hat{T}\tag{2.10}$$

of the interaction part of the Hamiltonian of the dot. Here

$$\hat{N} = \sum_{ns} d_{ns}^\dagger d_{ns},\quad \hat{\mathbf{S}} = \sum_{nss'} d_{ns}^\dagger \frac{\boldsymbol{\sigma}_{ss'}}{2} d_{ns'},\quad \hat{T} = \sum_n d_{n\uparrow}^\dagger d_{n\downarrow}^\dagger\tag{2.11}$$

are the operators of the total number of electrons in the dot, of the dot's spin, and the "pair creation" operator corresponding to the interaction in the Cooper channel.

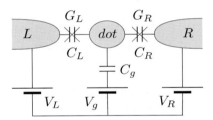

Fig. 1. Equivalent circuit for a quantum dot connected to two leads by tunnel junctions and capacitively coupled to the gate electrode. The total capacitance of the dot $C = C_L + C_R + C_g$.

The first term in Eq. (2.10) represents the electrostatic energy. In the conventional equivalent circuit picture, see Fig. 1, the charging energy E_C is related to the total capacitance C of the dot, $E_C = e^2/2C$. For a mesoscopic ($k_F L \gg 1$) conductor, the charging energy is large compared to the mean level spacing δE. Indeed, using the estimates $C \sim \kappa L$ and Eqs. (2.3) and (2.5), we find

$$E_C/\delta E \sim L/a_0 \sim r_s\sqrt{N}.\tag{2.12}$$

Except in the exotic case of an extremely weak interaction, this ratio is large for $N \gg 1$; for the smallest quantum dots formed in GaAs heterostructures, $E_C/\delta E \sim 10$ [4]. Note that Eqs. (2.4), (2.6), and (2.12) imply that

$$E_T/E_C \sim 1/r_s \gtrsim 1,$$

which justifies the use of RMT for the description of single-particle states with energies $|\epsilon_n| \lesssim E_C$, relevant for Coulomb blockade.

The second term in Eq. (2.10) describes the intra-dot exchange interaction, with the exchange energy E_S given by

$$E_S = \int dr\, dr'\, U(r - r') F^2(|r - r'|). \tag{2.13}$$

In the case of a long-range interaction the potential U here should properly account for the screening [23]. For $r_s \ll 1$ the exchange energy can be estimated with logarithmic accuracy by substituting $U(r) = (e^2/\kappa r)\theta(a_0 - r)$ into Eq. (2.13) (here we took into account that the screening length in two dimensions coincides with the Bohr radius a_0), which yields

$$E_S \sim r_s \ln(1/r_s)\,\delta E \ll \delta E. \tag{2.14}$$

The estimate Eq. (2.14) is valid only for $r_s \ll 1$. However, the ratio $E_S/\delta E$ remains small for the experimentally relevant[2] value $r_s \sim 1$ as long as the Stoner criterion for the absence of itinerant magnetism [26] is satisfied. This guarantees the absence of a macroscopic (proportional to N) magnetization of a dot in the ground state [22].

The third term in Eq. (2.10), representing interaction in the Cooper channel, is renormalized by higher-order corrections arising due to virtual transitions to states outside the energy strip of the width E_T about the Fermi level. For attractive interaction ($\Lambda < 0$) the renormalization enhances the interaction, eventually leading to the superconducting instability and formation of a gap Δ_Λ in the electronic spectrum. Properties of very small ($\Delta_\Lambda \sim \delta E$) superconducting grains are reviewed in, e.g., [27]; for properties of larger grains ($\Delta_\Lambda \sim E_C$) see [28]. Here we concentrate on the repulsive interaction ($\Lambda > 0$), in which case Λ is very small,

$$\Lambda \sim \frac{\delta E}{\ln(\epsilon_F/E_T)} \sim \frac{\delta E}{\ln N} \ll \delta E.$$

This estimate accounts for the logarithmic renormalization of Λ when the high-energy cutoff is reduced from the Fermi energy ϵ_F down to the Thouless energy E_T [23]. In addition, if time-reversal symmetry is lifted ($\beta = 2$) then the third term in Eq. (2.10) is zero to start with. Accordingly, hereinafter we neglect this term altogether by setting $\Lambda = 0$.

Obviously, the interaction part of the Hamiltonian, Eq. (2.10), is invariant with respect to a change of the basis of single-particle states $\phi_i(r)$. Choosing the

---

[2]For GaAs ($m^* \approx 0.07 m_e$, $\kappa \approx 13$) the effective Bohr radius $a_0 \approx 10$ nm, whereas a typical density of the two-dimensional electron gas, $n \sim 10^{11}$ cm^{-2} [4], corresponds to $k_F = \sqrt{2\pi n} \sim 10^6$ cm^{-1}. This gives $k_F a_0 \sim 1$.

basis in which the first term in Eq. (2.1) is diagonal, we arrive at the *universal Hamiltonian* [22, 23],

$$H_{\text{dot}} = \sum_{ns} \epsilon_n d_{ns}^\dagger d_{ns} + E_C (\hat{N} - N_0)^2 - E_S \hat{\mathbf{S}}^2. \qquad (2.15)$$

We included in Eq. (2.15) the effect of the capacitive coupling to the gate electrode: the dimensionless parameter N_0 is proportional to the gate voltage,

$$N_0 = C_g V_g / e,$$

where C_g is the capacitance between the dot and the gate, see Fig. 1. The relative magnitude of fully off-diagonal interaction terms in Eq. (2.1) (corresponding to h_{ijkl} with all four indices different), not included in Eq. (2.15), is of the order of $1/g \sim N^{-1/2} \ll 1$. Partially diagonal terms (two out of four indices coincide) are larger, of the order of $\sqrt{1/g} \sim N^{-1/4}$, but still are assumed to be negligible since $N \gg 1$.

As discussed above, in this limit the energy scales involved in Eq. (2.15) form a well-defined hierarchy

$$E_S \ll \delta E \ll E_C. \qquad (2.16)$$

If all the single-particle energy levels ϵ_n were equidistant, then the spin S of an even-N state would be zero, while an odd-N state would have $S = 1/2$. However, the level spacings are random. If the spacing between the highest occupied level and the lowest unoccupied one is accidentally small, then the gain in exchange energy associated with the formation of a higher-spin state, may be sufficient to overcome the loss of kinetic energy (cf. the Hund's rule in quantum mechanics). For $E_S \ll \delta E$ such deviations from the simple even-odd periodicity are rare [22, 29, 30]. This is why the last term in Eq. (2.15) is often neglected. Eq. (2.15) then reduces to the Hamiltonian of the *Constant Interaction Model*, widely used in the analysis of experimental data [1]. Finally, it should be emphasized that the SU(2)–invariant Hamiltonian (2.15) describes a dot in the absence of the spin-orbit interaction, which would destroy this symmetry.

Electron transport through the dot occurs via two dot-lead junctions. In a typical geometry, the confining potential forming a lateral quantum dot varies smoothly on the scale of the Fermi wavelength. Hence, the point contacts connecting the quantum dot to the leads act essentially as electronic waveguides. Potentials on the gates control the waveguide width, and, therefore, the number of electronic modes the waveguide support: by making the waveguide narrower one pinches the propagating modes off one-by-one. Each such mode contributes $2e^2/h$ to the conductance of a contact. Coulomb blockade develops when the conductances of the contacts are small compared to $2e^2/h$, i.e. when the very

last propagating mode approaches its pinch-off [31, 32]. Accordingly, in the
Coulomb blockade regime each dot-lead junction in a lateral quantum dot system supports only a single electronic mode [33].

As discussed below, for $E_C \gg \delta E$ the characteristic energy scale relevant to
the Kondo effect, the Kondo temperature T_K, is small compared to the mean level
spacing: $T_K \ll \delta E$. This separation of the energy scales allows us to simplify
the problem even further by assuming that the conductances of the dot-lead junctions are small. This assumption will not affect the properties of the system in
the Kondo regime. At the same time, it justifies the use of the tunneling Hamiltonian to describe the coupling between the dot and the leads. The microscopic
Hamiltonian of the system can then be written as a sum of three distinct terms,

$$H = H_{\text{leads}} + H_{\text{dot}} + H_{\text{tunneling}}, \qquad (2.17)$$

which describe free electrons in the leads, isolated quantum dot, and tunneling
between the dot and the leads, respectively. The second term in Eq. (2.17), the
Hamiltonian of the dot H_{dot}, is given by Eq. (2.15). We treat the leads as reservoirs of free electrons with continuous spectra ξ_k, characterized by constant density of states ν, equal for both leads. Moreover, since the typical energies $\omega \lesssim E_C$
of electrons participating in transport through a quantum dot in the Coulomb
blockade regime are small compared to the Fermi energy of the electron gas in
the leads, the spectra ξ_k can be linearized near the Fermi level, $\xi_k = v_F k$; here
k is measured from k_F. With only one electronic mode per junction taken into
account, the first and the third terms in Eq. (2.17) have the form

$$H_{\text{leads}} = \sum_{\alpha k s} \xi_k c^{\dagger}_{\alpha k s} c_{\alpha k s}, \quad \xi_k = -\xi_{-k}, \qquad (2.18)$$

$$H_{\text{tunneling}} = \sum_{\alpha k n s} t_{\alpha n} c^{\dagger}_{\alpha k s} d_{ns} + \text{H.c.} \qquad (2.19)$$

Here $t_{\alpha n}$ are tunneling matrix elements (tunneling amplitudes) "connecting" the
state n in the dot with the state k in the lead α ($\alpha = R, L$ for the right/left lead).

Tunneling leads to a broadening of discrete levels in the dot. The width $\Gamma_{\alpha n}$
that level n acquires due to escape of an electron to lead α is given by

$$\Gamma_{\alpha n} = \pi \nu \left| t^2_{\alpha n} \right| \qquad (2.20)$$

Randomness of single-particle states in the dot translates into a randomness of the
tunneling amplitudes. Indeed, the amplitudes depend on the values of the electron
wave functions at points r_α of the contacts, $t_{\alpha n} \propto \phi_n(r_\alpha)$. For $k_F |r_L - r_R| \sim$
$k_F L \gg 1$ the tunneling amplitudes [and, therefore, the widths (2.20)] in the

left and right junctions are statistically independent of each other. Moreover, the amplitudes to different energy levels are also uncorrelated, see Eq. (2.7):

$$\overline{t^*_{\alpha n} t_{\alpha' n'}} = \frac{\Gamma_\alpha}{\pi \nu} \delta_{\alpha\alpha'} \delta_{nn'}, \quad \overline{t_{\alpha n} t_{\alpha' n'}} = \frac{\Gamma_\alpha}{\pi \nu} \delta_{\beta,1} \delta_{\alpha\alpha'} \delta_{nn'}, \tag{2.21}$$

The average value $\Gamma_\alpha = \overline{\Gamma_{\alpha n}}$ of the width is related to the conductance of the corresponding dot-lead junction

$$G_\alpha = \frac{4e^2}{\hbar} \frac{\Gamma_\alpha}{\delta E}. \tag{2.22}$$

In the regime of strong Coulomb blockade ($G_\alpha \ll e^2/h$), the widths are small compared to the level spacing, $\Gamma_\alpha \ll \delta E$, so that discrete levels in the dot are well defined. Note that statistical fluctuations of the widths $\Gamma_{\alpha n}$ are large, and the corresponding distribution function is not Gaussian. Indeed, using Eqs. (2.20) and (2.21) it is straightforward [19, 20] to show that

$$P(\gamma) = \overline{\delta\left(\gamma - \Gamma_{\alpha n}/\Gamma_\alpha\right)} = \begin{cases} \dfrac{e^{-\gamma/2}}{\sqrt{2\pi\gamma}}, & \beta = 1 \\[2ex] e^{-\gamma}, & \beta = 2 \end{cases} \tag{2.23}$$

This expression is known as the Porter-Thomas distribution [34].

3. Thermally-activated conduction

At high temperatures, $T \gg E_C$, charging energy is negligible compared to the thermal energy of electrons. Therefore the conductance of the device in this regime G_∞ is not affected by charging and, independently of the gate voltage, is given by

$$\frac{1}{G_\infty} = \frac{1}{G_L} + \frac{1}{G_R}. \tag{3.1}$$

Dependence on N_0 develops at lower temperatures, $T \lesssim E_C$. It turns out that the conductance is suppressed for all gate voltages except narrow regions (*Coulomb blockade peaks*) around half-integer values of N_0. We will demonstrate this now using the method of rate equations [35, 36].

3.1. Onset of Coulomb blockade oscillations

We start with the regime of relatively high temperatures,

$$\delta E \ll T \ll E_C, \tag{3.2}$$

and assume that the gate voltage is tuned sufficiently close to one of the points of charge degeneracy,

$$|N_0 - N_0^*| \lesssim T/E_C \qquad (3.3)$$

(here N_0^* is a half-integer number).

Condition (3.2) enables us to treat the discrete single-particle levels within the dot as a continuum with a density of states $1/\delta E$. Condition 3.3, on the other hand, allows us to take into account only two charge states of the dot which are almost degenerate in the vicinity of the Coulomb blockade peak. For N_0 close to N_0^* these are states $|0\rangle$ with $N = N_0^* - 1/2$ electrons on the dot, and state $|1\rangle$ with $N = N_0^* + 1/2$ electrons. According to Eqs. (2.15) and (3.3), the difference of electrostatic energies of these states (the energy cost to add an electron to the dot) is

$$E_+(N_0) = E_{|1\rangle} - E_{|0\rangle} = 2E_C(N_0^* - N_0) \lesssim T. \qquad (3.4)$$

In addition to the constraints (3.2) and (3.3), we assume here that the inelastic electron relaxation rate within the dot $1/\tau_\epsilon$ is large compared to the escape rates $\Gamma_\alpha/\hbar$. In other words, transitions between discrete levels in the dot occur before the electron escapes to the leads[3]. Under this assumption the tunnelings across the two junctions can be treated independently of each other (this is known as *sequential tunneling* approximation).

With the help of the Fermi golden rule the current I_α from the lead α into the dot can be written as

$$I_\alpha = e \frac{2\pi}{\hbar} \sum_{kns} |t_{\alpha n}^2| \, \delta(\xi_k + eV_\alpha - \epsilon_n - E_+) \qquad (3.5)$$

$$\times \left\{ \mathcal{P}_0 f(\xi_k)[1 - f(\epsilon_n)] - \mathcal{P}_1 f(\epsilon_n)[1 - f(\xi_k)] \right\}.$$

Here $\mathcal{P}_i$ is the probability to find the dot in the charge states $|i\rangle$ $(i = 0, 1)$, $f(\omega) = [\exp(\omega/T) + 1]^{-1}$ is the Fermi function, and V_α is the electric potential on the lead α, see Fig. 1. In writing Eq. (3.5) we assumed that the distribution functions $f(\xi_k)$ and $f(\epsilon_n)$ are not perturbed. This is well justified provided that the relaxation rate $1/\tau_\epsilon$ exceeds the rate $\sim G_\infty |V_L - V_R|/e$ at which electrons pass through the dot. Replacing the summations over n and k in Eq. (3.5) by integrations over the corresponding continua, and making use of Eqs. (2.20) and (2.22), we find

$$I_\alpha = \frac{G_\alpha}{e}\left[\mathcal{P}_0 F(E_+ - eV_\alpha) - \mathcal{P}_1 F(eV_\alpha - E_+)\right], \quad F(\omega) = \frac{\omega}{e^{\omega/T} - 1}. \qquad (3.6)$$

---

[3]Note that a finite inelastic relaxation rate requires inclusion of mechanisms beyond the model Eq. (2.15), e.g., electron-phonon collisions.

In the steady state, the currents across the two junctions satisfy

$$I = I_L = -I_R. \tag{3.7}$$

Equations (3.6) and (3.7), supplemented by the obvious normalization condition $\mathcal{P}_0 + \mathcal{P}_1 = 1$, allow one to find the probabilities $\mathcal{P}_i$ and the current across the dot I in response to the applied bias $V = V_L - V_R$. This yields for the linear conductance across the dot [35]

$$G = \lim_{V \to 0} dI/dV = G_\infty \frac{E_C(N_0 - N_0^*)/T}{\sinh[2E_C(N_0 - N_0^*)/T]}. \tag{3.8}$$

Here $N_0 = N_0^*$ corresponds to the Coulomb blockade peak. At each peak, the conductance equals half of its high-temperature value G_∞, see Eq. (3.1). On the contrary, in the *Coulomb blockade valleys* ($N_0 \neq N_0^*$), the conductance falls off exponentially with the decrease of temperature, and all the valleys behave exactly the same way. Note that the sequential tunneling approximation disregards any interference phenomena for electrons passing through the dot. Accordingly, the result Eq. (3.8) is insensitive to a weak magnetic field.

3.2. Coulomb blockade peaks at low temperature

At temperatures below the single-particle level spacing in the dot δE, the activation energy for electron transport equals the difference between the ground state energies of the Hamiltonian (2.15) corresponding to two subsequent (integer) eigenvalues of N. Obviously, this difference includes, in addition to the electrostatic contribution $E_+(N_0)$, see Eq. (3.4), also a finite (and random) level spacing. As a result, the distance in N_0 between adjacent Coulomb blockade peaks is no longer 1, but contains a small fluctuating contribution of the order of $\delta E/E_C$. Mesoscopic fluctuations of spacings between the peaks are still the subject of significant disagreement between theory and experiments. We will not consider these fluctuations here (see [18] for a recent review), and discuss only the heights of the peaks.

We concentrate on the temperature interval

$$\Gamma_\alpha \ll T \ll \delta E, \tag{3.9}$$

which extends to lower temperatures the regime considered in the previous section, see Eq. (3.2), and on values of gate voltages tuned to the vicinity of the Coulomb blockade peak, see Eq. (3.3). Just as above, the latter condition allows us to neglect all charge states except the two with the lowest energy, $|0\rangle$ and $|1\rangle$. Due to the second inequality in Eq. (3.9), the thermal broadening of single-particle energy levels in the dot can be neglected, and the states $|0\rangle$ and

$|1\rangle$ coincide with the ground states of the Hamiltonian (2.15) with, respectively, $N = N_0^* - 1/2$ and $N = N_0^* + 1/2$ electrons in the dot. To be definite, consider the case when

$$N_0^* = N + 1/2 \tag{3.10}$$

with N an even integer; for simplicity, we also neglect the exchange term in Eq. (2.15). Then $|0\rangle$ (with an even number of electrons N) is the state in which all single particle levels below the Fermi level ($n < 0$) are doubly occupied. This state is, obviously, non-degenerate. The state $|1\rangle$ differs from $|0\rangle$ by the addition of a single electron on the Fermi level $n = 0$. The extra electron may be in two possible spin states, hence $|1\rangle$ is doubly degenerate; we denote the two components of $|1\rangle$ by $|s\rangle$ with $s = \uparrow, \downarrow$. As discussed below, the degeneracy eventually gives rise to the Kondo effect. However, at $T \gg \Gamma_\alpha$ [see Eq. (3.9)] the quantum coherence associated with the onset of the Kondo effect is not important, and the rate equations approach can still be used to study the transport across the dot [36].

Applying the Fermi golden rule, we write the contribution of electrons with spin s to the electric current $I_{\alpha s}$ from lead α to the dot as

$$I_{\alpha s} = e \frac{2\pi}{\hbar} |t_{\alpha 0}^2| \sum_k \delta \left(\xi_k + eV_\alpha - \epsilon_0 - E_+ \right) \left\{ \mathcal{P}_0 f(\xi_k) - \mathcal{P}_s \left[1 - f(\xi_k) \right] \right\}$$

We now neglect ϵ_0 as it is small compared to E_+ (thereby neglecting the mesoscopic fluctuations of the position of the Coulomb blockade peak) and replace the summation over k by an integration. This yields

$$I_{\alpha s} = \frac{2e}{\hbar} \Gamma_{\alpha 0} \left\{ \mathcal{P}_0 f \left(E_+ - eV_\alpha \right) - \mathcal{P}_s f \left(eV_\alpha - E_+ \right) \right\}. \tag{3.11}$$

In the steady state the currents $I_{\alpha s}$ satisfy

$$I_{Ls} = -I_{Rs} = I/2 \tag{3.12}$$

(here we took into account the fact that both projections of spin contribute equally to the total electric current I across the dot). Solution of Eqs. (3.11) and (3.12) subject to the normalization condition $\mathcal{P}_0 + \mathcal{P}_\uparrow + \mathcal{P}_\downarrow = 1$ results in [23]

$$G = \frac{4e^2}{\hbar} \frac{\Gamma_{L0}\Gamma_{R0}}{\Gamma_{L0} + \Gamma_{R0}} \left[\frac{-df/d\omega}{1 + f(\omega)} \right]_{\omega = E_+(N_0)}. \tag{3.13}$$

The case of odd N in Eq. (3.10) is also described by Eq. (3.13) after replacement $E_+(N_0) \to -E_+(N_0)$.

There are several differences between Eq. (3.13) and the corresponding expression Eq. (3.8) valid in the temperature range (3.2). First of all, the maximum of the conductance Eq. (3.13) occurs at the gate voltage slightly (by an amount of the order of T/E_C) off the degeneracy point $N_0 = N_0^*$, and, more importantly, the shape of the peak is not symmetric about the maximum. This asymmetry is due to correlations in transport of electrons with opposite spins through a single discrete level in the dot. In its maximum, the function (3.13) takes value

$$G_{\text{peak}} \sim \frac{e^2}{h} \frac{\Gamma_{L0}\Gamma_{R0}}{\Gamma_{L0} + \Gamma_{R0}} \frac{1}{T}. \tag{3.14}$$

Note that Eqs. (3.13) and (3.14) depend on the widths $\Gamma_{\alpha 0}$ of the energy level $n = 0$ rather than on the averages Γ_α over many levels in the dot, as in Eq. (3.8). As already discussed in Sec. 2, the widths $\Gamma_{\alpha 0}$ are related to the values of the electron wave functions at the position of the dot-lead contacts, and, therefore, are random. Accordingly, the heights G_{peak} of the Coulomb blockade peaks exhibit strong mesoscopic fluctuations. In view of Eq. (2.23), the distribution function of G_{peak}, see Eq. (3.14), is expected to be broad and strongly non-Gaussian, as well as very sensitive to the magnetic flux threading the dot. This is indeed confirmed by calculations [37, 38] and agrees with experimental data [25, 39]. The expression for the distribution function is rather cumbersome and we will not reproduce it here, referring the reader to the original papers [37, 38] and reviews [18, 20, 23]) instead.

An order-of-magnitude estimate of the average height of the peak can be obtained by replacing $\Gamma_{\alpha 0}$ in Eq. (3.14) by Γ_α, see Eq. (2.22), which yields

$$\overline{G_{\text{peak}}} \sim G_\infty \frac{\delta E}{T}. \tag{3.15}$$

This is by a factor $\delta E / T$ larger than the corresponding figure $G_{\text{peak}} = G_\infty/2$ for the temperature range (3.2), and may even approach the unitary limit ($\sim e^2/h$) at the lower end of the temperature interval (3.9). Interestingly, breaking of time-reversal symmetry results in an increase of the average conductance [23]. This increase is analogous to negative magnetoresistance due to weak localization in bulk systems [40], with the same physics involved.

4. Activationless transport through a blockaded quantum dot

According to the rate equations theory [35], at low temperatures, $T \ll E_C$, conduction through the dot is exponentially suppressed in the Coulomb blockade valleys. This suppression occurs because the process of electron transport through the dot involves a *real transition* to the state in which the charge of the dot differs by e from the thermodynamically most probable value. The probability of such fluctuation is proportional to $\exp\left(-E_C|N_0 - N_0^*|/T\right)$, which explains the conductance suppression, see Eq. (3.7). Going beyond lowest-order perturbation theory in conductances of the dot-leads junctions G_α allows one to consider processes in which states of the dot with a "wrong" charge participate in the tunneling process as *virtual states*. The existence of these higher-order contributions to the tunneling conductance was envisioned already in 1968 by Giaever and Zeller [41]. The first quantitative theory of this effect, however, was developed much later [42].

The leading contributions to the activationless transport, according to [42], are provided by the processes of *inelastic and elastic co-tunneling*. Unlike the sequential tunneling, in the co-tunneling mechanism, the events of electron tunneling from one of the leads into the dot, and tunneling from the dot to the other lead occur as a single quantum process.

4.1. Inelastic co-tunneling

In the inelastic co-tunneling mechanism, an electron tunnels from a lead into one of the vacant single-particle levels in the dot, while it is an electron occupying some other level that tunnels out of the dot, see Fig. 2(a). As a result, transfer of charge e between the leads is accompanied by a simultaneous creation of an electron-hole pair in the dot.

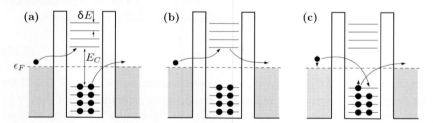

Fig. 2. Examples of the co-tunneling processes.
(a) inelastic co-tunneling: transferring of an electron between the leads leaves behind an electron-hole pair in the dot; (b) elastic co-tunneling; (c) elastic co-tunneling with a flip of spin.

Here we will estimate the contribution of the inelastic co-tunneling to the conductance deep in the Coulomb blockade valley, i.e. at almost integer N_0. Consider an electron that tunnels into the dot from the lead L. If the energy ω of the electron relative to the Fermi level is small compared to the charging energy, $\omega \ll E_C$, then the energy of the virtual state involved in the co-tunneling process is close to E_C. The amplitude of the inelastic transition via this virtual state to the lead R is then given by

$$A_{n,m} = \frac{t_{Ln}^* t_{Rm}}{E_C}. \tag{4.1}$$

The initial state of this transition has an extra electron in the single-particle state k in the lead L, while the final state has an extra electron in the state k' in the lead R and an electron-hole pair in the dot (state n is occupied, state m is empty). The conductance is proportional to the sum of probabilities of all such processes,

$$G_{in} \sim \frac{e^2}{h} \sum_{n,m} \nu^2 |A_{n,m}^2| \sim \frac{e^2}{h} \frac{1}{E_C^2} \sum_{n,m} \Gamma_{Ln} \Gamma_{Rm}, \tag{4.2}$$

where we made use of Eq. (2.20). Now we estimate how many terms contribute to the sum in (4.2). Given the energy of the initial state ω, the number of available final states can be found from the phase space argument, familiar from the calculation of the quasiparticle lifetime in the Fermi liquid theory [43]. For $\omega \gg \delta E$ the energies of the states n and m lie within a strip of the width $\sim \omega/\delta E$ about the Fermi level. The total number of the final states contributing to the sum in Eq. (4.2) is then of the order of $(\omega/\delta E)^2$. Since the typical value of ω is T, the average value of the inelastic co-tunneling contribution to the conductance can be estimated as

$$\overline{G_{in}} \sim \left(\frac{T}{\delta E} \right)^2 \frac{e^2}{h} \frac{\Gamma_L \Gamma_R}{E_C^2}.$$

Using now Eq. (2.22), we find [42]

$$\overline{G_{in}} \sim \frac{G_L G_R}{e^2/h} \left(\frac{T}{E_C} \right)^2. \tag{4.3}$$

The terms entering the sum in Eq. (4.2) are random and uncorrelated. As it follows from Eq. (2.23), fluctuation of each term in the sum is of the order of its average value. This leads to the estimate for the fluctuation $\delta G_{in} = G_{in} - \overline{G_{in}}$ of the inelastic co-tunneling contribution to the conductance

$$\overline{\delta G_{in}^2} \sim \left(\frac{T}{\delta E} \right)^2 \left(\frac{e^2}{h} \frac{\Gamma_L \Gamma_R}{E_C^2} \right)^2 \sim \left(\frac{\delta E}{T} \right)^2 (\overline{G_{in}})^2. \tag{4.4}$$

Accordingly, fluctuation of G_{in} is small compared to its average.

A comparison of Eq. (4.3) with the result of the rate equations theory (3.8) shows that inelastic co-tunneling overcomes the thermally-activated hopping at moderately low temperatures

$$T \lesssim T_{\text{in}} = E_C \left[\ln \left(\frac{e^2/h}{G_L + G_R} \right) \right]^{-1}. \tag{4.5}$$

The smallest energy of an electron-hole pair is of the order of δE. At temperatures below this threshold the inelastic co-tunneling contribution is exponentially suppressed. It turns out, however, that even at much higher temperatures this mechanism becomes less effective than the elastic co-tunneling.

4.2. Elastic co-tunneling

In the process of elastic co-tunneling, transfer of charge between the leads is not accompanied by the creation of an electron-hole pair in the dot. In other words, occupation numbers of single-particle energy levels in the dot in the initial and final states of the co-tunneling process are exactly the same, see Fig. 2(b). Close to the middle of the Coulomb blockade valley (at almost integer N_0) the average number of electrons on the dot, $N \approx N_0$, is also an integer. Both addition and removal of a single electron cost E_C in electrostatic energy, see Eq. (2.15). The amplitude of the elastic co-tunneling process in which an electron is transfered from lead L to lead R can then be written as

$$A_{el} = \sum_n t_{Ln}^* t_{Rn} \frac{\text{sign}(\epsilon_n)}{E_C + |\epsilon_n|} \tag{4.6}$$

The two types of contributions to the amplitude A_{el} are associated with the virtual creation of either an electron if the level n is empty ($\epsilon_n > 0$), or of a hole if the level is occupied ($\epsilon_n < 0$); the relative sign difference between the two types of contributions originates in the fermionic commutation relations.

As discussed in Sec. 2, the tunneling amplitudes $t_{\alpha n}$ entering Eq. (4.6) are Gaussian random variables with zero mean and variances given by Eq. (2.21). It is then easy to see that the second moment of the amplitude Eq. (4.6) is given by

$$\overline{|A_{el}^2|} = \frac{\Gamma_L \Gamma_R}{(\pi \nu)^2} \sum_n (E_C + |\epsilon_n|)^{-2}.$$

Since for $E_C \gg \delta E$ the number of terms making significant contribution to the sum over n here is large, and since the sum is converging, one can replace the summation by an integral which yields

$$\overline{|A_{el}^2|} \approx \frac{\Gamma_L \Gamma_R}{(\pi \nu)^2} \frac{1}{E_C \delta E}. \tag{4.7}$$

Substitution of this expression into

$$G_{el} = \frac{4\pi e^2 \nu^2}{\hbar} \left| A_{el}^2 \right| \tag{4.8}$$

and making use of Eq. (2.22) gives [42]

$$\overline{G_{el}} \sim \frac{G_L G_R}{e^2/h} \frac{\delta E}{E_C}. \tag{4.9}$$

for the average value of the elastic co-tunneling contribution to the conductance.

This result is easily generalized to gate voltages tuned away from the middle of the Coulomb blockade valley. The corresponding expression reads

$$\overline{G_{el}} \sim \frac{G_L G_R}{e^2/h} \frac{\delta E}{E_C} \left(\frac{1}{N_0 - N_0^*} + \frac{1}{N_0^* - N_0 + 1} \right). \tag{4.10}$$

and is valid when N_0 is not too close to the degeneracy points $N_0 = N_0^*$ and $N_0 = N_0^* + 1$ (N_0^* is a half-integer number):

$$\min\{\left| N_0 - N_0^* \right|, \left| N_0 - N_0^* - 1 \right|\} \gg \delta E / E_C$$

Comparison of Eq. (4.9) with Eq. (4.3) shows that the elastic co-tunneling mechanism dominates electron transport already at temperatures

$$T \lesssim T_{el} = \sqrt{E_C \delta E}, \tag{4.11}$$

which may exceed significantly the level spacing. However, as we will see shortly below, mesoscopic fluctuations of G_{el} are strong [44], of the order of its average value. Therefore, although $\overline{G_{el}}$ is always positive, see Eq. (4.10), the sample-specific value of G_{el} for a given gate voltage may vanish.

The key to understanding the statistical properties of the elastic co-tunneling contribution to the conductance is provided by the observation that there are many ($\sim E_C/\delta E \gg 1$) terms making significant contribution to the amplitude Eq. (4.6). All these terms are random and statistically independent of each other. The central limit theorem then suggests that the distribution of A_{el} is Gaussian [23], and, therefore, is completely characterised by the first two statistical moments,

$$\overline{A_{el}} = \overline{A_{el}^*}, \quad \overline{A_{el} A_{el}} = \overline{A_{el}^* A_{el}^*} = \delta_{\beta,1} \overline{A_{el} A_{el}^*} \tag{4.12}$$

with $\overline{A_{el}^* A_{el}}$ given by Eq. (4.7). This can be proven by explicit consideration of higher moments. For example,

$$\overline{\left| A_{el}^4 \right|} = 2 \left(\overline{A_{el} A_{el}^*} \right)^2 + \left| \overline{A_{el} A_{el}} \right|^2 + \delta \overline{\left| A_{el}^4 \right|}. \tag{4.13}$$

The non-Gaussian correction here, $\delta \overline{\left| A_{el}^4 \right|} \sim \overline{\left| A_{el}^2 \right|} (\delta E/E_C)$, is by a factor of $\delta E/E_C \ll 1$ smaller than the main (Gaussian) contribution.

It follows from Eqs. (4.8), (4.12), and (4.13) that the fluctuation of the conductance $\delta G_{el} = G_{el} - \overline{G_{el}}$ satisfies

$$\overline{\delta G_{el}^2} = \frac{2}{\beta} \left(\overline{G_{el}} \right)^2. \tag{4.14}$$

Note that breaking of time reversal symmetry reduces the fluctuations by a factor of 2, similar to conductance fluctuations in bulk systems, whereas the average conductance (4.10) is not affected by the magnetic field.

It is clear from Eq. (4.14) that the fluctuations of the conductance are of the order of the conductance itself, despite naive expectations that the large number of the contributing states results in self-averaging. The reason is that one has to add amplitudes, rather than probabilities, in order to compute the conductance. Because the fluctuations of the conductance are large, its distribution function is not Gaussian. Given the statistics (4.12) of the amplitudes, it is not quite surprising that the distribution of G_{el} normalized to its average coincides with the Porter-Thomas distribution (2.23). This result was obtained first in [44] by a different (and more general) method.

Finally, it is interesting to compare the elastic co-tunneling contribution to the conductance fluctuations Eq. (4.14) with that of inelastic co-tunneling, see Eq. (4.4). Even though the inelastic co-tunneling is the main conduction mechanism at $T \gtrsim T_{el}$, see Eq. (4.11), elastic co-tunneling dominates the fluctuations of the conductance throughout the Coulomb blockade regime $T \lesssim E_C$.

5. Kondo regime in transport through a quantum dot

In the above discussion of the elastic co-tunneling we made the tacit assumption that all single-particle levels in the dot are either empty or doubly occupied. This, however, is not the case when the dot has a non-zero spin in the ground state. A dot with an odd number of electrons, for example, would necessarily have a half-integer spin S. In the most important case of $S = 1/2$ the top-most occupied single-particle level is filled by a single electron and is spin-degenerate. This opens a possibility of a co-tunneling process in which the transfer of an electron between the leads is accompanied by a flip of the electron's spin with simultaneous flip of the spin on the dot, see Fig. 2(c).

The amplitude of such a process, calculated in the fourth order in tunneling matrix elements, diverges logarithmically when the energy ω of an incoming electron approaches 0. Since $\omega \sim T$, the logarithmic singularity in the transmission amplitude translates into a dramatic enhancement of the conductance G

across the dot at low temperatures: G may reach values as high as the quantum limit $2e^2/h$ [45, 46]. This conductance enhancement is not really a surprise. Indeed, in the spin-flip co-tunneling process a quantum dot with odd N behaves as an $S = 1/2$ magnetic impurity embedded into a tunneling barrier separating two massive conductors [47]. It is known [48] since the mid-60's that the presence of such impurities leads to zero-bias anomalies in tunneling conductance [49], which are adequately explained [50, 51] in the context of the Kondo effect [6].

5.1. Effective low-energy Hamiltonian

At energies well below the threshold $\Delta \sim \delta E$ for intra-dot excitations the transitions within the $(2S + 1)$-fold degenerate ground state manifold of a dot can be conveniently described by a spin operator $\mathbf{S}$. The form of the *effective Hamiltonian* describing the interaction of the dot with conduction electrons in the leads is then dictated by SU(2) symmetry[4],

$$H_{\text{eff}} = \sum_{\alpha ks} \xi_k c^\dagger_{\alpha ks} c_{\alpha ks} + \sum_{\alpha \alpha'} J_{\alpha \alpha'} (\mathbf{s}_{\alpha' \alpha} \cdot \mathbf{S}) \tag{5.1}$$

with $\mathbf{s}_{\alpha \alpha'} = \sum_{kk'ss'} c^\dagger_{\alpha ks}(\boldsymbol{\sigma}_{ss'}/2) c_{\alpha'k's'}$. The sum over k in Eq. (5.1) is restricted to $|\xi_k| < \Delta$. The exchange amplitudes $J_{\alpha \alpha'}$ form a 2×2 Hermitian matrix $\hat{J}$. The matrix has two real eigenvalues, the exchange constants J_1 and J_2 (hereafter we assume that $J_1 \geq J_2$). By an appropriate rotation in the $R - L$ space the Hamiltonian (5.2) can then be brought into the form

$$H_{\text{eff}} = \sum_{\gamma ks} \xi_k \psi^\dagger_{\gamma ks} \psi_{\gamma ks} + \sum_\gamma J_\gamma (\mathbf{s}_\gamma \cdot \mathbf{S}). \tag{5.2}$$

Here the operators ψ_γ are certain linear combinations of the original operators $c_{R,L}$ describing electrons in the leads, and

$$\mathbf{s}_\gamma = \sum_{kk'ss'} \psi^\dagger_{\gamma ks} \frac{\boldsymbol{\sigma}_{ss'}}{2} \psi_{\gamma k's'}$$

is the local spin density of itinerant electrons in the "channel" $\gamma = 1, 2$.

Symmetry alone is not sufficient to determine the exchange constants J_γ; their evaluation must rely upon a microscopic model. Here we briefly outline the derivation [33, 52, 53] of Eq. (5.1) for a generic model of a quantum dot system discussed in Section 2 above. For simplicity, we will assume that the gate voltage

---

[4]In writing Eq. (5.1) we omitted the potential scattering terms associated with the usual elastic co-tunneling. This approximation is well justified when the conductances of the dot-lead junctions are small, $G_\alpha \ll e^2/h$, in which case G_{el} is also very small, see Eq. (4.7).

N_0 is tuned to the middle of the Coulomb blockade valley. The tunneling (2.19) mixes the state with $N = N_0$ electrons on the dot with states having $N \pm 1$ electrons. The electrostatic energies of these states are high ($\sim E_C$), hence the transitions $N \to N \pm 1$ are virtual, and can be taken into account perturbatively in the second order in tunneling amplitudes [54].

For the Hamiltonian (2.15) the occupations of single-particle energy levels are good quantum numbers. Therefore, the amplitude $J_{\alpha\alpha'}$ can be written as a sum of partial amplitudes,

$$J_{\alpha\alpha'} = \sum_n J_{\alpha\alpha'}^n. \tag{5.3}$$

Each term in the sum here corresponds to a process in which an electron or a hole is created virtually on the level n in the dot, cf. Eq. (4.6). For $G_\alpha \ll e^2/h$ and $E_S \ll \delta E$ the main contribution to the sum in Eq. (5.3) comes from singly-occupied energy levels in the dot. A dot with spin S has $2S$ such levels near the Fermi level (hereafter we assign indices $n = -S, \ldots, n = S$ to these levels), each carrying a spin $S/2S$, and contributing

$$J_{\alpha\alpha'}^n = \frac{\lambda_n}{E_C} t_{\alpha n}^* t_{\alpha' n}, \quad \lambda_n = 2/S, \quad |n| \le S \tag{5.4}$$

to the exchange amplitude in Eq. (5.1). This yields

$$J_{\alpha\alpha'} \approx \sum_{|n| \le S} J_{\alpha\alpha'}^n. \tag{5.5}$$

It follows from Equations (5.3) and (5.4) that

$$\mathrm{tr}\hat{J} = \frac{1}{E_C} \sum_n \lambda_n \left(|t_{Ln}^2| + |t_{Rn}^2| \right). \tag{5.6}$$

By restricting the sum over n here to $|n| \le S$, as in Eq. (5.5), and taking into account that all λ_n in Eq. (5.4) are positive, we find $J_1 + J_2 > 0$. Similarly, from

$$\det \hat{J} = \frac{1}{2E_C^2} \sum_{m,n} \lambda_m \lambda_n |\mathcal{D}_{mn}^2|, \quad \mathcal{D}_{mn} = \det \begin{pmatrix} t_{Lm} & t_{Rm} \\ t_{Ln} & t_{Rn} \end{pmatrix} \tag{5.7}$$

and Equations (5.4) and (5.5), it follows that $J_1 J_2 > 0$ for $S > 1/2$. Indeed, in this case the sum in Eq. (5.7) contains at least one contribution with $m \ne n$; all such contributions are positive. Thus, both exchange constants $J_{1,2} > 0$ if the dot's spin S exceeds $1/2$ [33]. The peculiarities of the Kondo effect in quantum dots with large spin are discussed in Section 5.7 below.

We now turn to the most common case of $S = 1/2$ on the dot [4]. The ground state of such a dot has only one singly-occupied energy level ($n = 0$), so that $\det \hat{J} \approx 0$, see Eqs. (5.5) and (5.7). Accordingly, one of the exchange constants vanishes,

$$J_2 \approx 0, \tag{5.8}$$

while the remaining one, $J_1 = \operatorname{tr} \hat{J}$, is positive. Equation (5.8) resulted, of course, from the approximation made in Eq. (5.5). For the model (2.15) the leading correction to Eq. (5.5) originates in the co-tunneling processes with an intermediate state containing an extra electron (or an extra hole) on one of the empty (doubly-occupied) levels. Such a contribution arises because the spin on the level n is not conserved by the Hamiltonian (2.15), unlike the corresponding occupation number. Straightforward calculation [52] yields the partial amplitude in the form of Eq. (5.4), but with

$$\lambda_n = -\frac{2E_C E_S}{(E_C + |\epsilon_n|)^2}, \qquad n \neq 0.$$

Unless the tunneling amplitudes $t_{\alpha 0}$ to the only singly-occupied level in the dot are anomalously small, the corresponding correction

$$\delta J_{\alpha\alpha'} = \sum_{n \neq 0} J^n_{\alpha\alpha'} \tag{5.9}$$

to the exchange amplitude (5.5) is small,

$$\left| \frac{\delta J_{\alpha\alpha'}}{J_{\alpha\alpha'}} \right| \sim \frac{E_S}{\delta E} \ll 1,$$

see Eq. (2.16). To obtain this estimate, we assumed that all tunneling amplitudes $t_{\alpha n}$ are of the same order of magnitude, and replaced the sum over n in Eq. (5.9) by an integral. A similar estimate yields the leading contribution to $\det \hat{J}$,

$$\det \hat{J} \approx \frac{1}{E_C^2} \sum_n \lambda_0 \lambda_n |\mathcal{D}^2_{0n}| \sim -\frac{E_S}{\delta E} (\operatorname{tr} \hat{J})^2,$$

or

$$J_2/J_1 \sim -E_S/\delta E. \tag{5.10}$$

According to Eq. (5.10), the exchange constant J_2 is negative [55], and its absolute value is small compared to J_1. Hence, Eq. (5.8) is indeed an excellent approximation for large chaotic dots with spin $S = 1/2$ as long as the intra-dot exchange interaction remains weak[5], i.e. for $E_S \ll \delta E$. Note that corrections to the universal Hamiltonian (2.15) also result in finite values of both exchange constants, $|J_2| \sim J_1 N^{-1/2}$, and become important for small dots with $N \lesssim 10$ [46]. Although this may significantly affect the conductance across the system in the weak coupling regime $T \gtrsim T_K$, it does not lead to qualitative changes in the results for $S = 1/2$ on the dot, as the channel with the smaller exchange constant decouples at low energies [56], see also Section 5.7 below. With this caveat, we adopt the approximation (5.8) in our description of the Kondo effect in quantum dots with spin $S = 1/2$. Accordingly, the effective Hamiltonian of the system (5.2) assumes the "block-diagonal" form

$$H_{\text{eff}} = H_1 + H_2 \tag{5.11}$$

$$H_1 = \sum_{ks} \xi_k \psi_{1ks}^\dagger \psi_{1ks} + J(\mathbf{s}_1 \cdot \mathbf{S}) \tag{5.12}$$

$$H_2 = \sum_{ks} \xi_k \psi_{2ks}^\dagger \psi_{2ks} \tag{5.13}$$

with $J = \text{tr}\hat{J} > 0$.

To get an idea about the physics of the Kondo model (see [57] for recent reviews), let us first replace the fermion field operator $\mathbf{s}_1$ in Eq. (5.12) by a single-particle spin-1/2 operator $\mathbf{S}_1$. The ground state of the resulting Hamiltonian of two spins

$$\widetilde{H} = J(\mathbf{S}_1 \cdot \mathbf{S})$$

is obviously a singlet. The excited state (a triplet) is separated from the ground state by the energy gap J_1. This separation can be interpreted as the binding energy of the singlet. Unlike $\mathbf{S}_1$ in this simple example, the operator $\mathbf{s}_1$ in Eq. (5.12) is merely a spin density of the conduction electrons at the site of the "magnetic impurity". Because conduction electrons are freely moving in space, it is hard for the impurity to "capture" an electron and form a singlet. Yet, even a weak local exchange interaction suffices to form a singlet ground state [58, 59]. However, the characteristic energy (an analogue of the binding energy) for this singlet is given not by the exchange constant J, but by the so-called Kondo temperature

$$T_K \sim \Delta \exp(-1/\nu J). \tag{5.14}$$

---

[5]Equation (5.8) holds identically for the Anderson impurity model [51] frequently employed to study transport through quantum dots. In that model a quantum dot is described by a single energy level, which formally corresponds to the infinite level spacing limit $\delta E \to \infty$ of the Hamiltonian (2.15).

Using $\Delta \sim \delta E$ and Equations (5.6) and (2.22), one obtains from Eq. (5.14) the estimate

$$\ln \left(\frac{\delta E}{T_K} \right) \sim \frac{1}{\nu J} \sim \frac{e^2/h}{G_L + G_R} \frac{E_C}{\delta E}. \tag{5.15}$$

Since $G_\alpha \ll e^2/h$ and $E_C \gg \delta E$, the r.h.s. of Eq. (5.15) is a product of two large parameters. Therefore, the Kondo temperature T_K is small compared to the mean level spacing,

$$T_K \ll \delta E. \tag{5.16}$$

It is this separation of the energy scales that justifies the use of the effective low-energy Hamiltonian (5.1), (5.2) for the description of the Kondo effect in a quantum dot system. The inequality (5.16) remains valid even if the conductances of the dot-leads junctions G_α are of the order of $2e^2/h$. However, in this case the estimate (5.15) is no longer applicable [60].

In our model, see Equations (5.11)-(5.13), one of the channels (ψ_2) of conduction electrons completely decouples from the dot, while the ψ_1-particles are described by the standard single-channel antiferromagnetic Kondo model [6, 57]. Therefore, the thermodynamic properties of a quantum dot in the Kondo regime are identical to those of the conventional Kondo problem for a single magnetic impurity in a bulk metal; thermodynamics of the latter model is fully studied [61]. However, all the experiments addressing the Kondo effect in quantum dots test their transport properties rather than their thermodynamics. The electron current operator is not diagonal in the (ψ_1, ψ_2) representation, and the contributions of these two sub-systems to the conductance are not additive. Below we relate the linear conductance and, in some special case, the non-linear differential conductance as well, to the t-matrix of the conventional Kondo problem.

5.2. Linear response

The linear conductance can be calculated from the Kubo formula

$$G = \lim_{\omega \to 0} \frac{1}{\hbar \omega} \int_0^\infty dt \, e^{i\omega t} \langle [\hat{I}(t), \hat{I}(0)] \rangle, \tag{5.17}$$

where the current operator $\hat{I}$ is given by

$$\hat{I} = \frac{d}{dt} \frac{e}{2} (\hat{N}_R - \hat{N}_L), \quad \hat{N}_\alpha = \sum_{ks} c^\dagger_{\alpha ks} c_{\alpha ks} \tag{5.18}$$

Here $\hat{N}_\alpha$ is the operator of the total number of electrons in the lead α. Evaluation of the linear conductance proceeds similarly to the calculation of the impurity contribution to the resistivity of dilute magnetic alloys (see, e.g., [62]). In order to take full advantage of the decomposition (5.11)-(5.13), we rewrite $\hat{I}$ in terms of the operators $\psi_{1,2}$. These operators are related to the original operators $c_{R,L}$ representing the electrons in the right and left leads via

$$\begin{pmatrix} \psi_{1ks} \\ \psi_{2ks} \end{pmatrix} = \begin{pmatrix} \cos\theta_0 & \sin\theta_0 \\ -\sin\theta_0 & \cos\theta_0 \end{pmatrix} \begin{pmatrix} c_{Rks} \\ c_{Lks} \end{pmatrix}. \tag{5.19}$$

The rotation matrix here is the same one that diagonalizes matrix $\hat{J}$ of the exchange amplitudes in Eq. (5.1); the rotation angle θ_0 satisfies the equation $\tan\theta_0 = |t_{L0}/t_{R0}|$. With the help of Eq. (5.19) we obtain

$$\hat{N}_R - \hat{N}_L = \cos(2\theta_0)(\hat{N}_1 - \hat{N}_2) - \sin(2\theta_0)\sum_{ks}(\psi^\dagger_{1ks}\psi_{2ks} + \text{H.c.}) \tag{5.20}$$

The current operator $\hat{I}$ entering the Kubo formula (5.17) is to be calculated with the equilibrium Hamiltonian (5.11)-(5.13). Since both $\hat{N}_1$ and $\hat{N}_2$ commute with H_{eff}, the first term in Eq. (5.20) makes no contribution to $\hat{I}$. When the second term in Eq. (5.20) is substituted into Eq. (5.18) and then into the Kubo formula (5.17), the result, after integration by parts, can be expressed via 2-particle correlation functions such as $\langle\psi^\dagger_1(t)\psi_2(t)\psi^\dagger_2(0)\psi_1(0)\rangle$ (see Appendix B of [63] for further details about this calculation). Due to the block-diagonal structure of H_{eff}, see Eq. (5.11), these correlation functions factorize into products of the single-particle correlation functions describing the (free) ψ_2-particles and the (interacting) ψ_1-particles. The result of the evaluation of the Kubo formula can then be written as

$$G = G_0 \int d\omega \left(-\frac{df}{d\omega}\right)\frac{1}{2}\sum_s [-\pi \nu \operatorname{Im} T_s(\omega)]. \tag{5.21}$$

Here

$$G_0 = \frac{2e^2}{h}\sin^2(2\theta_0) = \frac{2e^2}{h}\frac{4|t^2_{L0}t^2_{R0}|}{\left(|t^2_{L0}| + |t^2_{R0}|\right)^2}, \tag{5.22}$$

$f(\omega)$ is the Fermi function, and $T_s(\omega)$ is the t-matrix for the Kondo model (5.12). The t-matrix is related to the exact retarded Green function of the ψ_1-particles in the conventional way,

$$G_{ks,k's}(\omega) = G^0_k(\omega) + G^0_k(\omega)T_s(\omega)G^0_{k'}(\omega), \quad G^0_k = (\omega - \xi_k + i0)^{-1}.$$

Here $G_{ks,k's}(\omega)$ is the Fourier transform of $G_{ks,k's}(t) = -i\theta(t)\langle\{\psi_{1ks}(t), \psi^\dagger_{1k's}\}\rangle$, where $\langle\ldots\rangle$ stands for the thermodynamic averaging with Hamiltonian (5.12). In writing Eq. (5.21) we took into account the conservation of the total spin (which implies that $G_{ks,k's'} = \delta_{ss'}G_{ks,k's}$, and the fact that the interaction in Eq. (5.12) is local (which in turn means that the t-matrix is independent of k and k').

5.3. Weak coupling regime: $T_K \ll T \ll \delta E$

When the exchange term in the Hamiltonian (5.12) is treated perturbatively, the main contribution to the t-matrix comes from the transitions of the type [64]

$$|ks, \sigma\rangle \rightarrow |k's', \sigma'\rangle. \tag{5.23}$$

Here the state $|ks, \sigma\rangle$ has an extra electron with spin s in the orbital state k whereas the dot is in the spin state σ. By SU(2) symmetry, the amplitude of the transition (5.23) satisfies

$$A_{|k's',\sigma'\rangle \leftarrow |ks,\sigma\rangle} = A(\omega) \frac{1}{4} (\sigma_{s's} \cdot \sigma_{\sigma'\sigma}) \tag{5.24}$$

Note that the amplitude is independent of k, k' because the interaction is local. However, it may depend on ω due to retardation effects.

The transition (5.23) is *elastic* in the sense that the number of quasiparticles in the final state of the transition is the same as that in the initial state (in other words, the transition (5.23) is not accompanied by the production of electron-hole pairs). Therefore, the imaginary part of the t-matrix can be calculated with the help of the optical theorem [65], which yields

$$-\pi \nu \, \mathrm{Im} \, T_s = \frac{1}{2} \sum_{\sigma} \sum_{s'\sigma'} \left| \pi \nu \, A^2_{|k's',\sigma'\rangle \leftarrow |ks,\sigma\rangle} \right|. \tag{5.25}$$

The factor $1/2$ here accounts for the probability to have spin σ on the dot in the initial state of the transition. Substitution of the tunneling amplitude in the form (5.24) into Eq. (5.25), and summation over the spin indices with the help of the identity (2.9) result in

$$-\pi \nu \, \mathrm{Im} \, T_s = \frac{3\pi^2}{16} \nu^2 \left| A^2(\omega) \right|. \tag{5.26}$$

In the leading (first) order in J one readily obtains $A^{(1)} = J$, independently of ω. However, as discovered by Kondo [6], the second-order contribution $A^{(2)}$ not only depends on ω, but is logarithmically divergent as $\omega \rightarrow 0$:

$$A^{(2)}(\omega) = \nu J^2 \ln |\Delta/\omega|.$$

Here Δ is the high-energy cutoff in the Hamiltonian (5.12). It turns out [64] that similar logarithmically divergent contributions appear in all orders of perturbation theory,

$$\nu A^{(n)}(\omega) = (\nu J)^n \left[\ln |\Delta/\omega| \right]^{n-1},$$

resulting in a geometric series

$$\nu A(\omega) = \sum_{n=1}^{\infty} \nu A^{(n)} = \nu J \sum_{n=0}^{\infty} [\nu J \ln |\Delta/\omega|]^n = \frac{\nu J}{1 - \nu J \ln |\Delta/\omega|}.$$

With the help of Eq. (5.14) this can be written as

$$\nu A(\omega) = \frac{1}{\ln |\omega/T_K|}. \qquad (5.27)$$

Substitution of Eq. (5.27) into Eq. (5.26) and then into Eq. (5.21), and evaluation of the integral over ω with logarithmic accuracy yield for the conductance across the dot

$$G = G_0 \frac{3\pi^2/16}{\ln^2(T/T_K)}, \quad T \gg T_K. \qquad (5.28)$$

Equation (5.28) is the leading term of the asymptotic expansion in powers of $1/\ln(T/T_K)$, and represents the conductance in the *leading logarithmic approximation*.

Eq. (5.28) resulted from summing up the most-diverging contributions in all orders of perturbation theory. It is instructive to re-derive it now in the framework of *renormalization group* [66]. The idea of this approach rests on the observation that the electronic states that give a significant contribution to observable quantities, such as conductance, are states within an interval of energies of width $\omega \sim T$ about the Fermi level, see Eq. (5.21). At temperatures of order T_K, when the Kondo effect becomes important, this interval is narrow compared to the width of the band $D = \Delta$ in which the Hamiltonian (5.12) is defined.

Consider a narrow strip of energies of width $\delta D \ll D$ near the edge of the band. Any transition (5.23) between a state near the Fermi level and one of the states in the strip is associated with high ($\sim D$) energy deficit, and, therefore, can only occur virtually. Obviously, virtual transitions via each of the states in the strip result in the second-order correction $\sim J^2/D$ to the amplitude $A(\omega)$ of the transition between states in the vicinity of the Fermi level. Since the strip contains $\nu \delta D$ electronic states, the total correction is [66]

$$\delta A \sim \nu J^2 \delta D/D.$$

This correction can be accounted for by modifying the exchange constant in the effective Hamiltonian $\widetilde{H}_{\text{eff}}$ which is defined for states within a narrower energy band of the width $D - \delta D$ [66],

$$\widetilde{H}_{\text{eff}} = \sum_{ks} \xi_k \psi_{ks}^\dagger \psi_{ks} + J_{D-\delta D}(\mathbf{s}_\psi \cdot \mathbf{S}), \quad |\xi_k| < D - \delta D, \qquad (5.29)$$

$$J_{D-\delta D} = J_D + \nu J_D^2 \frac{\delta D}{D}. \qquad (5.30)$$

Here J_D is the exchange constant in the original Hamiltonian. Note that $\widetilde{H}_{\text{eff}}$ has the same form as Eq. (5.12). This is not merely a conjecture, but can be shown rigorously [59, 67].

The reduction of the bandwidth can be considered to be the result of a unitary transformation that decouples the states near the band edges from the rest of the band [68]. In principle, any such transformation should also affect the operators that describe the observable quantities. Fortunately, this is not the case for the problem at hand. Indeed, the angle θ_0 in Eq. (5.19) is not modified by the transformation. Therefore, the current operator and the expression for the conductance (5.21) retain their form.

Successive reductions of the high-energy cutoff D by small steps δD can be viewed as a continuous process during which the initial Hamiltonian (5.12) with $D = \Delta$ is transformed to an effective Hamiltonian of the same form that acts within the band of reduced width $D \ll \Delta$. It follows from Eq. (5.30) that the dependence of the effective exchange constant on D is described by the differential equation [66, 67]

$$\frac{dJ_D}{d\zeta} = \nu J_D^2, \quad \zeta = \ln(\Delta/D). \tag{5.31}$$

With the help of Eq. (5.14), the solution of the RG equation (5.31) subject to the initial condition $J_\Delta = J$ can be cast into the form

$$\nu J_D = \frac{1}{\ln(D/T_K)}. \tag{5.32}$$

The renormalization described by Eq. (5.31) can be continued until the bandwidth D becomes of the order of the typical energy $|\omega| \sim T$ for real transitions. After this limit has been reached, the transition amplitude $A(\omega)$ is calculated in lowest (first) order of perturbation theory in the effective exchange constant (higher order contributions are negligible for $D \sim \omega$),

$$\nu A(\omega) = \nu J_{D \sim |\omega|} = \frac{1}{\ln|\omega/T_K|}.$$

Using now Eqs. (5.26) and (5.21), we recover Eq. (5.28).

5.4. Strong coupling regime: $T \ll T_K$

As temperature approaches T_K, the leading logarithmic approximation result Eq. (5.28) diverges. This divergence signals the failure of the approximation. Indeed, we are considering a model with single-mode junctions between the dot and the leads. The maximal possible conductance in this case is $2e^2/h$. To obtain a more precise bound, we discuss in this section the conductance in the strong coupling regime $T \ll T_K$.

We start with the zero-temperature limit $T = 0$. As discussed above, the ground state of the Kondo model (5.12) is a singlet [58], and, accordingly, is not degenerate. Therefore, the t-matrix of the conduction electrons interacting with the localized spin is completely characterized by the scattering phase shifts δ_s for electrons with spin s at the Fermi level. The t-matrix is then given by the standard scattering theory expression [65]

$$-\pi \nu\, T_s(0) = \frac{1}{2i}\,(\mathbb{S}_s - 1)\,, \quad \mathbb{S}_s = e^{2i\delta_s}, \tag{5.33}$$

where $\mathbb{S}_s$ is the scattering matrix for electrons with spin s, which for a single channel case reduces to its eigenvalue. Substitution of Eq. (5.33) into Eq. (5.21) yields

$$G(0) = G_0 \frac{1}{2} \sum_s \sin^2 \delta_s \tag{5.34}$$

for the conductance, see Eq. (5.21). The phase shifts in Eqs. (5.33), (5.34) are obviously defined only mod π (that is, δ_s and $\delta_s + \pi$ are equivalent). This ambiguity can be removed if we set to zero the values of the phase shifts at $J = 0$ in Eq. (5.12).

In order to find the two phase shifts δ_s, we need two independent relations. The first one follows from the invariance of the Kondo Hamiltonian (5.12) under the particle-hole transformation $\psi_{ks} \rightarrow s\psi^{\dagger}_{-k,-s}$ (here $s = \pm 1$ for spin-up/down electrons). The particle-hole symmetry implies the relation for the t-matrix

$$T_s(\omega) = -T^*_{-s}(-\omega), \tag{5.35}$$

valid at all ω and T. In view of Eq. (5.33), it translates into the relation for the phase shifts at the Fermi level ($\omega = 0$) [69],

$$\delta_\uparrow + \delta_\downarrow = 0. \tag{5.36}$$

The second relation follows from the requirement that the ground state of the Hamiltonian (5.12) is a singlet [69]. In the absence of exchange ($J = 0$) and at $T = 0$, an infinitesimally weak (with Zeeman energy $B \rightarrow +0$) magnetic field

$$H_B = B S^z \tag{5.37}$$

would polarize the dot's spin. Since the free electron gas has zero spin in the ground state, the total spin in a very large but finite region of space $\mathcal{V}$ surrounding the dot coincides with the spin of the dot, $\langle S^z \rangle_{J=0} = -1/2$. If the exchange with the electron gas is now turned on, $J > 0$, a very weak field will not prevent the formation of a singlet ground state. In this state, the total spin within $\mathcal{V}$ is zero.

Such change of the spin is possible if the numbers of spin-up and spin-down electrons within $\mathcal{V}$ have changed to compensate for the dot's spin: $\delta N_\uparrow - \delta N_\downarrow = 1$. By the Friedel sum rule, δN_s are related to the scattering phase shifts at the Fermi level, $\delta N_s = \delta_s / \pi$, which gives

$$\delta_\uparrow - \delta_\downarrow = \pi. \qquad (5.38)$$

Combining Eqs. (5.36) and (5.38), we find

$$\delta_s = s \frac{\pi}{2}. \qquad (5.39)$$

Equation (5.34) then yields for the zero-temperature and zero-field conductance across the dot [45]

$$G(0) = G_0. \qquad (5.40)$$

Thus, the growth of conductance as the temperature is lowered, is limited only by the value of G_0. This value, see Eq. (5.22), depends only on the ratio of the tunneling amplitudes $|t_{L0}/t_{R0}|$; if $|t_{L0}| = |t_{R0}|$, then the conductance at $T = 0$ will reach the maximal value $G = 2e^2/h$ allowed by quantum mechanics [45].

As explained above, screening of the dot's spin by itinerant electrons amounts to $\pm \pi/2$ phase shifts for electrons at the Fermi level ($\omega = 0$). However, the phase shifts at a finite ω deviate from the unitary limit value. Such deviation is to be expected, as the Kondo "resonance" has a finite width $\sim T_K$. There is also an inelastic component of scattering appearing at finite ω. Virtual transitions to excited states, corresponding to a broken up Kondo singlet, induce a local repulsive interaction between itinerant electrons [69], which is the cause of inelastic scattering. This interaction enters the fixed-point Hamiltonian, applicable when the relevant energies are small compared to the Kondo temperature T_K. The fixed point Hamiltonian takes a relatively simple form [69] when written in the basis of electronic states that incorporate an extra $\pm \pi/2$ phase shift [see Eq. (5.39)] compared to the original basis of Eq. (5.12). In the new basis,

$$H_{\text{fixed point}} = \sum_{ks} \xi_k \varphi_{ks}^\dagger \varphi_{ks} - \sum_{kk's} \frac{\xi_k + \xi_{k'}}{2\pi \nu T_K} \varphi_{ks}^\dagger \varphi_{k's} + \frac{1}{\pi \nu^2 T_K} \rho_\uparrow \rho_\downarrow. \qquad (5.41)$$

Here $\rho_s = \sum_{kk'} :\varphi_{ks}^\dagger \varphi_{k's}:$ (the colons denote normal ordering). The form of the last two terms on the r.h.s. of Eq. (5.41) is dictated by the particle-hole symmetry [69]. The ratio of their coefficients is fixed by the physical requirement that the Kondo singularity is tied to the Fermi level [69] (i.e. that the phase shifts depend only on the distance ω to the Fermi level), while the overall coefficient

$1/T_K$ can be viewed as a precise definition of the Kondo temperature[6]. Note that by construction, the t-matrix for φ−particles $\widetilde{T}$ vanishes for $\omega = T = 0$.

The second term in the r.h.s. of Eq. (5.41) describes a purely elastic scattering, and yields a small ω−dependent phase shift [69]

$$\widetilde{\delta}(\omega) = \frac{\omega}{T_K}. \tag{5.42}$$

The last term in Eq. (5.41) gives rise to an inelastic contribution to $\widetilde{T}$. Evaluation of this contribution in the second order of perturbation theory (see [62, 69] for details) results in

$$-\pi \nu \widetilde{T}_{in}(\omega) = i \frac{\omega^2 + \pi^2 T^2}{2 T_K^2}. \tag{5.43}$$

In order to obtain the t-matrix in the original basis $T_s(\omega)$, we note that the elastic scattering phase shifts here are obtained by simply adding $\pm \pi/2$ to $\widetilde{\delta}$, i.e.

$$\delta_s(\omega) = s \frac{\pi}{2} + \widetilde{\delta}(\omega). \tag{5.44}$$

The relation between $T_s(\omega)$ and $\widetilde{T}(\omega)$, accounting for small inelastic term $\widetilde{T}_{in}$, reads

$$-\pi \nu T_s(\omega) = \frac{1}{2i} \left[e^{2i\delta_s(\omega)} - 1 \right] + e^{2i\delta_s(\omega)} \left[-\pi \nu \widetilde{T}_{in}(\omega) \right]. \tag{5.45}$$

Note that the inelastic contribution enters this expression with an extra factor of $\exp(2i\delta_s)$. The appearance of this factor is a direct consequence of the unitarity of the scattering matrix in the presence of both elastic and inelastic scattering channels, see [69]. Taking into account Eqs. (5.42), (5.43), and (5.44), we find

$$-\pi \nu \operatorname{Im} T_s(\omega) = 1 - \frac{3\omega^2 + \pi^2 T^2}{2 T_K^2}. \tag{5.46}$$

Substitution of this expression into Eq. (5.21) then yields

$$G = G_0 \left[1 - (\pi T/T_K)^2 \right], \quad T \ll T_K. \tag{5.47}$$

Accordingly, corrections to the conductance are quadratic in temperature – a typical Fermi liquid result [69]. The weak-coupling ($T \gg T_K$) and the strong-coupling ($T \ll T_K$) asymptotes of the conductance have a strikingly different structure. Nevertheless, since the Kondo effect is a crossover phenomenon rather

---

[6]With this convention the RG expression Eq. (5.14) is regarded as an estimate of T_K with logarithmic accuracy.

than a phase transition [57, 58, 59, 61], the dependence $G(T)$ is a smooth and featureless [70] function throughout the crossover region $T \sim T_K$.

Finally, note that both Eqs. (5.28) and (5.47) have been obtained here for the particle-hole symmetric model (5.12). This approximation is equivalent to neglecting the elastic co-tunneling contribution to the conductance G_{el}. The asymptotes (5.28), (5.47) remain valid [33] as long as $G_{el}/G_0 \ll 1$. The overall temperature dependence of the linear conductance in the middle of the Coulomb blockade valley is sketched in Fig. 3.

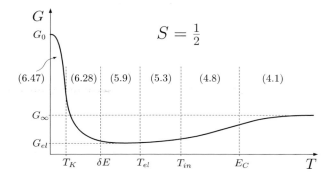

Fig. 3. Sketch of the temperature dependence of the conductance in the middle of the Coulomb blockade valley with $S = 1/2$ on the dot. The numbers in brackets refer to the corresponding equations in the text.

The fixed point Hamiltonian (5.41) also allows one to calculate the corrections to zero-temperature conductance due to a finite magnetic field, which enters Eq. (5.41) via a term $H_B = \sum_{ks}(s/2)B\,\varphi_{ks}^\dagger\varphi_{ks}$. The field polarizes the electron gas, and the operators ρ_s acquire non-zero ground state expectation value $\langle \rho_s \rangle = -\nu Bs/2$. In order to calculate the phase shifts at the Fermi energy we replace $\rho_\uparrow\rho_\downarrow$ in the third term in the r.h.s. (5.41) by $\sum_s \langle \rho_{-s} \rangle \rho_s$, and $\xi_k, \xi_{k'}$ in the second term by their values at the Fermi level $\xi_k = -Bs/2$. As a result, Eq. (5.41) simplifies to

$$H = \sum_{ks}(\xi_k + \frac{s}{2}B)\varphi_{ks}^\dagger\varphi_{ks} + \sum_{kk's}\frac{Bs}{\pi\nu T_K}\rho_s,$$

from which one can read off the scattering phase shifts for $\varphi-$particles,

$$\widetilde{\delta}_s = -sB/T_K.$$

The phase shifts in the original basis are given by [69]

$$\delta_s = s\frac{\pi}{2} + \tilde{\delta}_s = s\left(\frac{\pi}{2} - \frac{B}{T_K}\right),$$

cf. Eq. (5.44). Equation (5.34) then gives

$$G = G_0\left[1 - (B/T_K)^2\right], \quad B \ll T_K. \tag{5.48}$$

Note that the effect of a finite magnetic field on the zero-temperature conductance is very similar to the effect of a finite temperature has on the conductance at zero field, see Eq. (5.47). The similarity is not limited to the strong coupling regime. Indeed, the counterpart of Eq. (5.28) reads

$$G = G_0\frac{\pi^2/16}{\ln^2(B/T_K)}, \quad B \gg T_K. \tag{5.49}$$

5.5. *Beyond linear response*

In order to study transport through a quantum dot away from equilibrium we add to the effective Hamiltonian (5.11)-(5.13) a term

$$H_V = \frac{eV}{2}\left(\hat{N}_L - \hat{N}_R\right) \tag{5.50}$$

describing a finite voltage bias V applied between the left and right electrodes. Here we will evaluate the current across the dot at arbitrary V but under the simplifying assumption that the dot-lead junctions are strongly asymmetric:

$$G_L \ll G_R.$$

Under this condition the angle θ_0 in Eq. (5.19) is small, $\theta_0 \approx |t_{L0}/t_{R0}| \ll 1$. Expanding Eq. (5.20) to linear order in θ_0 we obtain

$$H_V(\theta_0) = \frac{eV}{2}\left(\hat{N}_2 - \hat{N}_1\right) + eV\theta_0\sum_{ks}(\psi_{1ks}^{\dagger}\psi_{2ks} + \text{H.c.}) \tag{5.51}$$

The first term in the r.h.s. here can be interpreted as the voltage bias between the reservoirs of 1 and 2 particles, cf. Eq. (5.50), while the second term has an appearance of k-conserving tunneling with very small (proportional to $\theta_0 \ll 1$) tunneling amplitude.

Similar to Eq. (5.51), the current operator $\hat{I}$, see Eq. (5.18), splits naturally into two parts,

$$\hat{I} = \hat{I}_0 + \delta\hat{I},$$

$$\hat{I}_0 = \frac{d}{dt}\frac{e}{2}(\hat{N}_1 - \hat{N}_2) = -ie^2 V\theta_0 \sum_{ks} \psi_{1ks}^\dagger \psi_{2ks} + \text{H.c.},$$

$$\delta\hat{I} = -e\,\theta_0 \frac{d}{dt} \sum_{ks} \psi_{1ks}^\dagger \psi_{2ks} + \text{H.c.}$$

It turns out that $\delta\hat{I}$ does not contribute to the average current in the leading (second) order in θ_0 [47]. Indeed, in this order

$$\langle \delta\hat{I} \rangle = i\theta_0^2 e^2 V \frac{d}{dt} \int_{-\infty}^{t} dt' \left\langle [\hat{O}(t), \hat{O}(t')] \right\rangle_0, \qquad \hat{O} = \sum_{ks} \psi_{1ks}^\dagger \psi_{2ks},$$

where $\langle \ldots \rangle_0$ denotes thermodynamic averaging with the Hamiltonian $H_{\text{eff}} + H_V$. The thermodynamic (equilibrium) averaging is well defined here, because the Hamiltonian $H_{\text{eff}} + H_V$ conserves separately the numbers of 1- and 2-particles. Taking into account that $\langle [\hat{O}(t), \hat{O}(t')] \rangle_0$ depends only on the difference $\tau = t - t'$, we find

$$\langle \delta\hat{I} \rangle = i\theta_0^2 e^2 V \frac{d}{dt} \int_0^{\infty} d\tau \left\langle [\hat{O}(\tau), \hat{O}(0)] \right\rangle_0 = 0.$$

The remaining contribution $I = \langle \hat{I}_0 \rangle$ corresponds to tunneling current between two bulk reservoirs containing 1- and 2-particles. Its evaluation follows the standard procedure [71] and yields [47]

$$\frac{dI}{dV} = G_0 \frac{1}{2} \sum_s [-\pi \nu \, \text{Im}\, T_s(eV)] \tag{5.52}$$

for the differential conductance across the dot at zero temperature. Here G_0 coincides with the small θ_0-limit of Eq. (5.22). Using now Eqs. (5.26), (5.27), and (5.46), we obtain

$$\frac{1}{G_0}\frac{dI}{dV} = \begin{cases} 1 - \dfrac{3}{2}\left(\dfrac{eV}{T_K}\right)^2, & |eV| \ll T_K \\[3mm] \dfrac{3\pi^2/16}{\ln^2(|eV|/T_K)}, & |eV| \gg T_K \end{cases} \tag{5.53}$$

Thus, a large voltage bias has qualitatively the same destructive effect on the Kondo physics as the temperature does. The result Eq. (5.53) remains valid as long as $T \ll |eV| \ll \delta E$. If temperature exceeds the bias, $T \gg eV$, the differential conductance coincides with the linear conductance, see Eqs. (5.28) and (5.47) above.

5.6. Splitting of the Kondo peak in a magnetic field

According to Eq. (5.53), the differential conductance exhibits a peak at zero bias. The very appearance of such a peak, however, does not indicate that we are dealing with the Kondo effect. Indeed, even in the absence of interactions, tunneling via a resonant level situated at the Fermi energy would also result in a zero-bias peak in dI/dV. These two situations can be distinguished by considering the evolution of the zero-bias peak with the magnetic field. In both cases, a sufficiently strong field (such that the corresponding Zeeman energy B appreciably exceeds the peaks width) splits the zero-bias peak in two smaller ones, located at $\pm V^*$. In the case of conventional resonant tunneling, $eV^* = B/2$. However, if the split peaks are of the Kondo origin, then $eV^* \approx B$. This doubling of the distance between the split peaks is viewed by many as a hallmark of the many-body physics associated with the Kondo effect [4].

In order to address the splitting of the zero-bias Kondo peak we add to the Hamiltonian (5.12) a term

$$H_B = B S_z + \sum_{ks} \frac{Bs}{2} \psi_{1ks}^\dagger \psi_{1ks} \tag{5.54}$$

describing the Zeeman effect of the magnetic field[7]. We concentrate on the most interesting case of

$$\Delta \gg B \gg T_K. \tag{5.55}$$

The first inequality here ensures that the zero-bias peak is split, while the second one allows the development of the Kondo correlations to a certain extent. For simplicity, we also consider here a strongly asymmetric setup, as in Section 5.5, which allows us to use Eq. (5.52) relating the differential conductance to the t-matrix. Note, however, that the results presented below remain qualitatively correct even when this assumption is lifted.

---

[7]In general, a magnetic field affects both the orbital and spin parts of the electronic wave function. However, as long as the spin of the dot remains equal to $1/2$ (the opposite case is considered in the next Section), the orbital effect of the field is unimportant. It modifies the exchange amplitude J and the bandwidth Δ in the Hamiltonian (5.12), which may eventually alter the values of G_0 and T_K in Eq. (5.53). However, the functional form of the dependence of dI/dV on V at a fixed B remains intact.

As long as the bias $|eV|$ is small compared to B, the differential conductance approximately coincides with the linear one, see Eq. (5.49). Similarly, at $|eV| \gg B$ the field has a negligible effect, and dI/dV is given by the second line in Eq. (5.53). The presence of the field, however, significantly affects the dependence of dI/dV on V at $|eV| \sim B \gg T$, and it is this region which we address here.

Clearly, for $B \sim |eV| \gg T_K$ the system is in the weak coupling Kondo regime, so that the perturbative RG procedure described in Section 5.3 is an adequate tool to study it. Let us suppose, for simplicity, that in the absence of magnetic field electrons fill up the lower half of a band characterized by width 2Δ and constant density of states ν, see Fig. 4(a). In the presence of the field, the energy of an itinerant electron consists of the orbital part ξ_k and the Zeeman energy $Bs/2$, see the second term in Eq. (5.54). Condition $\xi_k + Bs/2 = \epsilon_F$ cuts out different strips of orbital energies for spin-up and spin-down electrons, see Fig. 4(b). To set the stage for the RG, it is convenient to cut the initial band asymmetrically for spin-up and spin-down electrons. The resulting band structure, shown in Fig. 4(c), is the same for both directions of spin. Further symmetric reductions of the band in the course of RG leave this band structure intact.

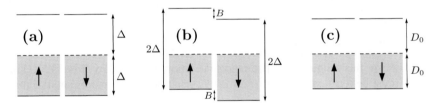

Fig. 4. The energy bands for spin-up and spin-down itinerant electrons. The solid lines denote the boundaries of the bands, the dashed lines correspond to the Fermi energy.
(a) Band structure without Zeeman splitting; (b) same with $B \neq 0$; (c) same after the initial asymmetric cutting of the bands ($D_0 = \Delta - B$).

After the initial cutting of the band, the effective Hamiltonian $H_{\text{eff}} + H_B$ takes the form

$$H = \sum_{ks} \xi_k \psi_{ks}^{\dagger} \psi_{ks} + J(\mathbf{s}_\psi \cdot \mathbf{S}) + \eta B S^z. \qquad (5.56)$$

Here k is measured from the Fermi momentum of electrons with spin s, and we suppressed the subscript 1 in the operators ψ_{1ks} etc. The Hamiltonian (5.56) is defined within a symmetric band

$$|\xi_k| < D_0 = \Delta - B \approx \Delta,$$

see Fig. 4(c). The parameter η in (5.56) is given by

$$\eta = 1 - \nu J/2 \tag{5.57}$$

with J the "bare" value of the exchange constant. The correction $-\nu J/2$ in Eq. (5.57) is nothing else but the Knight shift [72]. In a finite magnetic field, there are more spin-down electrons than spin-up ones, see Fig. 4(b). This mismatch results in a non-zero ground state expectation value $\langle s_1^z \rangle = -\nu B/2$ of the operator s_1^z in Eq. (5.12). Due to the exchange interaction, this induces an additional effective magnetic field $-\nu J B/2$ acting on the localized spin.

The form of the effective Hamiltonian (5.56) remains invariant under the RG transformation. The evolution of J and μ with the reduction of the bandwidth D is governed by the equations

$$\frac{dJ_D}{d\zeta} = \nu J_D^2, \tag{5.58}$$

$$\frac{d\ln \eta_D}{d\zeta} = -\frac{1}{2}(\nu J_D)^2, \tag{5.59}$$

where $\zeta = \ln(\Delta/D)$. Equation (5.58) was derived in Section 5.3 above; the derivation of Eq. (5.59) proceeds in a similar way. It is clear from the solution of Eq. (5.58), see Eq. (5.32), that νJ_D remains small throughout the weak coupling regime $D \gg T_K$. It then follows from Eqs. (5.57) and (5.59) that η_D is of the order of 1 in this range of D. The renormalization described by Eqs. (5.58) and (5.59) must be terminated at $D \sim 2B \gg T_K$, so that $\eta_B B$ remains smaller than D. At this point

$$\nu J_B = \frac{1}{\ln(B/T_K)}, \tag{5.60}$$

see Eq. (5.32). Solution of Eq. (5.58) in the scaling limit[8], defined as

$$\frac{\ln(B/T_K)}{\ln(\Delta/T_K)} \ll 1,$$

takes the familiar form [57]

$$\eta_B = 1 - \frac{1}{2\ln(B/T_K)}. \tag{5.61}$$

Now we are ready to evaluate the differential conductance. The Hamiltonian Eq. (5.56) with J and η given by Eqs. (5.60) and (5.61) is defined in a sufficiently

---

[8] In this limit the Knight shift in Eq. (5.57) can be neglected.

wide band $D \sim 2B$ to allow for excitations with energy of the order of the Zeeman splitting. Since the current operator remains invariant in the course of renormalization, the differential conductance can still be calculated from Eq. (5.52). We start with the lowest order in νJ_B (Born approximation). Calculation similar to that in Section 5.3 yields

$$\frac{1}{G_0}\frac{dI}{dV} = \frac{\pi^2/16}{\ln^2(B/T_K)}\Big[1 + 2\theta(|eV| - \eta_B B)\Big]. \tag{5.62}$$

The main new feature of Eq. (5.62) compared to (5.53) is a threshold behavior at $|eV| = \eta_B B$. The origin of this behavior is apparent: for $|eV| > \eta_B B$ an inelastic (spin-flip) scattering channel opens up, causing a step in dI/dV.

In the absence of Zeeman splitting, going beyond the Born approximation in the renormalized exchange amplitude Eq. (5.32) would be meaningless, as this would exceed the accuracy of the logarithmic RG. This largely remains true here as well, except when the bias is very close to the inelastic scattering threshold,

$$\Big||eV| - \eta_B B\Big| \ll B.$$

The existence of the threshold makes the t-matrix singular at $\omega = \eta_B B$; such logarithmically divergent contribution appears in the third order in J_B [50]. Retaining this contribution does not violate the accuracy of our approximations, and we find

$$\frac{1}{G_0}\frac{dI}{dV} = \frac{\pi^2/16}{\ln^2(B/T_K)}\Big[1 + 2\theta(|eV| - \eta_B B)\Big]$$
$$\times \left(1 + \frac{1}{\ln(B/T_K)}\ln\frac{B}{\big||eV| - \mu B\big|}\right). \tag{5.63}$$

We see now that the zero-bias peak in the differential conductance is indeed split in two by the applied magnetic field. The split peaks are located at $\pm V^*$ with

$$V^* = \eta_B B = \left[1 - \frac{1}{2\ln(B/T_K)}\right]B, \tag{5.64}$$

see Eq. (5.61).

Because of the logarithmic divergence in Eq. (5.63) this result needs some refinement. A finite temperature $T \gg T_K$ would of course cut the divergence [50]; this amounts to the replacement of $\big||eV| - \eta_B B\big|$ under the logarithm in Eq. (5.63) by $\max\{\big||eV| - \mu B\big|, T\}$. This is fully similar to the cut-off of the zero-bias Kondo singularity in the absence of the field.

The $T = 0$ case, however, is different, and turns out to be much simpler than the $B = 0$ Kondo anomaly at zero bias. The logarithmic singularity in Eq. (5.63) is brought about by the second-order in J_B contribution to the scattering amplitude. This contribution involves a transition to the excited ($S^z = +1/2$) state of the localized spin. Unlike the $B = 0$ case, now such a state has a finite lifetime: the spin of the dot may flip, exciting (a triplet) electron-hole pair of energy $\eta_B B$ in the band. Such relaxation mechanism was first considered by Korringa [73]. In our case the corresponding relaxation rate is easily found with the help of the Fermi Golden Rule,

$$\frac{\hbar}{\tau_K} = \frac{\pi}{2} \frac{\eta_B B}{\ln^2(B/T_K)}.$$

Due to the finite Korringa lifetime τ_K the state responsible for the logarithmic singularity is broadened by $\hbar/\tau_K$, which cuts off the divergence [74] in Eq. (5.63). It also smears the step-like dependence on V, which we found in the Born approximation, Eq. (5.62). As a result, the split peaks in the dependence of dI/dV on V are broad and low. The characteristic width of the peaks is

$$\delta V \sim \frac{\hbar}{\tau_K} \sim \frac{B}{\ln^2(B/T_K)}, \tag{5.65}$$

and their height is given by

$$\frac{G(V^*) - G(0)}{G(0)} \sim \frac{\ln\left[\ln^2(B/T_K)\right]}{\ln^2(B/T_K)} \ll 1. \tag{5.66}$$

Here we used the short-hand notation $dI/dV = G(V)$. When evaluating $G(V^*)$, we replaced $\big||eV| - \eta_B B\big|$ by $\hbar/\tau_K$ in the argument of the logarithm in Eq. (5.63). The right-hand side of Eq. (5.66) is small, and the higher-order in νJ_B corrections to Eq. (5.63) are negligible.

The problem of the field-induced splitting of the Kondo peak was revisited many times since the works of Appelbaum [50], see e.g. [75, 76, 77].

5.7. Kondo effect in quantum dots with large spin

If the dot's spin exceeds $1/2$, see Refs. [78, 79, 80], then, as discussed in Section 5.1 above, both exchange constants J_γ in the effective Hamiltonian (5.2) are finite and positive. This turns out to have a dramatic effect on the dependence of the conductance in the Kondo regime on temperature T and on Zeeman energy B. Unlike the case of $S = 1/2$ on the dot, see Fig. 3, now the dependences on T and B are *non-monotonic*: an initial increase of G is followed by a drop when the temperature is lowered [33, 81] at $B = 0$; the variation of G with B at $T = 0$ is similarly non-monotonic.

The origin of this peculiar behavior is easier to understand by considering the B-dependence of the zero-temperature conductance [33]. We assume that the magnetic field $H_\parallel$ is applied *in the plane* of the dot. Such field leads to the Zeeman splitting B of the spin states of the dot, see Eq. (5.37), but barely affects the orbital motion of electrons.

At any finite B the ground state of the system is not degenerate. Therefore, the linear conductance at $T = 0$ can be calculated from the Landauer formula

$$G = \frac{e^2}{h} \sum_s |\mathbb{S}^2_{s;RL}| , \tag{5.67}$$

which relates G to the amplitude of scattering $\mathbb{S}_{s;RL}$ of an electron with spin s from lead L to lead R. The amplitudes $\mathbb{S}_{s;\alpha\alpha'}$ form 2×2 scattering matrix $\hat{\mathbb{S}}_s$. In the basis of "channels", see Eq. (5.2), this matrix is obviously diagonal, and its eigenvalues $\exp(2i\delta_{\gamma s})$ are related to the scattering phase shifts $\delta_{\gamma s}$. The scattering matrix in the original $(R - L)$ basis is obtained from

$$\hat{\mathbb{S}}_s = \hat{U}^\dagger \text{diag}\left\{e^{2i\delta_{\gamma s}}\right\} \hat{U},$$

where $\hat{U}$ is a matrix of a rotation by an angle θ_0, see Eq. (5.19). The Landauer formula (5.67) then yields

$$G = G_0 \frac{1}{2} \sum_s \sin^2 (\delta_{1s} - \delta_{2s}) , \quad G_0 = \frac{2e^2}{h} \sin^2(2\theta_0), \tag{5.68}$$

which generalizes the single-channel expression (5.34).

Equation (5.68) can be further simplified for a particle-hole symmetric model (5.2). Indeed, in this case the phase shifts satisfy $\delta_{\gamma\uparrow} + \delta_{\gamma\downarrow} = 0$, cf. Eq. (5.36), which suggests a representation

$$\delta_{\gamma s} = s\delta_\gamma. \tag{5.69}$$

Substitution into Eq. (5.68) then results in

$$G = G_0 \sin^2 (\delta_1 - \delta_2) . \tag{5.70}$$

If the spin S on the dot, exceeds $1/2$, then both channels of itinerant electrons participate in the screening of the dot's spin [56]. Accordingly, in the limit $B \rightarrow 0$ both phase shifts δ_γ approach the unitary limit value $\pi/2$, see Fig. 5. However, the increase of phase shifts with lowering the field is characterized

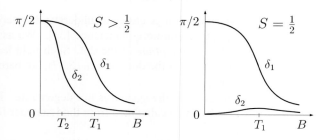

Fig. 5. Dependence of the scattering phase shifts at the Fermi level on the magnetic field for $S > 1/2$ (left panel) and $S = 1/2$ (right panel).

by two different energy scales. These scales, the Kondo temperatures T_1 and T_2, are related to the corresponding exchange constants in the effective Hamiltonian (5.2),

$$\ln\left(\frac{\Delta}{T_\gamma}\right) \sim \frac{1}{\nu J_\gamma},$$

so that $T_1 > T_2$ for $J_1 > J_2$. It is then obvious from Eq. (5.70) that the conductance across the dot is small both at weak ($B \ll T_2$) and strong ($B \gg T_1$) fields, but may become large ($\sim G_0$) at intermediate fields $T_2 \ll B \ll T_1$, see Fig. 6. In other words, the dependence of the zero-temperature conductance on the magnetic field is non-monotonic.

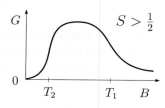

Fig. 6. Sketch of the magnetic field dependence of the Kondo contribution to the linear conductance at zero temperature. The conductance as a function of temperature exhibits a similar non-monotonic dependence.

This non-monotonic dependence is in sharp contrast with the monotonic increase of the conductance with lowering the field when $S = 1/2$. Indeed, in the latter case it is the channel whose coupling to the dot is the strongest that screens the dot's spin, while the remaining channel decouples at low energies [56], see Fig. 5.

The dependence of the conductance on temperature $G(T)$ is very similar[9] to $G(B)$. For example, for $S = 1$ one obtains [33]

$$
G/G_0 = \begin{cases} (\pi T)^2 \left(\dfrac{1}{T_1} - \dfrac{1}{T_2} \right)^2, & T \ll T_2 \\[4mm] \dfrac{\pi^2}{2} \left[\dfrac{1}{\ln(T/T_1)} - \dfrac{1}{\ln(T/T_2)} \right]^2, & T \gg T_1 \end{cases} \tag{5.71}
$$

The conductance reaches its maximun G_{max} at $T \sim \sqrt{T_1 T_2}$. The value of G_{max} can be found analytically for $T_1 \gg T_2$. For $S = 1$ the result reads [33]

$$
G_{max} = G_0 \left[1 - \frac{3\pi^2}{\ln^2(T_1/T_2)} \right]. \tag{5.72}
$$

Consider now a Coulomb blockade valley with $N =$ even electrons and spin $S = 1$ on the dot. In a typical situation, the dot's spin in two neighboring valleys (with $N \pm 1$ electrons) is $1/2$. Under the conditions of applicability of the approximation in Eq. (5.5), there is a single non-zero exchange constant $J_{N \pm 1}$ for each of these valleys. If the Kondo temperatures T_K are the same for both valleys with $S = 1/2$, then $J_{N+1} = J_{N-1} = J_{odd}$. Each of the two singly-occupied energy levels in the valley with $S = 1$ is also singly-occupied in one of the two neighboring valleys. It then follows from Eqs. (5.4)-(5.6) that the exchange constants $J_{1,2}$ for $S = 1$ satisfy

$$
J_1 + J_2 = \frac{1}{2} (J_{N+1} + J_{N-1}) = J_{odd}.
$$

Since both J_1 and J_2 are positive, this immediately implies that $J_{1,2} < J_{odd}$. Accordingly, both Kondo temperatures $T_{1,2}$ are expected to be smaller than T_K in the nearby valleys with $S = 1/2$.

This consideration, however, is not applicable when the dot is tuned to the vicinity of the singlet-triplet transition in its ground state [11, 12, 79, 80], i.e. when the energy gap Δ between the triplet ground state and the singlet excited state of an isolated dot is small compared to the mean level spacing δE. In this case the exchange constants in the effective Hamiltonian (5.2) should account for additional renormalization that the system's parameters undergo [82] when the high-energy cutoff (the bandwidth of the effective Hamiltonian) D is reduced from $D \sim \delta E$ down to $D \sim \Delta \ll \delta E$, see also [63]. The renormalization enhances the exchange constants $J_{1,2}$. If the ratio $\Delta/\delta E$ is sufficiently small, then the Kondo temperatures $T_{1,2}$ for $S = 1$ may become of the same order [78,

---

[9]Note, however, that $\langle \psi_1^\dagger(t) \psi_2(t) \psi_2^\dagger(0) \psi_1(0) \rangle \neq \langle \psi_1^\dagger(t) \psi_1(0) \rangle \langle \psi_2(t) \psi_2^\dagger(0) \rangle$ at finite T. Therefore, unlike Eq. (5.34), there is no simple generalization of Eq. (5.21) to the two-channel case.

80], or even significantly exceed [11, 12, 79] the corresponding scale T_K for $S = 1/2$.

In GaAs-based lateral quantum dot systems the value of Δ can be controlled by a magnetic field $H_\perp$ applied *perpendicular to the plane* of the dot [79]. Because of the smallness of the effective mass m^*, even a weak field $H_\perp$ has a strong orbital effect. At the same time, smallness of the quasiparticle g-factor in GaAs ensures that the corresponding Zeeman splitting remains small [12]. Theory of the Kondo effect in lateral quantum dots in the vicinity of the singlet-triplet transition was developed in [83], see also [84].

6. Concluding remarks

In the simplest form of the Kondo effect considered in these notes, a quantum dot behaves essentially as an artificial "magnetic impurity" with spin S, coupled via exchange interaction to two conducting leads. The details of the temperature dependence $G(T)$ of the linear conductance across the dot depend on the dot's spin S. In the most common case of $S = 1/2$ the conductance in the Kondo regime monotonically increases with decreasing temperature, potentially up to the quantum limit $2e^2/h$. Qualitatively (although not quantitatively), this increase can be understood from the Anderson impurity model in which the dot is described by a single energy level. On the contrary, when the spin on the dot exceeds $1/2$, the conductance evolution proceeds in two stages: the conductance first increases, and then drops again when the temperature is lowered.

In a typical experiment [4], one measures the dependence of the differential conductance on temperature T, Zeeman energy B, and dc voltage bias V. When one of these parameters is much larger than the other two, and is also large compared to the Kondo temperature T_K, the differential conductance exhibits a logarithmic dependence

$$\frac{1}{G_0}\frac{dI}{dV} \propto \left[\ln\frac{\max\{T, B, eV\}}{T_K}\right]^{-2}, \tag{6.1}$$

characteristic of the weak coupling regime of the Kondo system, see Section 5.3. Consider now a zero-temperature transport through a quantum dot with $S = 1/2$ in the presence of a strong field $B \gg T_K$. In accordance with Eq. (6.1), the differential conductance is small compared to G_0 both for $eV \ll B$ and for $eV \gg B$. However, the calculation in the third order of perturbation theory in the exchange constant yields a contribution that diverges logarithmically at $eV \approx \pm B$ [50]. The divergence is reminiscent of the Kondo zero-bias anomaly. However, the full development of resonance is inhibited by a finite lifetime of the excited spin state of the dot [74]. As a result, the peak in the differential conductance at $eV = \pm\mu B$ is broader and lower than the corresponding peak at zero bias in the absence of field, see Section 5.6.

One encounters similar effects in studies of the influence of a weak ac signal of frequency $\Omega \gtrsim T_K$ applied to the gate electrode [85] on transport across the dot. In close analogy with the usual photon-assisted tunneling [86], such perturbation is expected to result in the formation of satellites [87] at $eV = n\hbar\Omega$ (here n is an integer) to the zero-bias peak in the differential conductance. Again, the resonances responsible for the formation of the satellite peaks are limited by finite lifetime effects [88].

Spin degeneracy is not the only possible source of the Kondo effect in quantum dots. Consider, for example, a large dot connected by a single-mode junction to a conducting lead and tuned to the vicinity of the Coulomb blockade peak [31]. If one neglects the finite level spacing in the dot, then the two almost degenerate charge state of the dot can be labeled by a pseudospin, while real spin plays the part of the channel index [31, 89]. This setup turns out to be a robust realization [31, 89] of the symmetric (i.e. having equal exchange constants) two-channel $S = 1/2$ Kondo model [56]. The model results in a peculiar temperature dependence of the observable quantities, which at low temperatures follow power laws with manifestly non-Fermi-liquid fractional values of the exponents [90].

It should be emphasized that in the usual geometry consisting of two leads attached to a Coulomb-blockaded quantum dot with $S = 1/2$, only the conventional Fermi-liquid behavior can be observed at low temperatures. Indeed, in this case the two exchange constants in the effective exchange Hamiltonian (5.2) are vastly different, and their ratio is not tunable by conventional means, see the discussion in Section 5.1 above. A way around this difficulty was proposed in [91]. The key idea is to replace one of the leads in the standard configuration by a very large quantum dot, characterized by a level spacing $\delta E'$ and a charging energy E_C'. At $T \gg \delta E'$, particle-hole excitations within this dot are allowed, and electrons in it participate in the screening of the smaller dot's spin. At the same time, as long as $T \ll E_C'$, the number of electrons in the large dot is fixed. Therefore, the large dot provides a separate screening channel which does not mix with that supplied by the remaining lead. In this system, the two exchange constants are controlled by the conductances of the dot-lead and dot-dot junctions. A strategy for tuning the device parameters to the critical point characterized by the two-channel Kondo physics is discussed in [92].

Finally, we should mention that the description based on the universal Hamiltonian (2.15) is not applicable to large quantum dots subjected to a *quantizing* magnetic field $H_\perp$ [93, 94]. Such a field changes drastically the way the screening occurs in a confined droplet of a two-dimensional electron gas [95]. The droplet is divided into alternating domains containing compressible and incompressible electron liquids. In the metal-like compressible regions, the screening is almost perfect. On the contrary, the incompressible regions behave very much like insulators. In the case of lateral quantum dots, a large compressible domain is formed near the center of the dot. The domain is surrounded by a narrow incompressible region separating it from another compressible ring-shaped domain formed along the edges of the dot [96]. This system can be viewed as two concentric capacitively coupled quantum "dots" - the core dot and the edge dot [93, 96]. When the leads are attached to the edge dot, the measured conductance is sensitive to its spin state: when the number of electrons in the edge dot is odd, the conductance becomes large due to the Kondo effect [93]. Changing the field causes redistribution of electrons between the core and the edge, resulting in a striking checkerboard-like pattern of high- and low-conductance regions [93, 94]. This behavior persists as long as the Zeeman energy remains small compared to the Kondo temperature. Note that compressible regions are also formed around an *antidot* – a potential hill in a two-dimensional electron gas in the quantum Hall regime [97]. Both Coulomb blockade oscillations and Kondo-like behavior were observed in these systems [98].

The Kondo effect arises whenever a coupling to a Fermi gas induces transitions within otherwise degenerate ground state multiplets of an interacting system. Both coupling to a Fermi gas and interactions are naturally present in a nanoscale transport experiment. At the same time, many nanostructures can be easily tuned to the vicinity of a degeneracy point. This is why the Kondo effect in its various forms often influences the low temperature transport in meso- and nanoscale systems.

In these notes we reviewed the theory of the Kondo effect in transport through quantum dots. A Coulomb-blockaded quantum dot behaves in many aspects as an artificial "magnetic impurity" coupled via exchange interaction to two conducting leads. Kondo effect in transport through such "impurity" manifests itself by the lifting of Coulomb blockade at low temperatures, and, therefore, can be unambiguously identified. Quantum dot systems not only offer a direct access to transport properties of an artificial impurity, but also provide one with a broad arsenal of tools to tweak the impurity properties, unmatched in conventional systems. The characteristic energy scale for the intra-dot excitations is much smaller than the corresponding scale for natural magnetic impurities. This allows one to induce degeneracies in the ground state of a dot which are more exotic than just the spin degeneracy. This is only one out of many possible extensions of the simple model discussed in these notes.

Acknowledgements

This project was supported by NSF grants DMR02-37296 and EIA02-10736, and by the Nanoscience/Nanoengineering Research Program of Georgia Tech.

References

[1] L.P. Kouwenhoven et al., in: *Mesoscopic Electron Transport*, eds. L.L. Sohn et al., (Kluwer, Dordrecht, 1997), p. 105; M.A. Kastner, Rev. Mod. Phys. **64** (1992) 849; U. Meirav and E.B. Foxman, Semicond. Sci. Technol. **11** (1996) 255; L.P. Kouwenhoven and C.M. Marcus, Phys. World **11** (1998) 35.

[2] P. Joyez et al., Phys. Rev. Lett. **79** (1997) 1349; M. Devoret and C. Glattli, Phys. World **11** (1998) 29.

[3] B.L. Altshuler and A.G. Aronov, in: *Electron-Electron Interactions in Disordered Systems*, eds. A.L. Efros and M. Pollak (North-Holland, Amsterdam, 1985), p. 1.

[4] D. Goldhaber-Gordon et al., Nature **391** (1998) 156; S.M. Cronewett, T.H. Oosterkamp and L.P. Kouwenhoven, Science **281** (1998) 540; J. Schmid et al., Physica B **256-258** (1998) 182.

[5] L. Kouwenhoven and L. Glazman, Physics World **14** (2001) 33.

[6] J. Kondo, Prog. Theor. Phys. **32** (1964) 37.

[7] Furusaki A and K.A. Matveev Phys. Rev. B **52** (1995) 16676.

[8] I.L. Aleiner and L.I. Glazman, Phys. Rev. B **57** (1998) 9608.

[9] S.M. Cronenwett et al., Phys. Rev. Lett. **81** (1998) 5904.

[10] S. Tarucha et al., Phys. Rev. Lett. **84** (2000) 2485; L.P. Kouwenhoven, D.G. Austing and S. Tarucha, Rep. Prog. Phys. **64** (2001) 701.

[11] S. Sasaki et al., Nature **405** (2000) 764; S. Sasaki et al., Phys. Rev. Lett. **93** (2004) 017205.

[12] M. Pustilnik et al., Lecture Notes in Physics **579** (2001) 3 (cond-mat/0010336).

[13] J. Nygård, D.H. Cobden and P.E. Lindelof, Nature **408** (2000) 342; W. Liang, M. Bockrath and H. Park, Phys. Rev. Lett. **88** (2002) 126801; B. Babić, T. Kontos and C. Schönenberger, cond-mat/0407193.

[14] J. Park et al., Nature **417** (2002) 722; W. Liang et al., Nature **417** (2002) 725; L.H. Yu and D. Natelson, Nano Lett. **4** (2004) 79.

[15] L.T. Li et al., Phys. Rev. Lett. **80** (1998) 2893; V. Madhavan et al., Science **280** (1998) 567; H.C. Manoharan et al., Nature **403** (2000) 512.

[16] M.V. Berry, Proc. R. Soc. A **400** (1985) 229; B.L. Altshuler and B.I. Shklovskii, Zh. Eksp. Teor. Fiz. **91** (1986) 220 [Sov. Phys.–JETP **64** (1986) 127].

[17] C.W.J. Beenakker, Rev. Mod. Phys. **69** (1997) 731.

[18] Y. Alhassid, Rev. Mod. Phys. **72** (2000) 895.

[19] F. Haake, *Quantum Signatures of Chaos* (Springer-Verlag, New York, 2001).

[20] K. Efetov, *Supersymmetry in Disorder and Chaos* (Cambridge University Press, Cambridge, 1997).

[21] B.L. Altshuler et al.,Phys. Rev. Lett. **78** (1997) 2803; O. Agam et al., Phys. Rev. Lett. **78** (1997) 1956; Ya.M. Blanter, Phys. Rev. B **54** (1996) 12807; Ya.M. Blanter and A.D. Mirlin, Phys. Rev. B **57** (1998) 4566; Ya.M. Blanter, A.D. Mirlin and B.A. Muzykantskii, Phys. Rev. Lett. **78** (1997) 2449; I.L. Aleiner and L.I. Glazman, Phys. Rev. B **57** (1998) 9608.

[22] I.L. Kurland, I.L. Aleiner and B.L. Altshuler, Phys. Rev. B **62** (2000) 14886.

[23] I.L. Aleiner, P.W. Brouwer and L.I. Glazman, Phys. Rep. **358** (2002) 309.

[24] M.V. Berry, Journ. of Phys. A **10** (1977) 2083; Ya.M. Blanter and A.D. Mirlin, Phys. Rev. E **55** (1997) 6514 Ya.M. Blanter, A.D. Mirlin and and B.A. Muzykantskii, Phys. Rev. B **63** (2001) 235315; A.D. Mirlin, Phys. Rep. **326** (2000) 259.

[25] J.A. Folk et al., Phys. Rev. Lett. **76** (1996) 1699.

[26] J.M. Ziman, *Principles of the Theory of Solids* (Cambridge University Press, Cambridge, 1972), p. 339.

[27] J. von Delft and D.C. Ralph, Phys. Rep. **345** (2001) 61.

[28] K.A. Matveev, L.I. Glazman and R.I. Shekhter, Mod. Phys. Lett. B **8** (1994) 1007.

[29] P.W. Brouwer, Y. Oreg and B.I. Halperin, Phys. Rev. B **60** (1999) R13977; H.U. Baranger, D. Ullmo and L.I. Glazman, Phys. Rev. B **61** (2000) R2425.

[30] R.M. Potok et al., Phys. Rev. Lett. **91** (2003) 016802; J.A. Folk et al., Phys. Scripta **T90** (2001) 26; S. Lindemann et al., Phys. Rev. B **66** (2002) 195314.

[31] K.A. Matveev, Phys. Rev. B **51** (1995) 1743.

[32] K. Flensberg, Phys. Rev. B **48** (1993) 11156.

[33] M. Pustilnik and L.I. Glazman, Phys. Rev. Lett. **87** (2001) 216601.

[34] C.E. Porter and R.G. Thomas, Phys. Rev. **104** (1956) 483.

[35] I.O. Kulik and R.I. Shekhter, Zh. Eksp. Teor. Fiz. **62** (1975) 623 [Sov. Phys.–JETP **41** (1975) 308]; L.I. Glazman and R.I. Shekhter, Journ. of Phys.: Condens. Matter **1** (1989) 5811.

[36] L.I. Glazman and K.A. Matveev, Pis'ma Zh. Exp. Teor. Fiz. **48**, 403 (1988) [JETP Lett. **48**, 445 (1988)]; C.W.J. Beenakker, Phys. Rev. B **44** (1991) 1646; D.V. Averin, A.N. Korotkov and K.K. Likharev, Phys. Rev. B **44**, (1991) 6199.

[37] R.A. Jalabert, A.D. Stone and Y. Alhassid, Phys. Rev. Lett. **68** (1992) 3468.

[38] V.N. Prigodin, K.B. Efetov and S. Iida, Phys. Rev. Lett. **71** (1993) 1230.

[39] A.M. Chang et al., Phys. Rev. Lett. **76** (1996) 1695.

[40] B.L. Altshuler et al., Phys. Rev. B **22** (1980) 5142; S. Hikami, A.I. Larkin and Y. Nagaoka, Progr. Theor. Phys. **63** (1980) 707.

[41] I. Giaever and H.R. Zeller, Phys. Rev. Lett. **20** (1968) 1504; H.R. Zeller and I. Giaever, Phys. Rev. **181** (1969) 789.

[42] D.V. Averin and Yu.V. Nazarov, Phys. Rev. Lett. **65** (1990) 2446.

[43] A.A. Abrikosov, *Fundamentals of the Theory of Metals* (North-Holland, Amsterdam, 1988), p. 620.

[44] I.L. Aleiner and L.I. Glazman, Phys. Rev. Lett. **77** (1996) 2057.

[45] L.I. Glazman and M.E. Raikh, Pis'ma Zh. Eksp. Teor. Fiz. **47** (1988) 378 [JETP Lett. **47** (1988) 452]; T.K. Ng and P.A. Lee, Phys. Rev. Lett. **61** (1988) 1768.

[46] W.G. van der Wiel et al., Science **289** (2000) 2105; Y. Ji, M. Heiblum and H. Shtrikman, Phys. Rev. Lett. **88** (2002) 076601.

[47] L.I. Glazman and M. Pustilnik, in: *New Directions in Mesoscopic Physics (Towards Nanoscience)* eds. R. Fazio et al., (Kluwer, Dordrecht, 2003), p. 93 (cond-mat/0302159).

[48] C.B. Duke, *Tunneling in Solids* (Academic Press, New York, 1969); J.M. Rowell, in: *Tunneling Phenomena in Solids*, eds. E. Burstein and S. Lundqvist (Plenum Press, New York, 1969), p. 385.

[49] A.F.G. Wyatt, Phys. Rev. Lett. **13** (1964) 401; R.A. Logan and J.M. Rowell, Phys. Rev. Lett. **13** (1964) 404.

[50] J. Appelbaum, Phys. Rev. Lett. **17** (1966) 91; J.A. Appelbaum, Phys. Rev. **154** (1967) 633.

[51] P.W. Anderson, Phys. Rev. Lett. **17** (1966) 95.

[52] G.A. Fiete et al., Phys. Rev. B **66** (2002) 024431.

[53] M. Pustilnik and L.I. Glazman, Journ. of Physics: Condens. Matter **16** (2004) R513.

[54] J.R. Schrieffer and P.A. Wolff, Phys. Rev. **149** (1966) 491.

[55] P.G. Silvestrov and Y. Imry, Phys. Rev. Lett. **85** (2000) 2565.

[56] P. Nozières and A. Blandin, J. Physique **41** (1980) 193.

[57] P. Coleman, in: *Lectures on the Physics of Highly Correlated Electron Systems VI*, ed. F. Mancini (American Institute of Physics, New York, 2002), p. 79 (cond-mat/0206003); A.S. Hewson, *The Kondo Problem to Heavy Fermions* (Cambridge University Press, Cambridge, 1997).

[58] P.W. Anderson, *Basic Notions of Condensed Matter Physics* (Addison-Wesley, Reading, 1997).

[59] K.G. Wilson, Rev. Mod. Phys. **47** (1975) 773.

[60] L.I. Glazman, F.W.J. Hekking and A.I. Larkin, Phys. Rev. Lett. **83** (1999) 1830.

[61] A.M. Tsvelick and P.B. Wiegmann, Adv. Phys. **32** (1983) 453; N. Andrei, K. Furuya and J.H. Lowenstein, Rev. Mod. Phys. **55** (1983) 331.

[62] I. Affleck and A.W.W. Ludwig, Phys. Rev. B **48** (1993) 7297.

[63] M. Pustilnik and L.I. Glazman, Phys. Rev. B **64** (2001) 045328.

[64] A.A. Abrikosov, Physics **2** (1965) 5; A.A. Abrikosov, Usp. Fiz. Nauk **97** (1969) 403 [Sov. Phys.–Uspekhi **12** (1969) 168]; H. Suhl, Phys. Rev. **138** (1965) A515.

[65] R.G. Newton, *Scattering Theory of Waves and Particles* (Dover, Mineola, 2002).

[66] P.W. Anderson, Journ. of Physics C **3** (1970) 2436.

[67] P.W. Anderson, G. Yuval and D.R. Hamann, Phys. Rev. B **1** (1970) 4464.

[68] F. Wegner, Ann. Phys. **3** (1994) 77; Nucl. Phys. B **90** (2000) 141; S.D. Glazek and K.G. Wilson, Phys. Rev. D **49** (1994) 4214; Phys. Rev. D **57** (1998) 3558.

[69] P. Nozières, J. Low Temp. Phys. **17** (1974) 31; P. Nozières, in *Proceedings of the 14th International Conference on Low Temperature Physics* edited by M. Krusius and M. Vuorio (North Holland, Amsterdam, 1974), Vol. 5, pp. 339-374; P. Nozières, J. Physique **39** (1978) 1117.

[70] T.A. Costi, A.C. Hewson and V. Zlatić, Journ. of Physics: Condens. Matter **6** (1994) 2519.

[71] G.D. Mahan, *Many-Particle Physics* (Plenum, New York, 1990).

[72] W.D. Knight, Phys. Rev. **76** (1949) 1259; C. H. Townes, C. Herring, and W. D. Knight, Phys. Rev. **77** (1950) 736.

[73] J. Korringa, Physica (Amsterdam) **16** (1950) 601.

[74] D.L. Losee and E.L. Wolf, Phys. Rev. Lett. **23** (1969) 1457.

[75] Y. Meir, N.S. Wingreen and P.A. Lee, Phys. Rev. Lett. **70** (1993) 2601.

[76] J.E. Moore and X.-G. Wen, Phys. Rev. Lett. **85** (2000) 1722.

[77] J. Paaske, A. Rosch, and P. Wölfle, Phys. Rev. B **69** (2004) 155330.

[78] J. Schmid et al., Phys. Rev. Lett. **84** (2000) 5824.

[79] W.G. van der Wiel et al., Phys. Rev. Lett. **88** (2002) 126803.

[80] A. Kogan et al., Phys. Rev. B **67** (2003) 113309.

[81] W. Izumida, O. Sakai and Y. Shimizu, J. Phys. Soc. Jpn. **67** (1998) 2444.

[82] M. Eto and Yu.V. Nazarov, Phys. Rev. Lett. **85** (2000) 1306; M. Pustilnik and L.I. Glazman, Phys. Rev. Lett. **85** (2000) 2993.

[83] V.N. Golovach and D. Loss, Europhys. Lett. **62** (2003) 83; M. Pustilnik, L.I. Glazman and W. Hofstetter, Phys. Rev. B **68** (2003) 161303(R).

[84] M. Pustilnik, Y. Avishai and K. Kikoin, Phys. Rev. Lett. **84** (2000) 1756.

[85] J.M. Elzerman et al., J. Low Temp. Phys. **118** (2000) 375.

[86] P.K. Tien and J.P. Gordon, Phys. Rev. **129** (1963) 647.

[87] M.H. Hettler and H. Schoeller, Phys. Rev. Lett. **74** (1995) 4907.

[88] A. Kaminski, Yu.V. Nazarov and L.I. Glazman, Phys. Rev. Lett. **83** (1999) 384; A. Kaminski, Yu.V. Nazarov and L.I. Glazman, Phys. Rev. B **62** (2000) 8154.

[89] K.A. Matveev, Zh. Eksp. Teor. Fiz. **99** (1991) 1598 [Sov. Phys.–JETP **72** (1991) 892].

[90] D.L. Cox and A. Zawadowski, Adv. Phys. **47** (1998) 599.

[91] Y. Oreg and D. Goldhaber-Gordon, Phys. Rev. Lett. **90** (2003) 136602.

[92] M. Pustilnik et al., Phys. Rev. B **69** (2004) 115316.

[93] M. Keller et al., Phys. Rev. B **64** (2001) 033302; M. Stopa et al., Phys. Rev. Lett. **91** (2003) 046601.

[94] S.M. Maurer et al., Phys. Rev. Lett. **83** (1999) 1403; D. Sprinzak et al., Phys. Rev. Lett. **88** (2002) 176805; C. Fühner et al., Phys. Rev. B **66** (2002) 161305(R).

[95] C.W.J. Beenakker, Phys. Rev. Lett. **64** (1990) 216; A.M. Chang, Solid State Commun. **74** (1990) 871; D.B. Chklovskii, B.I. Shklovskii and L.I. Glazman, Phys. Rev. B **46** (1992) 4026.

[96] P.L. McEuen et al., Phys. Rev. B **45** (1992) 11419; A.K. Evans, L.I. Glazman and B.I. Shklovskii, Phys. Rev. B **48** (1993) 11120.

[97] V.J. Goldman and B. Su, *Science* **267** (1995) 1010.

[98] M. Kataoka et al., Phys. Rev. Lett. **83** (1999) 160; M. Kataoka et al., Phys. Rev. Lett. **89** (2002) 226803.

Seminar 3

TRANSPORT THROUGH QUANTUM POINT CONTACTS

Yigal Meir

Physics Department, Ben Gurion University, Beer Sheva 84105, ISRAEL

H. Bouchiat, Y. Gefen, S. Guéron, G. Montambaux and J. Dalibard, eds.
Les Houches, Session LXXXI, 2004
Nanophysics: Coherence and Transport

Contents

Abstract

Conductance through quantum point contacts, smoothly coupled to two dimensional leads, increases in units of $2e^2/h$. In addition, experiments on such point contacts have highlighted an anomalous conductance plateau at $0.7(2e^2/h)$. The physics behind this phenomenon is explained in terms of the Kondo effect, resulting from the binding of an electron in the vicinity of the point contact. Among the experimental predictions resulting from this model are a spin-polarized current, anomalous temperature-dependent noise, and increased dephasing.

1. Introduction

A quantum point contact (QPC) is a narrow constriction in a two-dimensional electron gas, where electrons are forced to flow through a quasi-1d section of the sample. The conductance G through a QPC is observed to be quantized in units of $2e^2/h$ (Fig. 1) [1, 2]. These steps can be understood from the Landauer conductance formula, $G = \frac{e^2}{h} \sum_{n\sigma} \mathcal{T}_{n\sigma}$, where $\mathcal{T}_{n\sigma}$ is the transmission coefficient through the n-th transverse channel (and spin σ). Thus for adiabatic transport, where each channel is either fully transparent or fully reflecting, the conductance is just $2e^2/h$ times the number of transparent channel, where the factor of 2 comes from spin degeneracy.

A surprising observation is that in addition to these integer conductance steps, there exists an extra conductance plateau around $0.7(2e^2/h)$, also evident in Fig. 1, and is expanded in Fig. 2, where the magnetic field dependence of the effect is depicted. With increasing magnetic field and the breaking of spin-degeneracy, the Landauer formula predicts steps in units of e^2/h, namely half of the zero field step size, which is indeed observed. Interestingly, the transition from the 0.7 to the 0.5 plateau is continuous. It has also been observed that the effect exists only for intermediate temperatures. When the temperature is lowered the 0.7 step gradually increases towards the $2e^2/h$ plateau.

This phenomenon has attracted considerable experimental [3, 4, 5, 6] and theoretical [7, 8, 9, 10] effort. A recent experiment[11] has highlighted features in QPC transport strongly suggestive of the Kondo effect: a zero-bias peak in the differential conductance which splits in a magnetic field, and a crossover to perfect transmission below a characteristic "Kondo" temperature T_K, consistent with the peak width. A puzzling observation was the large value of the residual conductance, $G > 0.5(2e^2/h)$, for $T \gg T_K$.

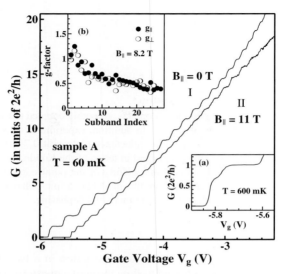

Fig. 1: Demonstration of the conductance steps in quantum point contacts [1]. Note the additional step at $G \simeq 0.7 \times 2e^2/h$ (inset).

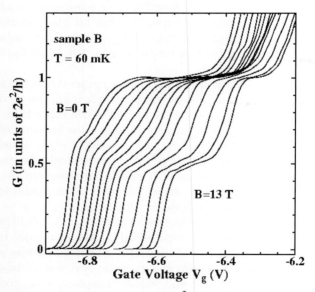

Fig. 2: The anomalous step around $0.7(2e^2/h)$ and its continuous evolution towards $0.5(2e^2/h)$ in large magnetic fields, as expected from the Landauer formula [3].

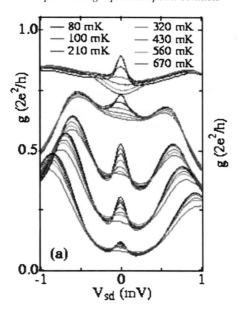

Fig.3: The zero bias anomaly in the first mode,[11] indicating Kondo physics.

In the next section I show that spin-density functional calculations [12] predict the binding of an electron in the vicinity of the the QPC, thereby underscoring the relevance of an Anderson model to the explanation of the phenomenon.[13] The model and a perturbative approach to its solution will be introduced in the following section. The last section will be devoted to predictions concerning the noise in such systems.

2. Spin-density-functional calculations

An intuitive picture for the formation of a quasi-bound states is motivated by transport across a square barrier. For a wide and tall barrier, in addition to the exponentially increasing transparency, there are narrow transmission resonances above the barrier. These result from multiple reflections from the edges of the barrier, and are associated with quasi-bound states, which can play the role of localized orbitals in an Anderson model. We present further evidence from spin-density-functional theory (SDFT) [14] for the formation of such a local moment (bound spin) at the center of a GaAs QPC, which supports our use of the Anderson model. SDFT is applied within the local-density approximation [15, 16].

The external potential consists of a clean quantum wire with a parabolic confining potential of $V_{\text{wire}}^0(y) = (1/2)m^*\omega_y^2 y^2$ and a QPC potential

$$V_{\text{QPC}}(x, y) = V(x)/2 + m^* [V(x)/\hbar]^2 \, y^2/2, \tag{2.1}$$

where $V(x) = V_0/\cosh^2(\alpha x)$, with $\alpha = \omega_x \sqrt{m^*/2V_0}$. A contour plot of the QPC potential $V_{\text{QPC}}(x, y)$ is shown in the left inset of Fig. 4(b).

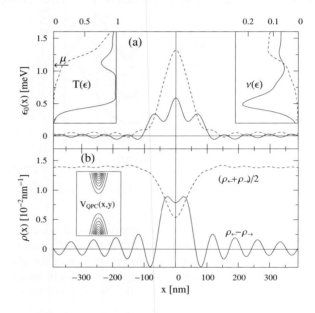

Fig. 4. Results of spin-density-functional theory. (a) Self-consistent "barrier", *i.e.* energy of the bottom of the lowest 1D subband as a function of position x in the direction of current flow through the QPC. The electrochemical potential μ is indicated by an arrow on the left. In this panel, solid curves are for spin-up electrons and dashed curves are for spin-down electrons. Left inset: transmission coefficient. Right inset: local density of states at center of QPC. (b) 1D electron density in QPC. The solid curve gives the net spin-up density and the dashed curve gives the spin-averaged density. Inset: contour plot of the QPC potential $V_{\text{QPC}}(x, y)$.

We solve the Kohn-Sham equations [15] using the material constants for GaAs, $m^* = 0.067m_0$ and $\kappa = 13.1$. The external confinement in the y-direction in the wire is fixed by $\hbar\omega_y = 2.0$ meV. The parameters for the QPC potential are taken to be $V_0 = 3.0$ meV and $\hbar\omega_x = 1.5$ meV.

Fig. 4(a) shows the spin-dependent, self-consistent QPC barriers at $T=0.1K$ obtained from SDFT [17]. Specifically, we plot the energy of the bottom of the lowest 1D subband $\epsilon_\sigma(x)$, relative to the value ϵ_0 far into the wire, for both spin-up and spin-down. The local density of states $\nu(\epsilon)$ at the center of the QPC is shown for both spin-up and spin-down in the right inset. Fig. 4(b) shows the average 1D electron density through the QPC and the net density of spin-up electrons. The integrated spin-up density is 0.96 electrons. The data from SDFT gives strong evidence for a quasi-bound state centered at the QPC: there is a resonance in the local density of states $\nu(\epsilon)$ for spin-up, with a net of one spin bound in the vicinity of the QPC. The transmission coefficient $T(\epsilon)$ for electrons in the lowest subband is shown in the left inset to Fig. 4(a). Transmission for spin-up is approximately 1 over a broad range of energies above the spin-up resonance. This implies an onset of strong hybridization at energies above the quasi-bound state.

3. The Anderson model

Our SDFT results indicate that even an initially smooth QPC potential can produce a narrow quasi-bound state, resulting in a spin bound at the center of the QPC. We thus model the QPC and its leads by the generalized Anderson Hamiltonian [18]

$$
\begin{aligned}
H &= \sum_{\sigma;k\in L,R} \varepsilon_{k\sigma} c_{k\sigma}^\dagger c_{k\sigma} + \sum_\sigma \varepsilon_\sigma d_\sigma^\dagger d_\sigma + U n_\uparrow n_\downarrow \\
&+ \sum_{\sigma;k\in L,R} [V_{k\sigma}^{(1)}(1-n_{\bar\sigma}) c_{k\sigma}^\dagger d_\sigma + V_{k\sigma}^{(2)} n_{\bar\sigma} c_{k\sigma}^\dagger d_\sigma + H.c.]
\end{aligned}
\tag{3.1}
$$

where $c_{k\sigma}^\dagger$ ($c_{k\sigma}$) creates (destroys) an electron with momentum k and spin σ in one of the two leads L and R, $d_\sigma^\dagger$ (d_σ) creates (destroys) a spin-σ electron on "the site", *i.e.* the quasi-bound state at the center of the QPC, and $n_\sigma = d_\sigma^\dagger d_\sigma$. The hybridization matrix elements, $V_{k\sigma}^{(1)}$ for transitions between 0 and 1 electrons on the site and $V_{k\sigma}^{(2)}$ for transitions between 1 and 2 electrons, are taken to be step-like functions of energy, mimicking the exponentially increasing transparency (the position of the step defines our zero of energy). Physically, we expect $V_{k\sigma}^{(2)} < V_{k\sigma}^{(1)}$, as the Coulomb potential of an electron already occupying the QPC will reduce the tunneling rate of a second electron through the bound state. In the absence of magnetic field the two spin directions are degenerate, $\varepsilon_\downarrow = \varepsilon_\uparrow = \varepsilon_0$.

For a noninteracting system, the conductance G will be a (temperature broadened) resonance of Lorentzian form, with a width proportional to V^2. If V rises abruptly to a large value, such that the width becomes larger than $\varepsilon_F - \varepsilon_0$, where ε_F is the Fermi energy, G saturates to a value of $2e^2/h$. For the interacting system, we similarly expect the *high-temperature* contribution from the $0 \leftrightarrow 1$ valence fluctuations to G to saturate at $0.5(2e^2/h)$ for $\varepsilon_F > 0 > \varepsilon_0$, because the probability of an opposite spin electron occupying the site in this regime is ≈ 0.5. Since $V_{k\sigma}^{(2)}$ may be significantly smaller than $V_{k\sigma}^{(1)}$, the contribution to the conductance from the $1 \leftrightarrow 2$ valence fluctuations may be small, until $\varepsilon_F \simeq \varepsilon_0 + U$. However, the Kondo effect will enhance this contribution with decreasing temperature, until at zero temperature the conductance will be equal to $2e^2/h$, due to the Friedel sum rule [19] for the Anderson model.

To obtain a quantitative estimate of the conductance we note that the relevant gate-voltage range corresponds to the Kondo regime (singly occupied site), a fact further supported by the observation of a zero-bias peak where the conductance first becomes measurable [11], so the Kondo limit of the Anderson Hamiltonian should be applicable. We therefore perform a Schrieffer-Wolff transformation [20] to obtain the Kondo Hamiltonian [21, 22]

$$
\begin{aligned}
H &= \sum_{\sigma;k\in L,R} \varepsilon_{k\sigma} \mathbf{c}_{k\sigma}^\dagger \mathbf{c}_{k\sigma} + \sum_{\sigma,\sigma';k,k'\in L,R} (J_{kk'\sigma\sigma}^{(1)} - J_{kk'\sigma\sigma}^{(2)}) \, \mathbf{c}_{k\sigma}^\dagger \mathbf{c}_{k'\sigma} \\
&+ 2\sum_{\sigma,\sigma',\alpha,\alpha';k,k'\in L,R} (J_{kk'\sigma\bar\sigma}^{(1)} + J_{kk'\sigma\bar\sigma}^{(2)})(\mathbf{c}_{k\sigma}^\dagger \vec\sigma_{\sigma\sigma'} \mathbf{c}_{k'\sigma'}) \cdot \vec S,
\end{aligned}
\tag{3.2}
$$

$$
J_{kk'\sigma\sigma'}^{(i)} = \frac{(-)^{i+1}}{4} \left(\frac{V_{k\sigma}^{(i)} V_{k'\sigma'}^{*(i)}}{\varepsilon_{k\sigma} - \varepsilon_\sigma^{(i)}} + \frac{V_{k\sigma}^{(i)} V_{k'\sigma'}^{*(i)}}{\varepsilon_{k'\sigma'} - \varepsilon_{\sigma'}^{(i)}} \right),
$$

where $\varepsilon_\sigma^{(1)} = \varepsilon_\sigma$ and $\varepsilon_\sigma^{(2)} = \varepsilon_\sigma + U$. The Pauli spin matrices are indicated by $\vec\sigma$, and the local spin due to the bound state is $\vec S \equiv \frac{1}{2}\mathbf{d}_\alpha^\dagger \vec\sigma_{\alpha\alpha'} \mathbf{d}_{\alpha'}$.

4. Results

Following Appelbaum [23], we treat the above Kondo Hamiltonian perturbatively in the couplings $J_{kk'\sigma\sigma'}^{(i)}$. The differential conductance to lowest order, J^2, is given by

$$
\begin{aligned}
G_2 &= \frac{4\pi e^2}{\hbar} \rho_L(\varepsilon_F)\rho_R(\varepsilon_F) \left\{ (J_{LR}^{(-)})^2 + (J_{LR}^{(+)})^2 \right. \\
&\times \left. \left[3 + 2\langle M\rangle \left(\tanh\frac{\Delta + eV}{2k_B T} + \tanh\frac{\Delta - eV}{2k_B T} \right) \right] \right\}
\end{aligned}
\tag{4.1}
$$

where, for simplicity, $J^{(i)}_{kk'\sigma\sigma'}$ are replaced by their (magnetic-field independent) values at the Fermi energy

$$J^{(i)}_{ll'} \equiv J^{(i)}_{k_F \in l\ k_F \in l'\ \sigma\sigma} = \frac{(-)^{i+1} V_i^2}{2(\varepsilon_F - \varepsilon_0^{(i)})} f_{FD}(-\varepsilon_F/\delta), \qquad (4.2)$$

where symmetric leads have been assumed, and the V_i and δ are constants. The $J^{(i)}$ increase in a step of the Fermi-Dirac form $f_{FD}(x) = 1/[1 + \exp(x)]$. We define the combinations $J^{(\pm)}_{ll'} = J^{(1)}_{ll'} \pm J^{(2)}_{ll'}$ for, respectively, the direct and exchange couplings in Eq. (2). In (4.1), $\Delta = g^* \mu_B B$ is the Zeeman splitting, $\langle M \rangle = -(1/2) \tanh(\Delta/2k_B T)$ is the magnetization for the uncoupled site, and $\rho_{L/R}(\varepsilon) = \sum_{k \in L/R} \delta(\varepsilon - \varepsilon_{k\sigma})$ is the single-spin electron density of states in the leads. We assume $\rho = \rho_L(\varepsilon) = \rho_R(\varepsilon)$.

At low temperatures the Kondo effect leads to a logarithmically diverging contribution G_3 to the differential conductance at order J^3, due to integrals running from the Fermi energy to either band edge. Because of the steplike increase of the $J^{(i)}$, the band integral for $J^{(1)}$ runs down from ε_F to the hybridization step at zero, but runs up from ε_F to $\varepsilon_0 + U$ for $J^{(2)}$. Since in the region of interest $\varepsilon_0 + U - \varepsilon_F \gg \varepsilon_F$, *the logarithmic contribution from $J^{(2)}$ dominates G_3.*

Fig. 5 depicts the linear-response conductance $(G_2 + G_3)$. Since G_2 depends only on the values of $J^{(i)}_{LR}$ at ε_F, it is dominated by $J^{(1)}$, while the Kondo enhancement is dominated by $J^{(2)}$. As argued above, the contribution due to $J^{(1)}$ is set around $0.5(2e^2/h)$ by construction, while the contribution due to $J^{(2)}$, resulting from the $1 \leftrightarrow 2$ valence fluctuations is small at high temperature, but grows with decreasing temperature in a form following the Kondo scaling function, $F(T/T_K)$, where $T_K \simeq U \exp(-1/4\rho J^{(2)}) = U \exp[(\varepsilon_F - \varepsilon_0 - U)/2\rho V_2^2]$, in agreement with the experimental observation of a Kondo temperature increasing exponentially with gate voltage $\sim \varepsilon_F$. Note that in perturbation theory the conductance is not bound by its physical limit: $2e^2/h$.

The dependence of conductance on magnetic field is shown in Fig. 5(b). The Kondo logarithms in G_3 are suppressed and the term in G_2 that depends on $\langle M \rangle$ gives a negative contribution $\propto \tanh^2(\Delta/2k_B T))$, leading to the evolution of the 0.7 plateau towards and below 0.5. In agreement with experiment[11], the conductance is no longer monotonically increasing with Fermi energy ε_F: the energy denominator causes the $J^{(1)}$ contribution to G_2 to decrease, and this is no longer compensated by an increase of G_3. Due to shortcomings of perturbation theory the conductance at large magnetic field reduces to a value smaller than $0.5(2e^2/h)$.

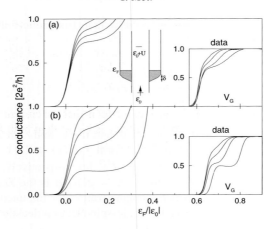

Fig. 5: (a) Conductance at four temperatures values, as a function of Fermi energy ε_F. Right inset: experimental conductance of a QPC at 4 different temperatures [11]. Center inset: Schematic of the band structure for our Anderson model. (b) Conductance in a magnetic field, for four values of Zeeman splitting. Inset: experimental conductance of a QPC at different magnetic fields.[11]

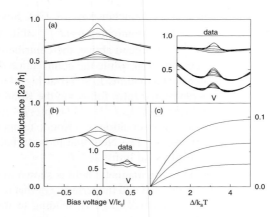

Fig. 6: Differential conductance dI/dV for the Kondo model. (a) dI/dV versus bias at different chemical potentials. For each chemical potential, curves are shown for four different temperatures. Inset: experimental differential conductance [11]. (b) dI/dV in different magnetic fields. Inset: experimental differential conductance at different magnetic fields [11]. (c) Spin conductance for several values of ε_F.

Fig. 6(a) shows the differential conductance as a function of bias voltage, for several values of ε_F and temperatures. Even at the lowest conductances (small ε_F) there is a clear Kondo peak, as is seen in experiment (inset). Due to the suppression of the Kondo effect by voltage, the large voltage traces are independent of temperature, again in agreement with experiment. Magnetic field splits the Kondo peak as shown in Fig. 6(b).

An important prediction of the Kondo model is that the current through a QPC will be spin polarized if the Zeeman splitting is larger than both $k_B T$ and $k_B T_K$ (Fig. 6(c)). The net spin conductance G_σ, is given, to second order in J, by

$$G_\sigma = \frac{16\pi e^2}{\hbar} \rho^2 \langle M \rangle \left[(J_{LR}^{(1)})^2 - (J_{LR}^{(2)})^2 \right]. \tag{4.3}$$

Therefore, at low temperatures and in the vicinity of the $0.7(2e^2/h)$ plateau where T_K is small, a QPC can be an effective spin filter at weak magnetic fields ($\Delta > k_B T_K, k_B T$).

5. Current noise

We have presented a microscopic Anderson model, supported by spin-density-functional theory, for transport through a quantum point contact. The anomalous $0.7(2e^2/h)$ plateau is attributed to a high background conductance plus a Kondo enhancement. Since the conductance consists practically of transport through two channels, one of which opens up first, then a value of the conductance of e^2/h corresponds to one channel perfectly transmitting ($\mathcal{T} = 1$) and one channel perfectly reflecting ($\mathcal{T} = 1$). Thus at that value of the conductance we predict a strong reduction in the current shot noise (which is proportional to $\mathcal{T}(1 - \mathcal{T})$), in contradiction with the traditional expectation of a transmission of $\mathcal{T} = 0.5$ per channel at e^2/h, and, consequently, maximal noise. As temperature is lowered, the conductance of the second channel is enhanced by the Kondo effect, eventually giving rise to two transmitting channels, and the reduction of the noise dip around $G = e^2/h$. Preliminary calculation [26] of the Fano factor, as a function of magnetic field, is depicted in Fig. 7, where the disappearance of the dip is clear. Indeed such a suppression of shot noise was reported recently.[27]

6. Conclusions

The presence of bound spins in QPCs near pinch-off has potentially profound effects on transport through quantum dots with QPCs as leads. In particular

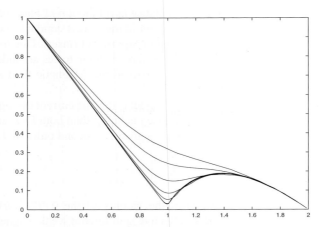

Fig. 7: The Fano factor, noise divided by current, as a function of the conductance for different magnetic fields. At high field the Kondo effect is suppressed, and the conductance for $G = e^2/h$ is dominated by one fully transmitting channel, and zero noise. As the magnetic field is reduced, the Kondo effect emerges, and the the conductance at $G = e^2/h$ is due to two transmitting channel, each with $\mathcal{T} = 1/2$ and maximal noise.[26]

the leads may act as magnetic impurities, and cause the apparent saturation of the dephasing time in transport through open semiconductor quantum dots at low temperatures [28], and may complicate attempts to measure the spin of dot electrons.[29]. Experimental attempts to directly probe the binding of this state are currently underway.[30]

Acknowledgements

This work was partly sponsored by the Israel Science Foundation.

References

[1] B. J. van Wees *et al.*, *Phys. Rev. Lett.* **60**, 848 (1988).

[2] D. A. Wharam *et al.*, *J. Phys. C* **21**, L209 (1988).

[3] K. J. Thomas *et al.*, *Phys. Rev. Lett.* **77**, 135 (1996); K. J. Thomas *et al.*, *Phys. Rev. B* **58**, 4846 (1998).

[4] A. Kristensen *et al.*, *Phys. Rev. B* **62**, 10950 (2000).

[5] D. J. Reilly *et al.*, *Phys. Rev. B* **63**, 121311 (2001).

[6] S. Nuttinck *et al.*, *Jap. J. of App. Phys* **39**, L655 (2000); K. Hashimoto *et al.*, *ibid* **40**, 3000 (2001).

[7] C.-K. Wang and K.-F. Berggren, Phys. Rev. B **57**, 4552 (1998).

[8] B. Spivak and F. Zhou, *Phys. Rev. B* **61**, 16730 (2000).

[9] H. Bruus, V. V. Cheianov, and K. Flensberg, *Physica E* **10**, 97 (2001).

[10] G. Seelig and K. A. Matveev, *Phys. Rev. Lett.* **90**, 176804 (2003); K. A. Matveev, *Phys. Rev. Lett.* **92**, 106801 (2004).

[11] S. M. Cronenwett *et al.*, *Phys. Rev. Lett.* **88**, 226805 (2002).

[12] K. Hirose, Y. Meir, and N. S. Wingreen, *Phys. Rev. Lett.* **90** 026804 (2003).

[13] Y. Meir, K. Hirose, and N. S. Wingreen, *Phys. Rev. Lett.* **89** 196802 (2002).

[14] P. Hohenberg and W. Kohn, *Phys. Rev.* **136**, B864 (1964).

[15] W. Kohn and L. J. Sham, *Phys. Rev.* **140**, A1133 (1965). We use the local-density approximation for the exchange-correlation energy $E_{xc} = \int \rho(\mathbf{r}) \, \varepsilon_{xc}[\rho(\mathbf{r})] \, d\mathbf{r}$, where $\varepsilon_{xc}[\rho(\mathbf{r})]$ is the parameterized form by Tanatar and Ceperley for the two-dimensional electron gas [B. Tanatar and D. M. Ceperley, *Phys. Rev. B* **39**, 5005 (1989)].

[16] J. Callaway and N. H. March, *Solid State Phys.* **38**, 135 (1984).

[17] The solution with broken spin-symmetry coexists with an unpolarized solution (K. Hirose, N. S. Wingreen, and Y. Meir, in preparation). See also A. M. Bychkov, I. I. Yakimenko, and K-F Berggren, *Nanotechnology* **11**, 318 (2000).

[18] P. W. Anderson, *Phys. Rev.* **124**, 41 (1961).

[19] D. C. Langreth, *Phys. Rev.* **150**, 516 (1966); See also T. K. Ng and P. A. Lee, *Phys. Rev. Lett.* **61**, 1768 (1988).

[20] J. R. Schrieffer and P. A. Wolff, *Phys. Rev.* **149**, 491 (1966).

[21] J. Kondo, *Prog. Th. Phys.* (Kyoto) **32**, 37 (1964).

[22] The parameters appearing in the Kondo Hamiltonian are not the bare parameters of the Anderson model (3.1), but renormalized parameters after the bandwidth has been reduced to U [F. D. M. Haldane, *Phys. Rev. Lett.* **40**, 416 (1978)].

[23] J. A. Appelbaum, *Phys. Rev.* **154**, 633 (1967). Appelbaum approximates the diverging integrals by $\log(|A| + k_b T)$. We use $\log[A^2 + (k_b T)^2]$ instead.

[24] R. M. Potok et al., *Phys. Rev. Lett.* **89**, 266602 (2002).

[25] T. Morimoto *et al.*, *Appl. Phys. Lett.* **82**, 3952 (2003).

[26] A. Golub, T Aono and Y. Meir, unpublished.

[27] N. Y. Kim *et al.*, cond-mat/0311435; P. Roche *et al.*, cond-mat/0402194.

[28] D.P. Pivin *et al.*, *Phys. Rev. Lett.* **82**, 4687 (1999); A.G. Huibers *et al.*, *Phys. Rev. Lett.* **83**, 5090 (1999).

[29] J. A. Folk *et al.*, *Phys. Scr.* **T 90**, 26 (2001).

[30] D. Goldhaber-Gordon, private communication.

Course 8

TRANSPORT AT THE ATOMIC SCALE: ATOMIC AND MOLECULAR CONTACTS

A. Levy Yeyati and J.M. van Ruitenbeek

Departamento de Física Teórica de la Materia Condensada C-V,
and Instituto Universitario de Ciencia de Materiales "Nicolás Cabrera",
Universidad Autónoma de Madrid, E-28049 Madrid, Spain
and
Kamerlingh Onnes Laboratorium, Universiteit Leiden, Postbus 9504, NL-2300 RA Leiden,
The Netherlands

H. Bouchiat, Y. Gefen, S. Guéron, G. Montambaux and J. Dalibard, eds.
Les Houches, Session LXXXI, 2004
Nanophysics: Coherence and Transport

Contents

1. Introduction

Atomic and molecular contacts have a special place in the field of nanoscience. From an application point of view they are the playground to investigate the possibilities and limitations of electronics at the very smallest size scale. From a fundamental science point of view they are simple physical systems to test principles of electron transport at the atomic scale. We will mostly address the fundamental issues in this field. Here, an important distinction can be made between atomic and molecular contacts. While simple and reliable experimental techniques are widely available for investigating single-atom contacts for nearly all metals, single-molecule junctions have been more difficult to produce and characterize reliably.

Metallic atomic-sized contacts have proven to provide a rich field of physical phenomena, that we have recently summarized in an extensive review, written together with Nicolás Agraït [1]. It is not our intention to rewrite the review on this topic. Instead, we refer to Ref. [1] for a detailed presentation of the field and mention here only the main results. The major part of this chapter will cover work of the last two years that was not included in our previous review and some aspects of superconducting junctions that did not receive sufficient attention at that time and that are useful for those interested in entering the field.

For contacts between two metal electrodes formed by just a single atom, or just a few atoms, the electron transport can be described in terms of a limited number, N, of conduction channels, each characterized by its own transmission probability τ_i. The conductance is given by the Landauer expression [2]

$$G = G_0 \sum_{i=1}^{N} \tau_i,$$

where $G_0 = 2e^2/h$ is the quantum unit of conductance for spin-degenerate channels. The conductance is easily obtained experimentally, but does not provide much information about the number of channels involved, other than the total sum of their transmission probabilities. Once we have obtained the full set of transmission probabilities τ_i, also referred to as the mesoscopic PIN code, the junction is fully characterized. Many properties can be predicted quantitatively once the PIN code is known, including the supercurrent [3], dynamical Coulomb blockade [4], shot noise [5, 6], and conductance fluctuations [7]. One of the attractive points of this field is that it is actually possible to determine the PIN code experimentally, principally through a measurement of the I-V characteristics in the superconducting state, exploiting the information contained in the subgap structure [8, 9]. Thus a detailed quantitative understanding of quantum transport has been obtained. As a central result it has been shown that the number of conductance channels for a single atom is determined by the number of its valence orbitals [9]. The properties of superconducting junctions will be discussed in Sect. 3. The role of the environment has proven to be of importance in analyzing current-voltage relations, and the associated problems will be addressed in Sect. 4.

For molecular contacts the level of sophistication of the experiments is much less. Where atomic contacts of metals are homogenously formed out of a single species of metallic atoms, organic molecules that are sandwiched between metallic leads form an inhomogeneous system with many difficulties associated with the formation and characterization of the interfaces. The attractive point that motivates much of this research is the wide scope of possible organic molecules that can be studied, with functions such as diode characteristics, switching, and memory engineered into the chemical design of the molecule. Although much of the initial optimism in this field has proven premature by the difficulties that surfaced related to the proper characterization an analysis of the observed current-voltage characteristics, more recently important progress has been made. These developments will be briefly summarized in Sect. 5.

One further development that we will address is associated with atomic chains of single metal atoms. Chains of atoms have been demonstrated for Au [10, 11], Pt and Ir [12]. Several groups have predicted that atomic chains for monovalent metals should show oscillations in the conductance with the parity of the number of atoms forming the chain [13–15]. Such oscillations as a function of the length of the chain have been observed, not only for gold as predicted, but also for Pt and Ir, as will be discussed in Sect. 2.

2. Parity oscillations in atomic chains

When a metallic contact is gradually stretched to the point that only a single atom bridges the gap one would expect that further stretching will break the contact at this weakest spot. This is indeed what one observes for most metals, with the noteworthy exception of three metals: Ir, Pt and Au. The linear bond between two atoms in a chain configuration for these metals is so strong that, rather than breaking this bond, new atoms are pulled out of the leads into the chain thus increasing its length. The effect has been observed directly in a high-resolution transmission electron microscope under ultra-high vacuum at room temperature [10]. At low temperatures the chain formation can be inferred from the variation of the conductance as a function of the stretching of the contact [11, 12]. The latter experiments have the advantage that the chains can be held stable over extended periods of time and their properties can be investigated in detail. Further details on the formation of chains of atoms can be found in our review paper [1].

As one of the interesting properties of linear chains of atoms we want to focus in this section on oscillations of the conductance as a function of the length of the chain. As argued above, the last plateau of conductance before rupture is typically due to a single atom contact. The formation of an atomic wire results from further pulling of this one-atom contact, and its length can be estimated from the length of the last conductance plateau [11, 12, 16]. A histogram made of those lengths (filled curves in Fig. 1) shows peaks separated by distances equal to the inter-atomic spacing in the chain. These peaks correspond to the lengths of stretching at which the atomic chain breaks, since at that point the strain to incorporate a new atom is higher than the one needed to break the chain [17]. This implies that the number of atoms in the chain increases by one going from one peak to the next.

For gold, a monovalent metal, both the one-atom contact and the chain have a conductance of about 1 G_0 with only small deviations from this value, suggesting that the single conductance channel has a nearly perfect coupling to the banks. However, small changes of conductance during the pulling of the wire can be observed [17, 18] and are suggestive of an odd-even oscillation. Small jumps in the conductance result from changes in the connection between the chain and the banks when new atoms are being pulled into the atomic wire. In order to uncover possible patterns hidden in these changes one can average many plateaus of conductance starting from the moment that a single-atomic contact is formed until the wire is broken. These points can be defined by a criterion, e.g. by the conductance dropping below 1.2 G_0 and 0.5 G_0, respectively. In the upper panel of Fig. 1 it can be seen that the thus obtained averaged conductance plateau for gold shows an oscillatory dependence of the conductance with the length of the wire. The amplitude of the oscillation is small and differs slightly between experiments, but the period and phase are quite reproducible. In the averaged curves of the experiment, Fig. 1, the conductance does not quite reach a maximum of 1 G_0. This is largely due to the averaging procedure, where for a given length there are contributions from n and $n + 1$-numbered chains and only the relative weight of these varies. In individual traces the maxima come much closer to full transmission. Further suppression of the maximum conductance may result from asymmetries in the connections to the leads. The relatively small amplitude of the oscillations is consistent with the fact that the average conductance is close

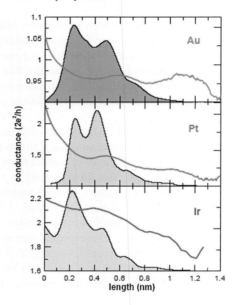

Fig. 1. Averaged plateaus of conductance for chains of atoms of the three metals investigated: Au, Pt, and Ir. Each of the curves are made by averaging individual traces of conductance while pulling atomic contacts or chains. Histograms of the plateau lengths for the three metals obtained from the the same set of data are shown by the filled curves. From Ref. [18].

to unity, implying that the contact between the chain and the banks is nearly adiabatic. The rise above $1\,G_0$ for short lengths in Figs. 1 can be attributed to tunneling contributions of additional channels.

Similar oscillations of even larger amplitude were observed for the other two metals forming chains of atoms, namely Pt and Ir [18]. These d metals have up to five channels of conductance and therefore the average plateau conductance is expected to show a more complicated behaviour. A one-atom Pt contact has a conductance of about $2\,G_0$ while for a Pt atomic chain it is slightly smaller, $\sim 1.5\,G_0$ [19] with variations during the pulling process that can be as large as $0.5\,G_0$. In the averaged curves one observes oscillations similar to those for Au, which are compared to the peak spacing in the length histogram in Fig. 1. Ir shows a similar behaviour although somewhat less pronounced. The periodicity p of the oscillation in the conductance for the three metals is about twice the inter-peak distance d of their corresponding plateau-length histogram. This behaviour agrees with an alternating odd-even evolution of the conductance with the number of atoms.

In addition to the oscillations for Pt and Ir, the mean conductance of the measurements in Fig.1 show an unexpected slope of about 0.3–0.4 G_0/nm. For a ballistic wire the conductance as a function of length is expected to be constant, apart from the oscillatory behaviour discussed above.

Theoretical models: Even-odd oscillations are already present at the level of simple tight-binding (TB) models. The model considered in Ref. [13] is a linear chain containing N atoms coupled to electrodes described as a Bethe lattice of a given coordination number Z. The model Hamiltonian can be decomposed as $\hat{H} = \hat{H}_L + \hat{H}_R + \hat{H}_{chain} + \hat{V}_L + \hat{V}_R$, where $\hat{H}_{L,R}$ correspond to the uncoupled electrodes, $\hat{H}_{chain} = \sum_{\sigma,i=1}^{N} v \hat{c}_{\sigma i}^{\dagger} \hat{c}_{\sigma i+1} + $ h.c. describes the electron states in the isolated linear chain and $\hat{V}_{L,R} = \sum_{\sigma} t_{L,R} \hat{c}_{\sigma L,R}^{\dagger} \hat{c}_{1,N} + $ h.c correspond to the coupling between the chain and the outermost sites of the left and right electrodes.

Electronic and transport properties within TB models are conveniently analyzed in terms of Green function techniques. For the linear response regime and in the non-interacting case it is sufficient to introduce the retarded and advanced Green operators, formally defined as

$$\hat{G}^{r,a}(\omega) = \lim_{\eta \to 0} \left[\omega \pm i\eta - \hat{H} \right]^{-1}. \qquad (2.1)$$

In terms of the matrix elements of $\hat{G}^{r,a}$ the transmission coefficient through the chain is then given by the following expression

$$\tau(E_F) = 4\Gamma_L(E_F)\,\Gamma_R(E_F)\,\left| G_{1,N}^r(E_F) \right|^2, \qquad (2.2)$$

where $\Gamma_\alpha(\omega) = t_\alpha^2 \mathrm{Im} g_\alpha^a(\omega)$ (with $\alpha = L, R$) are the tunneling rates from the chain to the leads. These rates are determined by both the hopping elements t_α coupling the chain to the leads and by the density of states on the outermost site of the uncoupled electrodes, determined by $\mathrm{Im} g_\alpha$. For the Bethe lattice model we have [20]

$$g_\alpha^{r,a}(\omega) = \frac{1}{\sqrt{Z} t_B} \left[\frac{\omega \pm i\eta}{2\sqrt{Z} t_B} - \sqrt{\left(\frac{\omega \pm i\eta}{2\sqrt{Z} t_B} \right)^2 - 1} \right], \qquad (2.3)$$

where t_B is the hopping element within the Bethe lattice.

The transmission coefficient and the linear conductance are related by the Landauer formula $G = \frac{2e^2}{h} \tau(E_F)$. As shown in Ref. [13] the conductance of the chain within this model exhibits an even-odd oscillation when $E_F \simeq 0$. This is essentially an interference phenomenon arising from the commensurability of

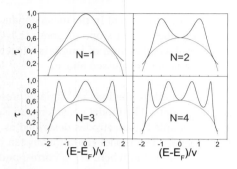

Fig. 2. Transmission through a linear chain of N atoms coupled to electrodes represented by Bethe lattices of coordination $Z = 4$. These curves correspond to $v = t_B = t_L = t_R$ (see text for the definition of parameters). The dotted lines correspond to the minimal envelope curve determined by Eq. (2.5).

the Fermi wavelength and the lattice spacing. It is a simple exercise [21] to show that for $E_F = 0$ one has

$$\tau = \frac{4v^2 \Gamma_L \Gamma_R}{(v^2 + \Gamma_L \Gamma_R)^2} \qquad \text{for even } N$$
$$\tau = \frac{4\Gamma_L \Gamma_R}{(\Gamma_L + \Gamma_R)^2} \qquad \text{for odd } N. \tag{2.4}$$

This result shows that perfect transmission is robust in the *odd* case as it only requires left-right symmetry, i.e. $\Gamma_L = \Gamma_R$. In the even case it requires a more stringent condition on the hopping elements (more precisely $v = \sqrt{\Gamma_L \Gamma_R}$) which is not necessarily fulfilled in an actual system. This elementary calculation thus provides a simple model for the appearance of parity oscillations.

When we analyze the transmission as a function of the Fermi energy for fixed N we observe an oscillatory behaviour, as depicted in Fig. 2 for a symmetric case. These oscillations are bound by the maximum value $\tau = 1$ and a minimal envelope curve determined by

$$\tau_{min}(x) = \frac{4\Gamma^2 v^2 (1 - x^2)}{\left(v^2 + \Gamma^2 (1 - 2x^2) \right)^2 + 4\Gamma^2 x^2 (1 - x^2)}, \tag{2.5}$$

where $\Gamma = \Gamma_{L,R}(0)$ and $x = \omega/2v$. This envelope curve is shown in Fig. 2 as a dotted line.

The even-odd effect has also been reported on the basis of ab-initio calculations for monovalent metals [14, 15, 22]. The calculations of Sim *et al.* [15] were

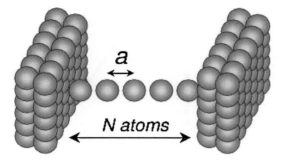

Fig. 3. Model geometry considered for the calculation of the conductance in Ref. [30]. The electrodes are represented by semi-infinite fcc crystals grown in the (111) direction.

based on the Friedel sum rule (FSR) which relates the phase of the transmission amplitude to the charge accumulated within the chain. For a single channel conductor the FSR states that $\delta Q = 2e\delta(E_F)/\pi$ (the factor 2 is due to spin degeneracy) [23]. For a monovalent metal and assuming local charge neutrality this relation implies that the Fermi energy should lie at the middle of a transmission resonance for odd N, i.e. $\delta(E_F) = (2n + 1)\pi/2$, and between two resonances for even N, in agreement with the simple TB calculations shown in Fig. 2. The charge neutrality argument for explaining the observed almost perfect quantization in monovalent one atom contacts was first introduced in Ref. [24].

Ab-initio calculations have also been presented for chains of non-monovalent metals like C, Si and Al [25–29]. These calculations indicate that not only the amplitude but also the actual periodicity can be extremely sensitive to the type of atoms in the chain. For instance, for the case of Al Ref. [28] showed that the conductance exhibits a four-atom period oscillation, which they explained using an effective single band tight-binding model with filling factor 0.25.

In spite of these theoretical efforts, and with the exception of Au, there are not many realistic calculations of the conductance for the actual $5d$ metals producing stable atomic chains. The simple explanation presented above can account qualitatively for the behaviour in the case of Au, characterized by a full $5d$ band and a nearly half-filled $6s$ band. However, for the case of Pt and Ir, in which the contribution of $5d$ orbitals to the conductance is important there is no reason why this simple picture should be valid.

The behaviour of the conductance for Au, Pt and Ir atomic chains was analyzed in detail in Ref. [30]. They considered model geometries like the one depicted in Fig. 3 in which the atomic chain is connected to bulk electrodes represented by two semi-infinite fcc perfect crystals grown along the (111) di-

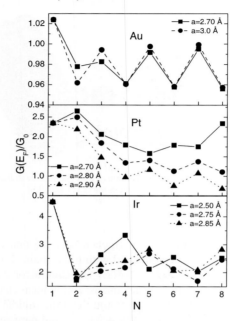

Fig. 4. Evolution of the conductance with N for different values of the interatomic distance a (from Ref. [30]).

rection. Using a parametrized self-consistent tight-binding model, which basically reproduces the bands around the Fermi energy for the infinite ideal chains, they obtained the evolution of the conductance with the number of atoms in the chain depicted in Fig. 4. This evolution is rather sensitive to the elongation, especially in the case of Pt and Ir (for Au the conductance exhibits small amplitude even-odd oscillations, $\sim 0.04G_0$, which remain basically unaffected upon stretching).

The main features and the differences between Au, Pt and Ir are more clearly understood by analyzing the local density of states and the energy dependent transmissions, shown in Fig. 5 for a $N = 5$ chain of these metals at an intermediate elongation. As has been shown in previous works (e.g., see Refs. [31,32]) Au chains are characterized by a single conducting channel around the Fermi energy with predominant s character. The transmission of this channel lies close to one and exhibits small oscillations as a function of energy resembling the behaviour of the single band TB model discussed above.

In the case of Pt the contribution from the almost filled $5d$ bands becomes important for the electronic properties at the Fermi energy. There are three conduction channels with significant transmission at E_F: one due to the hybridization

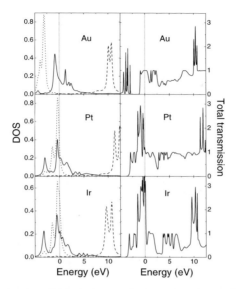

Fig. 5. Local density of states (LDOS) at the central atom and total transmission for Au, Pt and Ir chains with $N = 5$ at an intermediate elongation. The LDOS is decomposed in s (full line) d (dotted line) and p dashed line with the same normalization in the three cases (from Ref. [30]).

of $s - p_z$ and d_{z^2} orbitals, and another two almost degenerate, with $p_x - d_{xz}$ and $p_y - d_{yz}$ character respectively (here z corresponds to the chain axis). The contribution of the $5d$ orbitals is even more important in the case of Ir where a fourth channel exhibits a significant transmission.

As discussed in Ref. [30] more insight into these results can be obtained by analyzing the band structure of the infinite chains. The left panel in Fig. 6 shows the bands around the Fermi energy for Pt obtained from ab-initio calculations. Two main features are worth commenting on: 1) Symmetry considerations allow to classify the bands according to the projection of the angular momentum along the chain axis, m. 2) Close to the Fermi level there is an almost flat filled two-fold degenerate band with d_{xy} and $d_{x^2-y^2}$ ($m = \pm 2$) character. The other partially filled and more dispersing bands have $s - p_z - d_{z^2}$ ($m = 0$) and $p_x - d_{xz}$ or $p_y - d_{yz}$ ($m = \pm 1$) character (see labels in Fig. 6).

The close connection between this band structure and the conduction channels of the chains is realized when analyzing the evolution of the conductance and its channel decomposition for even longer chains than in Fig. 4 ($N > 8$). This is illustrated in the right panel in Fig. 6. As can be observed the decrease of the total conductance of Pt for $N < 7 - 8$ corresponds actually to a long period oscillation in the transmission of the two nearly degenerate channels associated

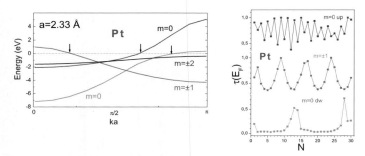

Fig. 6. Left panel: band structure of the infinite Pt chain The bands are classified by the quantum number m corresponding to the projection of the angular momentum on the chain axis. The arrows indicate the crossing of the Fermi level for the $m = 0$ and the $m = \pm 1$ bands. Right panel: channel decomposition for Pt chains as a function of N. The legends indicate the symmetry of the associated bands in the infinite chain.

with the $m = \pm 1$ bands. This period can be related to the small Fermi wave vector of these almost filled d bands, as indicated by the arrows in the left panel of Fig. 6. In addition, the upper $m = 0$ band crossing the Fermi level is close to half-filling giving rise to the even-odd oscillatory behaviour observed in the transmission of the channel with predominant s character. The lower $m = 0$ band tends to be completely filled and the corresponding channel is nearly closed for short chains. However, one can notice a very long period oscillation in its transmission, rising up to $\sim 0.5G_0$, for $N \sim 13 - 14$.

The general rule that emerges from the above analysis is that the transmission corresponding to each conduction channel oscillates as $\sim \cos^2(k_{F,i} N a)$ where $k_{F,i}$ is the Fermi wave vector of the associated band in the infinite chain. In the case of Pt the total conductance for short chains ($N < 7 - 8$) exhibits an overall decrease with superimposed even-odd oscillations in qualitative agreement with the experimental results of Ref. [18]. For even longer chains (not yet attainable in experiments) these calculations predict an increase of the conductance due to the contribution of conduction channels with d_{xz}, d_{yz} character.

3. Superconducting quantum point contacts

Atomic sized conductors have revealed to be unique systems to test predictions on superconducting transport in the strong quantum regime [1]. In order to understand the universal behaviour observed in their superconducting transport properties in terms of the PIN code $\{\tau_n\}$ it is important to consider the large difference between energy scales associated with superconductivity and the typical energy scale for the variation of the normal conductance. Thus, for instance, in the case of Al the superconducting gap Δ is of the order of 180 μeV, while for observing an appreciable variation in its normal conductance it is necessary to apply a bias voltage much larger than 10 meV. One can then safely assume that normal conduction channels are not affected by the superconducting transition and one is allowed to neglect the energy dependence in the channel transmissions when analyzing superconducting transport. As a consequence, in setting up a theoretical description one can concentrate on the properties of a superconducting quantum point contact (SQPC) with a single channel and with fixed transmission τ and describe the experimental results as a collection of independent channels.

3.1. The Hamiltonian approach

In the spirit of tight-binding models discussed in the previous section, the simplest model describing a single channel contact with arbitrary transmission can be written as $\hat{H} = \hat{H}_L + \hat{H}_R + \hat{H}_T$, where $H_{L,R}$ correspond to the left and right electrodes (which in the superconducting case are described by the usual BCS pairing model) that are coupled through $\hat{H}_T = \sum_\sigma v\hat{c}_{L\sigma}^\dagger \hat{c}_{R\sigma} e^{i\theta(t)/2} + \text{h.c.}$, where $\theta(t) = 2e/\hbar \int_0^t V(t)dt + \theta_0$ is the phase difference between the electrodes, which is determined by the imposed voltage bias. The normal transmission within this model is given by $\tau = 4v^2 W^2/(W^2 + v^2)^2$, where $1/W\pi$ is the density of states on the normal leads at the Fermi energy [33].

Determining the superconducting transport properties of this model requires the use of some field theoretical techniques. One needs to combine the Nambu formalism appropriate to describe the superconducting state [34] with the Keldysh formalism which allows to deal with a non-equilibrium situation [35]. In the Nambu formalism one introduces spinor field operators $\hat{\psi}_i$ and $\hat{\psi}_i^\dagger$ defined as

$$\hat{\psi}_i = \begin{pmatrix} \hat{c}_{i\uparrow} \\ \hat{c}_{i\downarrow}^\dagger \end{pmatrix} \qquad \text{and} \qquad \hat{\psi}_i^\dagger = \begin{pmatrix} \hat{c}_{i\uparrow}^\dagger & \hat{c}_{i\downarrow} \end{pmatrix}, \tag{3.1}$$

where the indices i, j denote the left and right electrodes.

In terms of these operators the usual retarded and advanced Green functions are given by

$$\begin{aligned} G_{i,j}^r(t, t') &= -i\theta(t - t') < \left[\hat{\psi}_i(t), \hat{\psi}_j^\dagger(t') \right]_+ > \\ G_{i,j}^a(t, t') &= i\theta(t' - t) < \left[\hat{\psi}_i(t), \hat{\psi}_j^\dagger(t') \right]_+ >, \end{aligned} \tag{3.2}$$

where $\left[\hat{A}, \hat{B} \right]_+$ denotes the anticommutator. Notice that these quantities are now 2×2 matrices in Nambu space. For the calculation of $G^{r,a}$ one needs to solve the corresponding Dyson equation $\mathbf{G}^{r,a} = \mathbf{g}^{r,a} + \mathbf{g}^{r,a} \circ \mathbf{v} \circ \mathbf{G}^{r,a}$, relating them to the Green functions of uncoupled electrodes $\mathbf{g}^{r,a}$. In this equation the product $\circ$ denotes integration over internal time arguments and summation over intermediate indexes while $\mathbf{v}$ denotes the hopping elements in Nambu space given by

$$\hat{v}_{LR}(t) = v \begin{pmatrix} e^{i\theta(t)/2} & 0 \\ 0 & -e^{-i\theta(t)/2} \end{pmatrix} = \left(\hat{v}_{RL}^a(t) \right)^{\dagger}. \tag{3.3}$$

The local BCS Green functions are given by $\hat{g}_{ij}^a = \hat{g}_{ij}^{r,*} = \delta_{ij} \left(g\hat{I} + f\hat{\sigma}_x \right)$, where $g = -\omega / W \sqrt{\Delta^2 - \omega^2} = -\omega/\Delta f$.

The retarded and advanced Green functions are not sufficient for dealing with a general non-equilibrium situation. It is in general necessary to determine in addition the Keldysh Green function defined as

$$G_{i,j}^{+-}(t, t') = i < \hat{\psi}_j^{\dagger}(t') \hat{\psi}_i(t) >, \tag{3.4}$$

which satisfies the equation [35]

$$\mathbf{G}^{+-} = \mathbf{g}^{+-} + \mathbf{G}^r \circ \mathbf{v} \circ \mathbf{g}^{+-} + \mathbf{g}^{+-} \circ \mathbf{v} \circ \mathbf{G}^a + \mathbf{G}^r \circ \mathbf{v} \circ \mathbf{g}^{+-} \circ \mathbf{v} \circ \mathbf{G}^a, \tag{3.5}$$

where $\mathbf{g}^{+-}$ corresponds to the uncoupled electrodes and are determined by $\hat{g}_{ij}^{+-} = \delta_{ij} n(\omega) \left(\hat{g}_{ii}^a - \hat{g}_{ii}^r \right)$, $n(\omega)$ being the Fermi distribution.

The various transport properties of the system can be expressed in terms of these quantities. Thus, for instance, the mean current can be written as

$$\left\langle \hat{I}(t) \right\rangle = \frac{e}{h} \mathrm{Tr} \left[\hat{\sigma}_z \left(\hat{v}_{LR}(t) \hat{G}_{RL}^{+-}(t, t) - \hat{v}_{RL}(t) \hat{G}_{LR}^{+-}(t, t) \right) \right]. \tag{3.6}$$

Obtaining analytical results in the superconducting case is (even for this simple model) a difficult task. The main difficulty lies in the intrinsic time dependence of the problem when $V \neq 0$. In the stationary case ($V = 0$) the Dyson equations for the retarded and advanced Green functions become 2×2 algebraic equations which can be readily solved. The main result for this case is the existence of two bound states inside the superconducting gap with energies given by [36]

$$\epsilon_{\pm}(\theta_0) = \pm \Delta \sqrt{1 - \tau \sin^2(\theta_0)/2}. \tag{3.7}$$

These are the so-called Andreev bound states, which result from multiple Andreev reflections (MAR) between the superconducting electrodes [37, 38]. The position of these states depends on the stationary phase difference θ_0 and they are

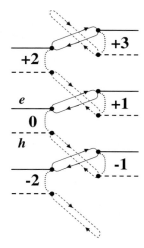

Fig. 7. Pictorial representation of the set of equations which determine the Green function components $\mathbf{G}_{n0}$.

the current-carrying states. In fact the Josephson current at zero temperature can be directly evaluated as

$$I(\theta_0) = \frac{2e}{\hbar} \frac{d\epsilon_-(\theta_0)}{d\theta_0}. \tag{3.8}$$

When a constant bias voltage is applied we have $\theta(t) = \omega_J t$, where $\omega_J = 2eV/\hbar$ is the Josephson frequency. To obtain a more tractable form of the Dyson equations one can perform a double Fourier transformation of the Green functions. Due to the special form of the phase factors in the hopping elements, the dependence on the two frequencies is linked by $\mathbf{G}(\omega, \omega') = \sum_n \mathbf{G}_{n0}(\omega)\delta(\omega - \omega' + n\omega_J)$. A pictorial representation of the Dyson equations determining the Fourier components $\mathbf{G}_{n0}$ is given in Fig. 7. In this figure we draw replicas of the left and right electrodes labeled by an integer number which determines the number of quanta eV which are emitted or absorbed in a given tunneling process. The horizontal lines represent the propagation inside the electrodes as an electron (full lines) or as a hole (dashed lines). The dotted lines indicate the coupling between electron and holes due to the anomalous propagator f. When a quasiparticle tunnels from left to right as an electron it absorbs an energy quantum eV, while when it tunnels as a hole a quantum is emitted. As a result when all processes up to infinite order in v are considered the replicas from $n = -\infty$ to $n = \infty$ are coupled like in a tight-binding one-dimensional chain with nearest-neighbour hopping. The different components $\mathbf{G}_{n0}$ can be expressed as a continued fraction

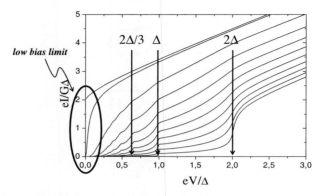

Fig. 8. Zero temperature dc current I_0 for a single channel superconducting quantum point contact for values of the transmission ranging from $\tau = 1$ (top) to tunneling at $\tau \ll 1$ (bottom).

in Nambu space. Thus, for instance

$$\hat{G}_{00}^r = \left[(\hat{g}_0^r)^{-1} - \hat{v}^+ \hat{\mathcal{G}}_1^r \hat{v}^- - \hat{v}^- \hat{\mathcal{G}}_{-1}^r \hat{v}^+ \right]^{-1}, \tag{3.9}$$

where the quantities $\hat{\mathcal{G}}_{\pm n}^r$ satisfy the recursive equations

$$\hat{\mathcal{G}}_{\pm n}^r = \left[(\hat{g}_{\pm n}^r)^{-1} - \hat{v}^\pm \hat{\mathcal{G}}_{\pm n \pm 1}^r \hat{v}^\mp \right]^{-1} \qquad \text{for even n,}$$

$$\hat{\mathcal{G}}_{\pm n}^r = \left[(\hat{g}_{\pm n}^r)^{-1} - \hat{v}^\mp \hat{\mathcal{G}}_{\pm n \pm 1}^r \hat{v}^\pm \right]^{-1} \qquad \text{for odd n.}$$

In the above expressions $\hat{g}_n^r = \hat{g}^r (\omega + neV)$ and $\hat{v}^\pm = v \left(\hat{\sigma}_z \pm \hat{I} \right) / 2$.

Once the components $\mathbf{G}_{n0}$ have been determined one can compute the mean current as $\left\langle \hat{I}(V, t) \right\rangle = \sum_m I_m(V)e^{im\omega_J t}$. The most interesting from the experimental point of view is the dc component I_0 which can be written as [33]

$$I_0 = -\frac{4e}{h} \int d\omega \sum_{n=odd} \mathrm{ReTr} \left[\hat{\sigma}_z \left(\hat{T}_n^\dagger \hat{g}_n^{+-} - \hat{T}_n \hat{g}_0^a \right) \right], \tag{3.10}$$

where $\hat{T}_n = \hat{v}^+ \hat{G}_{n+1,0}^a + \hat{v}^- \hat{G}_{n-1,0}^a$. The numerical evaluation of these equations yields the results shown in Fig. 8 for I_0 at zero temperature and for a range of values of the transmission τ. As can be observed $I_0(V)$ exhibits a highly non linear behaviour. The most remarkable feature is the appearance of current steps at $eV = 2\Delta/n$. This is the so-called subgap structure which is more pronounced

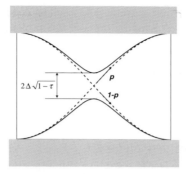

Fig. 9. Schematic representation of the Andreev states dynamics accounting for the transport properties of a SQPC in the limit of high tranmission and small bias voltage. The horizontal axis corresponds
to the phase evolution from 0 to 2π. The gray area above and below are the quasiparticle continuum
outside the superconducting gap.

in the tunnel limit $\tau \to 0$. Physically the subgap structure appears due to the occurrence of multiple Andreev reflection processes between the superconducting
electrodes. Associated with a MAR process of order n, a charge ne is transferred
coherently with a probability scaling as τ^n at low transmission. For the occurrence of these processes a minimum bias voltage $eV = 2\Delta/n$ is required, which
explains the observed jumps in the current. A rather simple expression can be
found for the current for $V \sim 2\Delta/n$ in this low transmission regime [33]

$$I_0 \simeq \frac{e\pi^2}{h}\tau^n \int_{\Delta-neV}^{-\Delta} d\omega\rho(\omega)\rho(\omega+neV)\Gamma_n(\omega), \tag{3.11}$$

where $\Gamma_n(\omega) = \prod_{k=1}^{n-1} |f(\omega+keV)|^2$ and $\rho(\omega)$ is the BCS density of states. The
scaling of the current steps with τ^n was first confirmed experimentally by van der
Post *et al.* using Nb break junctions [39].

As the transmission is increased the subgap structure becomes progressively
more rounded and eventually disappears for $\tau = 1$. Particularly interesting is the
behaviour of the current at low bias voltage when $\tau \to 1$ indicated in Fig. 8. In
this region the current exhibits an exponential increase with bias voltage. One can
understand this behaviour in terms of the low bias dynamics of the Andreev states
[40]. As commented above the zero bias limit is characterized by the presence
of Andreev states. Suppose that we have initially the equilibrium situation in
which only the lower Andreev state $\epsilon_-(\theta)$ is occupied. When a small bias voltage
is applied the state evolves adiabatically according to $\epsilon_-(\omega_J t)$ as illustrated in
Fig. 9. However, close to $\theta = \pi$ the gap between the lower and the upper
state gets smaller and a Landau-Zener transition can take place leading to the

appearance of a quasiparticle current. The Landau-Zener probability is given by $p = \exp[-\pi\Delta(1-\tau)/eV]$ and in terms of this probability the mean current is just

$$I_0(V) = \frac{4e\Delta}{h} \exp[-\pi\Delta(1-\tau)/eV].$$ (3.12)

This simple expression accounts for the behaviour of I_0 at low bias and large transmission.

3.2. Comparison to experimental results

The comparison between the theoretical predictions for a single channel SQPC and the experimental results for one atom contacts is particularly instructive. The top panel of Fig. 10 shows some typical results for Al one-atom contacts together with the set of theoretical curves for I_0. As can be observed, the theoretical results for one channel capture the main qualitative behaviour of the experimental

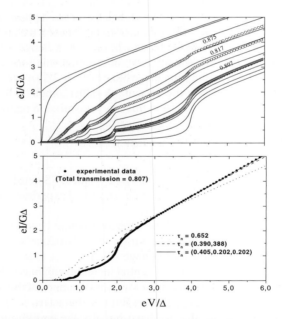

Fig. 10. Comparison of theoretical and experimental results for the dc component I_0. The symbols in the top panel are typical I-V curves recorded on the last plateau for Al atomic contacts at 30 mK. Three cases with similar normal conductance exhibit very different subgap structure. In the lower panel we show the fits of one of the experimental curves using one (dotted) two (dashed) and three channels (full curve). The corresponding sets of transmissions are given in the inset.

I-V curves. However, if we compare theoretical and experimental curves having the same normal conductance we see that they look very different. Moreover, we observe that experimental results for contacts with rather similar normal conductance exhibit large differences in their subgap structure.

The reason for this discrepancy, first pointed out in [8] is that i) even for the case of one atom contacts more than a single channel with significant transmission can be contributing to tranport and ii) the channel content or PIN code of the three experimental curves can be rather different even when their total conductance is similar. This hypothesis was confirmed by microscopic calculations for Al atomic contacts which predict the contribution of three channels for the one atom geometry [41]. In fact, when the theoretical results for three channels are superposed and their transmissions are varied as fitting parameters one can reach an excellent quantitative agreement with experiments, as shown in the lower panel of Fig.10.

Besides confirming the validity of the microscopic theory of MAR, this fitting procedure provides a powerful tool to determine the set of transmissions τ_n with high accuracy. Since its introduction in Ref. [8], the technique has been applied to many different materials [9, 42, 43] and to studying different properties of atomic contacts in well controlled conditions [3, 6]. All these studies have confirmed the reliability of the subgap structure analysis. Particularly remarkable are the studies of shot noise in Ref. [6] in which the determination of the PIN code by this technique allowed a direct comparison with the theory without any fitting parameter.

Shot noise is one of the most peculiar features of quantum transport in mesoscopic systems (see the contribution by T. Martin in this volume). It contains information both on the charge of the quasiparticles that are transferred and on their quantum correlations [44]. In the case of superconducting quantum point contacts shot noise studies offer new insight about the MAR mechanism. The quantity of interest is now the noise spectral density defined as

$$S(\omega) = \hbar \int dt e^{i\omega t} \left\langle \left[\delta \hat{I}(t), \delta \hat{I}(0) \right]_+ \right\rangle, \tag{3.13}$$

where $\delta \hat{I}(t) = \hat{I}(t) - \left\langle \hat{I}(t) \right\rangle$. The noise spectrum can be evaluated using the Keldysh-Nambu Green function techniques discussed above [45, 46]. Although we are not going to discuss this issue in detail let us just point out the more remarkable prediction of the theory, which is the quantization of the effective charge in the tunnel limit. This quantity is defined as $Q^* = S(0)/2I_0$. In a normal tunnel junction Q^* is just the electron charge. However, when the electrodes are superconducting the current in the subgap region is due to MAR processes in

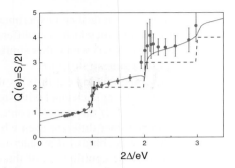

Fig. 11. Effective charge $Q^* = S(0)/2I_0$ as a function of the inverse reduced voltage for a super-conducting Al one-atom contact. The dashed line is the prediction of the theory for the tunnel limit. The experimental data (dots) are for a contact with PIN code $\{0.40, 0.27, 0.03\}$. The full line is the prediction of the microscopic theory for this set of transmission values.

which n quasiparticles are transferred coherently and this is reflected in the fact that $Q^* \to ne$, where $n = [1 + 2\Delta/V]$, when $\tau \to 0$ [45].

The experimental test of these predictions were provided by the work of Cron *et al.* [6]. Their results for the effective charge are shown in Fig. 11 together with the theoretical calculations. Although these results do not correspond to a very poorly transmitted channel ($\tau \sim 0.4$), one can already observe a tendency to a staircase behaviour as the bias voltage is reduced.

4. Environmental effects

A basic assumption in most theoretical descriptions of transport in atomic-sized conductors is that the nanoscale system can be ideally voltage biased, i.e. it can be connected directly to an ideal voltage source fixing a well defined chemical potential difference between the left and the right leads. The leads are thus assumed to behave as ideal electron reservoirs. An actual experimental situation can certainly deviate from this idealized picture. Our nanoscale device is always embedded in a macroscopic circuit which is determined mainly by the geometry of the metallic leads in the close vicinity of the atomic conductor. The actual situation would be more accurately described as illustrated in Fig. 12 in which the embedding circuit is characterized by a macroscopic impedance $Z(\omega)$. An important ingredient to be taken into account is the effective capacitance C associated with the atomic size conductor itself. Being a nanoscale object this capacitance can be rather small which could result in the appearance of observable charging effects when $e^2/C \gg kT$. In this section we briefly discuss how these environmental effects are manifested in different transport properties.

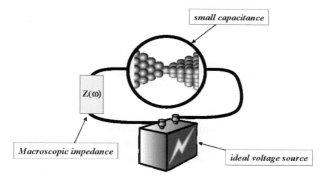

Fig. 12. Schematic representation of a typical circuit for transport measurements in atomic-sized conductors.

The theoretical description of these effects requires to analyze the influence on the transport properties of the fluctuations in the phase difference across the atomic conductor induced by the macroscopic impedance. One can distinguish between *classical* and *quantum* effects depending on the nature of these fluctuations. Thus, for sufficiently high temperatures, when charging effects are negligible, only thermal fluctuations have to be considered. These fluctuations have an important effect on the Josephson current, reducing strongly its maximum value from the theoretically predicted result, Eq. (3.8), which explains the quasi-absence of a supercurrent branch in many experiments. On the other hand, when the charging energy e^2/C is comparable to, or larger than, kT and the series impedance is not negligible compared to h/e^2 quantum fluctuations in the phase start to play an important role. This leads to a phenomenon called *dynamical Coulomb blockade*, i.e. a suppression of the current at low bias, which can be observed in normal atomic size conductors when the embedding circuit is designed in order to satisfy the above mentioned conditions.

4.1. Classical phase diffusion

To analyze the influence of thermal fluctuations on the Josephson current in superconducting atomic size contacts one can generalize the so-called resistively and capacitively shunted junction (RCSJ) model, traditionally used in the context of tunnel junctions [47]. The starting point of this approach is to write down a Langevin equation for the phase difference across the contact taking into account the non-sinuoidal behavior of the current-phase relation at high transmissions. This has been done in Refs. [3] and [48]. In the latter reference the model was generalized to include also the effect of a microwave field in order to study the influence of thermal fluctuations on the so-called Shapiro steps.

Within the RCSJ model the parallel combination of the atomic contact with a shunting resistance R and a capacitance C is current polarized by a current source I_b. The atomic contact is characterized by its current-phase relation $I(\theta)$. Current conservation then implies

$$I_b = C\frac{dv}{dt} + I(\theta) + \frac{v}{R} + L(t), \tag{4.1}$$

where $L(t)$ is the Johnson-Nyquist noise produced by dissipation in the resistance. The voltage across the contact and the phase are related by $v = \phi_0\dot{\theta}$, where $\phi_0 = \hbar/2e$ is the reduced flux quantum. Equation (4.1) is completely analogous to the one describing the motion of a brownian particle in a potential $U(\theta)$, determined by

$$U(\theta) = \phi_0\theta I_b - \phi_0 \int_0^\theta I(\theta)d\theta. \tag{4.2}$$

When the capacitance is negligible (overdamped regime) the problem can be translated into the study of the distribution function $\sigma(\theta, t)$ for the phase across the contact, which satisfies a Smoluchowski equation [49]

$$\frac{d\sigma}{dt} = \frac{R}{\phi_0^2}\frac{\partial}{\partial\theta}\left[-\frac{\partial U}{\partial\theta} + T\frac{\partial\sigma}{\partial\theta}\right]. \tag{4.3}$$

This equation can be written as $\frac{d\sigma}{dt} + \frac{\partial w}{\partial\theta} = 0$, where $w = \frac{R}{\phi_0^2}\left[-\frac{\partial U}{\partial\theta} + T\frac{\partial\sigma}{\partial\theta}\right]$. This quantity can thus be interpreted as a probability current which must be related to σ by $w = \sigma v/\phi_0$. From this relation one can obtain the mean voltage $\langle v \rangle$ across the contact as a function of the biasing current I_b. Extracting the current-voltage charateristic from the Smoluchowski equation is particularly simple for the case of a constant dc bias. In this case the stationary solution $\sigma(\theta)$ does not depend on time and w is just a constant. The equation can then be easily integrated yielding

$$\sigma(\theta) = \frac{w\phi_0^2}{TR}\frac{f(\theta)}{f(2\pi) - f(0)}\left[f(2\pi)\int_\theta^{2\pi}\frac{d\theta}{f(\theta)} + f(0)\int_0^\theta\frac{d\theta}{f(\theta)}\right], \tag{4.4}$$

where $f(\theta) = exp[-U(\theta)/T]$. The normalization condition $\int_0^{2\pi}d\theta\sigma(\theta) = 1$ then determines w and hence $\langle v \rangle$.

The supercurrent branch in superconducting atomic-sized contacts was analyzed experimentally in Ref. [3] using Al microfabricated break junctions. In order to have good control of the thermal and quantum fluctuations they designed an on-chip dissipative environment with small resistors of known value placed close

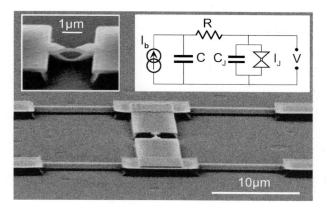

Fig. 13. Micrograph of the experimental setup used in Ref. [3] to study the supercurrent in an atomic contact. Each probe contains a AuCu resistor (thin lines of 10 μm length) and a large capacitor with the metallic substrate. The left inset shows a close-up of the microfabricated MCBJ and the right inset illustrates the equivalent circuit.

to the atomic contact, as illustrated in Fig. 13. With the appropriate choice of the environment parameters the current-voltage curve becomes hysteretic, which allows the measurement of the supercurrent and the quasi-particle branch for the same contact. The advantage is that this permits the determination of the PIN code by the subgap structure analysis as discussed in the previous section. A typical current voltage characteristic for this set-up is depicted in Fig. 14. The inset in this figure shows a blow-up of the supercurrent branch together with the result obtained from the Smoluchowski equation for the corresponding set of parameters. As can be observed the threshold or *switching* current, I_S, at which the jump from the supercurrent branch to the quasiparticle current branch takes place lies very close to the maximum in the current-voltage characteristic predicted by the RCSJ model. This value is very sensitive to thermal fluctuations and decreases with increasing temperatures as illustrated by the results shown in Fig. 15. The experimental results for I_S for different sets $\{\tau_n\}$ are in excellent agreement with the theoretical values except for contacts having an almost perfectly transmitted channel (with transmissions between 0.95 and 1.0). In this case it is found that the switching current is less sensitive to thermal fluctuations than predicted by the theoretical model. This discrepancy was attributed in Ref. [3] to the contribution due to Landau Zener transitions between Andreev states, not included in the above description. The development of a unified theory including classical phase diffusion and transitions between Andreev states constitutes one of the open problems in the theory of superconducting quantum point contacts.

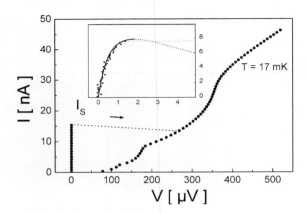

Fig. 14. Typical current-voltage characteristic for the set up of Fig. 13 at 17 mK. A jump from the supercurrent branch near zero voltage to the finite voltage dissipative branch is observed. The inset shows a blow-up of the supercurrent branch for another contact at a larger temperature together with a fit using the Smoluchowski equation (4.3) for the corresponding set of parameters.

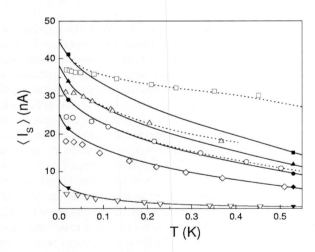

Fig. 15. Experimental (open symbols) and theoretical (lines) results for the switching current obtained in Ref. [3]. The results correspond to one atom contacts with different channel content. ($\triangledown$) $\{\tau_i\} = \{0.21, 0.07, 0.07\}$. From the fit a zero-temperature supercurrent of $I_0 = 8.0 \pm 0.1$ nA is obtained. ($\diamondsuit$) $\{\tau_i\} = \{0.52, 0.26, 0.26\}$, $I_0 = 25.3 \pm 0.4$ nA. ($\circ$) $\{\tau_i\} = \{0.925, 0.02, 0.02\}$, $I_0 = 33.4 \pm 0.4$ nA. ($\triangle$) $\{\tau_i\} = \{0.95, 0.09, 0.09, 0.09\}$, $I_0 = 38.8 \pm 0.2$ nA. (open squares) $\{\tau_i\} = \{0.998, 0.09, 0.09, 0.09\}$, $I_0 = 44.2 \pm 0.9$ nA. The full lines are the predictions of the Smoluchowski equation (4.3) while the dotted lines are the results of a numerical simulation of the circuit equation (4.1) allowing for Landau-Zener transitions between Andreev states. From [3].

4.2. Dynamical Coulomb blockade

Dynamical Coulomb blockade (DCB) in ultra-small tunnel junctions was extensively analyzed in the early 90's. The basic theory for describing this effect in tunnel junctions is called $P(E)$-theory. Within this theory the electromagnetic environment of the junction is described as a set of LC circuits in the spirit of Caldeira-Legget model for quantum dissipation [50]. One can in principle describe any impedance $Z(\omega)$ as a continuous distribution of harmonic oscillators which are coupled to the tunneling electrons. The tunneling then becomes an inelastic process: the electron can only tunnel if it can excite a mode in the environment. This necessarily reduces the current because there is a reduction in the phase space available for the final electron states. The name of the theory is due to the fact that the current is determined by a certain function $P(E)$ giving the probability of exciting a mode of energy E.

$P(E)$−theory is somewhat equivalent to the Fermi golden rule, i.e. lowest order perturbation theory in the coupling between the leads. This theory is clearly insufficient for describing the case of an atomic contact of arbitrary transmission. DCB in a normal QPC coupled to a macroscopic impedance was analyzed in Refs. [51, 52]. In Ref. [52] it was shown that there is a close connection between DCB and shot noise in these types of systems. The starting point of their theoretical analysis is again the simple model Hamiltonian of Sect. 3.1 which is modified to include the coupling to the environment. This is done by introducing an additional phase factor $\hat{\Lambda}_e = \exp[i\hat{\theta}]$ in the hopping term, where $\hat{\theta}$ is the phase operator which is a conjugate variable of the charge $\hat{Q}$ between the leads, i.e. $[\hat{\theta}, \hat{Q}] = ie$. The phase factor $\hat{\Lambda}_e$ then acts as a translation operator which shifts the charge between the leads by one each time an electron tunnels. The modified hopping term then reads $H_T = \sum_\sigma v\hat{c}^\dagger_{L\sigma}\hat{c}_{R\sigma}\hat{\Lambda}_e + \text{h.c.}$.

The information characterizing the electromagnetic environment is contained in the phase correlation function $J(t) = \left\langle\hat{\theta}(t)\hat{\theta}(0)\right\rangle - \left\langle\hat{\theta}^2\right\rangle$, which is related to the total impedance $Z_t(\omega)$ by $J(t) = G_0 \int d\omega \text{Re} Z_t(\omega)(e^{i\omega t} - 1)/\omega$ [53]. Here the total impedance is the parallel combination of $Z(\omega)$ and the contact capacitance, i.e $Z_t(\omega) = \left(Z(\omega)^{-1} + i\omega C\right)^{-1}$.

In Ref [52] the blockade in the current within this model was calculated using Keldysh formalism. However, in the limit of weak coupling with the environment (i.e. $Z \ll h/e^2$) the relation between DCB and noise can be obtained by more conventional methods [54]. To lowest order in $J(t)$ the current blockade is given by [54]

$$\left\langle\delta\hat{I}\right\rangle = -\frac{1}{2e^2} \int d\omega J(\omega) \int d\omega' \text{sign}(\omega - \omega')\frac{\partial S_I}{\partial V}(\omega', V), \qquad (4.5)$$

where $S_I(\omega, V)$ denotes the current noise spectrum of the QPC.

For a QPC at low frequencies and zero temperature $S_I = 2eG_0 \sum_n \tau_n (1 - \tau_n)(eV - \hbar\omega)\theta(eV - \hbar\omega)$ (see the chapter on noise by T. Martin in this volume) and Eq. (4.5) yields

$$\frac{\delta G}{G} = -G_0 \frac{\sum_n \tau_n (1 - \tau_n)}{\sum_n \tau_n} \int_{eV}^{\infty} d\omega \frac{\text{Re} Z_t(\omega)}{\omega}. \tag{4.6}$$

It should be noted that the correction to the conductance is affected by the same reduction factor that applies for shot noise. In the simple but realistic case for which the impedance is composed by the resistance R of the leads embedding the contact in parallel with the capacitance C of the contact itself, the blockade in the conductance reduces to

$$\frac{\delta G}{G} = -G_0 R \frac{\sum_n \tau_n (1 - \tau_n)}{\sum_n \tau_n} \ln \sqrt{1 + \left(\frac{\hbar\omega_R}{eV}\right)^2}, \tag{4.7}$$

where $\omega_R = 1/RC$. At finite temperature the logarithmic singularity in Eq. (4.7) at $V = 0$ becomes progressively rounded. The finite temperature expression of this equation can be found in Ref. [4].

In order to verify these predictions Cron *et al.* [4] fabricated an atomic contact embedded in an electromagnetic environment essentially equivalent to a pure ohmic resistor of the order of 1 kΩ defined by e-beam lithography. The material chosen for both the atomic contact and the series resistor was Al, which allowed to extract the channel decomposition or mesoscopic PIN code for the contacts, using the subgap structure analysis in the superconducting state. The environmental Coulomb blockade was then measured in the presence of a 0.2 T magnetic field which brings the sample to the normal state. The results for two contacts with very different transmissions are shown in Fig. 16. The standard $P(E)$-theory is able to account for the results obtained for the contact in the left panel, which has a single weakly transmitted channel. On the other hand, for the contact in the right panel, with a well transmitted channel ($\tau \simeq 0.83$) the amplitude of the dip at zero bias is markedly reduced with respect to the tunnel limit predictions (dashed line). The experimental results are in good agreement with the predictions of Eq. (4.6) (solid line).

5. Single-molecule junctions

More recently the focus of attention in the field of transport at the atomic scale is moving towards conducting bridges of individual molecules. As early as 1974 Aviram and Ratner [55] proposed that a molecule with donor and acceptor groups

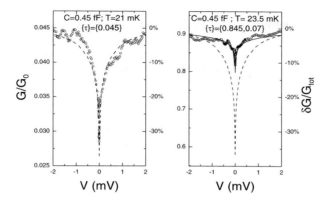

Fig. 16. Measured differential conductance curves of two atomic contacts (symbols referred to the left axes, G_0 being the conductance quantum), and comparison with the predictions for the dynamic Coulomb blockade (lines, right axes, relative reduction of the conductance). Dashed lines are the predictions for the tunnel case [53] and the full line is the prediction of the theory in [4], (Eq. (4.6)), summing the contributions of the two channels of the contact. The wiggles and asymmetry appearing on the experimental curves are reproducible conductance fluctuations due to interference effects depending on the detailed arrangement of the atoms in the vicinity of the contact. On the left panel, the contact consists of a single weakly transmitting channel, and it is well described by the standard theory of DCB valid for tunnel contacts, as expected. On the right pannel, the contact has a well transmitting channel with $\tau = 0.835$. In this case, the relative reduction of conductance is much smaller than for the tunnel case and is in agreement with the predictions of Eq. (4.6).

separated by a poorly conducting spacer group might act as a diode when connected between metallic electrodes. This and other ideas led to the concept of molecular electronics: the prospect of building electronics out of organic molecular components, that would ideally self-assemble into large integrated circuits. However, the first step would be to contact a single molecule, or at least a single monolayer. Although several initial attempts towards this goal seemed promising it has turned out to be difficult to obtain stable and reproducible results. For self-assembled monolayers of molecules many experimental results have been influenced by defects in the monolayers. Therefore, many attempts more recently focus on contacting single molecules only and some encouraging progress is now being made. For a more complete recent review, see Ref. [56].

Several experimental techniques have been developed that allow measuring single molecules. One approach is using scanning tunneling microscopes, or conducting-tip atomic force microscopes to find and contact a desired molecule. In order to avoid contacting many molecules in parallel the target molecules are often embedded at defect positions into a monolayer matrix of less conducting alkanethiol molecules [57, 58]. As metal contacts, almost exclusively gold has been used, in view of its low-reactive surfaces making it relatively easy to avoid reactions with other molecules. Strong bonds to organic molecules can nevertheless be made through the use of sulfur end-groups (often referred to as the alligator clips). Although a relatively stable junction of the molecules to a gold sample surface can be ensured by allowing a solution of the compounds to react over a period of many hours, the top contact is usually a weak-coupling tunnel junction to the STM or AFM tip. By using thiol alligator clips at both ends and binding a gold cluster to the top of the molecules Lindsay's group has achieved important progress in reproducibility of the measured conductance [59, 60]. Nevertheless, the conductance varies as a result of variations in the metal-molecule bonding. A particularly attractive method for obtaining ensemble averaged values for the conductance of a metal-molecule-metal junction was introduced by Tao and his group [61]. They repeatedly make and break contacts between a gold sample surface and a gold tip by forcing the STM tip into and out of contact. This is done at room temperature with tip and sample immersed in a solution of the molecules of interest, that have been prepared with the proper thiol end groups. One observes frequently occurring plateaus in the conductance traces recorded during breaking, that were shown to be associated with the molecules. The mean conductance for a molecular bridge configuration can then be obtained from the position of a peak in the histogram of conductance values.

Standard microfabrication approaches are not capable of reproducibly making electrode separations small enough that they can be bridged by a single molecule. Several solutions have been demonstrated that combine microfabrication with other tricks. One method uses electromigration of a microfabricated gold wire on a substrate [62–64]. Prior to the experiment the wire is allowed to interact with the thiol-ended molecules in solution, after which the wire is broken by controllably sending a large current through it. The combination of Joule heating and electron wind force breaks the wire, and it was found that the resulting junction is often bridged by one, or several, of the target molecules. The advantage of this approach is the additional experimental control parameter provided by the gate electrode that can be defined along with the wire. This method has recently been further improved by adding mechanical control over the electrode separation after breaking [65]. Other microfabrication approaches include the low-temperature shadow-evaporation technique introduced by Kubatkin *et al.* [66] and a combination with electrochemical etching/deposition as described by Kervennic *et al.* [67, 68].

A significant fraction of the experimental work on single molecule transport was inspired by the paper by Reed *et al.* [71]. The experiment was performed using a mechanically controllable break junction device [1] working at room tem-

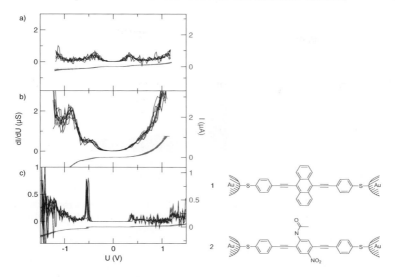

Fig. 17. Current voltage characteristics, reproducibly recorded for a stable junction in a mechanically controlled break junction (lower) and their numerical derivative dI/dU (upper curve). a) with molecule **1** a spatially symmetric (9,10-Bis((2'-para-mercaptophenyl)-ethinyl)-anthracene) at room temperature, b) with molecule **2** an asymmetric molecule (1,4-Bis((2'-para-mercaptophenyl)ethinyl)-2-acetyl-amino-5-nitro-benzene) at room temperature, and c) with molecule **2** at $T = 30$ K. From Refs. [69, 70].

perature, with the junction immersed in a solution of the organic compound of interest. The compound they selected was benzene-1,4-dithiol that has become the workhorse in this field of science. In the experiment the broken gold wire was allowed to interact with molecules for a number of hours so that a self-assembled monolayer covered the surface. Next, the junction was closed and re-opened a number of times and current-voltage ($I - V$) curves were recorded at the position just before contact was lost completely. The $I - V$ curves showed some degree of reproducibility with a fairly large energy gap feature of about 2 eV, that was attributed to a metal-molecule-metal junction. The zero-bias conductance was of order $10^9 \Omega$ which is one or two orders of magnitude larger than typically found in model calculations for a benzenedithiol bridging two metal electrodes [72].

Further experiments along these lines have been performed on larger organic molecules in the groups of Bourgoin [73] and Weber [69, 70]. In particular, the latter showed that the symmetry of the molecules is reflected in the symmetry of the $I - V$ curves, see Fig. 17. Dulić *et al.* [74] used a photochromic molecule that can be switched from a high-conductance state to a low-conductance state under the influence of visible light and back under near-UV irradiation. The

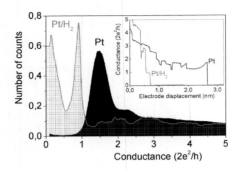

Fig. 18. Conductance curves and histograms for clean Pt, and Pt in a H_2 atmosphere. The curves for Pt in the inset and the histogram in the main panel were measured at a bias of 10 mV. The curve for Pt with H_2 in the inset was measured at 100 mV, and the histogram was obtained at 140 mV. All data were taken at 4.2 K under cryogenic vacuum. From Ref. [75].

two states had been investigated before in detail in solution. Dulić *et al.* used a lithographically fabricated break junction device to contact individual molecules of this kind, modified to have thiol-anchoring groups. They observed switching of the conductance of a molecule from the high-conductance state to the low-conductance state, but the reverse step was not obtained. They present evidence that the reverse step is suppressed as a result of the interaction of the UV light with surface plasmons in the gold electrodes.

The early experiments aimed at probing the electronic transport properties of individual organic molecules have shown that it is difficult to identify the number of molecules actually contacted and that the characteristics observed vary widely between experiments. Under such conditions it is not surprising that there is also very little agreement with calculations. This situation forms a strong motivation to study simple systems, that by themselves will not be useful as molecular devices, but that may provide a more viable test system to identify the problems in experiment and theory. The simplest molecule is dihydrogen, which has been shown can be contacted between platinum electrodes [75]. The discussion of the hydrogen experiments will occupy most of the remainder of this section.

Smit *et al.* [75] obtained molecular junctions of a hydrogen molecule between platinum leads using the mechanically controllable break junction technique. The inset to Fig. 18 shows a conductance curve for clean Pt (black) at 4.2 K, before admitting H_2 gas into the system. About 10,000 similar curves were used to build the conductance histogram shown in the main panel (black, normalized by the area). After introducing hydrogen gas the conductance curves were observed to change qualitatively as illustrated by the gray curve in the inset. The dramatic change is most clearly brought out by the conductance histogram (gray, hatched). Clean Pt contacts show a typical conductance of $1.5 \pm 0.2 \, G_0$ for a single-atom contact, as can be inferred from the position and width of the first peak in the Pt conductance histogram. Below $1 \, G_0$ very few data points are recorded, since Pt contacts tend to show an abrupt jump from the one-atom contact value into the tunneling regime towards tunnel conductance values well below $0.1 \, G_0$. In contrast, after admitting hydrogen gas a lot of structure is found in the entire range below $1.5 \, G_0$, including a pronounced peak in the histogram near $1 \, G_0$. The research to date on this system has focussed on the molecular arrangement responsible for this sharp peak. Clearly, many other junction configurations can be at the origin of the large density of data points at lower conductance, but they have not yet been studied in detail.

The interpretation of the peak at $1 \, G_0$ was obtained from a combination of measurements, including vibration spectroscopy and the analysis of conductance fluctuations, and Density Functional Theory (DFT) calculations. Experimentally, the vibration modes of the molecular structure were investigated by exploiting the principle of point contact spectroscopy for contacts adjusted to sit on a plateau in the conductance near $1 \, G_0$. The principle of point contact spectroscopy is similar to inelastic tunneling spectroscopy (IETS), but differs somewhat in a few important details. As for IETS, the differential conductance is measured using a small modulation superimposed on a dc bias that is slowly swept over a wide voltage range. When the bias increases from zero and crosses a voltage corresponding to the energy of a vibration mode in the contact, $eV = \hbar\omega$, a new channel for electron scattering opens. For an ideal one-channel contact the only option is backscattering since all forward propagating states are occupied. Thus, in contrast to IETS, to first approximation scattering by vibration modes leads to a drop in the conductance.

Figure 19 shows examples for Pt-H$_2$ and Pt-D$_2$ junctions at a plateau near $1 \, G_0$. The conductance is seen to drop by about 1 or 2%, symmetrically at positive and negative bias, as expected for electron-phonon scattering. The energies are in the range 50–60 meV, well above the Debye energy of ~ 20 meV for Pt metal. A high energy for a vibration mode implies that a light element is involved, since the frequency is given by $\omega = \sqrt{\kappa/M}$ with κ an effective spring constant and M the mass of the vibrating object. The proof that the spectral features are indeed associated with hydrogen vibration modes comes from further experiments where H$_2$ was substituted by the heavier isotopes D$_2$ and HD. The positions of the peaks in the spectra of d^2I/dV^2 vary within some range between measurements on different junctions, which can be attributed to variations in the atomic geometry of the leads to which the molecules bind. Figure 20 shows histograms for the vibration modes observed in a large number of spectra for each of the three isotopes.

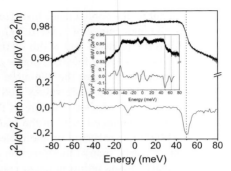

Fig. 19. Differential conductance curve for a molecule of D_2 contacted by Pt leads. The dI/dV curve (top) was recorded over 1 minute, using a standard lock-in technique with a voltage bias modulation of 1 meV at a frequency of 700Hz. The lower curve shows the numerically obtained derivative. The spectrum for H_2 in the inset shows two phonon energies, at 48 and 62 meV. All spectra show some, usually weak, anomalies near zero bias that can be partly due to excitation of modes in the Pt leads, partly due to two-level systems near the contact. From Ref. [76].

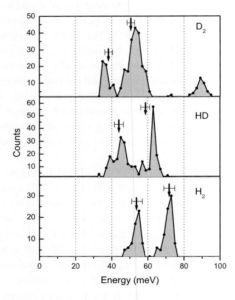

Fig. 20. Distribution of vibration mode energies observed for H_2, HD, and D_2 between Pt electrodes, with a bin size of 2 meV. The peaks in the distribution for H_2 are marked by arrows and their widths by error margins. These positions and widths were scaled by the expected isotope shifts, $\sqrt{2/3}$ for HD and $\sqrt{1/2}$ for D_2, from which the arrows and margins in the upper two panels have been obtained. From Ref. [76].

Two pronounced peaks are observed in each of the distributions, that scale approximately as the square root of the mass of the molecules, as expected. The two modes can often be observed together, as in the inset to Fig. 19. For D_2 an additional mode appears near 90 meV. This mode cannot easily be observed for the other two isotopes, since the lighter HD and H_2 mass shifts the mode above 100 meV where the junctions become very unstable. For a given junction with spectra as in Fig. 19 it is often possible to stretch the contact and follow the evolution of the vibration modes. The frequencies for the two lower modes were seen to increase with stretching, while the high mode for D_2 is seen to shift downward. This unambiguously identifies the lower two modes as transverse modes and the higher one as a longitudinal mode for the molecule. This interpretation agrees nearly quantitatively with DFT calculations for a configuration of a Pt-H-H-Pt bridge in between pyramidally shaped Pt leads [76, 77]. The conductance obtained in the DFT calculations [75–77] also reproduces the value of nearly 1 G_0 for this configuration. The number of conduction channels found in the calculations is one, which agrees with the analysis of conductance fluctuations in the experiment of Refs. [75, 78]. The fact that the conductance is carried by a single channel demonstrates that there is indeed just a single molecule involved.

Several other DFT calculations have been performed, see e.g. Refs. [79, 80], where the agreement is only partial. Although Cuevas *et al.* [80] find a similar high value for the conductance, the molecular orbitals responsible for the transport are the bonding orbitals, while Refs. [75–77] attribute the transport almost entirely to the antibonding orbitals. This difference implies that the sign of the charge transfer between the molecule and the metal leads differs between the two groups of calculations. Using a slightly different approach Garcia *et al.* [79] agree with Cuevas *et al.* on the bonding orbitals as the transport channel, but they obtain a conductance well below 1 G_0. They propose an alternative atomic arrangement to explain the high conductance for the Pt-H bridge, consisting of a Pt-Pt-bridge with two H atoms bonded to the sides. However, this configuration gives rise to three conductance channels, which is excluded based on the analysis of the conductance fluctuations as discussed above. The rather strong disagreement between various approaches in DFT calculations for this simple molecule shows that there is a need for a reliable set of experimental data against which the various methods can be tested. The hydrogen metal-molecule-metal bridge may provide a good starting point since it is the simplest and it can be compared in detail by virtue of the many parameters that have been obtained experimentally.

Conductance histograms recorded using Fe, Co or Ni electrodes in the presence of hydrogen also show a pronounced peak near 1 G_0 [81], indicating that many transition metals may form similar single-molecule junctions. Pd also seemed a good candidate, but Csonka *et al.* [78] did not observe the same suppression of conductance fluctuations as for Pt. There is an additional peak at

0.5 G_0 in the conductance histogram, and it was argued that hydrogen is incorporated into the bulk of the Pd metal electrodes.

Going beyond the simplest molecule using similar techniques, much work is still in progress. Preliminary results have been obtained for CO and for C_2H_2 between Pt electrodes [81, 82].

6. Concluding remarks

In this chapter we have presented an overview of some aspects of electron transport in atomic and molecular contacts. In the case of atomic contacts our fabrication capabilities and theoretical understanding are reaching a level of sophistication which has allowed accurate quantitative agreement between theory and experimental results for several properties. A key ingredient in this analysis is the concept of the mesoscopic PIN code, i.e. the set of transmission eigenvalues which determine all transport properties within the Landauer-Büttiker picture. The results for atomic contacts are consistent with the idea of this set being determined by the valence orbital structure of each element [9]. Superconducting transport has played a central role in these studies, providing a powerful tool to determine the PIN code by the subgap structure analysis. Conversely, atomic contacts have revealed themselves as an almost ideal system to verify the predictions of the microscopic theory of superconducting transport in the coherent MAR regime, which has been developed in the last decade.

As discussed in Sect. 4 some of the observable properties are fundamentally determined by the electromagnetic environment of the nano-junction. The combination of break junctions with lithographic techniques, such as pioneered in Refs. [3, 4], is opening the possibility to study some fundamental rules for circuits at the nanometer scale under well controlled conditions. In spite of this recent progress the influence of environmental effects on the transport properties of atomic size conductors is an issue which still deserves further analysis, specially regarding the superconducting state.

We have also discussed in some detail the question of the length dependence of the conductance through a chain of atoms. Atomic chains constitute an intermediate step between atomic-sized contacts and molecular junctions, and it is rather natural to ask whether the simple one-electron concepts which proved useful to analyze the former are also valid for the latter. The results that we have presented for the conductance oscillations as a function of the number of atoms in the chain seem to indicate that the connection between conduction channels and valence orbital structure remains valid. In the case of the chains the oscillations appear to be associated with the band structure around the Fermi energy of an infinite chain.

Finally we have summarized some developments towards the study of transport across single molecules. Although the level of sophistication in this research is still behind that of the metallic single-atom contacts significant progress has been made. We believe that it is worthwhile to start from simple model molecules and gradually work our way up to the more complicated ones. From the simple molecules we can learn how to analyze the problem and control the level of agreement with model calculations. The simple molecules are also simple in the sense that they are describable completely in a one-electron picture. Electron-electron interaction effects will gradually show up as we move towards more extended organic molecules, and the models will need to be extended to include these effects. There are many open questions regarding the proper description of non-equilibrium currents, on-site repulsion effects and strong electron-phonon coupling effects. There is already a significant number of theoretical works addressing these issues, but there are not too many experiments that they can be compared with. This poses an immediate challenge to the community of experimental physicists.

We thank our coworkers N. Agraït, J.C. Cuevas, R. Cron, M.H. Devoret, D. Djukic, F. Flores, M. Goffman, M.H. van Hemert, K.W. Jacobsen, P. Joyez, N. Lang, A. Martín-Rodero, Y. Noat, A. Saúl, E. Scheer, R.H.M. Smit, K. Thygesen, C. Untiedt, C. Urbina, E. Vecino and L. de la Vega for their contributions to the various works on atomic and molecular contacts which have been discussed in this paper. This work is part of the research program of the "Stichting FOM". and was partially supported by the EU through the DIENOW training network.

References

[1] N. Agraït, A. Levy Yeyati, and J. M. van Ruitenbeek. Quantum properties of atomic-sized conductors. *Phys. Rep.*, **377**, 81–279, (2003).

[2] R. Landauer. Electrical resistance of disordered one-dimensional lattices. *Phil. Mag.*, **21**, 863–867, (1970).

[3] M. Goffman, R. Cron, A. Levy Yeyati, P. Joyez, M. H. Devoret, D. Esteve, and C. Urbina. Supercurrent in atomic point contacts and Andreev states. *Phys. Rev. Lett.*, **85**, 170–173, (2000).

[4] R. Cron, E. Vecino, M. H. Devoret, D. Esteve, P. Joyez, A. Levy Yeyati, A. Martín-Rodero, and C. Urbina. Dynamical Coulomb blockade in quantum point contacts. In T. Martin, G. Montambaux, and J. Trân Thanh Vân, editors, *Electronic Correlations: from Meso- to Nano-Physics*, pages 17–22, Les Ulis, France, (2001). EDP Sciences.

[5] H. E. van den Brom and J. M. van Ruitenbeek. Quantum suppression of shot noise in atom-size metallic contacts. *Phys. Rev. Lett.*, **82**, 1526–1529, (1999).

[6] R. Cron, M. F. Goffman, D. Esteve, and C. Urbina. Multiple-charge-quanta shot noise in superconducting atomic contacts. *Phys. Rev. Lett.*, **86**, 4104–4107, (2001).

[7] B. Ludoph, M. H. Devoret, D. Esteve, C. Urbina, and J. M. van Ruitenbeek. Evidence for saturation of channel transmission from conductance fluctuations in atomic-size point contacts. *Phys. Rev. Lett.*, **82**, 1530–1533, (1999).

[8]   E. Scheer, P. Joyez, D. Esteve, C. Urbina, and M. H. Devoret. Conduction channel transmissions of atomic-size aluminum contacts. *Phys. Rev. Lett.*, **78**, 3535–3538, (1997).

[9]   E. Scheer, N. Agraït, J. C. Cuevas, A. Levy Yeyati, B. Ludoph, A. Martín-Rodero, G. Rubio Bollinger, J. M. van Ruitenbeek, and C. Urbina. The signature of chemical valence in the electrical conduction through a single-atom contact. *Nature*, **394**, 154–157, (1998).

[10] H. Ohnishi, Y. Kondo, and K. Takayanagi. Quantized conductance through individual rows of suspended gold atoms. *Nature*, **395**, 780–785, (1998).

[11] A. I. Yanson, G. Rubio Bollinger, H. E. van den Brom, N. Agraït, and J. M. van Ruitenbeek. Formation and manipulation of a metallic wire of single gold atoms. *Nature*, **395**, 783–785, (1998).

[12] R. H. M. Smit, C. Untiedt, A. I. Yanson, and J. M. van Ruitenbeek. Common origin for surface reconstruction and the formation of chains of metal atoms. *Phys. Rev. Lett.*, **87**, 266102, (2001).

[13] P. L. Pernas, A. Martín-Rodero, and F. Flores. Electrochemical-potential variations across a constriction. *Phys. Rev. B*, **41**, 8553–8556, (1990).

[14] N. D. Lang. Anomalous dependence of resistance on length in atomic wires. *Phys. Rev. Lett.*, **79**, 1357–1360, (1997).

[15] H.-S. Sim, H.-W. Lee, and K. J. Chang. Even-odd behavior of conductance in monatomic sodium wires. *Phys. Rev. Lett.*, **87**, 096803, (2001).

[16] C. Untiedt, A. I. Yanson, R. Grande, G. Rubio-Bollinger, N. Agraït, S. Vieira, and J. M. van Ruitenbeek. Calibration of the length of a chain of single gold atoms. *Phys. Rev. B*, **66**, 85418, (2002).

[17] G. Rubio-Bollinger, S. R. Bahn, N. Agraït, K. W. Jacobsen, and S. Vieira. Mechanical properties and formation mechanisms of a wire of single gold atoms. *Phys. Rev. Lett.*, **87**, 026101, (2001).

[18] R. H. M. Smit, C. Untiedt, G. Rubio-Bollinger, R. C. Segers, and J. M. van Ruitenbeek. Observation of a parity oscillation in the conductance of atomic wires. *Phys. Rev. Lett.*, **91**, 076805, (2003).

[19] S. K. Nielsen, Y. Noat, M. Brandbyge, R. H. M. Smit, K. Hansen, L. Y. Chen, A. I. Yanson, F. Besenbacher, and J. M. van Ruitenbeek. Conductance of single-atom platinum contatcs: voltage dependence of the conductance histogram. *Phys. Rev. B*, **67**, 245411, (2003).

[20] E. N. Economou. *Green's functions in Quantum Physics*. Springer, Berlin, (1983).

[21] Z.Y. Zeng and F. Claro. *Phys. Rev. B*, **65**, 193405, (2002).

[22] P. Havu, T. Torsti, M. J. Puska, and R. M. Nieminen. *Phys. Rev. B*, **66**, 075401, (2002).

[23] A. Levy Yeyati and M. Büttiker. Scattering phases in quantum dots: an analysis based on lattice models. *Phys. Rev. B*, **62**, 7307–7315, (2000).

[24] A. Levy Yeyati, A. Martín-Rodero, and F. Flores. Conductance quantization and electron resonances in sharp tips and atomic-size contacts. *Phys. Rev. B*, **56**, 10369–10372, (1997).

[25] N. D. Lang and Ph. Avouris. Oscillatory conductance of carbon-atom wires. *Phys. Rev. Lett.*, **81**, 3515–3518, (1998).

[26] N. D. Lang and Ph. Avouris. Carbon-atom wires: charge-transfer doping, voltage drop, and the effect of distortions. *Phys. Rev. Lett.*, **84**, 358–361, (2000).

[27] J. J. Palacios, A. J. Pérez-Jiménez, E. Louis, E. SanFabián, and J. A. Vergés. First-principles approach to electrical transport in atomic-scale nanostructures. *Phys. Rev. B*, **66**, 035322, (2002).

[28] K. S. Thygesen and K. W. Jacobsen. *Phys. Rev. Lett.*, **91**, 146801, (2003).

[29] S. Okano, K. Shiraishi, and A. Oshiyama. *Phys. Rev. B*, **69**, 045401, (2004).

[30] L. de la Vega, A. Martín-Rodero, A. Levy Yeyati, and A. Saúl. *Phys. Rev. B*, **70**, 113107, (2004).

[31] M. Okamoto and K. Takayanagi. Structure and conductance of a gold atomic chain. *Phys. Rev. B*, **60**, 7808–7811, (1999).

[32] H. Häkkinen, R. N. Barnett, and U. Landman. Gold nanowires and their chemical modifications. *J. Phys. Chem. B*, **103**, 8814–8816, (1999).

[33] J. C. Cuevas, A. Martín-Rodero, and A. Levy Yeyati. Hamiltonian approach to the transport properties of superconducting quantum point contacts. *Phys. Rev. B*, **54**, 7366–7379, (1996).

[34] Y. Nambu. Quasi-particles and gauge invariance in the theory of superconductivity. *Phys. Rev.*, **117**, 648–663, (1960).

[35] L. V. Keldysh. Diagram technique for nonequilibrium processes. *Sov. Phys. JETP*, **20**, 1018, (1965).

[36] A. Martín-Rodero, F. J. García-Vidal, and A. Levy Yeyati. Microscopic theory of Josephson mesoscopic constrictions. *Phys. Rev. Lett.*, **72**, 554–557, (1994).

[37] C. W. J. Beenakker and H. van Houten. Quantum transport in semiconductor nanostructures. *Solid State Physics*, **44**, 1–228, (1991).

[38] C. W. J. Beenakker. Three 'universal' mesoscopic Josephson effects. In H. Fukuyama and T. Ando, editors, *Proceedings of the 14th Taniguchi International Symposium on Transport Phenomena in Mesoscopic Systems*, pages 235–253, Berlin, (1992). Springer-Verlag.

[39] N. van der Post, E. T. Peters, I. K. Yanson, and J. M. van Ruitenbeek. Subgap structure as function of the barrier in atom-size superconducting tunnel junctions. *Phys. Rev. Lett.*, **73**, 2611–2613, (1994).

[40] D. Averin and A. Bardas. ac Josephson effect in a single quantum channel. *Phys. Rev. Lett.*, **75**, 1831–1834, (1995).

[41] J. C. Cuevas, A. Levy Yeyati, A. Martín-Rodero, G. Rubio Bollinger, C. Untiedt, and N. Agraït. Evolution of conducting channels in metallic atomic contacts under elastic deformation. *Phys. Rev. Lett.*, **81**, 2990–2993, (1998).

[42] B. Ludoph, N. van der Post, E. N. Bratus', E. V. Bezuglyi, V. S. Shumeiko, G. Wendin, and J. M. van Ruitenbeek. Multiple Andreev reflection in single atom niobium junctions. *Phys. Rev. B*, **61**, 8561–8569, (2000).

[43] M. Häfner, P. Konrad, F. Pauly, J.C. Cuevas, and E. Scheer. Conduction channels of one-atom zinc contacts. cond-mat/0407207.

[44] Ya. M. Blanter and M. Büttiker. Shot noise in mesoscopic conductors. *Phys. Rep.*, **336**, 2–166, (2000).

[45] J. C. Cuevas, A. Martín-Rodero, and A. Levy Yeyati. Shot noise and coherent multiple charge transfer in superconducting quantum point contacts. *Phys. Rev. Lett.*, **82**, 4086–4089, (1999).

[46] A. Martín-Rodero, J. C. Cuevas, A. Levy Yeyati, R. Cron, M. F. Goffman, D. Esteve, and C. Urbina. Quantum noise and multiple Andreev reflections in superconducting contacts. In Y. V. Nazarov, editor, *Quantum Noise in Mesoscopic Physics*, pages 51–71. Kluwer Academic Publishers, Dordrecht, (2003).

[47] M. Tinkham. *Introduction to superconductivity*. McGraw-Hill, New-York, (1996).

[48] R. Duprat and A. Levy Yeyati. Phase diffusion and fractional shapiro steps in superconducting quantum point contacts. *Phys. Rev. B*, **71**, 054510, (2005).

[49] F. Schwabl. *Statistical mechanics*. Springer, Berlin, (2002).

[50] A. O. Caldeira and A. J. Leggett. Quantum tunnelling in a dissipative system. *Annals of Physics*, **149**, 374–456, (1983).

[51] D. S. Golubev and A. D. Zaikin. Coulomb interaction and quantum transport through a coherent scatterer. *Phys. Rev. Lett.*, **86**, 4887–4890, (2001).

[52] A. Levy Yeyati, A. Martín-Rodero, D. Esteve, and C. Urbina. Direct link between Coulomb blockade and shot noise in a quantum coherent structure. *Phys. Rev. Lett.*, **87**, 46802, (2001).

[53] G.-L. Ingold and Yu. V. Nazarov. Charge tunneling rates in ultrasmall junctions. In H. Grabert and M. H. Devoret, editors, *Single Charge Tunneling*, chapter 2, pages 21–107. Plenum Press, New York, (1992).

[54] M.H. Devoret. (private communications).

[55] A. Aviram and M. A. Ratner. Molecular rectifiers. *Chem. Phys. Lett.*, **29**, 277–283, (1974).

[56] G. Cuniberti, G. Fagas, and K. Richter. *Introducing Molecular Electronics*. Springer, Heidelberg, (2005).

[57] J. Chen, M. A. Reed, C. L. Asplund, A. M. Cassell, M. L. Myrick, A. M. Rawlett, J. M. Tour, and P. G. van Patten. Placement of conjugated oligomers in an alkanethiol matrix by scanned probe microscope lithography. *Appl. Phys. Lett.*, **75**, 624–626, (1999).

[58] T. D. Dunbar, M. T. Cygan, L. A. Bumm, G. S. McCarty, T. P. Burgin, W. A. Reinerth, L. Stone II, J. J. Jackiw, J. M. Tour, P. S. Weiss, and D. L. Allara. Combined scanning tunneling microscopy and infrared spectroscopic characterization of mixed surface assemblies of linear conjugated guest molecules in host alkanethiolate monolayers on gold. *J. Phys. Chem.*, **104**, 4880–4893, (2000).

[59] X. D. Cui, A. Primak, X. Zarate, J. Tomfohr, O. F. Sankey, A. L. Moore, T. A. Moore, D. Gust, G. Harris, and S. M. Lindsay. Reproducible measurement of single-molecule conductivity. *Science*, **294**, 571–574, (2001).

[60] X. D. Cui, A. Primak, X. Zarate, J. Tomfohr, O. F. Sankey, A. L. Moore, T. A. Moore, D. Gust, L. A. Nagahara, and S. M. Lindsay. Changes in the electronic properties of a molecule when it is wired into a circuit. *J. Phys. Chem. B*, **106**, 8609–8614, (2002).

[61] B. Xu and N. J. Tao. Measurement of single-molecule resistance by repeaed formation of molecular junctions. *Science*, **301**, 1221–1223, (2003).

[62] H. Park, A. K. L. Lim, A. P. Alivisatos, J. Park, and P. L. Mc Euen. Fabrication of metallic electrodes with nanometer separation by electromigration. *Appl. Phys. Lett.*, **75**, 301–303, (1999).

[63] W. Liang, M. P. Shores, M. Bockrath, J. R. Long, and H. Park. Kondo resonance in a single-molecule transistor. *Nature*, **417**, 725–729, (2002).

[64] J. Park, A. N. Pasupathy, J. I. Goldsmith, C. Chang, Y. Yaish, J. R. Petta, M. Rinkoski, J. P. Sethna, H. D. Abruña, P. L. McEuen, and D. Ralph. Coulomb blockade and the Kondo effect in single atom transistors. *Nature*, **417**, 722–725, (2002).

[65] A. R. Champagne, A. N. Pasupathy, and D. C. Ralph. Mechanically adjustable and electrically gated single-molecule transistors. *Nano Lett.*, **5**, 305–308, (2005).

[66] S. Kubatkin, A. Danilov, M. Hjort, J. Cornil, J.-L. Brédas, N. Stuhr-Hansen, P. Hedegård, and T. Bjørnholm. Single-electron transistor of a single organic molecule with access to several redox states. *Nature (London)*, **425**, 698–701, (2003).

[67] Y. V. Kervennic, D. Vanmaekelbergh, L. P. Kouwenhoven, and H. S. J. van der Zant. Planar nanocontacts with atomically controlled separation. *Appl. Phys. Lett.*, **83**, 3782–3784, (2003).

[68] Y. V. Kervennic, J. M. Thijssen, D. Vanmaekelbergh, C. A. van Walree, L. W. Jenneskens, and H. S. J. van der Zant. Orbital determined transport and gat-effect in a molecular junction. unpublished.

[69] J. Reichert, R. Ochs, D. Beckmann, H. B. Weber, M. Mayor, and H. von Löhneysen. Driving current through single organic molecules. *Phys. Rev. Lett.*, **88**, 176804, (2002).

[70] J. Reichert, H. B. Weber, M. Mayor, and H. von Löhneysen. Low-temperature conductance measurements on single molecules. *Appl. Phys. Lett.*, **82**, 4137–4139, (2003).

[71] M. A. Reed, C. Zhou, C. J. Muller, T. P. Burgin, and J. M. Tour. Conductance of a molecular junction. *Science*, **278**, 252–254, (1997).

[72] E. G. Emberly and G. Kirczenow. Comment on "first-principles calculation of transport properties of a molecular device". *Phys. Rev. Lett.*, **87**, 269701, (2001).

[73] C. Kergueris, J.-P. Bourgoin, S. Palacin, D. Esteve, C. Urbina, M. Magoga, and C. Joachim. Electron transport through a metal-molecule-metal junction. *Phys. Rev. B*, **59**, 12505–12513, (1999).

[74] D. Dulić, S. J. van der Molen, T. Kudernac, H. T. Jonkman, J. J. D. de Jong, T. N. Bowden, J. van Esch, B. L. Feringa, and B. J. van Wees. One-way optoelectronic switching of photochromic molecules on gold. *Phys. Rev. Lett.*, **91**, 207402, (2003).

[75] R. H. M. Smit, Y. Noat, C. Untiedt, N. D. Lang, M. C. van Hemert, and J. M. van Ruitenbeek. Measurement of the conductance of a hydrogen molecule. *Nature*, **419**, 906–909, (2002).

[76] D. Djukic, K. S. Thygesen, C. Untiedt, R. H. M. Smit, K. W. Jacobsen, and J. M. van Ruitenbeek. Stretching dependence of the vibration modes of a single-molecule Pt-H_2-Pt bridge. Preprint, cond-mat/0409640.

[77] K. S. Thygesen and K. W. Jacobsen. Conduction mechanism in a molecular hydrogen contact. *Phys. Rev. Lett.*, **94**, 036807, (2005).

[78] Sz. Csonka, A. Halbritter, G. Mihály, O. I. Shklyarevskii, S. Speller, and H. van Kempen. Conductance of Pd-H nanojunctions. *Phys. Rev. Lett.*, **93**, 016802, (2004).

[79] Y. García, J. J. Palacios, E. SanFabián, J. A. Vergés, A. J. Pérez-Jiménez, and E. Louis. Electronic transport and vibrational modes in a small molecular bridge: H_2 in Pt nanocontacts. *Phys. Rev. B*, **69**, 041402, (2004).

[80] J. C. Cuevas, J. Heurich, F. Pauly, W. Wenzel, and G. Schön. Theoretical description of the electrical conduction in atomic and molecular junctions. *Nanotechnology*, **14**, R29–R38, (2003).

[81] C. Untiedt, D. M. T. Dekker, D. Djukic, and J. M. van Ruitenbeek. Absence of magnetically induced fractional quantization in atomic contacts. *Phys. Rev. B*, **69**, 081401(R), (2004).

[82] D. Djukic and J. M. van Ruitenbeek. unpublished.

Course 9

SOLID STATE QUANTUM BIT CIRCUITS

Daniel Estève and Denis Vion

Quantronics, SPEC, CEA-Saclay,
91191 Gif sur Yvette, France

H. Bouchiat, Y. Gefen, S. Guéron, G. Montambaux and J. Dalibard, eds.
Les Houches, Session LXXXI, 2004
Nanophysics: Coherence and Transport

Contents

1. Why solid state quantum bits?

Solid state quantum bit circuits are a new type of electronic circuit that aim to implement the building blocks of quantum computing processors, namely the quantum bits or qubits. Quantum computing [1] is a breakthrough in the field of information processing because quantum algorithms could solve some mathematical tasks presently considered as intractable, such as the factorisation of large numbers, exponentially faster than classical algorithms operated on sequential Von Neumann computers. Among the various implementations envisioned, solid state circuits have attracted a large interest because they are considered as more versatile and more easily scalable than qubits based on atoms or ions, despite worse quantumness. The 2003 Les Houches School devoted to *Quantum Coherence and Information Processing* [2] has covered many aspects of quantum computing [3], including solid state qubits [4–7]. Superconducting circuits were in particular thoroughly discussed. Our aim is to provide in this course a rational presentation of all solid state qubits. The course is organised as follows: we first introduce the basic concepts underlying quantum bit circuits. We classify the solid state systems considered for implementing quantum bits, starting with semiconductor circuits, in which a qubit is encoded in the quantum state of a single particle. We then discuss superconducting circuits, in which a qubit is encoded in the quantum state of the whole circuit. We detail the case of the quantronium circuit that exemplifies the quest for quantum coherence.

1.1. From quantum mechanics to quantum machines

Quantum Computing opened a new field in quantum mechanics, that of quantum machines, and a little bit of history is useful at this point. In his seminal work, Max Planck proved that the quantisation of energy exchanges between matter and the radiation field yields a black-body radiation law free from the divergence previously found in classical treatments, and in good agreement with experiment. This success led to a complete revision of the concepts of physics. It took nevertheless about fifty years to tie together the new rules of physics in what is now called quantum mechanics. The most widely accepted interpretation of quantum mechanics was elaborated by a group physicists around Niels Bohr

541

in Copenhagen. Whereas classical physics is based on Newtonian mechanics for the dynamics of any system, and on fields, such as the electromagnetic field described by Maxwell's equations, quantum mechanics is based on the evolution of a system inside a Hilbert space associated to all its physically possible states. For example, localised states at all points in a box form a natural basis for the Hilbert space of a particle confined in this box. Any superposition of the basis states is a possible physical state. The evolution inside this Hilbert space follows a unitary operator determined by the Hamiltonian of the system. Finally, when a measurement is performed on a system, an eigenvalue of the measured variable (operator) is found, and the state is projected on the corresponding eigenspace. Although these concepts seem at odds with physical laws at our scale, the quantum rules do lead to the classical behavior for a system coupled to a sufficiently complex environment. More precisely, the theory of decoherence in quantum mechanics predicts that the entanglement between the system and its environment suppresses coherence between system states (interferences are no longer possible), and yields probabilities for the states that can result from the evolution. Classical physics does not derive from quantum mechanics in the sense that the state emerging from the evolution of the system coupled to its environment is predicted only statistically. As a result, quantum physics has been mainly considered as relevant for the description of the microscopic world, although no distinction exists in principle between various kinds of degrees of freedom: their underlying complexity does not come into play within the standard framework of quantum mechanics.

This blindness explains the fifty years delay between the establishment of quantum mechanics, and the first proposals of quantum machines in the nineteen-eighties. On the experimental side, the investigation of quantum effects in electronic circuits carried out during the last thirty years paved the way to this conceptual revolution. The question of the quantumness of a collective variable involving a large number of microscopic particles, such as the current in a superconducting circuit, was raised. The quantitative observation of quantum effects such as macroscopic quantum tunneling [8] contributed to establishing the confidence that quantum mechanics can be brought in the realm of macroscopic objects.

Before embarking on the description of qubits, it is worth noticing that quantum machines offer a new direction to probe quantum mechanics. Recently, the emphasis has been put on the entanglement degree rather than on the mere size of a quantum system. Probing entanglement between states of macroscopic circuits, or reaching quantum states with a high degree of entanglement are now major issues in quantum physics. This is the new border, whose exploration started by the demonstration of the violation of Bell's inequalities for entangled pairs of photons [9]. This research direction, confined for a long time in Byzantine discussions about the EPR and Schrödinger cat paradoxes, is now accesible to experimental tests [10].

First proposals of quantum machines   Commonly accepted quantum machines such as the laser only involve quantum mechanics at the microscopic level, atom-field interactions in this case. A true quantum machine is a system in which machine-state variables are ruled by quantum mechanics. One might think that quantum machines more complex than molecules could not exist because the interactions between any complex system and the numerous degrees of freedom of its environment tend to drive it into the classical regime. Proposing machines that could benefit from the quantum rules was thus a bold idea. Such propositions appeared in the domain of processors after Deutsch and Josza showed that the concept of algorithmic complexity is hardware dependent. More precisely, it was proved that a simple set of unitary operations on an ensemble of coupled two level systems, called qubits, is sufficient to perform some specific computing tasks in a smaller number of algorithmic steps than with a classical processor [1].

Although the first problem solved "more efficiently" by a quantum algorithm was not of great interest, it initiated great discoveries. Important results were obtained [1], culminating with the factorisation algorithm discovered by Shor in 1994, and with the quantum error correction codes [1] developed by Shor, Steane, Gottesman and others around 1996. These breakthroughs should not hide the fact that the number of quantum algorithms is rather small. But since many problems in the same complexity class can be equivalent, solving one of them can provide a solution to a whole class of problems. Pessimists see in this lack of algorithms a major objection to quantum computing. Optimists point out that simply to simulate quantum systems, it is already worthwhile to develop quantum processors, since this task is notoriously difficult for usual computers. A more balanced opinion might be that more theoretical breakthroughs are still needed before quantum algorithms are really worth the effort of making quantum processors. How large does a quantum processor need to be to perform a useful computation? It is considered that a few tens of robust qubits would already be sufficient for performing interesting computations. Notice that the size of the Hilbert space of such a processor is already extremely large.

1.2. Quantum processors based on qubits

A sketch of a quantum processor based on quantum bits is shown in Fig. 1. It consists of an array of these qubits, which are two level systems. Each qubit is controlled independently, so that any unitary operation can be applied to it. Qubits are coupled in a controlled way so that all the two qubit gate operations required by algorithms can be performed. As in Boolean logic, a small set of

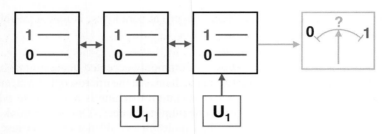

Fig. 1. Sketch of a quantum processor based on qubits. Each qubit is here a robust qubit, with its error correction circuitry. The detailed architecture of a quantum processor strongly depends on the set of gates that can be implemented. The single two qubit gates, combined with single qubit gates, should form a universal set of gates, able to process any quantum algorithm.

gates is sufficient to form a universal set of operations, and hence to operate a quantum processor. A two-qubit gate is universal when, combined with a subset of single qubit gates, it allows implementation of any unitary evolution [1]. For instance, the control-not gate (C-NOT), which applies a not operation on qubit 2 when qubit1 is in state 1, is universal.

Criteria required for qubits   Not all two level systems are suitable for implementing qubits. A series of points, summarised by DiVicenzo, need to be addressed (see chapter 7 in [1]):

1) The level spectrum should be sufficiently anharmonic to provide a good two level system.

2) An operation corresponding to a 'reset' is needed.

3) The quantum coherence time must be sufficient for the implementation of quantum error correction codes.

4) The qubits must be of a scalable design with a universal set of gates.

5) A high fidelity readout method is needed.

These points deserve further comments:

The requirement on the coherence is measured by the number of gate operations that can be performed with an error small enough so that error correcting codes can be used. This requirement is extremely demanding: less than one error in 10^4 gate operations. Qubits rather better protected from decoherence than those available today will be needed for this purpose.

If a readout step is performed while running the algorithm, a perfect readout system should provide answers with the correct probabilities, and project the register on the state corresponding to the outcome read.The state can then be stored for other purpose. This is the definition of a quantum non demolition

(QND) measurement. Such a QND readout would be useful to measure quantum correlations in coupled qubit circuits and to probe whether or not Bell's inequalities are violated as predicted by quantum mechanics like in the microscopic world [9]. However, non QND readout systems could provide answers with the correct probabilities, but fail to achieve the projection afterwards. Note that QND readout is not essential for quantum algorithms; although the factorization algorithm is often presented with intermediate projection step, it is not necessary.

1.3. Atom and ion versus solid state qubits

On the experimental side, implementing quantum processors is a formidable task, and no realistic scalable design presently exists. The activity has been focused on the operation of simple systems, with at most a few qubits. Two main roads have been followed. First, microscopic quantum systems like atoms [10] and ions [11] have been considered. Their main advantage is their excellent quantumness, but their scalability is questionable. The most advanced qubit implementation is based on ions in linear traps, coupled to their longitudinal motion [11] and addressed optically. Although the trend is to develop atom-chips, these implementations based on microscopic quantum objects still lack the flexibility of microfabricated electronic circuits, which constitute the second main road investigated. Here, quantumness is limited by the complexity of the circuits that always involve a macroscopic number of atoms and electrons. We describe in the following this quest for quantumness in electronic solid state circuits.

1.4. Electronic qubits

Two main strategies based on quantum states of either single particles or of the whole circuit, have been followed for making solid-state qubits.

In the first strategy, the quantum states are nuclear spin states, single electron spin states, or single electron orbital states. The advantage of using microscopic states is that their quantumness has already been probed and can be good at low temperature. The main drawback is that qubit operations are difficult to perform since single particles are not easily controlled and read out.

The second strategy has been developed in superconducting circuits based on Josephson junctions, which form a kind of artificial atoms. Their Hamiltonian can be tailored almost at will, and a direct electrical readout can be incorporated in the circuit. On the other hand, the quantumness of these artificial atoms does not yet compare to that of natural atoms or of spins.

2. Qubits in semiconductor structures

Microscopic quantum states suitable for making qubits can be found in semiconductor nanostructures, but more exotic possibilities such as Andreev states at a superconducting quantum point contact [12] have also been proposed. Single particle quantum states with the best quantumness have been selected, and a few representative approaches are described below. Two families can be distinguished: the first one being based on quantum states of nuclear spins, or of localised electrons, while the second one is based on propagating electronic states (flying qubits).

2.1. Kane's proposal: nuclear spins of P impurities in silicon

The qubits proposed by Kane are the S=1/2 nuclear spins of P^{31} impurities in silicon [13]. Their quantumness is excellent, and rivals that of atoms in vacuum. In the ref. [13], the author has proposed a scheme to control, couple and readout such spins. A huge effort has been started in Australia in order to implement this proposal sketched in Fig. 2. The qubits are controlled through the hyperfine interaction between the nucleus of the P^{31} impurity and the bound electron around it. The effective Hamiltonian of two neighboring nucleus bound electron spins:

$$H = A_1 \sigma^{1n} \sigma^{1e} + A_2 \sigma^{1n} \sigma^{1e} + J \sigma^{1e} \sigma^{2e},$$

where the subscripts n and e refer to nuclei and bound electrons respectively. The transition frequency of each qubit is determined by the magnetic field applied to it, and by its hyperfine coupling A controlled by the gate voltage applied to the A gate electrode, which displaces the wavefunction of the bound electron. Single qubit gates would be performed by using resonant pulses, like in NMR, while two qubit gates would be performed using the J gates, which tune the exchange interaction between neighboring bound electrons. The readout would be performed by transfering the information on the qubit state to the charge of a quantum dot, which would then be read using an rf-SET. Although the feasibility of Kane's proposal has not yet been demonstrated, it has already yielded significant progress in high accuracy positioning of a single impurity atom inside a nanostructure.

2.2. Electron spins in quantum dots

Using electron spins for the qubits is attractive because the spin is weakly coupled to the other degrees of freedom of the circuit, and because the spin state

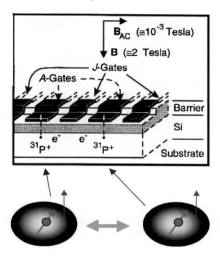

Fig. 2. Kane's proposal: nuclear spins of phosphorus impurities form the qubits. The control is provided by the hyperfine interaction with a bound electron around each impurity. Each qubit level scheme is controlled by applying a voltage to an A gate (labelled A) electrode that displaces slightly the wavefunction of the bound electron, and thus modifies the hyperfine interaction. Single qubit gates are performed by applying an ac field on resonance, like in Nuclear Magnetic Resonance. The two qubit gates are performed using the J gates (labelled J), which control the exchange interaction between neighboring bound electrons. The exchange interaction mediates an effective interaction between the qubits. The readout is performed by transfering the information on the qubit state to the charge of a quantum dot (not shown), which is then read using an rf-Single Electron Transistor *(picture taken from [13].)*

can be transferred to a charge state for the purpose of readout (see [14] and refs. therein). Single qubit operations can be performed by applying resonant magnetic fields (ESR), and two qubit gates can be obtained by controlling the exchange interaction between two neighboring electrons in a nanostructure. The device shown in Fig. 3 is a double dot in which the exchange interaction between the single electrons in the dots is controlled by the central gate voltage. The readout is performed by monitoring the charge of the dot with a quantum point contact transistor close to it. The measurement proceeds as follows: first, the dot gate voltage is changed so that an up spin electron stays in the dot, while a down spin electron leaves it. In that case, another up spin electron from the reservoir can enter the dot. The detection of changes in the dot charge thus provides a measurement of the qubit state. Note that such a measurement can have a good fidelity as required, but is not QND because the quantum state is destroyed afterwards.

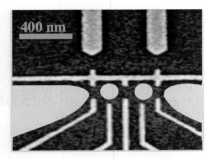

Fig. 3. Scanning Electron Micrograph of a double dot implementing two qubits. The qubits are based on the spin of a single electron in the ground state of each dot (disks). The two qubits are coupled through the exchange energy between electrons, which is controlled by the central gate. Single qubit gates are obtained by applying local resonant ac magnetic fields. Readout is performed by monitoring each dot charge with a point contact transistor, after a sudden change of the dot gate voltage. An electron with the up spin state stays in the dot, whereas a down spin exits, and is replaced by an up spin electron. A change in the dot charge thus signals a down spin. *(Courtesy of Lieven Vandersypen, T.U. Delft).*

2.3. Charge states in quantum dots

The occupation of a quantum dot by a single electron is not expected to provide an excellent qubit because the electron strongly interacts with the electric field. Coherent oscillations in a semiconductor qubit circuit [15] were nevertheless observed by measuring the transport current in a double dot charge qubit repetitively excited by dc pulses, as shown in Fig. 4.

2.4. Flying qubits

Propagating electron states provide an interesting alternative to localised states. Propagating states in wires with a small number of conduction channels have been considered, but edge states in Quantum hall Effect structures offer a better solution [4] . Due to the absence of back-scattering, the phase coherence time at low temperature is indeed long: electrons propagate coherently over distances longer than 100 μm. Qubit states are encoded using electrons propagating in opposite directions, along the opposite sides of the wires. The qubit initialisation can be performed by injecting a single electron in an edge state. As shown in Fig. 5, single qubit gates can be obtained with a quantum point contact that transmits or reflects incoming electrons, and two qubit gates can be obtained by coupling edge states over a short length. The readout can be performed by detecting the passage of the electrons along the wire, using a corrugated edge in order to increase the readout time. This system is not easily scalable because of

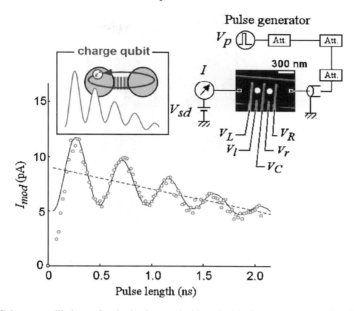

Fig. 4. Coherent oscillations of a single electron inside a double dot structure, as a function of the duration of a dc pulse applied to the transport voltage. These oscillations are revealed by the average current when the pulse is repeated at a large rate *(picture taken from Hayashi et al. [15])*

its topology, but is well suited for entangling pairs of electrons and measuring their correlations.

3. Superconducting qubit circuits

The interest of using the quantum states of a whole circuit for implementing qubits is to benefit from the wide range of Hamiltonians that can be obtained when inductors and capacitors are combined with Josephson junctions. These junctions are necessary because a circuit built solely from inductors and capacitors constitutes a set of harmonic modes. A Josephson junction [16] has a Hamiltonian which is not quadratic in the electromagnetic variables, and hence allows to obtain an anharmonic energy spectrum suitable for a qubit. Josephson qubits can be considered as artificial macroscopic atoms, whose properties can be tailored. The internal and coupling Hamiltonians can be controlled by applying electric or magnetic fields, and bias currents. The qubit readout can also be performed electrically.

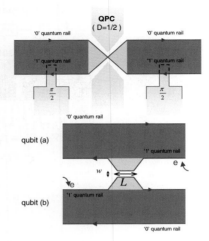

Fig. 5. Single qubit gate (top) and two qubit gate (bottom) for flying qubits based on edge states in QHE nanostructures *(Courtesy of C. Glattli).*

3.1. Josephson qubits

A direct derivation of the Hamiltonian can often be performed for simple circuits. There are however systematic rules to derive the Hamiltonian of a Josephson circuit [17, 18], and different forms are possible depending on the choice of variables. When branch variables are chosen, the contribution to the Hamiltonian of a Josephson junction in a given branch is:

$$h(\theta) = -E_J \cos(\theta),$$

where $E_J = I_0 \phi_0$ is the Josephson energy, with I_0 the critical current of the junction, and θ the superconducting phase difference between the two nodes connected by the branch. The phase θ is the conjugate of the number N of Cooper pairs passed across the junction. In each quantum state of the circuit, each junction is characterised by the fluctuations of θ and of N. Often, the circuit junctions are either in the phase or number regimes, characterised by small and large fluctuations of the phase, respectively. Qubit circuits can be classified according to the regime to which they belong.

3.1.1. Hamiltonian of Josephson qubit circuits

In the case of a single junction, the electromagnetic Hamiltonian of the circuit in which the junction is embedded adds to the junction Hamiltonian. The phase

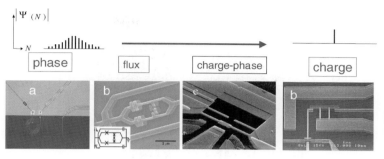

Fig. 6. Josephson qubits can be classified along an axis ranging from the phase regime to the charge regime: the current biased large junction(a), the flux qubit (b), the quantronium charge-phase qubit (c), the Cooper pair box with small Josephson energy (d). In the phase regime, the number of Cooper pairs transferred across each junction has large fluctuations, whereas these fluctuations are small in the charge regime. *(Courtesy of NIST, T.U. Delft, CEA-Saclay, and Chalmers).*

biased junction is in the phase regime, whereas the charge biased junction, a circuit called the Single Cooper Pair Box, can be in a charge regime, phase regime, or intermediate charge-phase regime, depending on the circuit parameters. The Cooper-pair box in the charge regime was the first Josephson qubit in which coherent behavior was demonstrated [19].

In practice, all Josephson qubits are multi junction circuits in order to tailor the Hamiltonian, to perform the readout, and to achieve the longest possible coherence times. The main types of superconducting qubit circuits can be classified along a phase to charge axis, as shown in Fig. 6. The phase qubit [20] developed at NIST (Boulder) consists of a Josephson junction in a flux biased loop, with two potential wells. The qubit states are two quantized levels in the first potential well, and the readout is performed by resonantly inducing the transfer to the second well, using a monitoring SQUID to detect it. The flux qubit [21, 22] developed at T.U. Delft consists of three junctions in a loop, placed in the phase regime. Its Hamiltonian is controlled by the flux threading the loop. The flux qubit can be coupled in different ways to a readout SQUID. The quantronium circuit [7, 23–25], developed at CEA-Saclay is derived from the Cooper pair box, but is operated in the intermediate charge-phase regime. A detailed description of all Josephson qubits, with extensive references to other works, is given in [5–7].

3.1.2. The single Cooper pair box

The single Cooper pair box [7] consists of a single junction connected to a voltage source across a small gate capacitor, as shown in Fig. 7. Its Hamiltonian is the

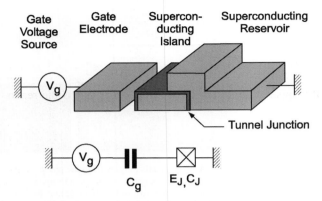

Fig. 7. Schematic representation and electrical circuit of a Single Cooper pair box: A small super-conducting island is connected to a voltage source across a capacitor on one side, and a Josephson junction on the other side. In the schematic circuit, the cross in a box represents a small Josephson junction.

sum of the Josephson Hamiltonian and of an electrostatic term:

$$\widehat{H}(N_g) = E_C(\widehat{N} - N_g)^2 - E_J \cos\widehat{\theta}, \qquad (3.1)$$

where $E_C = (2e)^2/2C_\Sigma$ is the charging energy, and $N_g = C_g V_g/(2e)$ the reduced gate charge with V_g the gate voltage. The operators $\widehat{N}$ and $\widehat{\theta}$ obey the commutation relation $\left[\widehat{\theta}, \widehat{N}\right] = i$. The eigenstates and eigenenergies can be analytically determined, or calculated numerically using a restriction of the Hamiltonian in a subspace spanned by a small set of $|N\rangle$ states. They are $2e$ periodic with the gate charge. The two lowest energy levels provide a quantum bit because the eigenenergy spectrum is anharmonic for a wide range of parameters. When $E_J \ll E_C$, the qubit states are two successive $|N\rangle$ states away from $N_g \equiv 1/2 \bmod[1]$, and symmetric and antisymmetric combinations of successive $|N\rangle$ states in the vicinity of $N_g \equiv 1/2 \bmod[1]$.

3.1.3. Survey of Cooper pair box experiments
The most direct way to probe the Cooper pair box is to measure the island charge. Following this idea, the island charge was measured in its ground state in 1996 [26] by capacitively coupling the box island to an electrometer based on a Single Electron Transistor (SET) [27]. This readout method could not be used however for time resolved experiments because its measuring time was too long. The first Josephson qubit experiment was performed in 1999 at NEC [19], by monitoring the current through an extra junction connected on one side to the box island and on the other side to a voltage source. When the box gate charge

is suddenly (i.e. non adiabatically) moved from $N_g \approx 0$ to $N_g = 1/2$, the initial ground state $|0\rangle$ state is no longer an eigenstate, and coherent oscillations between states take place between $|0\rangle$ and $|1\rangle$ at the qubit transition frequency. When N_g is suddenly moved back to its initial value ≈ 0, the probability for the qubit to be in the excited state $|1\rangle$ is conserved. The readout takes advantage of the available energy in the upper state to transfer a Cooper pair across the readout junction. When the experiment is repeated, the average current through the readout junction provides a measurement of the qubit state at the end of the gate charge pulse. This method of readout provides a continuous average measurement of the box. It proved extremely well suited to many experiments. However, it cannot provide a single shot readout of the qubit. The evolution of qubit design was then driven by the aim of achieving a better quantumness and a more efficient readout. Better quantumness means a longer coherence time, with a controlled influence of the environment to avoid decoherence. More efficient readout means single shot readout, with a fidelity as high as possible, and ideally quantum non demolition (QND). The quantronium operated in 2001 at Saclay was the first qubit circuit combining a single shot readout with a long coherence time [7, 23–25]. In 2003, the charge readout of a Cooper pair box was achieved at Chalmers [29] using an rf-SET [30], which is a SET probed at high frequency. A sample and hold charge readout was operated in 2004 at NEC [31], with a fidelity approaching 90%. In 2004, a Cooper pair box embedded in a resonant microwave cavity was operated at Yale [32] using the modification of the cavity transmission by the Cooper pair box, similar to the effect of a single atom in cavity-QED experiments [10].

3.2. How to maintain quantum coherence?

When the readout circuit measures the qubit, its backaction results in full qubit decoherence during the time needed to get the outcome, and even faster if the readout efficiency is below the quantum limit. The readout should thus be switched off when the qubit is operated, and switched on just at readout time. Furthermore, even when the readout is off, the qubit is subject to decoherence, partly due to the connection of the qubit to the readout circuitry. How could one possibly limit the influence of the environment and of the readout even when it is off, and to switch on the readout when needed?

Before explaining a possible strategy to circumvent this major problem, we expose the basic concepts underlying decoherence in qubit circuits. The interaction between a qubit and the degrees of freedom of its environment entangles both parties. This entanglement takes a simple form in the weak coupling regime, which is usually the case in qubit circuits [28]. The coupling arises from the fact that the control parameters of the qubit Hamiltonian (such as N_g for the Cooper pair box), are in fact fluctuating variables of the qubit environment.

3.2.1. Qubit-environment coupling Hamiltonian

We call λ the set of control variables entering the Hamiltonian of a qubit. At a given working point λ_0, the qubit space is analogous to a fictitious spin $1/2$ with σ_z eigenstates $|0\rangle$ and $|1\rangle$. Using the Pauli matrix representation of spin operators, the expansion of the Hamiltonian around λ_0 yields the coupling Hamiltonian:

$$\widehat{H}_X = -1/2 \left(\overrightarrow{D}_\lambda \cdot \overrightarrow{\sigma} \right) \left(\widehat{\lambda} - \lambda_0 \right), \tag{3.2}$$

where $\overrightarrow{D}_\lambda \cdot \overrightarrow{\sigma}$ is the restriction of $-2\widehat{\partial H/\partial \lambda}$ to the $\{|0\rangle, |1\rangle\}$ space. This coupling Hamiltonian determines the qubit evolution when a control parameter is varied at the qubit transition frequency, and the coupling to decoherence noise sources.

In the weak coupling regime, the fluctuations of the qubit environment are characterised by the spectral density:

$$S_{\lambda_0}(\omega) = \frac{1}{2\pi} \int_{-\infty}^{+\infty} d\tau \left\langle \left(\widehat{\lambda}(t) - \lambda_0 \right) \left(\widehat{\lambda}(t + \tau) - \lambda_0 \right) \right\rangle \exp(-i\omega\tau) \tag{3.3}$$

This spectral density is defined for positive and negative ω's, proportional to the number of environmental modes that can absorb and emit a quantum $\hbar\omega$, respectively. In the case of the Cooper pair box, the fluctuations of the gate charge N_g arise from the impedance of the biasing circuitry and from microscopic charge fluctuators in the vicinity of the box island [7, 25].

3.2.2. Relaxation

The decay of the longitudinal part of the density matrix in the eigenstate basis $\{|0\rangle, |1\rangle\}$ involves $|1\rangle \rightarrow |0\rangle$ qubit transitions, with the energy transferred to the environment. Such an event resets the qubit in its ground state. The decay is exponential, with a rate:

$$\Gamma_1 = \frac{\pi}{2} \left(\frac{D_{\lambda,\perp}}{\hbar} \right)^2 S_{\lambda_0}(\omega_{01}). \tag{3.4}$$

The symbol $\perp$ indicates that only transverse fluctuations at positive frequency ω_{01} induce downward transitions. Upward transitions, which involve $S_{\lambda_0}(-\omega_{01})$, occur at a negligible rate for experiments performed at temperatures $k_B T \ll \hbar\omega_{01}$, provided the environment is at thermal equilibrium. The relaxation time is thus $T_1 = 1/\Gamma_1$.

3.2.3. Decoherence: relaxation + dephasing

When a coherent superposition $a\,|0\rangle + b\,|1\rangle$ is prepared, the amplitudes a and b evolve in time, and the non diagonal part of the density matrix oscillates at the qubit frequency ω_{01}. The precise definition of decoherence is the decay of this part of the density matrix. There are two distinct contributions to this decay. Relaxation contributes to decoherence by an exponential damping factor with a rate $\Gamma_1/2$. Another process, called dephasing, often dominates. When the qubit frequency Ω_{01} fluctuates, an extra phase factor $\exp[i\,\Delta\varphi(t)]$ with $\Delta\varphi(t) = \frac{D_{\lambda,z}}{\hbar} \int_0^t \left(\widehat{\lambda}(t') - \lambda_0\right) dt'$ builds up between both amplitudes, the coupling coefficient $D_{\lambda,z}$ being:

$$D_{\lambda,z} = \langle 0|\,\widehat{\partial H/\partial\lambda}\,|0\rangle - \langle 1|\,\widehat{\partial H/\partial\lambda}\,|1\rangle = \hbar\partial\omega_{01}/\partial\lambda.$$

Dephasing involves longitudinal fluctuations, and contributes to decoherence by the factor:

$$f_X(t) = \langle \exp[i\,\Delta\varphi(t)]\rangle. \tag{3.5}$$

Note that the decay of this dephasing factor $f_X(t)$ is not necessarily exponential. When $D_{\lambda,z} \neq 0$, and assuming a gaussian process for $\left(\widehat{\lambda}(t') - \lambda_0\right)$, one finds using a semi-classical approach:

$$f_X(t) = \exp\left[-\frac{t^2}{2}\left(\frac{D_{\lambda,z}}{\hbar}\right)^2 \int_{-\infty}^{+\infty} d\omega\, S_{\lambda_0}(\omega)\mathrm{sinc}^2(\frac{\omega t}{2})\right], \tag{3.6}$$

A full quantum treatment of the coupling to a bath of harmonic oscillators justifies using the quantum spectral density in the above expression [28]. When the spectral density $S_{\lambda_0}(\omega)$ is regular at $\omega = 0$, and flat at low frequencies, $f_X(t)$ decays exponentially at long times, with a rate $\Gamma_\varphi \approx \pi\left(\frac{D_{\lambda,z}}{\hbar}\right)^2 S_{\lambda_0}(\omega \approx 0)$. When the spectral density diverges at $\omega = 0$, like for the ubiquitous $1/f$ noise, a careful evaluation has to be performed [33, 34].

3.2.4. The optimal working point strategy

The above considerations on decoherence yield the following requirements for the working point of a qubit:

-In order to minimize the relaxation, the coefficients $D_{\lambda,\perp}$ should be small, and ideally $D_{\lambda,\perp} = 0$.

-In order to minimize dephasing, the coefficients $D_{\lambda,z} \propto \partial\Omega_{01}/\partial\lambda$ should be small. The optimal case is when the transition frequency is stationary with respect to all control parameters: $D_{\lambda,z} = 0$. At such optimal points, the qubit is decoupled from its environment and from the readout circuitry in particular. This means that the two qubit states cannot be discriminated at an optimal point. One must therefore depart in some way from the optimal point in order to perform the readout. The first application of the optimal working point strategy was applied to the Cooper pair box, with the quantronium circuit [7, 23, 24].

4. The quantronium circuit

The quantronium circuit is derived from a Cooper pair box. Its Josephson junction is split into two junctions with respective Josephson energies $E_J(1 \pm d)/2$, with $d \in [0, 1]$ a small asymmetry coefficient (see Fig. 8). The reason for splitting the junction into two halves is to form a loop that can be biased by a magnetic flux Φ. The split box, which we first explain, has two degrees of freedom, which can be chosen as the island phase $\widehat{\theta}$ and the phase difference $\widehat{\delta}$ across the two box junctions. In this circuit, the phase $\widehat{\delta}$ is a mere parameter $\delta = \Phi/\phi_0$.

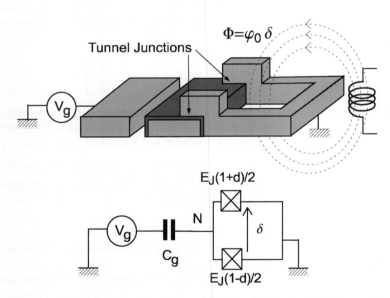

Fig. 8. *Schematic representation of a split Cooper pair box showing its island, its two Josephson junctions connected to form a grounded superconducting loop, its gate circuit, and its magnetic flux bias. Bottom: Corresponding electrical drawing.*

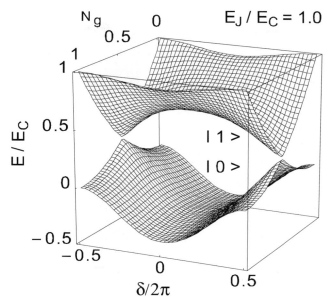

Fig. 9. Two lowest energy levels of a split Cooper pair box having an E_J/E_C ratio equal to 1, as a function of the two external parameters N_g and δ. Energies are normalized by the Cooper pair Coulomb energy. The asymetry coefficient used here is $d = 2\%$. The principal effect of d is to control the gap at $N_g = 1/2$, $\delta = \pi$.

The Hamiltonian of the split box, which depends on the two control parameters N_g and δ, is:

$$\widehat{H} = E_C(\widehat{\mathbf{N}} - N_g)^2 - E_J \cos\left(\frac{\delta}{2}\right)\cos(\widehat{\theta}) + dE_J \sin\left(\frac{\delta}{2}\right)\sin(\widehat{\theta}) . \qquad (4.1)$$

The two lowest energies of this Hamiltonian are shown in Fig. 9 as a function of the control parameters. The interest of the loop is to provide a new variable to probe the qubit: the loop current. The loop current is defined by the operator:

$$\widehat{I}(N_g, \delta) = (-2e)\left(-\frac{1}{\hbar}\frac{\partial \widehat{H}}{\partial \delta}\right).$$

The average loop current $\langle i_k \rangle$ in state $|k\rangle$ obeys a generalized Josephson relation: $\langle i_k(N_g, \delta)\rangle = \langle k |\widehat{I}| k\rangle = \frac{1}{\varphi_0}\partial E_k(N_g, \delta)/\partial\delta$. The difference between the loop currents of the two qubit states is $\Delta i_{10} = \langle i_1\rangle - \langle i_0\rangle = 2e\partial\omega_{10}/\partial\delta$. As expected, the difference Δi_{10} vanishes at an optimal point.

The variations of the qubit transition frequency with the control parameters are shown in Fig.10. Different optimal points where all derivatives $\partial\Omega_{01}/\partial\lambda_i$

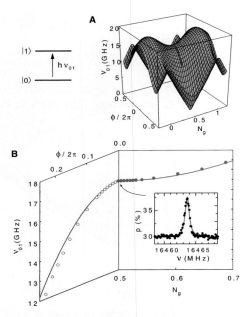

Fig. 10. A: Calculated transition frequency as a function of the control parameters N_g and δ for the parameters $E_J = 0.86\, k_B K$, $E_C = 0.68\, k_B K$. Three optimal points where the frequency is stationary, are visible. The optimal point used in the experiments is the saddle point $(N_g = 1/2, \delta = 0)$. B: cuts along the planes $N_g = 1/2$ and $\delta = 0$. Symbols: position of the resonance of the switching probability in CW excitation; lines: predictions. The lineshape at the optimal point is plotted in inset. *(Taken from Vion et al. [24])*.

vanish are present. The charge difference $\Delta N_{10} = \langle N_1 \rangle - \langle N_0 \rangle$ also vanishes at these points. The optimal point $\{N_g = 1/2, \delta = 0\}$ was first used.

4.1. Relaxation and dephasing in the quantronium

The split box is unavoidably coupled to noise sources affecting the gate charge N_g and the phase δ [7, 25]. The coupling to these noise sources $D_{\lambda,\perp}$ and $D_{\lambda,z}$ for relaxation and dephasing are obtained from the definition 3.2.

The coupling vector $D_{\lambda,\perp}$ for relaxation is:

$$D_{\lambda,\perp} = \left\{ 4E_C \left| \langle 0 | \widehat{N} | 1 \rangle \right|, 2\varphi_0 \left| \langle 0 | \widehat{I} | 1 \rangle \right| \right\}.$$

Relaxation can thus proceed through the charge and phase ports, but one finds that the phase port does not contribute to relaxation at $N_g = 1/2$ when the asymmetry factor d vanishes. Precise balancing of the box junctions is thus important in the quantronium.

The coupling vector for dephasing is directly related to the derivatives of the transition frequency:

$$\lambda_{,z} = \hbar \left(\partial \omega_{01} / \partial N_g, \ \partial \omega_{01} / \partial \delta \right).$$

The charge noise arises from the noise in the gate bias circuit and from the background charge noise due to microscopic fluctuators in the vicinity of the box tunnel junctions. This background charge noise has a $1/f$ spectral density at low frequency, with a rather universal amplitude. The phase noise also has a $1/f$ spectral density, but its origin in Josephson junction circuits is not well understood and is not universal.

4.2. Readout

The full quantronium circuit, shown in the top of Fig. 11, consists of a split-box with an extra larger junction inserted in the loop for the purpose of readout.

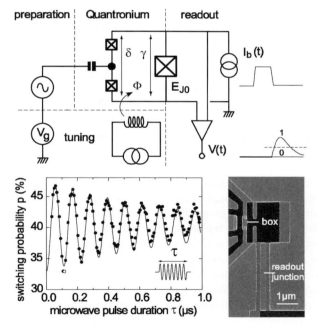

Fig. 11. Top: Schematic circuit of the quantronium qubit circuit. The quantronium consists of a readout junction inserted in the loop of a split-junction Cooper pair box. When a trapezoidal current pulse is applied, the readout junction switches to the voltage state with a larger probability for state $|1\rangle$ than for state $|0\rangle$. Bottom right: Scanning electron picture of a quantronium sample made using double angle shadow-mask evaporation of aluminum. Bottom left: Rabi oscillations of the switching probability as a function of the duration of a resonant microwave pulse.

The Hamiltonian of the whole circuit is the sum of the split-box Hamiltonian4.1 and of the Hamiltonian of a current-biased Josephson junction [7, 25]. The phase difference $\widehat{\delta}$ in the split-box Hamiltonian is related to the phase difference across the readout junction by the relation $\widehat{\delta} = \widehat{\gamma} + \Phi/\phi_0$. The phase $\widehat{\delta}$ is still an almost classical variable, except at readout time, when the qubit gets entangled with the readout junction. This readout junction can be used in different ways in order to discriminate the qubit states, as we now show.

4.2.1. Switching readout

The simplest readout method consists in using the readout junction to perform a measurement of the loop current after adiabatically moving away from the optimal point. For this purpose, a trapezoidal readout pulse with a peak value slightly below the readout junction critical current is applied to the circuit. Since this bias current adds to the loop current in the readout junction, the switching of the readout junction to a finite voltage state can be induced with a large probability for state $|1\rangle$ and with a small probability for state $|0\rangle$. This switching method is in principle a single shot readout. It has been applied to the quantronium [24] and to the flux qubit [22], with switching probability difference up to 40% and 70%, respectively. The lack of fidelity is attributed to spurious relaxation during the readout bias current pulse.This switching method also destroys the qubit after measurement: this is not a QND readout.

4.2.2. AC methods for QND readout

Recently, microwave methods measuring the phase of a microwave signal reflected or transmitted by the circuit have been used with various superconducting qubits in order to attempt a non destructive readout. A QND readout should also lead to a better readout fidelity. Although correlated measurements on coupled qubits and quantum algorithms do not require QND readout, achieving this goal seems essential for probing quantum mechanics in macroscopic objects. In general, with these rf methods, the working point stays, on average, at the optimal point, and undergoes small amplitude oscillations at a frequency different than the qubit frequency. The case of the Cooper pair box embedded in a resonant microwave cavity is an exception because the cavity frequency is comparable with the qubit frequency [32]. Avoiding moving far away from the optimal point might thus reduce the spurious relaxation observed with the switching method, and thus improve the readout fidelity. These rf methods, proposed for the flux qubit [35], the quantronium [36, 37], and the Cooper pair box [38], give access to the second derivative of the energy of each qubit state with trespect to the control parameter that is driven. In the quantronium, this parameter is the phase $\widehat{\delta}$. The qubit slightly modifies the inductance of the whole circuit [37], with opposite changes for the two qubit states. The readout of the inductance change is obtained by measuring the reflected signal at a frequency slightly below the plasma frequency of the readout junction. The discrimination between the two qubit states is furthermore greatly helped there by the non-linear resonance of the junction and the consequent dynamical transition from an in phase oscillation regime, to an out of phase oscillation regime when the drive amplitude is increased [36].

5. Coherent control of the qubit

Coherent control of a qubit is performed by driving the control parameters of the Hamiltonian. This evolution of the qubit state can be either adiabatic, or non adiabatic. A slow change of the control parameters yields an adiabatic evolution of the qubit that can be useful for some particular manipulations. Note however that the adiabatic evolution of the ground state of a quantum system can be used to perform certain quantum computing tasks [39]. Two types of non-adiabatic evolutions have been performed, with dc-pulses and with resonant ac-pulses.

5.1. Ultrafast 'DC' pulses versus resonant microwave pulses

In the dc-pulse method [19], a sudden change of the Hamiltonian is performed. The qubit state does not in principle evolve during the change, but evolves afterwards with the new Hamiltonian. After a controlled duration, a sudden return to the initial working point is performed in order to measure the qubit state. In this method, the qubit manipulation takes place at the qubit frequency, which allows time-domain experiments even when the coherence time is very short. Its drawbacks are its lack of versatility, and the extremely short pulse rise-time necessary to reach the non-adiabatic regime.

In the second method, a control parameter is varied sinusoidally with a frequency matching the resonance requency of the qubit. This method is more versatile and more accurate than the dc pulse method, but is slower. When the gate voltage of a Cooper pair box is modulated by a resonant microwave pulse with amplitude δN_G, the Hamiltonian 3.2 contains a term $h(t) = -2E_C \langle 0 | \widehat{N} | 1 \rangle \sigma_X$, which induces Rabi precession at frequency $\omega_R = 2E_C \, \delta N_G / \hbar \, |\langle 0 | \widehat{N} | 1 \rangle|$, as shown in Fig. 12. The fictitious spin representing the qubit rotates around an axis located in the equatorial plane of the Bloch sphere. The position of this axis is defined by the relative phase χ of the microwave with respect to the microwave carrier that defines the X axis. A single resonant pulse with duration τ induces a rotation by an angle $\omega_R \tau$, which manifests itself by oscillations of the switching probability, as shown in Fig. 11. When the pulse is not resonant, the detuning adds a z component to the rotation vector.

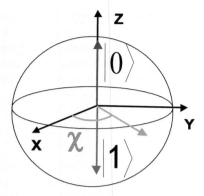

Fig. 12. Bloch sphere in the rotating frame. On resonance, the ac excitation corresponds to a static magnetic field in the equatorial plane for the fictitious spin representing the qubit (thin arrow). The angle between the field and the x axis is the phase χ of the excitation with respect to the reference phase that determines the x axis of the Bloch sphere.

5.2. NMR-like control of a qubit

Rabi precession, which is the basic coherent control operation, has been demonstrated for several Josephson qubits [19,22,24,29]. More complex manipulations inspired from NMR [40,41,43] have also been applied in order to perform arbitrary single qubit gates, and to probe decoherence processes [44,45].

Although it is possible to rotate around an out of plane axis by detuning the microwave, it is more convenient to combine on-resonance pulses. Indeed, three sequential rotations around two orthogonal axes, for instance the x and y axes on the Bloch sphere, should allow to perform any desired rotation. It is thus important to test whether or not two subsequent rotations combine as predicted. The result is shown in Fig. 13. A two pulse sequence was also used to probe rotations around the z axis and was performed using adiabatic pulses applied to the gate charge or to the phase port. Indeed, varying the qubit frequency during a short time results in an extra phase factor between the two components of the qubit, which is equivalent to a rotation around the Z axis by an angle $\varsigma = \int \delta\omega_{01}(t)dt$. As discussed further below, the two pulse sequence also probes decoherence during the free evolution of the qubit between the two pulses.

The issue of gate robustness is also extremely important because the needs of quantum computing are extremely demanding. In NMR, composite pulse methods have been developed in order to make transformations less sensitive to pulse imperfections [41–43]. In these methods, a single pulse is replaced by a series of pulses that yield the same operation, but with a decreased sensitivity to pulse imperfections. In the case of frequency detuning, a particular sequence named

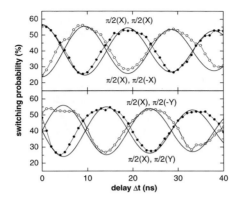

Fig. 13. Switching probability after two $\pi/2$ pulses around x, y, $-x$, or $-y$ axes, as a function of the delay between the pulses. The phase of the oscillating signal at the detuning frequency 50 MHz depends as predicted for the different combinations of rotation axes. The solid lines are theoretical fits *(taken from [45])*.

CORPSE (Compensation for Off-Resonance with a Pulse Sequence) has proved to be extremely efficient [42]. The sensitivity to detuning is indeed strongly reduced, the error starting at fourth order in detuning instead of second order for a single pulse. This sequence has been probed in the case of a π rotation around the X axis. As shown in Fig. 14, it is significantly more robust against detuning than a single π pulse. This robustness was probed starting from state 0, but also from any state with a representative vector in the YZ plane (see inset).

6. Probing qubit coherence

We discuss now decoherence during the free evolution of the qubit, which induces the decay of the qubit density matrix. As explained in section 3.2.3, decoherence is characterised by relaxation, affecting the diagonal and off diagonal parts of the density matrix, and by dephasing, which affects only its off diagonal part.

6.1. Relaxation

Relaxation is readily obtained from the decay of the signal after a π pulse, as shown in Fig. 15. The relaxation time in the quantronium ranges from a few hundreds of nanoseconds up to a few microseconds. These relaxation times are

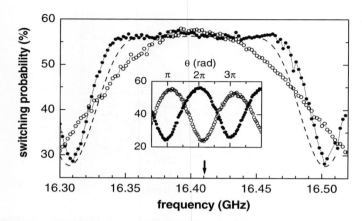

Fig. 14. Demonstration of the robustness of a composite pulse with respect to frequency detuning: switching probability after a $CORPSE\ \pi(X)$ sequence (disks), and after a single $\pi(X)$ pulse (circles). The dashed line is the prediction for the $CORPSE\ \pi(X)$ sequence, the arrow indicates the qubit transition frequency. The $CORPSE$ sequence works over a larger frequency range. The Rabi frequency was 92 MHz. Inset: oscillations of the switching probability after a single pulse $\theta(-X)$ followed (disks) or not (circles) by a $CORPSE\ \pi(X)$ pulse. The patterns are phase shifted by π, which shows that the $CORPSE$ sequence is indeed equivalent to a π pulse (Taken from [45]).

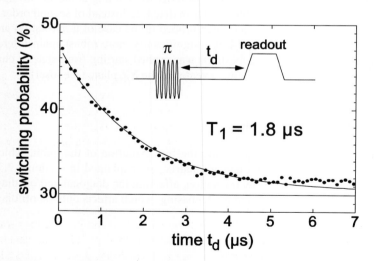

Fig. 15. Decay of the switching probability of the quantronium's readout junction as a function of the delay t_d between a π pulse that prepares state $|1\rangle$ and the readout pulse.(Taken from [24])

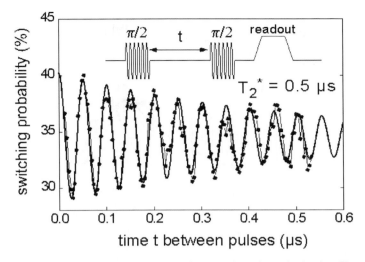

Fig. 16. Ramsey fringe experiment on a quantronium sample at the optimal point. Two $\pi/2$ microwave pulses slightly out of resonance and separated by a time delay t are applied to the gate, The oscillations of the switching probability (dots) at the detuning frequency give direct access to decoherence. In this experiment, their decay time was 500 ns, as estimated by the fit to an exponentially decaying cosine (full line). Coherence times have been measured to be in the range 200 − 500 ns for the quantronium *(Quantronics group)*.

shorter than those calculated from the coupling to the external circuit using an estimated value for the asymmetry factor d. Note however that the electromagnetic properties of the circuit are difficult to evaluate at the qubit transition frequency. Since a similar discrepancy is found in all Josephson qubits, this suggests that qubits with a simple microwave design are preferable, and that microscopic relaxation channels may be present in all these circuits, as suggested by recent experiments on phase qubits [46]. A confirmation of this would imply the necessity of a better junction technology.

6.2. *Decoherence during free evolution*

The most direct way to probe decoherence is to perform a Ramsey fringe experiment, as shown in Fig. 16, using two $\pi/2$ pulses slightly out of resonance. The first pulse creates a superposition of states, with an off diagonal density matrix. After a period of free evolution, during which decoherence takes place, a second pulse transforms the off-diagonal part of the density matrix into a longitudinal component, which is measured by the subsequent readout pulse. The decay of the obtained oscillations at the detuning frequency characterise decoher-

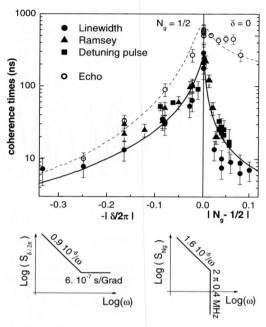

Fig. 17. Coherence times T_2 and T_{Echo} in a quantronium sample extracted from the decay of free evolution signals. The full and dashed lines are calculated using the the spectral densities depicted by the bottom graphs for the phase noise (left) and for the charge noise (right), respectively. *(Quantronics group)*.

ence. This experiment was first performed in atomic physics, and it corresponds to the free induction decay (FID) in NMR. When the decay is not exponential, we define the coherence time as the time corresponding to a decay factor $\exp(-1)$. Other more sophisticated pulse methods have been developed to probe coherence. When the operating point is moved away from the optimal point at which decoherence is weak during a fraction of the delay between the two pulses of a Ramsey sequence, the signal gives access to decoherence at this new working point. The interest of this 'detuning' method is to perform qubit manipulations at the optimal working point without being hindered by strong decoherence. When the coherence time is too short for time domain experiments, the lineshape, which is the Fourier transform of the Ramsey signal, gives access to the coherence time. Coherence times obtained with all these methods on a single sample away from the optimal point in the charge and phase directions are indicated by full symbols in Fig. 17.

It is possible to shed further light on the decoherence processes and to fight them using the echo technique well known in NMR [40]. An echo sequence is a two $\pi/2$ pulse Ramsey sequence with a π pulse in the middle, which causes the phase accumulated during the second half to be subtracted from the phase accumulated during the first half. When the noise-source producing the frequency fluctuation is static on the time scale of the pulse sequence, the echo does not decay. The observed echo decay times, indicated by open disks in Fig. 17, thus set constraints on the spectral density of the noise sources. In particular, these data indicate that the charge noise is significantly smaller than expected from the low frequency $1/f$ spectrum, at least in the two samples in which echo experiments were performed. Bang-bang suppression of dephasing, which generalises the echo technique, could fight decoherence more efficiently [47].

6.3. Decoherence during driven evolution

During driven evolution, the density matrix is best defined using the eigenstate basis in the rotating frame. On resonance, these eigenstates are the states $|X\rangle$ and $|-X\rangle$ on the Bloch sphere. As in the laboratory frame, the decay of the density matrix involves relaxation and dephasing. The measurement of the relaxation time can be performed using the so-called spin locking technique in NMR [40], which allows one to measure the qubit polarisation after the preparation of the state $|X\rangle$. The coherence time during driven evolution is easily obtained from Rabi oscillations. Indeed, the initial state $|0\rangle$ is a coherent superposition of the eigenstates during driven evolution since $|0\rangle = (|X\rangle + |-X\rangle)/\sqrt{2}$. The Rabi signal measured after a pulse of duration t thus probes decoherence during driven evolution. The corresponding coherence time is longer than the coherence time during free evolution because the driving field quenches the effect of the low frequency fluctuations that dominate dephasing during free evolution.

7. Qubit coupling schemes

Single qubit control and readout has been achieved for several Josephson qubits. Although the control accuracy and readout fidelity do not yet meet the requirements for quantum computing, the demonstration on such 'working' qubits of logic gates is now the main goal. Presently, only a few experiments have been performed on coupled qubits. A logic $C - NOT$ gate was operated in 2003 on charge qubits [48], but without a single shot readout. The correlations between coupled phase qubits have been measured recently using a single-shot readout [49]. However, the entanglement between two coupled qubits has not yet been investigated with sufficient accuracy to probe the violation of Bell inequalities predicted by quantum mechanics. Only such an experiment can indeed test if collective degrees of freedom obey quantum mechanics, and whether or not the entanglement decays as predicted from the known decoherence processes.

7.1. *Tunable versus fixed couplings*

In a processor, single qubit operations have to be supplemented with two qubit logic gate operations. During a logic gate operation, the coupling between the two qubits has to be controlled with great accuracy. For most solid state qubits, there is however no simple way to switch on and off the coupling and to control its amplitude. In the cases of the implementations based on P impurities in silicon and on electrons in quantum dots, the exchange energy between two electrons, which can be varied with a gate voltage, provides a tunable coupling. In the case of the superconducting qubits, controllable coupling circuits have been proposed, but fixed coupling Hamiltonians have been mostly considered: capacitive coupling for phase, charge-phase and charge qubits, and inductive coupling for flux qubits. It is nevertheless possible to use a constant coupling Hamiltonian provided that the effective qubit-qubit interaction induced by this coupling Hamiltonian is controlled by other parameters. We now discuss all these coupling schemes.

7.2. *A tunable coupling element for Josephson qubits*

The simplest way to control the coupling between two Josephson qubits is to use a Josephson junction as a tunable inductance. For small phase excursions, a Josephson junction with phase difference δ behaves as an effective inductance $L = \varphi_0/(I_0 \cos \delta)$. Two Josephson junctions in parallel form an effective junction whose inductance can be controlled by the magnetic flux through the loop. When an inductance is placed in a branch shared by two qubit loop circuits, which is possible for phase, charge-phase and flux qubits, the coupling between the two qubits is proportional to the branch inductance. Note that, in this tunable coupling scheme, the qubits have to be moved slightly away from their optimal working point, which deteriorates quantum coherence. The spectroscopy of two flux qubits whose loops share a common junction has been performed [51], and been found to be in close agreement with the predictions. In the case of two charge-phase qubits sharing a junction in a common branch [52], the coupling takes a longitudinal form in the qubit eigenstate basis: $H_{cc} = -\hbar\omega_C \, \hat{\sigma}_{Z1}\hat{\sigma}_{Z2}$, where ω_C is the coupling frequency. The amplitude of the effective z field acting on each fictitious spin is changed proportionally to the z component of the other spin. This coupling allows one to control the phase of each qubit, conditional upon the state of the other one. It thus allows the implementation of the Controlled Phase gate, from which the controlled not *CNOT* gate can be obtained.

7.3. Fixed coupling Hamiltonian

The first demonstration of a logic gate was performed using a fixed Hamiltonian. The system used consisted of two Cooper pair boxes with their islands connected by a capacitance C_C. The coupling Hamiltonian is

$$H_{cc} = -E_{CC}(\widehat{N}_1 - N_{G1})(\widehat{N}_2 - N_{G2}) \tag{7.1}$$

where $E_{CC} = -E_{C1}E_{C2}C_C/(2e)^2$ is the coupling energy, smaller than the charging energy of the Cooper pair boxes. This Hamiltonian corresponds to changing the gate charges by $(E_{CC}/2E_{C1})/(\widehat{N}_2 - N_{G2})$ for qubit 1, and by $(E_{CC}/2E_{C2})/(\widehat{N}_1 - N_{G1})$ for qubit 2. The correlations between the two qubits predicted for this Hamiltonian have been probed, as shown in Fig. 18. A C-NOT logic gate was operated with this circuit [48].

In the uncoupled eigenstate basis, The Hamiltonian (7.1) contains both longitudinal terms of type $\widehat{\sigma}_{Z1}\widehat{\sigma}_{Z2}$ and transverse terms of type $\widehat{\sigma}_{X1}\widehat{\sigma}_{X2}$. At the double optimal point $N_{G1} = N_{G2} = 1/2$, $\delta_1 = \delta_2 = 0$, the Hamiltonian (7.1) is transverse $H_{CC} = \hbar\Omega_C\widehat{\sigma}_{X1}\widehat{\sigma}_{X2}$, with $\Omega_C = E_{CC}/\hbar |\langle 0_1 \widehat{N}_1 1_1\rangle| |\langle 0_2 |\widehat{N}_2| 1_2\rangle|$. When the two qubits have the same resonance frequency ω_{01}, and when $\Omega_C \ll \omega_{01}$, the non-secular terms in H_{CC} that do not commute with the single qubit Hamiltonian are ineffective, and the effective Hamiltonian reduces to:

$$H_{CC}^{\text{sec}} = (\hbar\Omega_C)\,(\widehat{\sigma}_{+1}\widehat{\sigma}_{-2} + \widehat{\sigma}_{-1}\widehat{\sigma}_{+2}). \tag{7.2}$$

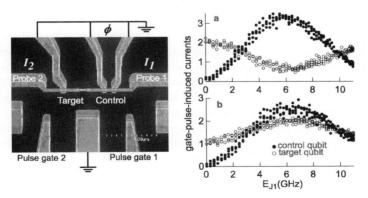

Fig. 18. Demonstration of the correlation between two capacitively coupled charge qubits. Pulse-induced current as a function of the Josephson energy of the control qubit. The control qubit is prepared in a superposition of states that depends on its Josephson energy E_{J1}. A pulse applied on the target qubit yields a π rotation only when the control qubit is in state $|0\rangle$. The currents through the two probe junctions can be anticorrelated (a) or correlated (b) when E_{J1} is varied. *(Courtesy of T. Yamamoto et al. [48], NEC, Japan).*

The evolution of the two qubits corresponds to swapping them periodically. More precisely, a swap operation is obtained at time π/Ω_C. This gate is called *ISWAP* because of extra factors i:

$ISWAP\,|00\rangle = |00\rangle$; $ISWAP\,|10\rangle = -i\,|01\rangle$;

$ISWAP\,|01\rangle = -i\,|10\rangle$; $ISWAP\,|11\rangle = |11\rangle$.

At time $\pi/4\Omega_C$, the evolution operator corresponds to the gate $\sqrt{ISWAP}$, which is universal.

7.4. Control of the interaction mediated by a fixed Hamiltonian

The control of the qubit-qubit interaction mediated by a fixed Hamiltonian depends on the form of this Hamiltonian.

For a coupling of the form 7.2, the effective interaction can be controlled by varying the qubit frequencies since the qubits are affected only when their frequency difference is smaller than Ω_C. This tuning strategy was recently applied to capacitively coupled phase qubits, in which the qubit frequency is directly controlled by the bias current of the junctions [6]. The correlations predicted by quantum mechanics between the readouts of the two qubits were observed [49]. The tuning strategy would be also well suited for coupling many qubits together through an oscillator [28]. The virtual exchange of photons between each qubit and the oscillator indeed yields a coupling of the form 7.2, which is efficient only when the two qubits are tuned. This coupling scheme yields truly scalable designs, whereas most of other schemes are limited to 1D qubit arrays, with nearest neighbor couplings. The coupling between a qubit and a resonator has been already demonstrated for the charge and flux qubits [32, 50].

Another method proposed recently consists in maintaining the qubits out of resonance, but in reaching an equivalent resonance condition in the presence of resonant microwave pulses applied to each one [53]. This method is based on an NMR protocol developed by Hartmann and Hahn in order to place two different spin species 'on speaking terms'. In this scheme, the energy difference between the two qubits is exchanged with the microwave fields.

The case of the longitudinal coupling $H_{CC} \propto -\hbar\omega_C \widehat{\sigma}_{Z1}\widehat{\sigma}_{Z2}$ has not been considered yet. Although the control of this coupling is commonly performed in high resolution NMR, adaptation to qubits has not been attempted.

7.5. Running a simple quantum algorithm

Despite the fact that no quantum processor is yet available, running a simple quantum algorithm in a Josephson qubit circuit is nonetheless presently within reach. Let us consider Grover's search algorithm, which is able to retrieve an object among N in $\sqrt{N}$ algorithmic steps [1]. In the simple case of 4 objects, it requires a single algorithmic step. Let us consider a two-qubit system $\{1, 2\}$

with an *ISWAP* gate. The object to be retrieved is an operator $\widehat{O}$ taken among the four operators $R_{1Z}(\pm\pi/2)R_{2Z}(\pm\pi/2)$, where $R_U(\alpha)$ denotes a rotation around the u axis by an angle α. A simplified version of Grover's algorithm proceeds as follows:

-first, a superposition of all eigenstates is prepared by applying single qubit rotations around the y axis:

$$|\Psi\rangle = 1/2(|00\rangle + |10\rangle + |01\rangle + |11\rangle)$$

We assume that single qubit rotations are fast enough to neglect the effect of the two qubit interaction during their duration.

-A single algorithm step is then applied, with the operator:

$$U = R_{1X}(\pi/2)R_{2X}(\pi/2)\ ISWAP\ \widehat{O}\ ISWAP.$$

-The state $U|\Psi\rangle$ is then read, and the outcome determines which operator had been selected. For instance, the outcome $|11\rangle$ corresponds to the operator $\widehat{O} = R_{1Z}(\pi/2)R_{2Z}(\pi/2)$.

With more qubits, more sophisticated quantum manipulations and algorithms become possible. Note in particular that teleportation is possible with 3 qubits [1].

8. Conclusions and perspectives

Many solid state qubits have been proposed, and several of them have already demonstrated coherent evolution.

For semiconductor qubits, the coherent transfer of an electron between two dots has been demonstrated, and other promising designs are under investigation.

For superconducting qubits, single qubit control, single-shot readout, and a two-qubit logic gate have been achieved. Methods inspired from NMR have been applied to qubit manipulation in order to improve its robustness, and to probe decoherence processes. The lack of an efficient readout scheme and of robust two qubit gates still hinders the development of the field. New QND readout schemes are presently investigated in order to reach a higher readout fidelity. Different two qubit gates have been proposed, but none of them is as robust as the *NAND* gate used in ordinary classical processors. Currently, the coherence time, the readout fidelity, and the gate accuracy are insufficient to envision quantum computing. But how far from this goal are solid state qubits?

In order to use quantum error correcting codes, an error rate of the order of 10^{-4} for each logic gate operation is required. Presently, the gate error rates can be estimated at about a few % for single qubit gates, and at about 20% at best for two qubit gates. The present solid state qubits thus miss the goal by many orders of magnitude. When decoherence and readout errors are taken into account, quantum computing appears even more unrealistic. This is not, however, a reason to give up because conceptual and technical breakthroughs can be expected in this rather new field, and because no fundamental objection has been found. One should not forget that, in physics, everything which is possible is eventually done. Furthermore, quantum circuits provide new research directions in which fundamental questions on quantum mechanics can be addressed. The extension of quantum entanglement out of the microscopic world, and the location and nature of the frontier between quantum and classical worlds, are two of these essential issues. For instance, the accurate measurement of the correlations between two coupled qubits would indeed probe whether or not the collective variables of qubit circuits do follow quantum mechanics.

Our feeling is that, whatever the motivation, complex quantum systems and quantum machines are a fascinating field worth the effort.

Acknowledgements

The qubit research is a collective effort carried out by groups worldwide. We acknowledge discussions with many colleagues from all these groups. We thank in particular the participants to the european project SQUBIT, whose input has been instrumental. The research on the Quantronium has been carried in the Quantronics group at CEA-Saclay. We warmly thank all the group members and all the visitors for maintaining a demanding but friendly research atmosphere. We thank P. Meeson and N. Boulant for their help with the manuscript. We acknowledge the support from the CEA, the CNRS, and of the european project SQUBIT. Last but not least, we thank all those who contribute to make the Les Houches school a so lively place. Teaching there is a unique experience.

References

[1] M.A. Nielsen and I.L. Chuang, "*Quantum Computation and Quantum Information*" (Cambridge University Press, Cambridge, 2000).
[2] *Quantum Coherence and Information Processing* , edited by D. Estève, J.M. Raimond, and J. Dalibard (Elsevier, 2004).
[3] I. Chuang, course 1 in ref. 2.
[4] C. Glattli, course 11 in ref. 2.
[5] M.H. Devoret and J. Martinis, course 12 in ref. 2.
[6] J. Martinis, course 13 in ref. 2.
[7] D. Vion, course 14 in ref. 2.

[8] M.H. Devoret, D.Estève, C. Urbina, J.M. Martinis, A.N. C leland, and J.Clarke, in *"Quantum Tunneling in Condensed Media"*, Kagan N Yu., Leggett A.J., eds. (Elsevier Science Publishers, 1992) pp. 313-345.

[9] A. Aspect, P. Grangier, and Gérard Roger, Phys. Rev. Lett **49**, 91 (1982).

[10] S. Haroche, course 2 in ref. 2; M. Brune, course 3 in ref. 2.

[11] R. Blatt, H. Häffner, C.F. Ross, C. Becher, and F. Schmidt-Kaler, course 5 in ref. 2; D.J. Wineland, course 6 in ref. 2.

[12] A. Zazunov, V. S. Shumeiko, E. N. Bratus', J. Lantz, and G. Wendin, Phys. Rev. Lett. 90, 087003 (2003).

[13] B. E. Kane, Nature.393, 133 (1998).

[14] J. M. Elzerman, R. Hanson, L. H. Willems van Beveren, B. Witkamp, J. S. Greidanus, R. N. Schouten, S. De Franceschi, S. Tarucha, L. M. K. Vandersypen, and L.P. Kouwenhoven, *Quantum Dots: a Doorway to Nanoscale Physics, in Series: Lecture Notes in Physics,* **667**, Heiss, WD. (Ed.), (2005), and refs. therein.

[15] T. Hayashi, T. Fujisawa, H. D. Cheong, Y. H. Jeong, and Y. Hirayama, Phys. Rev. Lett. 91, 226804 (2003).

[16] A. Barone and G. Paternò, *Physics and applications of the Josephson effect* (Wiley, New York, 1982).

[17] M.H. Devoret, in *"Quantum Fluctuations"*, S. Reynaud, E. Giacobino, J. Zinn-Justin, eds. (Elsevier, Amsterdam, 1996), p.351.

[18] Guido Burkard, Roger H. Koch, and David P. DiVincenzo, Phys. Rev. B 69, 064503 (2004).

[19] Y. Nakamura, Yu. A. Pashkin and J. S. Tsai, Nature **398**, 786, (1999).

[20] J. M. Martinis, S. Nam, J. Aumentado, and C. Urbina, Phys. Rev. Lett. 89, 117901 (2002).

[21] J. E. Mooij, T. P. Orlando, L. Levitov, Lin Tian, Caspar H. van der Wal, and Seth Lloyd, Science 285, 1036 (1999).

[22] I. Chiorescu, Y. Nakamura, C. J. P. M. Harmans, and J. E. Mooij, Science 299, 1869 (2003).

[23] A. Cottet, D. Vion, P. Joyez, P. Aassime, D. Estève, and M.H. Devoret, Physica C **367**, 197 (2002).

[24] D. Vion *et al.*, Science **296**, 886 (2002).

[25] A. Cottet, *Implementation of a quantum bit in a superconducting circuit*, PhD thesis, Université Paris VI, (2002); www-drecam.cea.fr/drecam/spec/Pres/Quantro/.

[26] V. Bouchiat, D. Vion, P. Joyez, D. Estève and M.H. Devoret, Physica Scripta, **76**, 165 (1998); V. Bouchiat, PhD thesis, Université Paris VI, (1997), www-drecam.cea.fr/drecam/spec/Pres/Quantro/.

[27] *Single Charge Tunneling*, edited by H. Grabert and M. H. Devoret (Plenum Press, New York, 1992).

[28] Y. Makhlin, G. Schön and A. Shnirman, Rev. Mod. Phy, **73**, 357 (2001).

[29] T. Duty, D. Gunnarsson, K. Bladh, and P. Delsing, Phys. Rev. **B 69**, 140503 (2004).

[30] R.J. Schoelkopf *et al.*, Science, **280**, 1238 (1998).

[31] O. Astafiev, Yu. A. Pashkin, Y. Nakamura, T. Yamamoto, and J. S. Tsai, Phys. Rev. Lett. **93**, 267007 (2004).

[32] A. Wallraff, D. Schuster,.-I.; A. Blais; L. Frunzio; R.-S. Huang,- J. Majer, S. Kumar, S.M.Girvin, R.J. Schoelkopf, Nature **431**, 162 (2004); and p. 591 in ref. 2.

[33] E. Paladino, L. Faoro, G. Falci, and Rosario Fazio, Phys. Rev. Lett. 88, 228304 (2002).

[34] Y. Makhlin and A. Shnirman, Phys. Rev. Lett. 92, 178301 (2004).

[35] A. Lupascu, J.M.Verwijs, R.N. Schouten, C.J.P.M. Harmans, and J.E. Mooij, Phys. Rev. Lett. **93**, 177006 (2004).

[36] I. Siddiqi, R. Vijay, F. Pierre, C. M. Wilson, M. Metcalfe, C. Rigetti, L. Frunzio,R.J. Schoelkopf, M. H. Devoret, D. Vion, and D. Estève, Phys. Rev. Lett. **94**, 027005 (2005).

[37] I. Siddiqi, R. Vijay, F. Pierre, C. M. Wilson, M. Metcalfe, C. Rigetti, L. Frunzio, and M. H. Devoret, Phys. Rev. Lett. **93**, 207002 (2004).

[38] Mika A. Sillanpää, Leif Roschier, and Pertti J. Hakonen, Phys. Rev. Lett. **93**, 066805 (2004).

[39] E. Fahri, J. Goldstone, S. Gutmann, and M. Sipser. Science **292**, 472 (2001).

[40] C.P. Slichter, *Principles of Magnetic Resonance*, Springer-Verlag (3rd ed: 1990).

[41] J. Jones, course 10 in ref. 2.

[42] H.K. Cummins, G. Llewellyn, and J.A. Jones, Phys. Rev. A **67**, 042308 (2003).

[43] L.M.K. Vandersypen and I.L. Chuang, quant-ph/0404064.

[44] D. Vion *et al.*, Fortschritte der Physik, **51**, 462 (2003).

[45] Collin E., Ithier G., Aassime A., Joyez P., Vion D., Estève D. Phys. Rev. Lett. **93**, 157005 (2004).

[46] K. B. Cooper, Matthias Steffen, R. McDermott, R. W. Simmonds, Seongshik Oh, D. A. Hite, D. P. Pappas, and John M. Martinis, Phys. Rev. Lett. **93**, 180401 (2004).

[47] G. Falci, A. D'Arrigo, A. Mastellone, and E. PaladinoPhys. Rev. A **70**, 040101 (2004); H. Gutmann, F.K. Wilhelm, W.M. Kaminsky, and S. Lloyd, Quantum Information Processing, Vol. **3**, 247 (2004).

[48] T. Yamamoto et al., Nature **425**, 941 (2003), and Yu. Pashkin *et al.*, Nature **421**, 823 (2003).

[49] R. McDermott, R. W. Simmonds, Matthias Steffen, K. B. Cooper, K. Cicak, K. D. Osborn, Seongshik Oh, D. P. Pappas, and John M. Martinis, Science **307**, 1299 (2005).

[50] I. Chiorescu, P. Bertet, K. Semba, Y. Nakamura, C. J. P. M. Harmans, and J. E. Mooij, Nature **431**, 159 (2004).

[51] Hans Mooij, private communication.

[52] J. Q. You, Y. Nakamura, and F. Nori, Phys. Rev. B 71, 024532 (2005); J. Lantz, M. Wallquist, V. S. Shumeiko, and G. Wendin, Phys. Rev. B **70**, 140507 (2004).

[53] C. Rigetti and M.H. Devoret, quant-ph/0412009.

ABSTRACTS OF SEMINARS PRESENTED AT
THE SCHOOL

CLASSICAL SPINS MADE BY FERROMAGNETIC π JUNCTIONS

A. Bauer[1], C. Strunk[2], M.L. Della Rocca [2,3], T. Kontos [4]
and M. Aprili [3,4]

[1] *Institut fur Experimentelle und Angewandte Physik, Universitat Regensburg, 93040 Regensburg, Germany.* [2] *Dipartimento di Fisica "E.R. Caianiello", Università degli Studi di Salerno,via S.Allende, 84081 Baronissi (Salerno),Italy.* [3]*Ecole Supérieure de Physique et Chimie Industrielles (ESPCI), 10 rue Vaquelin, 75005 Paris,France.* [4] *CSNSM-CNRS, Bât.108 Université Paris-Sud, 91405 Orsay Cedex, France.*

The exchange field inside a ferromagnetic weak link strongly suppresses the conventional spin singlet pairing known from bulk superconductors. However, in 1982 it was predicted by Buzdin et al. [1] that the pair amplitude F inside a ferromagnet in contact with a superconductor should oscillate as a function of position. The origin of the oscillation is the energy splitting of the spin up and spin down conduction bands in the ferromagnet, in which unconventional Cooper pairs with nonzero total momentum are formed [2]. The oscillation period is determined by the exchange field of the ferromagnet. Thus, by choosing appropriate values of the exchange field and the thickness of the ferromagnetic layer the ratio of the pair amplitude on both sides of a Josephson junction can become negative, and with it the current through the junction [3]. Such junctions are called π-junctions [4]. A π-junction in a superconducting loop behaves as a phase bias generator producing a spontaneous current and hence a magnetic flux [5]. More precisely a phase transition occurs at $2\pi L I_c = \phi_0$, where L is the loop inductance and I_c the critical current. In the limit $2\pi L I_c < \phi_0$ the system gains energy by minimizing its magnetic energy against the junction energy. The system maintains a constant phase everywhere and a shift of $\phi_0/2$ in the current-phase relationship of the junction is expected with no current in the loop. On the other hand, when $2\pi L I_c > \phi_0$ the system's minimum energy corresponds to that of the junction while maximizing its magnetic energy. A phase gradient is maintained by generating a spontaneous superconducting current, which sustains exactly half a flux quantum, $\phi_0/2$ [6], [7]. The ground state is degenerate since the spontaneous supercurrent can circulate clockwise and counterclockwise with exactly the same probability. Applying a small magnetic field can lift the degeneracy and define an easy magnetization direction.The local magnetic flux of a Nb loop interrupted by a 7.5nm thick PdNi thin film is measured by a micro-Hall sensor in the ballistic limit. The device is shown in Fig. 1a [5]. Fig. 1b shows the temperature dependence of the magnetic moment detected in the loops. Below the Nb critical temperature the loop shows a small but clearly visible Hall signal as the temperature of the junction is reduced in zero applied field, which results from the magnetic flux generated by the spontaneous current. This signal is compensated when cooling down in an applied magnetic field corresponding to half a flux quantum in the loop. This data can be taken as a direct evidence for the half integer flux quantization induced by the π−junction.

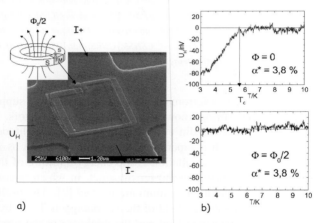

a)   b)

Fig. 1. (a) Scanning electron micrograph of the sample after lift-off. The loop is placed on top of the flux-sensitive area of the micro-Hall cross made of a GaAs/AlGaAs heterostructure.. The samples are niobium (Nb) loops with a planar ferromagnetic palladium-nickel (PdNi) Josephson-junction. (b). Temperature dependence of the spontaneous current. The measurements show the temperature dependent magnetic flux produced by a π-loop when cooling in zero field and in a magnetic field equal to half a flux quantum in the loop [5].

The degeneracy of the ground state is probed using the device shown in fig. 2a. Namely a π-junction (Nb/PdNi/Nb) shorted by a superconducting weak-link [8]. In the junction either a half quantum vortex or a half quantum antivortex are spontaneously generated. The spontaneous supercurrent associated to such vortices produces a phase gradient measured by a Josephson junction [9]. The ferromagnetic junction (source junction) and the detection junction (detector junction) are coupled by sharing one electrode, as schematically reported in Fig. 2(a), i.e. the top electrode of the conventional Josephson junction is simultaneously the bottom electrode of the ferromagnetic one. If half a quantum vortex is spontaneously generated in the ferromagnetic junction, the spontaneous supercurrent sustaining it circulates in the common electrode producing a phase variation equal to $\pi/2$. As a consequence, a $\phi_0/4$−shift of the diffraction pattern of the detection junction is expected. Since the detector has a higher critical temperature, this $\phi_0/4$−shift appears below the critical temperature of the source-junction as shown in Fig. 2 b. When cooling down from room temperature to 2 K, the shift, while reproducible in magnitude, becomes random in sign. This is shown in Fig. 3 c, where the histogram for 26 cooling-down from room temperature on different samples is plotted. The calculated gaussian distribution functions show mean values of $+0.24\phi_0$ and $-0.26\phi_0$ with equal dispersions of $\pm0.03\phi_0$. The same distribution would be expected for a magnetic mono-domain, for instance a small magnetic nano-particle, in the presence of an anisotropy.

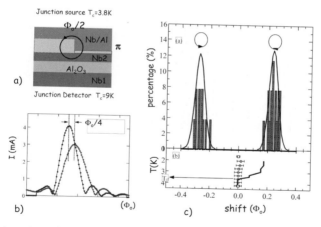

Fig. 2. a) Schematic section of the device: a Nb based Josephson junction (detector) is coupled to a ferromagnetic (PdNi) junction (source) by sharing one electrode (Nb2). The closed loop indicates the half quantum vortex location. b) Detector diffraction patterns for a source junction with π−coupling. Measurements are taken at $T = 4.2$ K and $T = 2$ K. A $\phi_0/4$−shift of the maximum critical current occurs at low temperature. c) Gaussian distributions (black curves) of the spontaneous shift for 26 cool-downs from room temperature on different samples. Inset: Temperature dependence of the spontaneous shift for the sample presented in b). The spontaneous flux appears in the ferromagnetic junction below the superconducting transition temperature of the source junction. The bold markers show that no spontaneous flux is measured in 0-junctions as expected.

A. Bauer et al.

References

[1] A. I. Buzdin, L. N. Bulaevski and S. V. Panyukov, *JETP Letters* **35**, 178 (1982).

[2] T. Kontos, M. Aprili, J. Lesueur and X. Grison, *Phys. Rev. Lett.* **86**, 304(2001).

[3] T. Kontos, et al., *Phys. Rev. Lett.* **89**, 137007(2002).

[4] W. Guichard et al., *Phys. Rev. Lett.* **90**, 167001(2003).

[5] A. Bauer, et al. *Phys. Rev. Lett.* **92**, 217001(2004).

[6] L. N. Bulaevskii, V. V. Kuzii and A. A. Sobyanin,*Solid.State.Com.* **25**, 1053 (1978).

[7] J. R. Kirtley et al., cond-mat/0411380.

[8] M. Aprili et al., Proceeding of the NATO Advanced Research Workshop on "Nanoscale Devices-Fundamentals and Applications NDFA-2004", September 18-22, 2004.

[9] M. L. Della Rocca et al., cond-mat/0501459.

ELECTRONIC TRANSPORT IN CARBON NANOTUBES

R. Egger

Institut für Theoretische Physik, Heinrich-Heine-Universität, D-40225 Düsseldorf, Germany

In this seminar, theoretical issues pertaining to carbon nanotubes, in particular to correlation effects in electronic transport, have been reviewed. The concept of a Luttinger liquid can be applied to metallic single-wall nanotubes, and the field-theoretic justification has been outlined [1]. Experimental evidence based on the power-law form of the tunneling density of states has already been accumulated. Additional evidence has recently been accumulated for transport in crossed nanotube geometries. A zero-bias anomaly in one tube has been observed which is suppressed by a current flowing through the other nanotube. The phenomenon is shown to be consistent with the picture of strongly correlated electrons within the Luttinger liquid model. The most relevant coupling between the nanotubes is the electrostatic interaction generated via crossing-induced backscattering processes. Explicit solution of a simplified model is able to describe qualitatively the observed experimental data with only one adjustable parameter [2]. The consequences of disorder on transport, e.g. in multiwall nanotubes, are discussed in some detail following Ref. [3]. In the last part of the talk, the low-energy theory of superconductivity in carbon nanotube ropes has been discussed. A rope is modelled as an array of metallic nanotubes, taking into account phonon-mediated as well as Coulomb interactions, and arbitrary Cooper pair hopping amplitudes (Josephson couplings) between different tubes. We use a systematic cumulant expansion to construct the Ginzburg-Landau action including quantum fluctuations. The regime of validity is carefully established, and the effect of phase slips is assessed. Quantum phase slips are shown to cause a depression of the critical temperature T_c below the mean-field value, and a temperature-dependent resistance below T_c. The theoretical results can be compared to recent experimental data [4] for the sub-T_c resistance, with good agreement with only one free fit parameter [5]. Ropes of nanotubes therefore represent superconductors in the one-dimensional limit.

References

[1] R. Egger and A.O. Gogolin, Eur. Phys. J. B **3**, 281 (1998).

[2] B. Gao *et al.*, Phys. Rev. Lett. **92**, 216804 (2004).

[3] R. Egger and A.O. Gogolin, Phys. Rev. Lett. **87**, 066401 (2001).

[4] A. Kasumov *et al.*, Phys. Rev. B **68**, 214521 (2003).

[5] A. De Martino and R. Egger, Phys. Rev. B **67**, 235419 (2003); *ibid.* **70**, 014508 (2004).

QUANTUM PHYSICS IN QUANTUM DOTS

Klaus Ensslin

ETH Zurich, 8093 Zurich, Switzerland

Quantum dots are fabricated in AlGaAs/GaAs heterostructures by AFM nano-lithography [1-5]. In the single level regime the broadening of a quantum state originating from the coupling of the dot to source and drain contacts is much less than the single-particle level separation. In this regime spin pairs, i.e. the successive population of one orbital state with spin-up and spin-down electrons can be experimentally detected [6]. It is demonstrated that the orbital states of quantum rings [7] reveal a magnetic field dispersion which is periodic with a flux quantum per ring area. In this regime a detailed understanding of the orbital properties of single-particle states in quantum rings can be obtained. This is extended to spin states, and in particular to an electric field driven singlet-triplet transition [8]. This becomes relevant in the context of using electron spins in quantum dots for quantum information processing. A multi-terminal dot is used to determine the tunnel coupling of the system to each lead [9]. By placing a quantum point contact close to a quantum dot, the charging of the dot can be determined in a time-resolved fashion [10] using the quantum point contact as a detector for single charge read-out. In this way the discrete energy spectrum of the dot allows to measure the distribution function in the source contact coupled to the dot. The coupling of an open phase-coherent ring located close to a quantum dot is measured on the single electron level [11]. Electronic transport through quantum dots is at least partially coherent. We detect the phase evolution of a composite system consisting of a ring with a quantum dot in each of the two arms[12]. As electrons are added to one or both of the dots the phase change of the corresponding Aharonov-Bohm signal is measured.

References

[1]   R. Held, T. Heinzel, P. Studerus, K. Ensslin, and M. Holland, *Fabrication of a semiconductor quantum point contact by lithography with an atomic force microscope*, Appl. Phys. Lett. **71**, 2689 (1997).

[2]   R. Held, T. Vancura, T. Heinzel, K. Ensslin, M. Holland, and W. Wegscheider, *In-plane gates and nanostructures fabricated by direct oxidation of semiconductor heterostructures with an atomic force microscope*, Appl. Phys. Lett. **73**, 262 (1998).

[3]   R. Held, S. Lüscher, T. Heinzel, K. Ensslin, and W. Wegscheider, *Fabricating tunable semiconductor devices with an atomic force microscope*, Appl. Phys. Lett. **75**, 1134 (1999).

[4]   S. Lüscher, A. Fuhrer, R. Held, T. Heinzel, K. Ensslin, and W. Wegscheider, *In-plane gate single electron transistor fabricated by scanning probe lithography*, Appl. Phys. Lett. **75**, 2452 (1999), cond-mat/9909340.

[5]   A. Fuhrer, A. Dorn, S. Lüscher, T. Heinzel, K. Ensslin, W. Wegscheider, and M. Bichler, *Electronic Properties of nanostructures defined in Ga[Al]As heterostructures by local oxidation*, Superl. Microstr. **31**, 19 (2002).

[6]   S. Lüscher, T. Heinzel, K. Ensslin, W. Wegscheider, and M. Bichler, *Signatures of spin pairing in a quantum dot in the Coulomb blockade regime*, Phys. Rev. Lett. **86**, 2118 (2001), cond-mat/0002226.

[7]   A. Fuhrer, S. Lüscher, T. Ihn, T. Heinzel, K. Ensslin, W. Wegscheider, and M. Bichler, *Energy spectra in quantum rings*, Nature **413**, 822 (2001), cond-mat/0109113.

[8]   A. Fuhrer, T. Ihn, K. Ensslin, W. Wegscheider, and M. Bichler, *Singlet-Triplet Transition Tuned by Asymmetric Gate Voltages in a Quantum Ring*, Phys. Rev. Lett. **91**, 206802 (2003), cond-mat/0306667.

[9]   R. Leturcq, D. Graf, T. Ihn, K. Ensslin, D. D. Driscoll, and A. C. Gossard, *Multi-terminal transport through quantum dots*, Europhys. Lett. **67**, 439 (2004), cond-mat/0406046.

[10]   R. Schleser, E. Ruh, T. Ihn, K. Ensslin, D. C. Driscoll and A. C. Gossard, *Time-Resolved Detection of Individual Electrons in a Quantum Dot*, Appl. Phys. Lett. **85**, 2005 (2004), cond-mat/0406568.

[11]   L. Meier, A. Fuhrer, T. Ihn, K. Ensslin, W. Wegscheider, and M. Bichler, *Single-Electron Effects in a Coupled Dot-Ring System*, Phys. Rev. B **69**, 41302 (2004), cond-mat/0406118.

[12]   M. Sigrist, A. Fuhrer, T. Ihn, K. Ensslin, S. E. Ulloa, W. Wegscheider, and M. Bichler, *Magnetic field dependent transmission phase of a double dot system in a quantum ring*, Phys. Rev. Lett. **93**, 66802 (2004), cond-mat/0308223.

Interplay of Coulomb and proximity effects in S-I-N nanostructures

M.V. Feigel'man

L. D. Landau Institute for Theoretical Physics, Moscow 119334, Russia

I discuss a number of effects which are due to the interplay and competition between the superconductive proximity effect and Coulomb blockade in superconductor-insulator-metal hybrid structures. The discussed results are obtained in collaboration with P. M. Ostrovsky and M. A. Skvortsov [2, 2].

First we consider the proximity effect in a normal dot coupled to a bulk superconducting reservoir by the tunnel contact with large normal conductance in the presence of Coulomb interaction [2]. In spite of the large conductance between the dot and the lead, the Coulomb blockade effect is still possible due to the superconducting gap in the reservoir provided that the capacitance of the junction is sufficiently low. The proximity induced minigap is suppressed by the Coulomb interaction. We find exact expressions for the thermodynamic and tunneling minigaps as functions of the capacitance. The tunneling minigap interpolates between its proximity-induced value in the regime of weak Coulomb interaction and the Coulomb gap in the regime of Coulomb blockade. In the intermediate case a non-universal two-step structure of the tunneling density of states is predicted while the temperature dependence of the minigap is nonmotonic and may even be reentrant. The charge quantization in the dot is also studied. In the Coulomb blockade regime the charge of the grain is a step-like function of the gate voltage while in the opposite limit the staircase is exponentially smeared.

Next, we generalize the above results for the case of S-I-N-I-S structure [2]. Namely, we study the system of two superconductors connected by a small normal grain. We consider the modification of the Josephson effect by the Coulomb interaction on the grain. Coherent charge transport through the junction is suppressed by Coulomb repulsion. An optional gate electrode may relax the charge blocking and enhance the current leading to the single Cooper pair transistor effect. The temperature dependences of the critical current and of the minigap

M.V. Feigel'man

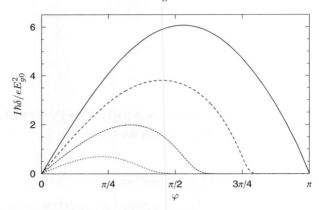

Fig. 3. Current vs. phase for the symmetric junction. The solid curve illustrates the case $E_C = 0$. The other curves correspond to $E_C \delta/E_{g0}^2 = 0.5$, 1.5, 2.5. At large φ charging effects are always strong and the current is exponentially suppressed. We assume $G_L = G_R = 20$ and $\log(\Delta/E_{g0}) = 5$.

induced in the normal grain by the proximity with the superconductor are studied. Both temperature and Coulomb interaction suppress the critical current and the minigap but their interplay may lead to a non-monotonous and even reentrant temperature dependence. The Current versus phase dependence acquires a strongly non-sinusoidal form, with the current vanishing exponentially fast as the phase difference approaches π, see Fig. 3.

References

[1] P. M. Ostrovsky, M. A. Skvortsov and M. V. Feigel'man, Phys. Rev. Lett. **92** (2004), 176805.

[2] P. M. Ostrovsky and M. V. Feigel'man, JETP Letters, **79** (2004) 602.

GEOMETRIC PHASES IN SUPERCONDUCTING NANOCIRCUITS

Rosario Fazio

NEST- INFM & Scuola Normale Superiore, piazza dei Cavalieri 7, I-56126 Pisa, Italy

When a quantum mechanical system undergoes an adiabatic cyclic evolution, it acquires a geometrical phase factor in addition to the dynamical one. This remarkable effect, found by Berry, has been demonstrated in a variety of systems [1]. Since recent years all the attention in detecting geometric phases have been confined to microscopic systems. More recently, however, due the advances in nanofabrication the possibility to observe geometric phases in mesoscopic systems has become an interesting possibility. Quantum dynamics has been already observed in superconducting nanocircuits [2], therefore these systems have been indicated as promising candidates to measure geometric interference.

The simplest configuration in which to observe the Berry phase consists of a superconducting electron box formed by an asymmetric SQUID, pierced by a magnetic flux Φ and with an applied gate voltage V_x [3]. See also the course by D. Estève in this volume. The Hamiltonian is

$$H = E_{\text{ch}}(n - n_x)^2 - E_J(\Phi) \cos(\theta - \alpha) \tag{1}$$

where

$$\tan \alpha = \frac{E_{J_1} - E_{J_2}}{E_{J_1} + E_{J_2}} \tan\left(\pi \frac{\Phi}{\Phi_0}\right),$$

and $E_J(\Phi)$ is the effective Josephson coupling of the loop and $\Phi_0 = h/2e$ is the (superconducting) quantum of flux. The phase difference across the junction θ and the number of Cooper pairs n are canonically conjugate variables $[\theta, n] = i$. Both external parameters of the Hamiltonian can be controlled. The offset charge $2en_x$ can be tuned by changing V_x and the coupling $E_J(\Phi)$ depends on Φ. The device operates in the regime where the Josephson couplings $E_{J_{1(2)}}$ of the junctions are much smaller than the charging energy E_{ch}. At temperatures much lower than E_{ch}, if n_x varies around the value $1/2$, only two charge states, $n = 0, 1$, are important. The effective Hamiltonian is obtained by projecting Eq.(1) on the computational Hilbert space, and reads $H_B = -(1/2)\vec{B} \cdot \vec{\sigma}$, where we have defined the fictitious field $\vec{B} \equiv (E_J(\Phi)\cos\alpha, -E_J(\Phi)\sin\alpha, E_{ch}(1-2n_x))$. Charging couples the system to B_z whereas the Josephson term determines the projection in the xy plane. By changing V_x and Φ the qubit Hamiltonian H_B describes a cylindroid in the parameter space $\{\vec{B}\}$.

By means of a more complicated circuit it is also possible to obtain a degenerate subspace and hence to observe non-Abelian holonomies [4].

Besides the interest in itself, the detection of geometric phases in superconducting nanocircuits may have an impact in the area of solid state quantum computation. Quantum computers are usually analysed in terms of qubits and gates. In most of the implementations proposed quantum gates are realized by varying in time in a controlled way the Hamiltonian of the individual qubits as well as their mutual coupling. An alternative design [5, 6] makes use of quantum geometric phases, obtained by adiabatically varying the qubits' Hamiltonian in such a way to describe a suitably chosen closed loop in its parameter space. It is believed that geoemtric quantum computation may be more robust to certain type of errors. It has already been proposed that geometric computation can be implemented with Josephson nanocircuits.

As a final remark it is worth to mention that the detection of geometric phases is intimately related to coherent pumping of Cooper pairs [7].

Additional informations concerning work done on geometric phases in superconducting nanocircuits can be found in the reference list of the papers quoted in this abstract. An important topic related to the role of the environment on geometric interferometry is discussed by Yu. Makhlin in this Volume.

References

[1] *Geometric phases in physics*, Shapere A. and Wilczek F., Eds. World Scientific, (Singapore, 1989).

[2] Yu. Makhlin, G. Schön, and A. Shnirman, Rev. Mod. Phys. **73**, 357 (2001).

[3] G. Falci, R. Fazio, G.M. Palma, J. Siewert, and V. Vedral, Nature **407**, 355 (2000).

[4] L. Faoro, J. Seiwert, and R. Fazio, Phys. Rev. Lett. **90**, 028301 (2003).

[5] P. Zanardi and M. Rasetti, Phys. Lett. A **264**, 94 (1999).

[6] J. Jones, V. Vedral, A. K. Ekert and G. Castagnoli, Nature **403**, 869 (2000).

[7] J. Pekola *et al.*, Phys. Rev. B **60**, R9931 (1999).

DECOHERENCE IN DISORDERED CONDUCTORS AT LOW TEMPERATURES, THE EFFECT OF SOFT LOCAL EXCITATIONS

Y. Imry

The Weizmann Institute, Rehovot, Israel

The conduction electrons' dephasing rate, τ_φ, is expected to vanish with temperature [2]. However, very intriguing apparent saturation of this dephasing rate in several systems was recently reported [2] at very low temperatures. It was demonstrated, by a direct determination of the electrons' temperature from the interaction conductivity correction, that the electrons were not heated in this experiment. The suggestion that this represents dephasing by zero-point fluctuations has generated both theoretical and experimental controversies.

We start by proving that the dephasing rate must vanish at the $T \to 0$ limit, unless a large ground state degeneracy exists. The thermodynamic proof employs well-known properties of equilibrium correlators. It therefore includes most systems of relevance and it is valid for any determination of τ_φ from *linear* transport measurements, because linear transport is redetermined by *equilibrium* correlators. In fact, further experiments [3] demonstrate unequivocally that indeed when strictly linear transport is used, the apparent low-temperature saturation of τ_φ is eliminated. However, the conditions to be in the linear transport regime were found to be more strict than hitherto expected, based just on the absence of heating of the electrons. We briefly discuss this qualitatively.

Another novel result of the experiments is that introducing heavy nonmagnetic impurities (gold) in the samples produces, even in linear transport, a shoulder in the dephasing rate at very low temperatures [3]. We then show theoretically that low-lying local defects may produce a relatively large dephasing rate at low temperatures. However, as expected, this rate in fact vanishes when $T \to 0$, in agreement with the experimental observations [3]. The same low-lying local defects may also provide the mechanism for dephasing without heating.

591

References

[1] Y. Imry, Z. Ovadyahu and A. Schiller, *Proceedings of the Euresco Conference on Fundamental Problems of Mesoscopic Physics*, Granada, September 2003, Kluwer, cond-mat/0312135.

[2] P. Mohanty, E.M. Jariwala and R.A. Webb, Phys. Rev. Lett. **77**, 3366 (1997).

[3] Z. Ovadyahu, Phys. Rev. B **63**, 235403 (2001).

Proximity Induced and Intrinsic Superconductivity in Long and Short Molecules

A.Yu. Kasumov[1,2,4], V.T. Volkov[2], I.I. Khodos[2],
Yu.A. Kasumov[2], D.V. Klinov[3], M. Kociak[4], R. Deblock[4],
S. Guéron[4], H. Bouchiat[4]

[1]*RIKEN, Hirosawa 2-1, Wako, Saitama 351-0198, Japan.* [2]*Institute of Microelectronics Technology RAS, Chernogolovka, Moscow district, 142432, Russia.* [3]*Shemyakin-Ovchinnikov Institute of Bioorganic Chemistry, RAS, Miklukho-Maklaya 16/10, Moscow 117871, Russia.* [4]*Laboratoire de Physique des Solides UMR 8502 - Universite Paris-Sud, Bat. 510, 91405 Orsay cedex, France.*

Proximity induced superconductivity has emerged as one of the most efficient tools of investigation of phase coherent transport at mesoscopic or nanoscopic scales. Observing proximity effect in long molecular wires implies indeed not only that these molecules are conducting and form a low-resistance contact with the superconducting electrodes but also, which is more fundamental, that both the thermal length and the phase coherence length are of the order of the length of the molecules.

This is especially interesting in the case of DNA molecules where very little is known about the nature of electronic transport. Observation of the proximity effect yields the order of magnitude of typical time- and length- scales involved in the charge transfer mechanism along the molecules. We have performed conductivity measurements on double-stranded DNA molecules deposited by a combing process across a submicron slit between rhenium carbon metallic contacts. Conduction is ohmic between room temperature and 1 K. The resistance per molecule is less than 100 kΩ and varies very slowly with temperature. Below 1K, which is the superconducting transition temperature of the contacts, we observe proximity induced superconductivity. This implies in particular that DNA molecules can be metallic down to mK temperature, and furthermore that phase coherence is achieved over several hundred nanometers [4]. We also emphasized the importance of the interaction of DNA molecules with the underlying substrate. For most commonly used substrates like mica or silicon oxide the interaction between the molecule and the surface is very strong and induces a very large compression deformation of deposited DNAs. The thickness of such compressed DNAs is 2-4 times less than the diameter (about 2nm) of native Watson-Crick B-DNA. We confirm the insulating character of DNA on such substrates. On the other hand when the substrate is treated (functionalized) so that deposited molecules keep their original thickness we observe a conducting behavior, both from conducting AFM and transport measurements on molecules connected to platinum electrodes [3].

In the case of carbon nanotubes on superconducting contacts, the observation of high supercurrents strongly suggests the existence of intrinsic superconducting fluctuations. This is corroborated by experiments on long ropes of carbon nanotubes on normal contacts. Intrinsic superconductivity is then only observed when the distance between the normal electrodes is large enough, otherwise superconductivity is destroyed by the (inverse) proximity effect. These experiments indicate the presence of attractive interactions in carbon nanotubes which overcome the repulsive Coulomb interaction at low temperature and opens the way to the investigation of superconductivity in a 1D limit never explored before. The temperature and bias dependences of the resistance reveal the presence quantum phase slips (QPS). A QPS is a topological vortex-like excitation of the superconducting phase field, which only exists in 1D superconductors. In addition to the T_c depression, QPSs produce a finite sub-T_c resistance in addition to the usual temperature-independent contact resistance. These observations [1], can be compared in a quantitative way to theory [2].

We also report the first study of transport through a nanometer size molecule in contact with superconducting electrodes and direct observation of the molecule by high resolution transmission electron microscopy (HRTEM) (fig. 1a). As a molecule we used a metallofullerene molecule, Gd@C82, which has a diameter of about 1 nm. Metallofullerene molecules have the same mechanical stability as fullerenes, but their doping by a metallic atom (in our case Gd) acting as a donor favors charge transfer through the molecule [5].

We find that a junction containing a single metallofullerene dimer between superconducting electrodes (fig. 1a and 1b) displays signs of proximity-induced superconductivity. In contrast, no superconductivity remains in junctions containing a cluster of dimers. These results can be understood, taking into account multiple Andreev reflections (MAR) [6], and the spin states of Gd atoms [8] (fig. 1 c).

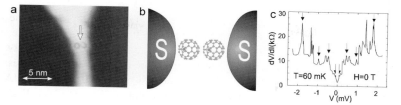

Fig. 4. (a) Actual HRTEM image of Gd@C82 molecular dimer between electrodes. (b) Schematic picture of the molecular dimer between superconducting electrodes. The black dots symbolize the Gd atoms inside the fullerene cage. (c) Voltage dependence of the differential resistance; The arrows indicate MAR differential resistance peaks at voltage $V_n = 2\Delta/ne$ [6]. Additional peaks are due to a Quantum Dot effect [7].

References

[1] A. Yu. Kasumov et al., Phys.Rev. B **68**, 214521 (2003).

[2] A. De Martino & R. Egger, Phys.Rev. B (2004), to be published.

[3] A. Yu. Kasumov et al., Appl.Phys.Lett. **84**, 1007 (2004).

[4] A.Yu. Kasumov et al., Science **291**, 280 (2001).

[5] H. Shinohara, Rep. Prog. Phys. **63**, 843 (2000).

[6] M. Octavio, M. Tinkham, G.E. Blonder and T.M. Klapwijk, Phys. Rev. B **27**, 6739 (1983).

[7] G. Johansson, E. N. Bratus, V. S. Shumeiko and G. Wendin, Phys. Rev. B **60**, 1382 (1999).

[8] A. Yu. Kasumov et al., cond-mat/0402312.

GEOMETRIC PHASES IN DISSIPATIVE QUANTUM

Yuriy Makhlin

Landau Institute for Theoretical Physics, Kosygin st. 2, 119334 Moscow, Russia and Institut für Theoretische Festkörperphysik, Universität Karlsruhe, D-76128 Karlsruhe, Germany

Substantial progress has been achieved recently in the studies of solid-state quantum-coherent systems (see the course by D. Estève at this School); these investigations were additionally motivated by the advancement of quantum information theory. In particular, recent experiments demonstrate a high level of coherence achieved in superconducting qubits. Coherent phenomena are affected by the environment, especially in solid-state systems due to a host of microscopic modes. Here we discuss its influence on the Berry phase (cf. the seminar by R. Fazio). We focus on the simplest case of a two-level system — spin-1/2 in a slowly varying magnetic field and anisotropically coupled to an environment (which thus produces a fluctuating field, the noise).

The dynamics can be described by the Bloch equations [1]. The Bloch-Redfield formalism [2] is suitable for systems subject to weak short-correlated noise [3]. For stationary field the effect of the environment on the coherent dynamics (the Larmor precession) amounts to relaxation of the longitudinal spin component and the decay of the transverse spin component (phase relaxation or dephasing), with rates denoted $1/T_1$ and $1/T_2$, respectively [4]. It is important to note that the noise also modifies the precession frequency (the 'Lamb shift' of the level splitting). If the field is varied, these rates — T_1, T_2, and the Lamb shift — are modified, and one can find the corrections using the Keldysh formalism.

On one hand, the Berry phase acquired after varying the field along a closed loop is modified by the noise. Similarly to the case of an isolated system, the modification is of geometric origin. While the original Berry phase may be interpreted as the flux of a monopole at origin, the modification is given by the flux of a quadrupole-like field [5].

Moreover, the field's variation modifies also the effects of the noise, giving rise to *geometric* relaxation and dephasing. Unlike the phase, these are well-defined also for open paths [5].

For a two-level system the modifications are sensitive only to the (symmetrized) noise correlator and may be obtained by considering a classical stochastic field. Of special interest are the limits of slow and fast fluctuations. This consideration provides simple heuristic arguments to interpret the results [5,6]. We also discuss the conditions on the parameters for observation of the effects discussed.

References

[1] F. Bloch, Phys. Rev. **70**, 460 (1946).

[2] F. Bloch, Phys. Rev. **105**, 1206 (1957); A.G. Redfield, IBM J. Res. Dev. **1**, 19 (1957).

[3] Yu. Makhlin, G. Schön, and A. Shnirman in *New Directions in Mesoscopic Physics*, V.F. Gantmakher, R. Fazio, and Y. Imry, eds. (Kluwer, Dordrecht, 2003); [cond-mat/0309049].

[4] C. Slichter, *Principles of Magnetic Resonance*. Springer, Berlin, 2nd ed., 1978.

[5] R. S. Whitney, Yu. Makhlin, A. Shnirman, and Y. Gefen, Phys. Rev. Lett., to appear; [cond-mat/0405267].

[6] R. S. Whitney, Yu. Makhlin, A. Shnirman, and Y. Gefen, in: *Proceedings of the ARW on Theory of Quantum Transport in Metallic and Hybrid Nanostructures*; [cond-mat/0401376].

INTERACTING ELECTRONS IN METAL NANOSTRUCTURES

D.C. Ralph

Physics Department, Cornell University, Ithaca, NY, USA

In my lectures, I focused primarily on two topics: how interactions affect the "electrons-in-a-box" states in metal nanoparticles, and how a current of spin-polarized electrons interacting with a nanomagnet can apply a torque to the magnet by the direct transfer of spin angular momentum.

The spectrum of discrete electronic states inside a nanoscale metal particle can be measured using electron tunneling if the particle is connected to electrical leads in a single-electron-transistor configuration [1, 2]. Measurements show that the spectra even in simple metals like aluminum, copper, or gold cannot be described by simple models of free, non-interacting electrons. In fact, all of the different types of forces and interactions that affect the electrons inside a metal affect the discrete spectrum in different ways, so that these spectra can serve as a sensitive probe into the nature of the interactions [3]. The situation is closely analogous to the physics of electrons in semiconductor quantum dots, except that in metals a wider variety of different types of interactions can be explored, including those that give rise to correlated-electron states such as superconductivity and ferromagnetism. My lectures surveyed the consequences of a number of different types of interactions. For example, spin-orbit interactions in heavy metals can lead to a reduction of the g-factors for Zeeman spin splitting below the free-electron value of 2, pronounced spatial anisotropies in the value of individual g-factors as a function of the angle of the applied magnetic field, and avoided crossings between predominantly spin-up and spin-down energy levels [4–8]. When considering electron-electron interactions in metallic quantum dots, theory suggests that the form of the interactions takes a particularly simple form, the "Universal Hamiltonian" [9, 10]. Even relatively weak Coulomb repulsion, which produces exchange interactions with strength less than the Stoner criterion for bulk ferromagnetism, can be capable of producing electronic ground states with non-minimal spin values in quantum dots [11–13]. Attractive effective electron-electron interactions in materials such as aluminum can lead to a super-conducting ground state. Despite the fact that it is not possible to measure either a Meissner effect or a zero-resistance state in a nanoparticle, a pair-correlated superconducting state can still be detected through correlation-induced gaps in the electronic spectra that are sensitive to whether the particle contains an even or odd number of electrons and that are suppressed by spin pair breaking in an applied magnetic field [3]. Strong repulsive Coulomb interactions lead to fer-romagnetic electronic states, with rather complicated behavior that is affected by magnetic anisotropy forces as well as the strong electron-electron interac-tion [14, 15]. Some progress in understanding the strongly-correlated electronic states inside a nanoscale ferromagnet has been achieved using both numerical simulations [16] and effective spin Hamiltonians [17–19].

Spin-transfer torques represent a mechanism for manipulating the orientation of the magnetic moments in small magnetic elements that does not involve a mag-netic field. The effect originates from the fact that thin layers of ferromagnets can act as filters for spins, producing partially spin-polarized currents. If a spin-polarized current is incident on a ferromagnetic layer, with a spin-polarization angle that is not collinear with the magnetic moment of the layer, then in the fil-tering process the current can transfer spin-angular momentum to the layer, in this way applying a torque to the layer's magnetic moment [20–22]. This torque has been measured in multilayered ferromagnet/normal metal/ferromagnet samples in which a current flows perpendicular to the layers [23–25]. For samples with a sufficiently narrow diameter, less than about 0.25 microns, the spin-transfer ef-fect provides much stronger torques per unit current than do current-generated magnetic fields. The response of a magnet to the spin-transfer torque can take two forms, depending on the magnitudes of the applied magnetic field and the current. For small fields, the spin-transfer effect can produce reversible switch-ing, in which one sign of current drives two magnetic layers to an antiparallel configuration, and the reversed current can drive them back parallel [25]. This effect is of interest as a means to write information in non-volatile magnetic ran-dom access memory elements, more efficiently than is possible using magnetic-field writing. The second type of magnetic dynamics that can be driven by the spin-transfer effect is steady-state precession – a DC current of spin-polarized electrons can excite a nanomagnet into a state of dynamical equilibrium in which the magnet precesses at GHz-scale frequencies that are tunable using either the current or by changing the value of an external magnetic field. These oscillations have been measured in the frequency domain [26,27] and also directly in the time domain [28], and they are under investigation for applications such as nanoscale oscillators and microwave sources.

References

[1]   D. C. Ralph, C. T. Black and M. Tinkham, Phys. Rev. Lett. **74** (1995) 3241.

[2]   D. C. Ralph, C. T. Black and M. Tinkham, Phys. Rev. Lett. **78** (1997) 4087.

[3]   J. von Delft and D. C. Ralph, Phys. Rep. **345** (2001) 61.

[4]   P. W. Brouwer, X. Waintal, and B. I. Halperin, Phys. Rev. Lett. **85** (2000) 369.

[5]   K. A. Matveev, L. I. Glazman, and A. I. Larkin, Phys. Rev. Lett. **85** (2000) 2789.

[6]   D. A. Gorokhov and P. W. Brouwer, Phys. Rev. B **69** (2004) 155417.

[7]   J. R. Petta and D. C. Ralph, Phys. Rev. Lett. **87** (2001) 266801.

[8]   J. R. Petta and D. C. Ralph, Phys. Rev. Lett. **89** (2002) 156802.

[9]   I. L. Kurland, I. L. Aleiner, and B. L. Altshuler, Phys. Rev. B **62** (2000) 14886.

[10] I. L. Aleiner, P. W. Brouwer, and L. I. Glazman, Phys. Rep. **358** (2002) 309.

[11] P. W. Brouwer, Y. Oreg, and B. I. Halperin, Phys. Rev. B **60** (1999) R13977.

[12] H. U. Baranger, D. Ullmo, and L. I. Glazman, Phys. Rev. B **61** (2000) R2425.

[13] G. Usaj and H. U. Baranger, Phys. Rev. B **67** (2003) 121308.

[14] S. Guéron *et al.*, Phys. Rev. Lett **83** (1999) 4148.

[15] M. M. Deshmukh *et al.*, Phys. Rev. Lett. **87** (2001) 226801.

[16] A. Cehovin, C. M. Canali, and A. H. MacDonald, Phys. Rev. B **66** (2002) 094430.

[17] C. M. Canali and A. H. MacDonald, Phys. Rev. Lett. **85** (2000) 5623.

[18] C. M. Canali and A. H. MacDonald, Solid State Comm. **119** (2001) 253.

[19] S. Kleff *et al.*, Phys. Rev. B **64** (2001) 220401.

[20] J. Slonczewski, J. Magn. Magn. Mater. **159** (1996) L1.

[21] L. Berger, Phys. Rev. B **54** (1996) 9353.

[22] M. D. Stiles and A. Zangwill, Phys. Rev. B **66** (2002) 014407.

[23] M. Tsoi *et al.*, Phys. Rev. Lett **80** (1998) 4281.

[24] E. B. Myers *et al.*, Science **285** (1999) 867.

[25] J. A. Katine *et al.*, Phys. Rev. Lett. **84** (2000) 3149.

[26] S. I. Kiselev *et al.*, Nature **425** (2003) 380.

[27] W. H. Rippard *et al.*, Phys. Rev. Lett. **92** (2004) 027201.

[28] I. N. Krivorotov *et al.*, Science **307** (2005) 228.

INVERSION OF DNA CHARGE BY A POSITIVE POLYMER AND FRACTIONALIZATION OF THE POLYMER CHARGE

B.I. Shklovskii

Theoretical Physics Institute, University of Minnesota, Minneapolis MN 55455, USA

Gene delivery requires inversion of DNA charge in order to facilitate its contact with negative cell membranes and penetration inside the cytoplasm. Charge inversion of a DNA double helix by a positively charged flexible polymer (polyelectrolyte) is used for this purpose including trials on cancer patients. Effectiveness of charge delivery due to complexation is known to grow thousand times.

We consider charge inversion in terms of discrete charges of DNA and concentrate on the worst scenario case when in the neutral state of the DNA-polyelectrolyte complex, matching of DNA and polyelectrolyte structure is so perfect that all DNA charges are locally compensated by a polyelectrolyte charge, so that one can suspect that charge inversion is impossible. We show that charge inversion exists even in this case. When additional polyelectrolyte molecule is adsorbed by DNA, its charge gets fractionalized into monomer charges of defects (tails and arches) on the background of the perfectly neutralized DNA. These charges spread all over the DNA helix eliminating the self-energy of the polyelectrolyte molecule. This phenomenon is similar to what happens when additional electron is added to the 1/3 filled Landau level in the fractional Hall effect (FQHE). The charge of such an electron is known to split in three excitations with charge $e/3$ each.

Elimination of the self-energy of polyelectrolyte due to fractionalization is the driving force of charge inversion. Fractionalization leads to a substantial positive charge of DNA-polyelectrolyte complex. It was observed in electrophoresis experiments.

We show that fractionalization driven charge inversion is also possible when polyelectrolyte is adsorbed on a surface with a two-dimensional lattice of opposite charges. This is the first classical example of fractionalization of charge beyond one dimension.

References

[1] T. T. Nguyen, A. Yu. Grosberg, B. I. Shklovskii, *Physics of charge inversion in chemical and biological systems*, Rev. Mod. Phys. **74**, 329 (2002).

[2] T. T. Nguyen, B. I. Shklovskii, *Inversion of DNA charge by a positive polymer via fractionalization of the polymer charge*, Physica A **310**, 197 (2002).

SPIN - CHARGE SEPARATION IN QUANTUM WIRES

A. Yacoby

Department of Condensed Matter Physics, Weizmann Institute of Science, Rehovot 76100, Israel

One-dimensional (1D) electronic systems are expected to show unique transport behavior as a consequence of the Coulomb interaction between carriers [1]. Unlike in two and three dimensions [2], where the Coulomb interaction affects the transport properties only perturbatively, in 1D they completely modify the ground state from its well-known Fermi-liquid form. The success of Landau Fermi-liquid theory in two and three dimensions lays in its ability to lump the complicated effects of the Coulomb interaction into the Fermi surface properties (i.e. mass and velocity) of some newly defined particles known as quasi-particles [3]. Within this new description the quasi-particles are interacting only weakly and, thus the underlying transport properties may still be described in terms of single-particle physics. However, in 1D the Fermi surface is qualitatively altered even for weak interactions [4] and, hence, Landau Fermi-liquid theory breaks down. Today, it is well established theoretically that the low temperature transport properties of interacting 1D electron systems are described in terms of a Luttinger-liquid (LL) rather than a Fermi-Liquid. This state is characterized by strong correlated electron behavior similar to the behavior of a Wigner crystal [5]. Of coarse, there can be no true long-range order in 1D due to the large quantum mechanical zero-point fluctuations of the electrons. The correlation functions thus decay algebraically in space and in time with exponents that depend continuously on the interaction strength [6–9].

The validity of Fermi liquid theory in 2D and 3D assures that even in the presence of Coulomb interaction between the electrons, the low-lying excitations are quasi particles with charge e and spin $\frac{1}{2}$. 1D electronic systems, on the other hand, have only collective excitations [1]. A unique property of 1D systems is that these collective modes decouple into two kinds: collective spin modes and collective charge modes. Coulomb interactions couple primarily to the latter, and thus strongly influence their dispersion. Conversely, the excitation spectrum of the spin modes is typically unaffected by interactions, and therefore remains similar to the non-interacting case. However, this unique excitation spectrum is not manifested in the transport properties of clean 1D systems. Furthermore, the decoupling of the spin and charge degrees of freedom will have only subtle effects on the transport properties of disordered wires such as to modify the power laws in the $I - V$ characteristics and modify the excitation spectrum of a quantum dot embedded in a LL.

Using momentum resolved tunneling between two clean parallel quantum wires in a AlGaAs/GaAs heterostructure we directly measure the dispersion of elementary excitation and follow its dependence on carrier density. A voltage bias between the wires determines the energy of the tunneling electrons and a magnetic field applied perpendicular to the plane formed by the two wires determines their momentum. We find clear signatures of three excitation modes in the data: The anti symmetric charge mode of the coupled wire system and two spin modes. The density dependence of the anti symmetric charge mode agrees well with Luttinger liquid theory. As the density of electrons is lowered, the Coulomb interaction is seen to become increasingly dominant leading to excitation velocities that are up to a factor of 2.5 faster than the bare Fermi velocity, determined experimentally from the carrier density. The symmetric charge excitation also expected from theory is however not visible in the data. The observed spin velocities are found to be 25% slower than the bare Fermi velocities and depend linearly on carrier density. Below a critical electron density the system abruptly looses translation invariance and becomes localized. This localized phase is characterized by Coulomb blockade and a broad momentum content of the many body wave functions.

References

[1] D. C. Mattis, in 'The Many Body Problem', World Scientific Publishing Co., 1992.

[2] B. L. Altshuler, A. G. Aronov in: Electron- Electron Interactions in Disordered Systems, Edited by A. L. Efros and M. Pollak.

[3] P. Nozieres, in 'Interacting Fermi Systems', Adison Willey Publishing Co., 1997.

[4] F. D. M. Haldane, J. Phys. C 14, 2585 (1981).

[5] L. I. Glazman, I. M. Ruzin, and B. I. Shklovskii, Phys. Rev. B 45, 8454 (1992).

[6] W. Apel and T. M. Rice, Phys. Rev. B 26, 7063 (1992).

[7] C. L. Kane and M. P. A. Fisher, Phys. Rev. Lett. 68, 1220 (1992); C. L. Kane and M. P. A. Fisher, Phys. Rev. B 46, 15233 (1992).

[8] D. Yue, L. I. Glazman, and K. A. Matveev, Phys. Rev. B 49, 1966 (1994).

[9] M. Ogata and H. Fukuyama, Phys. Rev. Lett. 73, 468 (1994).